DATE DUE

8-3-91	
12-15-92	

Library of Congress Cataloging-in-Publication Data

Calabrese, Edward J., 1946–
 Air toxics and risk assessment / Edward J. Calabrese and Elaina M.
Kenyon.
  p. cm.
 Includes bibliographical references and index.
 1. Air—Pollution—Toxicology—United States. 2. Health risk
assessment—United States. 3. Air—Pollution—Standards—United
States. I. Kenyon, Elaina M. II. Title.
 [DNLM: 1. Air Pollutants, Environmental—analysis.
2. Environmental Exposure. 3. Environmental Monitoring—methods.
4. Environmental Monitoring—standards—United States. 5. Risk
Factors—United States. WA 754 C141a]
RA576.5.C35   1991
615.9′02—dc20
DNLM/DLC 90-13404
ISBN 0-87371-165-3

LEWIS PUBLISHERS, INC.
121 South Main Street, Chelsea, Michigan 48118

PRINTED IN THE UNITED STATES OF AMERICA

Edward J. Calabrese:

*To Dr. Kenneth Howe, for all his
efforts to make me a better
student and researcher*

Elaina M. Kenyon:

To my friend and companion, John

Toxicology and Environmental Health Series

Series Editor: Edward J. Calabrese

Preface for the Series

Given the complex and ever-expanding body of information in toxicology and environmental health, the purpose of the Toxicology and Environmental Health Series is to present a genuine synthesis of information that not only will offer rational organization to rapidly evolving developments but also will provide significant insight into and evaluation of critical issues. In addition to its emphasis on assessing and assimilating the technical aspects of the field, the series will offer leadership in the area of environmental health policy, including international perspectives. Thus, the intention of this series is not only to provide a careful and articulate review of critical areas in toxicology and environmental health but to influence the directions of this field as well.

The Editorial Board will oversee and shape the series, while individual works will be peer-reviewed by appropriate experts in the field.

Edward J. Calabrese
University of Massachusetts
Amherst

Acknowledgments

Thanks are due to several individuals who were very helpful in the practical aspects of manuscript preparation. Particular thanks are due Leonard Adams and the librarians in government documents at the University of Massachusetts for their frequent help in tracking down hard-to-find documents. Special thanks are also extended to Alenka Chadwick and the librarians at Biological Sciences Library for their help with literature searches, and to the staff at the Environmental Protection Agency (EPA) Region I Library for the loan of numerous EPA documents and help with IRIS access. We would also like to acknowledge excellent secretarial support provided in the early phases of manuscript preparation by Claire Meissner and later by Paula Goodhind. Finally, this work was initially supported by Rohm and Haas, Inc.

Preface

The status of air toxics programs in the United States is a confusing one, especially in the area of risk assessment. The 1980s witnessed the lack of strong federal leadership in the air toxics area along with a concomitant drift toward decentralization of federal responsibilities to the state governments. While decentralization of various activities may often enhance efficiency of operation at reduced expense, its application to the field of environmental health in general, and to air toxics in particular, has resulted in a plethora of approaches to assessing potential risks to public health from exposure to air toxics. These approaches, while often in direct scientific conflict with each other, as well as reflecting various policies with respect to acceptable health risks, also clearly reflect the wide range of resources, both technical and financial, that individual states can direct to the challenge of air toxics. Therefore, the net result of the decentralized approach to air toxics regulation has been the creation of a polysyllabic approach to assessing risk at the state level. The resultant state of confusion in which we find ourselves represents a regression in public policy that flies in the face of major recommendations of the federal government and the National Academy of Sciences (NAS) to seek greater consistency in approaches to assessing public health risks of exposure to environmental contaminants.

This book was written with the intention of trying to be a vehicle to enhance that greater consistency. It presents a thoroughly researched, carefully constructed, yet simple to follow decision tree methodology for all major endpoints of public health concern in the derivation of acceptable levels of ambient air contaminants for use in air toxics programs at the state and other levels. This decision tree framework is consistent with major developments in the fields of toxicology, epidemiology, and risk assessment, as well as with the reports of specific expert committees of the NAS. However, in addition to providing a decision tree framework that could be applied to any compound, the book then applies the methodology to over 100 specific toxic agents covering a wide range of chemical classes and the spectrum of major toxicological endpoints. Thus, the information contained in this book will be of both theoretical and practical value to regulatory officials involved in the standard-setting process, as well as for those on the receiving end of such decisions. It is hoped that the methodologies developed and applied in this book will serve as a vehicle not only to bring the respective states closer to speaking in a similar language with respect to air toxics, but also to encourage productive dialogue between government regulators and industry.

Finally, we realize that the field of risk assessment is a rapidly changing one and that new and important advances will be forthcoming that will impact present recommendations. Thus, while we look forward to improving our methodologies in the future, we hope that this volume will be a useful addition and a handy reference for those actively engaged in the challenging domain of air toxics.

Chapter 1 sets the stage for the methodology presented in Chapter 3 by describing the historical reasons for the initiation of state air toxics programs and how this has resulted in a confusing diversity of approaches to regulating air toxics at the state level. This is illustrated by a review and critique of the general approaches used by the states in developing ambient air level (AAL) guidelines. Finally, a general analysis of some of the problems that may result from such divergent approaches is presented.

Chapter 2 supports the methodology presented in Chapter 3 by describing and reviewing the strengths and limitations of various secondary data sources used in the application of the methodology. Emphasis is also given to how the data source is used most efficiently in the context of the methodology in Chapter 3 and how it may be most easily accessed.

The methodology used to derive the AALGs presented in the chemical-specific assessments in Part II is described in Chapter 3. This methodology draws heavily on existing risk assessment techniques, assembled from a variety of sources to form a systematic means to rapidly evaluate a large number of chemicals and develop scientifically defensible air guidelines in a setting of limited organizational resources and a narrowly defined time frame, as would be typical for most state governmental agencies, individual industries, and consulting companies.

The second part of the book is devoted to over 100 chemical-specific assessments produced by the application of the methodology presented in Chapter 3. These assessments not only illustrate the application of the methodology, but also demonstrate the application of toxicologic principles of data evaluation to a diverse range of substances with varying levels of overall database quality, ranging from those agents with a substantial toxicologic database to those for which there has been minimal study.

Edward J. Calabrese is a board certified toxicologist who is professor of toxicology at the University of Massachusetts School of Public Health, Amherst. Dr. Calabrese has researched extensively in the area of host factors affecting susceptibility to pollutants, and is the author of more than 270 papers in scholarly journals, as well as 12 books, including *Principles of Animal Extrapolation, Nutrition and Environmental Health*, Vols. I and II, *Ecogenetics, Safe Drinking Act: Amendments, Regulations and Standards, Petroleum Contaminated Soils,* Vols. I, II, and III, *Ozone Risk Communication and Management*, and *Multiple Chemical Interactions*. He has been a member of the U.S. National Academy of Sciences and NATO Countries Safe Drinking Water committees, and most recently has been appointed to the Board of Scientific Counselors for the Agency for Toxic Substances and Disease Registry (ATSDR). Dr. Calabrese also serves as Chairman of the International Society of Regulatory Toxicology and Pharmacology's Council for Health and Environmental Safety of Soils (CHESS).

Elaina M. Kenyon is currently a postdoctoral fellow in the Environmental Sciences Program at the University of Massachusetts Medical School Department of Family and Community Medicine in Worcester. She received her BS in Resource Development from the University of Rhode Island, an MS in Epidemiology from Texas A&M University, and her PhD in Public Health concentrating in toxicology from the University of Massachusetts at Amherst. Dr. Kenyon worked for several years at the U. S. Environmental Protection Agency Marine Water Quality Laboratory in Narragansett, Rhode Island. Her research interests and publications are in the area of animal extrapolation and risk assessment.

Table of Contents

Preface   vii

List of Tables   xv

List of Figures   xvii

Abbreviations   xviii

PART I: BACKGROUND AND METHODOLOGY

1. Introduction to the Problem and Review of Current Approaches   3

Air Toxics Regulation at the Federal Level   4
Approaches to Regulating Air Toxics at the State and Local Level   5
Issues Surrounding the Use of OELs to Derive AALs   7
Problems with Current Approaches to AAL Derivation   8
References   12

2. Secondary Data Sources   15

Registry of Toxic Effects of Chemical Substances (RTECS)   15
National Air Toxics Information Clearinghouse (NATICH)   18
Integrated Risk Information System (IRIS)   18
Gene-Tox Program   20
National Toxicology Program (NTP)   21
REPROTOX Database   22
Documentation for Occupational Limits   23
Drinking Water Documents   24
Chemical-Specific Documentation   25
Conclusions   30
Contacts for Information Sources   30
References   32

3. An Approach for the Rapid Development of Scientifically Defensible AALs   35

Scope and Limitations   35
Classification of Toxicity   36
General Issues   37
AALG Derivation for Carcinogens   44
Use of Genotoxicity Data   49
AALG Derivation for Noncarcinogens: The Uncertainty
   Factor Approach   51
Issues Specific to Certain Noncarcinogenic Endpoints   63
Options for AALG Derivation When Only Acute Lethality
   Data Are Available   71

Multimedia Exposure and the Relative Source Contribution   73
Guidelines for the Selection of Averaging Times   75
References   75

PART II: CHEMICAL-SPECIFIC ASSESSMENTS

Acetic acid   81
Acetone   85
Acetone cyanohydrin   89
Acetonitrile   91
Acetylene   95
Acrolein   97
Acrylamide   101
Acrylic acid   107
Acrylonitrile   113
Allyl chloride   119
Ammonia   125
Antimony (and compounds)   129
Arsenic (and compounds)   135
Asbestos   141
Benzene   145
Benzoyl peroxide   151
Benzyl alcohol   155
Benzyl chloride   159
Bis (2-chloroethyl) ether (BCEE)   163
Bis (chloromethyl) ether (BCME)   167
Boron trifluoride   171
Bromine   173
1,3-Butadiene   177
Butyl acrylate   183
Butyl benzyl phthalate   187
n-Butyl alcohol   191
s-Butyl alcohol   195
t-Butyl alcohol   199
Cadmium (and compounds)   203
Caprolactam   209
Carbon disulfide   213
Carbon tetrachloride   219
Chlorine   223
Chlorobenzene   229
Chloroform   235
Chloromethyl methyl ether   241
Chromium (and compounds)   245
Cobalt (and compounds)   251
Copper (and compounds)   257
Cumene   261

Cyclohexanone   263
Dibutyl phthalate   267
1,2-Dichloropropane   273
Diethanolamine   277
Diisocyanate diphenyl methane   281
Dimethylamine   285
Dimethyl phthalate   289
1,4-Dioxane   293
Epichlorohydrin   297
Ethanediamine   303
Ethyl acrylate   305
Ethyl benzene   311
Ethylene dichloride (1,2-dichloroethane)   315
Ethylene glycol   321
Ethylene oxide   325
Ethylene thiourea   331
2-Ethylhexyl acrylate   337
Formaldehyde   341
n-Hexane   345
Hydrazine   349
Hydrochloric acid   353
Hydrogen cyanide   357
Hydroquinone   361
Isobutyraldehyde   365
Isophorone   369
Isopropyl alcohol   375
Lead (and compounds)   379
Maleic anhydride   385
Mancozeb   389
Maneb   393
Manganese (and compounds)   401
Mercury (and compounds)   407
Mesityl oxide   411
Methyl acrylate   415
Methyl alcohol   419
Methyl bromide   423
Methyl chloride   431
Methylene chloride (dichloromethane)   437
Methyl ethyl ketone   443
Methyl isobutyl ketone   447
Methyl methacrylate   451
Methyl styrene (vinyl toluene)   455
Naphthalene   459
Nickel (and compounds)   463
Nitric acid   473
Nitrofen   477

Phenol   483
Phosphoric acid   489
Phthalic anhydride   491
Propionic acid   495
Propylene   499
Propylene oxide   503
Resorcinol   507
Sodium hydroxide   511
Sodium sulfate   515
Styrene   519
Sulfuric acid   525
Tetrachloroethylene (perchloroethylene)   531
Titanium dioxide   537
Toluene   541
1,1,1-Trichloroethane (methylchloroform)   545
Trichloroethylene (TCE)   551
Vanadium pentoxide   557
Vinyl chloride   561
Vinylidene chloride (1,1-dichloroethylene)   567
Xylenes   573
Zinc acetate   579
Zinc oxide   583
Zinc sulfate   587
Zineb   591

APPENDICES

A.   Drinking Water Limits   597
B.   Chemical-Specific Worksheet   607
C.   Dose Calculations, Scaling, and Physiological Parameters   609
D.1. Criteria for Assessing the Quality of Individual Human
        Epidemiological Studies   617
D.2. Criteria for Assessing the Quality of Individual Animal
        Toxicity Studies   621
E.   Bibliography of NTP/NIEHS Papers Relevant to Evaluation of Toxicology
        and Carcinogenesis Studies   625
F.   U.S. Environmental Protection Agency Genetic Toxicology (Gene-Tox)
        Program Publication List   629
G.   Brief Description of the EPA RfC   635
H.   Examples of LOAEL or NOAEL Selection When Multiple
        Responses Are Available   643

CAS Number Index   645
Subject Index   649
Name Index   657

List of Tables

1.1 Substances with National Ambient Air Quality Standards 5
1.2 Disadvantages in the Use of OELs in AAL Derivation 7
1.3 Comparison of AALs for Benzene by S/L Agencies 9
1.4 Comparison of Highest and Lowest AAL Standardized to a 24-hour Averaging Time by Compound for Some Representative Known and Probable Human Carcinogens 10
1.5 Comparison of Highest and Lowest AAL Standardized to a 24-hour Averaging Time by Compound for Some Representative Noncarcinogens 11
2.1 Summary of Information in a Typical RTECS Entry 16
2.2 Examples of Various Types of NATICH Publications 19
2.3 ATSDR Toxicological Profiles Published in 1988 and 1989 27
2.4 Substances for Which HEADs Have Been Published 28
2.5 Titles in the Environmental Health Criteria (EHC) Series 29
3.1 Preliminary Categorization of Health Effects Using RTECS 37
3.2 Sequence with Options for Chemicals Flagged in the Carcinogen Class During Preliminary Screening 45
3.3 EPA Weight-of-Evidence Classification for Carcinogens and Basis for Suitability of Agents for Quantitative Risk Assessment 46
3.4 IARC Weight-of-Evidence Classification for Carcinogenicity with Definitions for Degree of Evidence 47
3.5 IARC Scheme for the Assessment of Evidence of Activity in Short-Term Tests 50
3.6 Endpoints for Short-Term Tests Recognized under the IARC Scheme for the Assessment of Evidence of Activity in Short-Term Tests 51
3.7 U.S. EPA Classification of Weight of Evidence for Potential Human Germ-Cell Mutagenicity 52
3.8 General Scheme with Options for AALG Derivation for Substances Identified as Developmental, Reproductive, or Systemic Toxicants During Preliminary Screening 53
3.9 NOAELs and Related Responses Defined 55
3.10 Guidelines Used by U.S. EPA in Selection of Animal Toxicity Data for Deriving Criteria Based on Noncarcinogenic Responses 56
3.11 General Considerations and Scheme for AALG Derivation Based on an Occupational Limit 58
3.12 Criteria and Guidelines for the Application of Uncertainty Factors 60
3.13 Guidelines for Application of the Database Factor 63
3.14 Definitions of the Major Endpoints of Developmental Toxicity and Other Important Terms 64
3.15 Examples of Indices and Endpoints Indicative of Male or Female Reproductive Toxicity Which Might Be Evaluated in Various Study Types 69

3.16 Sequence of Steps for the Evaluation of Compounds Identified as Sensory Irritants During Preliminary Screening with Options for AALG Derivation   70

3.17 Guidelines for Selection of the Relative Source Contribution Allowed from Ambient Air in the AALG Derivation Process   74

A.1 Three-Category Approach for Developing MCLGs   600

A.2 Derivation of MCLGs Based on Noncarcinogenic Effects   601

A.3 Contaminants Scheduled for Regulation by 1989 Under the 1986 Amendments to the Safe Drinking Water Act   602

A.4 Contaminants for Which Completed Health Advisories Have Been Issued as of 3/31/87   604

A.5 Chemicals and Classes of Chemicals for Which Ambient Water Quality Criteria Documents were Originally Published in 1980   605

C.1 Conversion of Exposure Levels in Units of ppm to Units of mg/m^3 610

C.2 Calculation of the Animal Breathing Rate (ABR) on the Basis of Animal Body Weight (ABW) Using a Body Weight–Surface Area Relationship   611

C.3 Default or Reference Values for Various Parameters in Humans and Laboratory Animals   612

C.4 Assumptions Operative in Dosimetric Adjustments Described in This Appendix   613

C.5 Conversion of Exposure Levels in ppm in the Diet or Water to mg of Test Article per kg of Body Weight per Day (mg/kg/day)   614

G.1 Equations for Calculation of the HEC for Particulates That Exert Respiratory Effects   638

G.2 Equations for Calculation of the HEC for Particulates That Exert Extrarespiratory Effects   639

G.3 Equations for Calculation of the HEC for Gases That Are More Reactive and Exert Respiratory Effects   640

G.4 Equation for Calculation of the HEC for Gases That Are More Soluble and Exert Respiratory Effects and for Gases That Exert Extrarespiratory Effects Where Periodicity Is Attained   641

G.5 Equation for Calculation of the HEC for Gases That Exert Extrarespiratory Effects Where Periodicity Is Not Attained   642

List of Figures

3.1 Prioritization scheme and organizational framework for application of the AALG derivation methodology   38

3.2 Schematic diagram of possible outcomes of chemical exposure in developmental toxicity studies   65

A.1 Former NPDWR regulatory scheme   598

A.2 Regulatory and criteria document development processes for MCL/MCLGs   599

A.3 Health advisory development process   603

G.1 Flowchart for HEC calculation   637

Abbreviations

AAL	Ambient Air Levels
AALG	Ambient Air Level Goals
ABR	Animal Breathing Rate
ABW	Animal Body Weight
ACGIH	American Conference of Governmental Industrial Hygienists
ACS	American Chemical Society
ADI	Acceptable Daily Intake
Air RISC	Air Risk Information Support Center
ALAPCO	Association of Local Air Pollution Control Officials
ANPRAM	Advanced Notice of Proposed Rulemaking
ATSDR	Agency for Toxic Substances and Disease Registry
AWQCD	Ambient Water Quality Criteria Document
BACT	Best Available Control Technology
CAG	Carcinogen Assessment Group (EPA)
CAP	Criteria Air Pollutant
CAS	Chemical Abstracts Service
CDC	Centers for Disease Control
CD-ROM	Compact Disk—Read Only Memory
CERCLA	Comprehensive Environmental Response, Compensation, and Liability Act
CIB	Current Intelligence Bulletin
CIS	Chemical Information Systems
CPSC	Consumer Product Safety Commission
2,4-D	2,4-Dichlorophenoxyacetic acid
DHHS	Department of Health and Human Services
DOE	Department of Energy
DOT	Department of Transportation
DWC	Drinking Water Contribution
DWCD	Drinking Water Criteria Document
DWEL	Drinking Water Equivalent Level
EHC	Environmental Health Criteria (WHO)
ELF	Extremely Low Frequency
EPA	Environmental Protection Agency
ETIC	Environmental Teratogen Information Center
FAO/WHO	Food and Agricultural Organization/World Health Organization
FASEB	Federation of American Societies for Experimental Biology
FDA	Food and Drug Administration
FEL	Frank Effect Level
FR	Federal Register
HA	Health Advisory
HAD	Health Assessment Document
HEAD	Health Effects Assessment Document

HEC	Human Equivalent Concentration (EPA, RfD)
IARC	International Agency for Research on Cancer
IPCS	International Program on Chemical Safety
IRIS	Integrated Risk Information System
kg	kilogram
LC_{50}	Lethal Concentration 50%
LD_{50}	Lethal Dose 50%
LOAEL	Lowest Observed Adverse Effect Level
LOEL	Lowest Observed Effect Level
MAC	Maximum Allowable Concentration
MAOL	Most Appropriate Occupational Limit (Massachusetts)
MCLG	Maximum Contaminant Level Goal
MCL	Maximum Contaminant Level
mg	milligrams
MSHA	Mine Safety and Health Administration
MW	Molecular Weight
NAAQS	National Ambient Air Quality Standard
NAS	National Academy of Sciences
NATICH	National Air Toxics Information Clearinghouse
NCAP	Noncriteria Air Pollutant
NCI	National Cancer Institute
NCTR	National Center for Toxicologic Research
NESHAPs	National Emission Standards for Hazardous Air Pollutants
NIEHS	National Institute of Environmental Health Sciences
NIH	National Institutes of Health
NIOSH	National Institute for Occupational Safety and Health
NLM	National Library of Medicine
NOAEL	No Observed Adverse Effect Level
NOEL	No Observed Effect Level
NOHS	National Occupational Hazard Survey
NPDWR	National Primary Drinking Water Regulations
NRC	National Research Council
NTIS	National Technical Information Service
NTP	National Toxicology Program
ODW	Office of Drinking Water (EPA)
OEL	Occupational Exposure Level
OPTS	Office of Pesticides and Toxic Substances (EPA)
OSHA	Occupational Safety and Health Administration
OTE	Office of Testing and Evaluation (EPA)
PAH	Polynuclear Aromatic Hydrocarbon
PB-PK	Physiologically Based Pharmacokinetic (modeling)
PCBs	Polychlorinated Biphenyls
PEL	Permissible Exposure Limit (OSHA)
PHN	Public Health Network
PHS	Public Health Service

ppb	parts per billion
ppm	parts per million
QRA	Quantitative Risk Assessment
RACT	Reasonably Available Control Technology
REL	Recommended Exposure Level (NIOSH)
RfC	Inhalation Reference Concentration
RfD	Reference Dose
RfD_i	Inhalation Reference Dose
RMCL	Recommended Maximum Contaminant Level
RQ	Reportable Quantity
RQD	Reportable Quantity Document
RSC	Relative Source Contribution
RTC	Reproductive Toxicology Center
RTECS	Registry of Toxic Effects of Chemical Substances
RTI	Relative Teratogenic Index
SARA	Superfund Amendments and Reauthorization Act
SCE	Sister-Chromatid Exchange
SDWC	Safe Drinking Water Committee (NAS)
SIC	Standard Industrial Classification
S/L	State/Local
SNARL	Suggested No Adverse Response Level
STAPPA	State and Territorial Air Pollution Program Administrators
TLVs	Threshold Limit Values
TLV-C	Threshold Limit Value—Ceiling
TLV-STEL	Threshold Limit Value—Short-Term Exposure Limit
TLV-TWA	Threshold Limit Value—Time-Weighted Average
TSCA	Toxic Substances Control Act
TSD	Technical Support Document
UFs	Uncertainty Factors
U.S. EPA	United States Environmental Protection Agency (EPA)
WHO	World Health Organization

AIR TOXICS
and
RISK ASSESSMENT

PART I

Background and Methodology

1   INTRODUCTION TO THE PROBLEM AND REVIEW OF CURRENT APPROACHES

This introductory chapter will provide an overview of how the current decentralized process of air toxics risk assessment developed. It will also establish the public health implications of the diverse risk assessment methodologies developed by states, in the absence of strong federal leadership, in terms of estimates of environmentally induced disease associated with projected possible exposures to air toxics. As will be seen, the end result of this decentralized process is societal acquiescence to differential levels of acceptable health risk in different states, a condition that was rejected by Congress two decades ago for occupationally induced disease.

This chapter will also illustrate the basic methodologies currently used by various states and provide a critique of their strengths and limitations, thus establishing the need for greater consistency in the use of risk assessment methodologies when dealing with air toxics issues at the state level. In addition, this critique will provide the basis for Chapter 3, which offers a critical and practical decision tree approach for the application of risk assessment to air toxics for those organizations, such as states, that must make health-based decisions with extreme time and resource limitations.

Over the past decade there has been a major effort at the state level to develop what are now called air toxics programs. These programs are essentially designed to address health concerns resulting from exposure to contaminants other than the seven that the Environmental Protection Agency (EPA) regulates with ambient air quality standards.[1] The need for additional regulatory action to control toxic air pollutants has been based on a number of arguments, most notably the occurrence of a higher incidence of lung cancer in the urban environment and ambient air monitoring studies, especially in urban areas, showing a large number of contaminants, including carcinogens such as benzene, vinyl chloride, and chloroform.

The attribution of the higher incidence of lung cancer in urban settings to air pollution is highly controversial given the multifactorial nature of complex diseases such as cancer. Factors such as dietary differences, stress, smoking patterns, indoor air pollutant levels and other important independent variables that may vary between

3

urban and rural areas contribute to uncertainties in seeking a consensus on the role of the urban factor in lung cancer. Biostatistical/epidemiological approaches of different investigators have reflected the lack of consensus on this point. At the high end, Karch and Schneiderman have estimated that at least 11%, and possibly as much as 21%, of lung cancer may be attributable to air pollution after correction for smoking and occupational effects; Doll and Peto estimate that the contribution of all forms of pollution (air, water, and food) to cancer incidence is 2%, with a range of 1 to 5%.[2,3] They also indicate that the contribution of air pollution to the observed incidence of lung cancer after correction for smoking is minimal and that it does not account, to a large extent, for differences in observed lung cancer rates between rural and urban areas.

Despite the unresolved issue over the causes of the urban factor, there is little dispute over the observation that urban air contains numerous pollutants, many of which are mutagens and carcinogens in animal systems. What is in dispute is whether the levels observed pose significant health risks to the general population. In the absence of an unequivocal answer, many states sought guidance from the U.S. EPA to develop new ambient standards. The response from the EPA during the Reagan years was to provide data on exposure levels based on specific studies and technical information transfer.

In addition to concern at the state level over the public health impact of air pollution, the states were also requested by EPA to develop their own air toxics programs.[4] This process has essentially led to a bewildering array of various approaches for deriving ambient air levels (AALs). This approach by EPA of encouraging the states to develop their own methodologies for deriving AALs for exposure to air toxics has evolved into decentralized risk assessment.

Air Toxics Regulation at the Federal Level

The first federal air pollution legislation in the United States was adopted in 1955. This legislation granted the federal government the authority to conduct research, training, and technical assistance programs. Further legislation in the form of the Clean Air Act of 1963, subsequently amended several times, was ultimately passed. The concept of national ambient air quality standards (NAAQS) did not come into being until the Clean Air Amendments of 1970 were passed; these amendments required the EPA to develop and promulgate NAAQS for pollutants for which health criteria had been issued. This requirement represented a significant change from previous amendments (1967) to the Clean Air Act, which had provided that the states develop their own air quality standards.[1]

The process of NAAQS development begins with the identification of air pollutants that pose the most serious threat to human health based on known health effects and extent of use, occurrence, and release. This is followed by an extensive and exhaustive review of the literature on the substance in question with respect not only to health effects in humans, but effects on other biota as well as analytical

Table 1.1 Substances with National Ambient Air Quality Standards

Sulfur dioxide
Carbon monoxide
Nitrogen dioxide
Hydrocarbons
Oxidants, e.g., ozone
Particulate matter (PM_{10})
Lead

techniques and environmental fate and distribution. Before a final criteria document is published, the information collected and the conclusions drawn are subject to extensive internal and external review. Since NAAQS are standards that are legally enforceable (as opposed to guidelines, which are not), considerations such as economic and technical feasibility also apply. Partially because the review process is extensive, the development of NAAQS has proceeded very slowly and at present there are only seven substances for which NAAQS have been developed. These are referred to as criteria pollutants and are listed in Table 1.1.

While efforts to develop NAAQS are lagging at the federal level, many states have been faced with the problem of mounting public and legislative pressure to address perceived problems with noncriteria air pollutants. As a result, and with the encouragement of the U.S. EPA, most states have developed, or are in the process of developing, air toxics programs.

The EPA strategy for reducing risks to public health from air toxics is embodied in a four-part plan:

1. maintaining a regulatory posture for contaminants of national concern
2. assisting states in dealing with the control of air toxics of a more local nature
3. improving the information base on multimedia contaminants from multiple sources
4. developing an improved program for emergency preparedness and responses, including the enhancement of information systems and training of state and local response teams

With particular reference to the development of air toxic programs, the EPA serves in the role of a technical advisor and facilitator of information exchange through mechanisms such as the National Air Toxics Information Clearinghouse (NATICH), described in Chapter 2.

Approaches to Regulating Air Toxics at the State and Local Level

The scope and general approach to air toxics regulation vary widely among state/local (S/L) agencies with respect to the pollutants regulated, whether emissions guidelines or ambient air guidelines or both have been or are being developed, and whether emissions standards or guidelines will be applied to existing or modified industrial sources emitting a given pollutant. In addition, requirements for the application of BACT (best available control technology) or RACT (reasonably available control technology) vary considerably in terms of the types of sources covered (i.e., new, existing, or modified), what controls are required, and when they must be applied.[5,6]

A number of state and local entities have chosen to develop AALs as part of their overall approach to air toxics control. Three general approaches have been used. In the case of substances classified as carcinogens, some S/L agencies have chosen to perform quantitative risk assessment, usually using the linearized multistage model applied to animal bioassay data and setting the AAL at a preselected *de minimus* risk level, usually one in a million (10^{-6}). More often however, S/L agencies have used preexisting risk assessments performed by the Carcinogen Assessment Group (CAG) of the U.S. EPA.[5,6]

Another approach, used mainly for noncarcinogens, but in some cases applied to carcinogens as well, has been to select a no observed adverse effect level (NOAEL)—or lowest observed adverse effect level (LOAEL) if a NOAEL is unavailable—from a suitable animal or human study and divide it by a series of multiplicative uncertainty factors (UFs). The uncertainty factors typically used are intended to account for factors such as animal to human extrapolation (interspecies variation), sensitivity of high-risk individuals (interindividual variation), use of a LOAEL rather than a NOAEL, and for extrapolation from less-than-chronic exposure to chronic exposure. This is the general approach used by the EPA and other federal agencies for the development of health-based guidelines for noncarcinogenic substances in a variety of media (e.g., food, water, air).[1]

The third, and most common, approach involves the application of uncertainty factors to occupational exposure levels (OELs). The OELs typically used are:

1. recommended exposure limits (RELs), developed by the National Institute of Occupational Safety and Health (NIOSH)
2. permissible exposure limits (PELs), developed by the Occupational Safety and Health Administration (OSHA)
3. threshold limit values (TLVs), developed by the American Conference of Governmental Industrial Hygienists (ACGIH)

PELs, unlike RELs, are standards rather than guidelines and, as such, must consider issues such as economic and technical feasibility in addition to health effects. However, while both RELs and TLVs are primarily health based, consideration is also given to the analytical methods and practical detection limits for the particular substance in question.

Since OELs are designed to protect workers during a typical 40-hour workweek, when modifying them to protect the general population they are often divided by a factor of 4.2 (168 hours/40 hours) to adjust for continuous exposure.[7] Additional uncertainty factors, usually in the range of 10 to 100, are incorporated to account for interindividual variation and/or relative severity of effect (e.g., carcinogens versus noncarcinogens).[6]

Table 1.2 Disadvantages in the Use of OELs for AAL Derivation

1. As typically used by S/L agencies, it is implicitly assumed that TLVs are the equivalent of human NOAELs, when in fact TLVs may be based on human or animal experimental data, industrial experience, or chemical analogy based on similar structure. The degree to which the type and quality of data are factored into the limit is inherently variable (thereby implying a variable margin of safety) since the final TLV is based on consensus judgment by the Chemical Agents TLV committee.
2. TLVs are intended to prevent or minimize a given effect in workers who are generally healthy individuals between the ages of 18 and 65. However, the health effect of major concern in workers may not be the same as that for the general population, which contains high-risk subpopulations, e.g., individuals who are very old or young or those with preexisting disease states, particularly respiratory disease.
3. TLVs are set assuming a "zero-exposure" recovery period, i.e., a time period during which there is no exposure. While it is possible that correction factors, e.g., 4.2, may appropriately correct for this in the case of cumulative effects, it may be overly conservative for noncumulative threshold activity agents such as primary irritants. In addition, recovery times allowed by different OELs vary, e.g., TLVs allow 16 hours between workdays and 64 hours on weekends and NIOSH RELs allow 14 hours between workdays and 68 hours on weekends; it is unclear how an AAL based on an OEL can account for this.
4. It is difficult when using TLVs or other OELs to account for the influence of factors such as environmental fate of the compound, multiple sources of exposure, and exposure to multiple agents, which may be important considerations in setting a final limit.
5. Use of OELs does not give the S/L agency the flexibility to change AALs to reflect new data until the TLV is changed, and substances for which OELs have not been set cannot be dealt within a system based on occupational limits.

Source: This table was assembled from material in Rowan et al.[7] and CMA.[9]

Issues Surrounding the Use of OELs to Derive AALs

Since they are the most commonly used OELs in AAL derivation, the discussion here will focus primarily on TLVs, although many of the points made apply to other OELs as well. While there are several types of TLVs, e.g., TLV–ceiling limits, TLV–short-term exposure limits (TLV-STELs), and TLV–time-weighted averages (TLV-TWAs), TLV-TWAs are the ones most commonly used by states as a starting point in AAL derivation. The TLV-TWA is defined as "the time-weighted average concentration for a normal 8-hour workday and a 40-hour workweek, to which nearly all workers may be repeatedly exposed day after day, without adverse effect."[8] With respect to the appropriateness of using TLVs to develop AALs, the ACGIH has stated the following:

These limits are intended for use in the practice of industrial hygiene as guidelines or recommendations in the control of potential health hazards and for no other use, e.g., in the evaluation or control of community air pollution nuisances, in estimating the toxic potential of continuous uninterrupted exposures . . .[8]

The use of TLVs as a starting point in AAL derivation is clearly inconsistent with their intended use, and this has been cited as one reason to avoid this practice. Table 1.2 summarizes other problems with the use of TLVs for AAL derivation.[7,9]

While ideally it would be best to perform complete risk assessments on a chemical-by-chemical basis to develop AALs, this remains a nonviable option for many S/L agencies faced with resource limitations and a pressing need for a concerted program to deal with their respective air toxics concerns.

In spite of the problems associated with the use of OELs as a starting point in AAL derivation, it remains an attractive option for many states for several reasons. OELs and their documentation constitute the largest database available on toxic substances in air that has been peer reviewed by highly qualified professionals from a number of relevant disciplines. Thus, they are an extremely valuable resource for agencies with limited staff and fiscal resources. In fact, to completely ignore the wealth of data that has been collected, analyzed, and critiqued for the purpose of health risk assessment for occupationally exposed workers by committees of nationally and internationally recognized experts would be a waste of limited organizational resources. In addition, the existence of a large number of OELs provides a convenient starting point for agencies faced with the need to regulate a large number of chemicals in a relatively short period of time.

It is our recommendation that OELs not be used as a basis for AAL derivation when the critical toxic effect is carcinogenicity. The reason for the latter distinction is the presumed nonthreshold dose-response relationship for carcinogenic effects compared to the generally accepted threshold dose-response relationship for most other toxic endpoints. However, if examination of the background documentation reveals that the numerical basis for the OEL is sufficiently documented and based on health effects and that the endpoint on which it is based is appropriate for the general population, then the OEL may reasonably serve as a starting point for AAL derivation. The need for careful examination of the background documentation of the particular TLV prior to using it as a basis for AAL derivation is reinforced by recently published analyses that indicate that TLVs are sometimes poorly correlated with levels known to produce adverse effects, but well correlated with measured exposure levels or levels thought to be reasonably achievable at the time the limit was adopted.[10,11] The operative question to answer before using a given OEL as a basis for AAL derivation is: Does available experimental data indicate that the OEL is a reasonable surrogate human NOAEL or LOAEL?

Problems with Current Approaches to AAL Derivation

The diversity of approaches to AAL derivation has resulted in a large variation among AALs between S/L agencies. Benzene, considered a known human carcinogen by both IARC (Group 1) and the U.S. EPA (Class A) weight-of-evidence classifications, is used to illustrate the wide range of approaches, and consequently the large variation of AALs, between different S/L agencies (Table 1.3). The large disparity in derived values is characteristic of AALs for carcinogens and is reflective of the use of a risk-based approach, which tends to result in more conservative AALs, versus an OEL-based or uncertainty factor approach, which tends to result in less conservative AALs. Further examples of the range of AALs suggested for known and probable human carcinogens (as classified by IARC) are presented in

6. U.S. EPA. 1989. "NATICH Data Base Report on State, Local and EPA Air Toxics Activities." EPA-450/3–89–29.

7. Rowan, C. A., W. M. Connolly, and H. S. Brown. 1984. "Evaluating the Use of Occupational Standards for Controlling Toxic Air Pollutants." *J. Environ. Sci. Health* B19:618–48.

8. ACGIH. 1988. "Threshold Limit Values and Biological Exposure Indices for 1988–89" (Cincinnati, OH: ACGIH).

9. Chemical Manufacturers Association (CMA). 1988. *Chemicals in the Community: Methods to Evaluate Airborne Chemical Levels* (Washington, DC: CMA).

10. Ziem, G. E., and B. I. Castleman. 1989. "Threshold Limit Values: Historical Perspectives and Current Practice." *J. Occ. Med.* 31:910–18.

11. Roach, S. A., and S. M. Rappaport. 1990. "But They Are Not Thresholds: A Critical Analysis of the Documentation of the Threshold Limit Values." *Am. J. Industrial Med.* 17:727–753.

12. Bierma, T. 1982. *The Use of Occupational Exposure Standards to Estimate Safe Ambient Concentrations for Air Pollutants: A Comparative Study.* Illinois Environmental Protection Agency.

13. National Research Council. 1983. *Risk Assessment in the Federal Government: Managing the Process* (Washington, DC: National Academy Press).

2   SECONDARY DATA SOURCES

The intent of this chapter is to review and comment on some of the secondary data sources that were used extensively in the preparation of the chemical-specific assessments presented in Part II. The primary advantage of secondary data sources is that the material has already been collected and usually screened to some extent. Thus they are convenient to access and can result in substantial savings in costs and time compared to chemical-specific searches of the primary scientific literature. In addition, since secondary data sources are screened and in some cases extensively peer reviewed, this can compensate somewhat for limitations in available organizational expertise and time. The data sources reviewed here include those that are most accessible and relevant to air toxics and have been most widely used by state and local (S/L) agencies in AAL derivation.

Registry of Toxic Effects of Chemical Substances (RTECS)

In the application of this methodology, the RTECS database produced by the National Institute for Occupational Safety and Health (NIOSH) was used as the initial starting point for obtaining information about specific compounds. The format and information provided in a typical RTECS entry are shown in Table 2.1. The reasons for choosing the RTECS database were primarily related to format, accessibility, and utility as a point of reference to other information sources. The toxicity information in RTECS is organized primarily by type of effect (e.g., skin and eye irritation, genotoxicity, tumorigenicity, reproductive and developmental toxicity, and acute and chronic systemic toxicity), which fits the endpoint-based format of the methodology presented in Chapter 3. In addition, the information provided on species or test system and route of exposure facilitates rapid referral to the appropriate studies in the primary literature. For example, in the case of this methodology, the most relevant route of exposure is inhalation.

Another strength of the RTECS database is its accessibility, which is related not only to the variety of media in which it is available, but also to the breadth of chemicals covered and the system by which they are cross-referenced. The current

Table 2.1 Summary of Information in a Typical RTECS Entry

Type of Information	Comments
Identification	—substance prime name, derived from American Chemical Society (ACS) Chemical Abstract Service (CAS) nomenclature —RTECS number —CAS registry number —molecular formula —molecular weight —synonyms and trade names
Update[a]	—when last change to file was made
Skin and Eye Irritation	—primary skin and eye irritation data; in both cases data on species, amount, exposure duration, severity of response, and literature citation are given
Mutation data and References	—positive results from in vivo and in vitro tests including species and/or test system, exposure route and duration if applicable, lowest dose or level causing effect
Reproductive Effects Data and References	—developmental and reproductive effects in mammalian species including type of effect, species, duration and route of exposure, and dose information —data from multigeneration studies also
Tumorigenic Data and References	—positive and equivocal results (as defined by RTECS criteria) including species, organs affected, exposure route, and duration and dose information
Toxicity Data and References	—toxic effects from acute and chronic exposures including effect, route and duration of exposure, species and dose information
Reviews	—ACGIH TLV documentation reference with numerical limits provided —IARC monograph references —general toxicology references
Standards and Regulations	—notations indicating regulation by DOT, EPA, MSHA, and OSHA (including transitional and final PEL data)
NIOSH REL	—actual figures with reference to appropriate criteria document, CIB, or other documentation —NIOSH occupational exposure survey data from National Occupational Hazard Surveys (NOHS) conducted and published in 1974 and 1983
Status Lines	—listed for EPA, NIOSH and NTP EPA: related to Toxic Substances Control Act and Gene-Tox database entries NIOSH: related to existence of NIOSH analytical method and CIBs NTP: tested or on test in NTP program with NTP classification of results for carcinogenicity bioassays and notation if listed in NTP *4th Annual Report on Carcinogens*

Note: This description most closely resembles RTECS microfiche format. Format for other media will vary somewhat. All types of information are not available for all substances.
[a]Microfiche can only be updated quarterly; the on-line version of RTECS is updated on a continuing basis.

RTECS file contains information on approximately 250,000 chemicals, listed by primary chemical name and cross-referenced by synonyms (including common and trade names where applicable) and CAS (Chemical Abstracts Service) number. RTECS is available in several forms:

1. books, published periodically (most recently as the 1985–86 edition)
2. microfiche, updated quarterly and available in most federal depository libraries
3. a computerized information file, available on line through the TOXNET system developed by the Toxicology Information Program (TIP) of the National Library of Medicine (NLM)
4. a single CD-ROM (read only memory) disk

The computerized version of RTECS has also been integrated into the Chemical Information System (CIS).[1] Contact information for these sources is given at the end of this chapter. In the application of the methodology presented in Chapter 3, RTECS on line was used since it is updated on an ongoing basis and is therefore the most current source.

Another feature of RTECS that made it uniquely useful in the air toxics methodology presented in this book is that its format provided for rapid reference to studies from the primary literature and relevant secondary sources as well. Some of these secondary sources include general toxicology review articles, IARC (International Agency for Research on Cancer) monographs, NTP (National Toxicology Program) bioassay reports, NIOSH criteria documents, NIOSH Current Intelligence Bulletins (CIBs), and TLV documentation.

While RTECS is a highly useful referral source to primary and secondary literature sources, it has certain limitations that should be recognized if it is to be used efficiently. References to studies reporting toxic effects that are cited in RTECS have been screened to a certain extent in the sense that only positive results (in some cases statistically significant—$p < 0.05$) are included, and in some cases, specifically tumorigenic data, other criteria must be met. However, this also poses a limitation because it does not enable one to distinguish between negative data and no data, which can be very important, especially for carcinogenic and mutagenic effects. In addition, RTECS does not attempt to resolve contradictory data, but "strives to accurately reflect the literature as it exists."[1] Another feature of RTECS, significant in the context of the methodology in Chapter 3, is that the irritation data provided refer to primary skin and eye irritation and do not include references to respiratory irritation. Given this information, it is particularly important to check the availability of secondary data sources described in this chapter and, if no secondary sources are available, to consider the given chemical as a possible candidate for primary literature search and review.

Another caution that should be observed is to avoid using the toxicity figures presented in RTECS as a basis for criteria derivation without consulting the original reference. It is explicitly cautioned in the introduction to the RTECS database:

It is not the purpose of this registry to quantitate a hazard through the use of the toxic concentration or dose data that are presented with each substance. Under no circumstances can the toxic dose values presented with these chemical substances be considered definitive values for describing safe versus toxic levels.[1]

National Air Toxics Information Clearinghouse (NATICH)

NATICH is a medium through which EPA acts as a technical advisor and facilitator of information exchange for air toxics information between state and local governments and the federal government. It was established by the EPA Office of Air Quality Planning and Standards in cooperation with the State and Territorial Air Pollution Program Administrators (STAPPA) and the Association of Local Air Pollution Control Officials (ALAPCO). The clearinghouse collects, classifies, and disseminates air toxics information (via both hardcopy reports and an on-line database) submitted by federal, state, local, and other agencies. A number of these reports with a brief description of each are cited in Table 2.2.

NATICH is not generally useful as a reference to the primary literature, but it is an excellent source of information on the activities of state and local air toxics programs, including such information as substances for which ambient air limits have been derived by various S/L agencies together with applicable methodologic information. Information, such as the names and addresses of contact persons, is also available on other aspects of S/L air toxics programs, including general regulatory programs, permitting of new, modified, and existing sources, source testing, emission inventory, indoor air quality programs, compliance and enforcement, dispersion and exposure modeling, risk communication, ambient air monitoring programs, health effects assessment, and criteria derivation. NATICH also provides ready access to citations to, and in some cases summaries of, information and publications of federal and state agencies. Thus NATICH, while not directly providing references to the primary literature on specific chemicals, is a valuable resource because it provides a mechanism for obtaining recent information and potential guidance from the activities of a variety of regulatory agencies at both the federal and state level on specific chemicals and on a variety of air toxics–related issues.

Integrated Risk Information System (IRIS)

IRIS is an on-line database established by the EPA that is organized on a chemical-specific basis and provides information related to risk assessment (hazard identification and dose-response assessment) and EPA standards, guidelines, and regulatory activities. The data provided in IRIS—specifically the basis for reference dose (RfD) and unit risk estimates—have been reviewed by working groups within the EPA and represent agency consensus.[2] Currently there is information on over 300 chemicals in IRIS,[3] and it is continuously updated. IRIS is available through several channels depending on the affiliation of the user:

1. EPA's electronic mail system for individuals within the agency
2. the Public Health Network (PHN) of the Public Health Foundation for individuals in state and local health departments

Table 2.2 Examples of Various Types of NATICH Publications

A. General NATICH User Information
1. How the Clearinghouse Can Help to Answer Your Air Toxics Questions. July 1986 450/5-86-009.
   —general information on the functions and type of assistance provided by NATICH
2. NATICH Database User's Guide for Data Entry and Editing. February 1987. EPA 450/5-87-002. NTIS: PB87 175576/AS.
   —information on use of the NATICH on-line database
3. NATICH Database User's Guide for Data Viewing. September 1985. EPA 450/5-85-008. NTIS: PB86 123601/AS
   —information on use of the NATICH on-line database
B. Periodic Publications
1. NATICH Report on State and Local Air Toxics Activities. Published annually in July as two or more volumes.
   —descriptions and summaries of activities of S/L agencies related to air toxics (including air toxics contacts; regulatory program information; permitting and regulation of new, modified, and existing sources; health effects assessment; ambient air monitoring; source testing; etc.)
2. NATICH Newsletter. Published bi-monthly.
   —updates on ongoing NATICH activities and activities of S/L and federal agencies relevant to air toxics
3. Bibliography of Selected Reports and Federal Register Notices Related to Air Toxics. Approximately annually.
   Volume 1—cumulative citations from before 1974 through March 1987, EPA-450/5-87-005, NTIS: PB88 136601/REB
   Volume 2—citations from April 1987 to March 1988, EPA-450/5-88-005, NTIS: PB89 103436/REB
   Volume 3—citations from April 1988 to January 1989, EPA-450/3-89-25
   —citations to reports and federal register notices useful in developing and operating air toxics control programs including those from the following: U.S. EPA, NAS NCI, NIEHS, NTP, NIOSH, ATSDR, CPSC, and WHO (also IARC)
4. Ongoing Research and Regulatory Development Projects. Annually in June or July.
   —brief description of projects in progress at EPA, NIOSH, ATSDR, and S/L agencies indexed by agency, project type, chemical name, CAS number, and source category Standard Industrial Classification (SIC) Code
C. Reports and Methodologic Information Publications
1. Methods for Pollutant Selection and Prioritization. July 1986. EPA/450/5-86-010. NTIS: PB87 124079/AS.
2. Case Studies in Risk Communication. February 1988. EPA-450/5-88-003. NTIS: PB89 104277/AS.
3. Rationale for Air Toxics Control in Seven State and Local Agencies. August 1985. EPA/450/5-86-005. NTIS: PB86 181179/AS.
4. Qualitative and Quantitative Carcinogenic Risk Assessment. June 1987. EPA-450/5-87-003. NTIS: PB88 113188/AS.
   —basic principles and assumptions associated with qualitative and quantitative cancer risk assessment with examples and case histories; intended audience is risk managers and staff in S/L agencies

Note: Contact information for NATICH is provided in the list at the end of this chapter.

3. the DIALCOM, Inc., electronic mail system and NLM's TOXNET system for users not affiliated with EPA or S/L health departments

IRIS is also available on diskette from NTIS with quarterly updates. Contact information concerning IRIS access through various media is provided at the end of this chapter.

IRIS is an excellent source of chemical-specific information on EPA regulatory activities and referral to EPA-based reports (e.g., criteria documents, health advisories, health assessment documents, and reportable quantity documents) and selected studies from the primary literature that support the RfD and unit risk estimates, i.e., principal and supporting studies (see description and definition of these terms in Chapter 3). IRIS is particularly useful if it is necessary to quickly obtain screened qualitative and quantitative estimates of risk accompanied by confidence ratings of the overall database for a specific substance. The general categories of data in IRIS are as follows:

- substance identification and use information
- chemical and physical properties
- noncarcinogenic assessment for lifetime exposure (RfD)
- carcinogenicity assessment for lifetime exposure (unit risk estimates)
- drinking water health advisories and acute toxicity information
- aquatic toxicity assessment
- EPA regulatory actions (including exposure guidelines and standards) under the Clean Air Act; Safe Drinking Water Act; Clean Water Act; Federal Insecticide, Fungicide, and Rodenticide Act; Toxic Substances Control Act; Resource Conservation and Recovery Act; and Superfund
- references

Another useful feature of IRIS is that names and contact numbers for EPA personnel with more detailed information on a given aspect of a chemical (e.g., quantitative risk estimates for carcinogens, RfD estimates, and different regulatory activities) are provided for each substance in IRIS.

Gene-Tox Program

The Gene-Tox program was established by the U.S. EPA in 1980 and is sponsored and directed by the Office of Testing and Evaluation (OTE) within the Office of Pesticides and Toxic Substances (OPTS). The two major activities of the Gene-Tox program are (1) evaluation of the existing literature on selected bioassays for mutagenicity and presumptive carcinogenicity and (2) establishment of a database organized on a chemical-specific basis that summarizes the results of various short-term tests designed to detect genotoxic effects.

The Gene-Tox program effort focusing on evaluation of the existing literature on various short-term genotoxicity assays was undertaken to aid EPA in "establishing standard genetic testing and evaluation procedures for the regulation of toxic substances and determining the direction of research and development in the area of genetic toxicology."[4] The structure and review process of this part of the Gene-Tox program has been described by Green and Auletta and by Waters and Auletta.[4,5] The work group reports on specific assays include sections on general test description and protocols, interpretation of data, test performance (number of chemicals and chemical classes tested and discussion of results), conclusions on the strengths and limitations of the assay with suggestions for future research, and references. These reports are usually published in the journal *Mutation Research*, and a bibliography of these reports is provided in Appendix F.

The chemical-specific Gene-Tox database is a peer-reviewed information file that summarizes the results of one or more of 73 short-term genetic toxicology bioassays on over 2500 chemicals covering some 30 different chemical classes.[6] The chemical-specific Gene-Tox database has been described by Palajda and Rosenkranz, and more recently by Ray et al.[6,7] The greatest strengths of the Gene-Tox database are that data on both positive and negative assay results are published (in contrast to RTECS) and the format is summarized and abbreviated, which allows for rapid accessing of relevant information. In the context of the methodology presented in Chapter 3, the Gene-Tox database is a useful source in cases where there is equivocal or limited data on carcinogenic effects or heritable genetic effects. The Gene-Tox database has been integrated into the Chemical Information System (CIS) and is expected to be available through the NLM TOXNET system in the near future.[3] Relevant contact information is provided in the list at the end of this chapter.

The Gene-Tox Carcinogen Database is an integral part of the Gene-Tox chemical-specific database and has been published in hardcopy form. It includes data from three sources:[8]

1. chemicals determined by IARC to have sufficient evidence of carcinogenic activity in experimental animals
2. selected chemicals from NCI/NTP bioassays (see the next section)
3. selected chemicals evaluated in genetic toxicology assays and abstracted from previous Gene-Tox reports

Since it draws data from two of the other sources described here, there will be an overlap of information. However, the summary form in which the information is given (i.e., a few lines for each chemical including species, strain, sex, route of administration, duration of study, significant tumors or organs affected, and an overall evaluation of the individual studies) provides a useful rapid referral point to other studies for the 506 chemicals listed.

National Toxicology Program (NTP)

The NTP is best described as "an interagency program of the federal government which coordinates toxicological programs at the NIH (NIEHS), FDA (NCTR), and CDC (NIOSH) with input from NCI, NIH, OSHA, CPSC, EPA, and ATSDR."[9] The activities of these agencies with respect to toxicologic research and testing programs on specific chemicals are summarized in two annual reports: *Review of Current DHHS, DOE, and EPA Research Related to Toxicology* and the *NTP Annual Plan*. The status of particular chemicals with respect to NTP Division of Toxicology Research and Testing (e.g., selected for general toxicology study, assigned to laboratory for toxicology study, prechronic studies in progress, two-year studies in progress, technical reports printed) as well as summary information on species, route of exposure, and results (single line descriptions for each chemical) for published bioassay reports are given in the *Chemical Status Report*, published quarterly. Contact information for obtaining these publications is given in the contact list at the end of this chapter.

Another valuable source of information from NTP are the NCI/NTP technical reports, which describe the results of long-term cancer bioassays on specific chemicals carried out under the auspices of the NTP carcinogenesis testing program. Results of 14-day and 13-week general toxicity studies and short-term tests for genetic activity are also reported in the NTP technical reports. One advantage of NTP bioassays compared to studies from other sources is that these bioassay reports are not published until the entire study in all its various phases has been subject to an extensive and rigorous review process. In addition, positive results reported in these bioassays are often used as the starting point for EPA/CAG quantitative risk assessments. The process by which chemicals are nominated and selected for testing in the NTP has been described by Heindel;[9] information on the availability of bioassay reports is given in the *Chemical Status Report* mentioned above.

REPROTOX Database

REPROTOX is an on-line computerized database developed by the Reproductive Toxicology Center (RTC) located at the Columbia Hospital for Women Medical Center. Information contained in this database is organized by substance and is broad in scope, including information on both developmental and reproductive toxicology in animal models and humans, and in vitro data. With respect to the degree of review and preparation of summaries on specific agents listed in this database, it is stated in the REPROTOX user's manual that:

> All of the information in the RTC database has been prepared using the RTC computer link with the medical-toxicology-pharmacology databases in this country (e.g., MEDLINE, ETIC, TOXLINE, CHEMLINE, etc.) and the specialized library of the RTC. Information obtained from these references is summarized by RTC professional staff and reviewed for completeness and accuracy by the Director of the RTC and by members of the Scientific Advisory Board of the RTC.[10]

Given the extensiveness of data review, frequency of updating, and the broad scope of information accessible through REPROTOX, it is a good source of summarized comprehensive information on developmental and reproductive effects of specific chemicals, including references to the primary literature. As of 1989, there were over 1500 chemicals, drugs, and agents listed in the REPROTOX database agent list, published annually and updated quarterly in the journal *Reproductive Toxicology*.[11] Access to REPROTOX is obtained through the RTC; contact information is provided in the list at the end of this chapter.

Documentation for Occupational Limits

Three forms of occupational limit documentation are included in this section: the book *Documentation of the Threshold Limit Values*,[12] which supports the ACGIH TLVs, and NIOSH criteria documents and current intelligence bulletin (CIBs), which support NIOSH RELs. It should be noted that there may be other documents that support the RELs and that are referenced in RTECS including Special Hazard Reviews, Occupational Hazard Assessments, and Technical Guidelines, and these should be consulted where appropriate. The annual *NIOSH Publications Catalog,* available in most federal depository libraries, is the most useful reference to consult concerning the availability of NIOSH publications. While the sources reviewed in this section vary in the depth and comprehensiveness of the information they provide, they are useful in the context of the methodology presented in Chapter 3 because they are a reviewed and summarized chemical-specific data source that emphasizes the inhalation route of exposure. In addition, if any consideration is given to basing an ambient air limit on an occupational limit, then review and evaluation of the documentation and the studies cited as the basis for the OEL are essential.

TLV Documentation

The basis for the individual TLVs is published by ACGIH in the form of a book, *Documentation of the Threshold Limit Values*, containing summaries of the relevant studies with references as well as the reasoning used in setting the TLV and is organized alphabetically by chemical name.[12] Since new TLVs are added and older ones modified each year (see the "Notice of Intended Changes" in the summary booklet), it is important to use the most recent edition together with the summary booklet for the current year (e.g., *Threshold Limit Values and Biological Exposure Indices for 1989–90*).[13] Proposed changes are considered trial limits for the first two years, and only TLVs that have been adopted and accepted as final limits are referenced in RTECS.[1] The address for ACGIH is provided in the contacts section at the end of the chapter.

The primary value of TLV documentation in the context of the methodology presented in Chapter 3 is to provide information on the numerical basis for any given TLV. TLV documentation for any given compound is usually two or three pages in length and therefore can provide only a limited description and evaluation of relevant studies.

NIOSH Criteria Documents

The basis for NIOSH-recommended limits are published in the form of criteria documents, which are critical evaluations of all relevant scientific information on a particular chemical or industrial process including preventive measures designed to limit exposure and prevent adverse health effects. These published documents are intended to provide background information and recommendations to OSHA or the

Mine Safety and Health Administration (MSHA) for use in promulgating legal standards.[1]

The criteria documents provide much more extensive documentation on a particular chemical than does TLV documentation. However, their availability is limited compared to TLV documentation: there are over 700 TLVs,[13] but approximately only 130 criteria documents have been published, and some of these are revisions of previous documents. In addition, although criteria documents represent a thorough evaluation of the literature on a given chemical, backed up by extensive expert peer review, many of them are over a decade old and therefore may not contain highly relevant information from more recent studies. Therefore, the best use of most of these documents is as a source of summarized information on the older chemical-specific literature of particular relevance to the occupational exposure situation and as support documents for the NIOSH RELs.

Current Intelligence Bulletins (CIBs)

NIOSH CIBs are one means that is used to partially compensate for the age of many criteria documents since they evaluate "new and emerging information on occupational hazards." In addition, they may also be used by NIOSH as a vehicle to "draw attention to a formerly unrecognized hazard, report new data on a known hazard or disseminate information on hazard control."[1] CIBs are referenced in RTECS, and information on their availability may be obtained from the NIOSH publications dissemination office, listed in the contacts section at the end of the chapter.

Drinking Water Documents

The documents described in this section are published by the EPA and, if available for a particular chemical, are referenced in IRIS. While they are published to support a particular drinking water guideline or standard and therefore do not emphasize the inhalation route of exposure, they are also an excellent review, summary, and critique of the toxicology literature, which also includes inhalation studies, for a specific chemical up to the date of publication. The various forms of drinking water standards and guidelines referred to in these documents have been described in Appendix A.

Drinking Water Criteria Documents (DWCDs)

DWCDs are the technical support documents for the health-based guidelines known as maximum contaminant level goals (MCLGs). They are typically organized into sections covering topics such as the chemical and physical properties of the substance, sources of the chemical in the environment, environmental fate, transport and distribution, typical environmental levels, effects of the substance on biological organisms, metabolism and toxicology of the substance in experimental animals and humans including occupational and epidemiologic studies, and the basis for the proposed MCLG. Since these documents are designed to support drinking water criteria, the major emphasis is on health effects of the agent in humans and animal models,[14] and for this reason much of the information is relevant to the methodology presented in Chapter 3.

Health Advisories (HAs)

Health advisory documents are prepared as part of the Office of Drinking Water's Health Advisory Program to support the drinking water guidelines known as health advisories, described more fully in Appendix A. Health advisories typically provide information similar to that in drinking water criteria documents, only in a much more abbreviated form and with the major emphasis on review of the principal and supporting studies for the various HAs (e.g., one-day, ten-day, longer-term, and lifetime) and description of the basis for derivation of the numerical criterion.[15] While multiple copies of HAs and entire sets of the documents are available through NTIS, single copies of individual HAs are usually available by calling the EPA Safe Drinking Water Hotline, listed at the end of the chapter.

Ambient Water Quality Criteria Documents (AWQCDs)

AWQCDs are published to support criteria developed to be applied to surface and ground waters under the Clean Water Act rather than finished drinking water under the Safe Drinking Water Act. While they contain information similar to that found in DWCDs and at a comparable level of detail, there is a much greater emphasis on the adverse effects of the chemical on ecosystems and aquatic plants and organisms.[16] Most of the AWQCDs that have been issued were published as a group in 1980 (see Table A.5), and although a few others have been subsequently published, AWQCDs are generally older than DWCDs or HAs for a given chemical.

Chemical-Specific Documentation

IARC Monographs

Monographs are published annually as hardcopy volumes by the International Agency for Research on Cancer (IARC) of the United Nations World Health Organization (WHO) and are referenced in RTECS and available in most university-associated libraries. The stated goal of the IARC monographs program, initiated in 1969, is:

> to prepare, with the help of international working groups of experts, and to publish in the form of monographs, critical reviews and evaluations of evidence on the carcinogenicity of a wide range of agents to which humans are or may be exposed. The monographs may also indicate where additional research efforts are needed.[17]

The general organization of the IARC monographs is as follows:

1. chemical and physical data
2. production, use, occurrence, and analysis
3. biological data relevant to the evaluation of carcinogenic risk to humans, including carcinogenicity studies in animals; human epidemiologic studies; data on other toxic effects; information on absorption, metabolism, and excretion; and genotoxic effects based on short-term tests
4. summary of data reported and evaluation
5. references

Due to the extensive nature and quality of the peer review associated with the monographs, they provide an excellent source of critically reviewed information on the potential carcinogenicity of a variety of chemicals and processes as well as referral to the relevant primary literature. An additional strength of these reviews is that weight-of-evidence ratings are provided on the strength of evidence for carcinogenic and genotoxic effects, based on clearly stated and rigorously defined and applied criteria. Recently, supplements (numbers 6 and 7) to the IARC monographs have been published that update these ratings for substances that had previously been evaluated as part of the IARC monographs program.

Health Assessment Documents (HADs)

HADs are prepared by the EPA Office of Health and Environmental Assessment and organized in a similar manner to the DWCDs. However the focus of these documents is broader than the drinking water documents since they were prepared to serve as source documents for agencywide (EPA) use. They are particularly strong in the area of evaluating critical and supporting studies that serve as the basis for RfD—formerly ADIs (acceptable daily intakes)—and unit risk estimates. In addition, they also provide a wealth of detail on the methods used by the EPA to derive these estimates. HADs are usually referenced in IRIS and are available in most federal depository libraries or through the NTIS.

ATSDR Toxicological Profiles

The background of the toxicological profiles as it is stated in the foreword to all the profiles is as follows:

> The Superfund Amendments and Reauthorization Act of 1986 (Public Law 99–499) extended and amended the Comprehensive Environmental Response, Compensation and Liability Act of 1980 (CERCLA or Superfund). This public law (also known as SARA) directed the Agency for Toxic Substances and Disease Registry (ATSDR) to prepare toxicological profiles for hazardous substances which are most commonly found at facilities on the CERCLA National Priorities List and which pose the most significant potential threat to human health, as determined by the ATSDR and the Environmental Protection Agency.[18]

The toxicological profiles are intended to be succinct characterizations of the key toxicologic and health effects information on the chosen substances. They are organized into general sections that cover the following areas:

Table 2.3 ATSDR Toxicological Profiles Published in 1988 and 1989

Substance	
Arsenic	1,2-Dichloropropane
Aldrin/dieldrin	Di(2-ethylhexyl)phthalate
Benzene	2,4-Dinitrotoluene and 2,6-dinitrotoluene
Benzidine	Heptachlor/heptachlor epoxide
Benzo(a)anthracene	α-, β-, γ-, and δ-Hexachlorocyclohexane
Benzo(a)pyrene	Isophorone
Benzo(b)fluoranthene	Lead
Beryllium	Mercury
Bis(2-chloroethyl)ether	Methylene chloride
Bis(chloromethyl)ether	Nickel
Bromodichloromethane	n-Nitrosodimethylamine
Cadmium	n-Nitrosodi-n-propylamine
Carbon tetrachloride	n-Nitrosodiphenylamine
Chlordane	Pentachlorophenol
Chloroethane	Phenol
Chloroform	Selected PCBs
Chromium	Selenium
Chrysene	2,3,7,8-Tetrachlorobenzo-p-dioxin
Cyanide	1,1,2,2-Tetrachloroethane
p,p'-DDT,p,p'-DDE, and p,p'-dinitrotoluene	Tetrachloroethylene
Dibenzo(a,h)anthracene	Toluene
1,4-Dichlorobenzene	1,1,2-Trichloroethane
3,3'-Dichlorobenzidine	Trichloroethylene
1,2-Dichloroethane	Vinyl chloride
1,1'-Dichloroethene	Zinc

1. public health statement
2. health effects
3. chemical and physical information
4. production, import, use, and disposal
5. potential for human exposure
6. analytical methods
7. regulations and advisories
8. references

Substances for which toxicological profiles have been published in 1988 and 1989 are listed in Table 2.3. The most recent complete list of substances for which toxicological profiles will be prepared was published on October 20, 1988, in the *Federal Register*.[19] The toxicologic profiles can generally be found in most federal depository libraries or may be purchased through the NTIS. Contact information is provided at the end of this chapter.

The toxicological profiles are probably one of the most generally useful secondary information sources reviewed in this chapter due to the comprehensive nature of the information that is brought together in a single document. Although similar in content to the information provided in documents such as HADs and drinking water documents, the profiles are more generally useful since their focus is broader (e.g., multiple route of exposure emphasized) and they were published more recently.

Table 2.4 Substances for Which HEADs Have Been Published

Acetone	Lindane
Arsenic	Manganese (and compounds)
Asbestos	Mercury
Barium	Methylene chloride
Benzene	Methyl ethyl ketone
Benzo(a)pyrene	Naphthalene
Cadmium	Nickel
Carbon Tetrachloride	Pentachlorophenol
Chlordane	Phenanthrene
Chlorobenzene	Phenol
Chloroform	Polycyclic biphenyls (PCBs)
Coal Tars	Polychlorinated aromatic hydrocarbons (PAHs)
Copper	Pyrene
Cresols	Selenium (and compounds)
Cyanide	Sodium cyanide
DDT	Sulfuric acid
1,1-Dichloroethane	2,3,7,8-Tetrachlorodibenzo-p-dioxin
1,2-Dichloroethane	1,1,2,2-Tetrachloroethane
1,1-Dichloroethylene	Tetrachloroethylene
1,2-cis-Dichloroethylene	Toluene
1,2-trans-Dichloroethylene	1,1,1-Trichloroethane
Ethylbenzene	1,1,2-Trichloroethane
Glycol ethers	Trichloroethylene
Hexachlorobenzene	2,4,5-Trichlorophenol
Hexachlorobutadiene	2,4,6-Trichlorophenol
Hexachlorcyclopentadiene	Trivalent chromium
Hexavalent chromium	Vinyl chloride
Iron (and compounds)	Xylene
Lead	Zinc (and compounds)

Additional Chemical Specific Documents

Certain other chemical-specific documents, while not as generally applicable in the application of the methodology presented in Chapter 3 as the ones previously reviewed, still have some utility and merit brief description. These include Reportable Quantity Documents (RQDs), Environmental Health Criteria (EHC), and Health Effects Assessment Documents (HEADs).

RQDs are published pursuant to the requirements of CERCLA (Superfund) to provide technical support for the criteria known as reportable quantities (RQs). RQs are environmental release limits in units of pounds for chemicals designated as hazardous substances under certain sections of CERCLA; if release of a substance in amounts equal to, or in excess of, the RQ occurs from a vessel or facility, then notification of the National Response Center is required.[20] RQDs are referenced in IRIS and generally available through NTIS.

The HEADs are a series of 58 short, quantitatively oriented documents prepared by the EPA Environmental Criteria and Assessment Office for the Office of Emergency and Remedial Response and were published in 1986.[21] In general these documents draw heavily from previously published EPA documents such as HADs, DWCDs, AWQCDs and RQDs. Thus there is considerable overlap in the information presented and reviewed. A list of substances for which HEADs were prepared is presented in Table 2.4; these documents are available through NTIS.

EHCs are published as part of the International Program on Chemical Safety

Table 2.5 Titles in the Environmental Health Criteria (EHC) Series

1. Mercury	37. Aquatic (Marine and Freshwater)
2. Polychlorinated Biphenyls and	Biotoxins
Terphenyls	38. Heptachlor
3. Lead	39. Paraquat and Diquat
4. Oxides of Nitrogen	40. Endosulfan
5. Nitrates, Nitrites and N-Nitroso	41. Quintozene
Compounds	42. Tecnazene
6. Principles and Methods for Evaluating	43. Chlordecone
the Toxicity of Chemicals, Part I	44. Mirex
7. Photochemical Oxidants	45. Camphechlor
8. Sulfur Oxides and Suspended	46. Guidelines for the Study of Genetic
Particulate Matter	Effects in Human Populations
9. DDT and Its Derivatives	47. Summary Report on the Evaluation of
10. Carbon Disulfide	Short-Term Tests for Carcinogens
11. Mycotoxins	(Collaborative Study on In Vitro Tests)
12. Noise	48. Dimethyl Sulfate
13. Carbon Monoxide	49. Acrylamide
14. Ultraviolet Radiation	50. Trichloroethylene
15. Tin and Organotin Compounds	51. Guide to Short-Term Tests for Detecting
16. Radiofrequency and Microwaves	Mutagenic and Carcinogenic Chemicals
17. Manganese	52. Toluene
18. Arsenic	53. Asbestos and Other Natural Mineral
19. Hydrogen Sulfide	Fibers
20. Selected Petroleum Products	54. Ammonia
21. Chlorine and Hydrogen Chloride	55. Ethylene Oxide
22. Ultrasound	56. Propylene Oxide
23. Lasers and Optical Radiation	57. Principles of Toxicokinetic Studies
24. Titanium	58. Selenium
25. Selected Radionuclides	59. Principles for Evaluating Health Risks
26. Styrene	from Chemicals During Infancy and
27. Guidelines on Studies in Environmental	Early Childhood
Epidemiology	60. Principles for the Assessment of
28. Acrylonitrile	Neurobehavioral Toxicology
29. 2,4-Dichlorophenoxy-acetic Acid (2,4-D)	61. Chromium
30. Principles for Evaluating Health Risks to	62. 1,2-Dichloroethane
Progeny Associated with Exposure to	63. Organophosphorus Insecticides: A
Chemicals During Pregnancy	General Introduction
31. Tetrachloroethylene	64. Carbamate Insecticides: A General
32. Methylene Chloride	Introduction
33. Epichlorohydrin	65. Butanols: Four Isomers
34. Chlordane	66. Kelevan
35. Extremely Low Frequency (ELF) Fields	67. Tetradifon
36. Fluorides and Fluorine	

(IPCS), which is a joint venture of the United Nations Environment Program, the International Labor Organization, and the WHO. These documents represent the collective views of international expert committees and present information to support recommended environmental criteria. They are organized in the following general format:[22]

1. summary and conclusions
2. identity, physical and chemical properties, and analytical methods
3. sources, environmental transport, and distribution
4. environmental levels and human exposure

5. kinetics and metabolism
6. effects on organisms in the environment
7. effects on experimental animals and in vitro test systems
8. effects on man
9. evaluation of risks for human health and effects on the environment
10. recommendations
11. previous evaluations by international bodies
12. references

A list of titles in the series is provided in Table 2.5; availability information is in the contacts section at the end of the chapter.

Conclusions

The secondary data sources reviewed in this chapter are those that were both used most extensively in the application of the methodology in Chapter 3 and judged to be most generally useful. It must be emphasized that the state and organization of information in toxicology is dynamic and constantly growing and useful new information sources are continually becoming available. In acknowledgment of this trend several recent articles have reviewed the organization and state of information in toxicology and are worthy of mention.[3,23,24]

Contacts for Information Sources

ACGIH (information related to TLVs):
    American Conference of Government Industrial Hygienists
    6500 Glenway Avenue, Building D-5
    Cincinnati, OH 45211
    phone (513) 661–7881
Air Risk Information Support Center:
    AIR RISC Hotline (919) 541–0888
Chemical Information Systems (RTECS, on line):
    7215 York Road
    Baltimore, MD 21212
    phone (800) CIS-USER
EPA Safe Drinking Water Hotline:
    phone (800) 426–4791
Gene-Tox Database via CIS:
    Chemical Information Systems
    Fein Marquart Associates
    7215 York Road
    Baltimore, MD 21212

Gene-Tox Database (general information and current status):
John S. Wassom
Environmental Mutagen, Carcinogen, and
    Teratogen Information Program
Oak Ridge National Laboratory
P.O. Box Y
Oak Ridge, TN 37831
IRIS Access via NLM:
IRIS Representative
Specialized Information Services
National Library of Medicine
8600 Rockville Pike
Bethesda, MD 20894
phone (301) 496–6531
IRIS Access via DIALCOM, Inc.:
DIALCOM, Inc.
Federal Systems Division
6120 Executive Blvd.
Suite 150
Rockville, MD 20852
IRIS on 5¹/₄-Inch Floppy Disks:
National Technical Information Service
U.S. Department of Commerce
5285 Port Royal Road
Springfield, VA 22161
phone (703) 487–4807
IRIS User Support (for general assistance and information):
phone (513) 569–7254
NATICH Information:
U.S. Environmental Protection Agency
826 Mutual Plaza (MD-13)
Research Triangle Park, NC 27711
phone (919) 541–0850
National Technical Information Service (NTIS):
5285 Port Royal Road
Springfield, VA 22161
phone (800) 336–4700
NIOSH Publications Availability (CIBs, etc.):
Publications Dissemination
Division of Standards Development and Technology Transfer
National Institute of Occupational Safety and Health
4676 Columbia Parkway
Cincinnati, OH 45226
NTP Information and Publications Availability:
Public Information Office, NTP

P.O. Box 12233
Research Triangle Park, NC 27709
phone (919) 541–3991
REPROTOX Database Access Information:
Reproductive Toxicology Center
Columbia Hospital for Women Medical Center
2425 L Street, N.W.
Washington, DC 20037–1485
phone (202) 293–5137
RTECS Database in Hardcopy:
Superintendent of Documents
U.S. Government Printing Office
Washington, DC 20402
phone (202) 783–3238
    document no. 017–033–00431–5
    (also source for RTECS in microfiche)
RTECS Database on CD/ROM Disk:
SilverPlatter Information, Inc.
37 Walnut St.
Wellesley, MA 02181
phone (617) 239–0306
WHO Publications Center (EHC documents):
49 Sheridan Avenue
Albany, NY 12210
General Directory of Information Sources in Toxicology:
Wexler, P. 1988. *Information Resources in Toxicology*
(New York: Elsevier).

References

1. NIOSH. 1990. "Registry of Toxic Effects of Chemical Substances (RTECS)." D. V. Sweet, Ed., quarterly microfiche (Cincinnati, OH: Dept. Health and Human Services, NIOSH), DHHS (NIOSH) 90–101–1.
2. National Library of Medicine (NLM). 1990. "IRIS—Integrated Risk Information System—Fact Sheet" (Bethesda, MD: DHHS, NLM).
3. Wexler, P. 1990. "The Framework of Toxicology Information." *Toxicology* 60:67–98.
4. Waters, M. D., and A. Auletta. 1981. "The GENE-TOX Program: Genetic Activity Evaluation." *J. Chem. Inform. Comp. Sci.* 21:35–38.
5. Green, S., and A. Auletta. 1980. Editorial introduction to the reports of "The GENE-TOX Program." *Mutat. Res.* 76:165–68.
6. Ray, V. A., L. D. Kier, K. L. Mannan, R. T. Haas, A. E. Auletta, J. S. Wassom, S. Nesnow, and M. D. Waters. 1987. "An Approach to Identifying Specialized Batteries of Bioassays for Specific Classes of Chemicals: Class Analysis Using Mutagenicity and Carcinogenicity Relationships and Phylogenetic Concordance and Discordance Patterns. 1. Composition and Analysis of the Overall Data Base." *Mutat. Res.* 185:197–241.

7. Palajda, M., and H. S. Rosenkranz. 1985. "Assembly and Preliminary Analysis of a Genotoxicity Database for Predicting Carcinogens." *Mutat. Res.* 153:79–134.

8. Nesnow, S., M. Argus, H. Bergman, K. Chu, C. Frith, T. Helmes, R. McGaughy, V. Ray, T. J. Slaga, R. Tennant, and E. Weisburger. 1986. "Chemical Carcinogens: A Review and Analysis of the Literature of Selected Chemicals and Establishment of the Gene-Tox Carcinogen Database." *Mutat. Res.* 185:1–195.

9. Heindel, J. J. 1988. "The National Toxicology Program Chemical Nomination Selection and Testing Process." *Reproductive Toxicol.* 2:273–79.

10. Reproductive Toxicology Center (RTC). 1986. *REPROTOX User's Manual* (Washington, DC: RTC, Columbia Hospital for Women).

11. Reproductive Toxicology Center (RTC). 1989. "The REPROTOX Agent List." *Reproductive Toxicol.* 2:63–77.

12. ACGIH. 1986. *Documentation of the Threshold Limit Values, 5th ed.* (Cincinnati, OH: American Conference of Governmental Industrial Hygienists).

13. ACGIH. 1989. *Threshold Limit Values and Biological Exposure Indices for 1989–90* (Cincinnati, OH: American Conference of Governmental Industrial Hygienists).

14. U.S. EPA. 1985. "National Primary Drinking Water Regulations; Synthetic Organic Chemicals, Inorganic Chemicals and Microorganisms; Proposed Rule." *Fed. Reg.* 50: 46936–7025.

15. Ohanian, E. V. 1989. "Office of Drinking Water's Health Advisory Program." In E. J. Calabrese, C. E. Gilbert, and H. Pastides (eds.). *Safe Drinking Water Act—Amendments, Regulations and Standards* (Chelsea, MI: Lewis Publishers), pp. 85–103.

16. U.S. EPA. 1980. "Water Quality Criteria Documents; Availability." *Fed. Reg.* 45: 79318–79.

17. IARC. 1987. "IARC Monographs Program on the Evaluation of Carcinogenic Risks to Humans—Preamble." *IARC Monog.* suppl. 7:17–34.

18. ATSDR. 1989. "Toxicological Profile for Phenol." ATSDR/TP-89/20.

19. ATSDR. 1988. "Hazardous Substances Priority List, Toxicological Profiles; Second List." *Fed. Reg.* 53:41280–85.

20. U.S. EPA. 1987. "Reportable Quantity Adjustments." *Fed. Reg.* 52:8140–94.

21. U.S. EPA. 1986. "Project Summary: Health Effects Assessment Documents." EPA/ 540/S1–86/059.

22. World Health Organization (WHO). 1986. *EHC 67—Tetradifon* (Geneva, Switzerland: WHO).

23. Wexler, P. 1990. "Toxicological Information Series; II. A Survey of Toxicology Information." *Fund. Appl. Toxicol.* 14:649–57.

24. Cosmides, G. J. 1990. "Toxicological Information Series; I. Toxicological Information." *Fund. Appl. Toxicol.* 14:439–43.

3

AN APPROACH FOR THE
RAPID DEVELOPMENT OF
SCIENTIFICALLY DEFENSIBLE AALs

Scope and Limitations

It is our contention that any methodology for the development of ambient air guidelines should carefully consider the risk assessment methodologies recommended by National Academy of Sciences expert committees and used by federal regulatory agencies (e.g., EPA, FDA). There are two reasons for this. First, the risk assessment methodologies used by federal agencies have been extensively researched, often by the NAS expert committees, subjected to extensive peer review from leading scientists in a variety of relevant disciplines as well as receiving public comment and revision before final adoption. Second, although efforts to develop ambient air quality standards and guidelines are presently lagging at the federal level (i.e., EPA), if this should change it is reasonable to expect that risk assessment methodologies for air toxics will be consistent with previous institutionalized risk assessment practice, given the explicit desire of the federal government, including EPA, to seek greater inter- and intraagency consistency in risk assessment methodologies.[1]

However, the specific application of the methodology may be modified by the resources of the organization undertaking the program. For example, the approach selected has to be consistent with limited organizational resources and the desired time frame for completing the task. This may at times place extreme constraints on the selection of the risk assessment methodology and database utilized. In circumstances where resources are limited and the time constraints are extreme, the state or private sector entity may, for example, be forced to use secondary literature sources rather than primary and may have to utilize and modify preexisting guidelines (e.g., occupational limits), among other time- and resource-conserving options. Therefore, although a preferred method is indicated at the various steps in the AAL derivation process outlined in this methodology, options will be provided at various decision points in order to allow maximum flexibility. In addition, when applicable, the limitations for a given option or recommendation are discussed and the implicit assumptions stated.

The numerical values derived utilizing the methodology presented in this chapter are referred to as ambient air level goals (AALGs). The reason for this designation is that these values are based only on health effects and do not include consideration of technical, economic, and analytical feasibility or any of the other issues that are within the realm of risk management. In this sense the AALGs derived using this methodology are somewhat analogous to the maximum contaminant level goals (MCLGs), derived by the U.S. EPA for drinking water, and like MCLGs are intended for use only as health-based guidelines.

Classification of Toxicity

In order for any AALG to be toxicologically defensible it should be endpoint-based; that is, an AALG should be based on the prevention of a specific adverse health effect. Thus, it is first necessary to categorize chemicals with respect to the endpoint(s) of concern prior to proceeding with the derivation of AALGs. In this section, a health effects categorization based upon information obtainable from the RTECS database is described and its criteria and justification presented.

The RTECS database, described in detail in Chapter 2, was selected for use for practical considerations since it provides the most comprehensive toxicity information available in a form that is readily accessible via a variety of media, is updated frequently, and is at least partially screened—generally only studies in which statistically significant ($p < 0.05$) toxic effects were manifest are listed for most endpoints. In addition, it is possible to use RTECS as a source for referral to studies in the primary literature and to useful secondary sources such as IARC monographs, ACGIH TLV documentation, NIOSH criteria documents and CIBs, NTP carcinogen technical reports, and the EPA Gene-Tox database to a limited extent. However, due to the limitations of the RTECS database described in Chapter 2, it should be used in conjunction with additional secondary data sources such as those described above. Those recommended and used in the chemical-specific assessments in Part II include ambient water quality criteria documents, drinking water criteria documents, drinking water health advisories, ATSDR toxicological profiles, EPA health assessment documents (HADs) and health effects assessment documents (HEADs), and the IRIS database. These data sources, and their advantages and limitations, were described in Chapter 2.

A preliminary health effects categorization scheme based on use of RTECS is outlined in Table 3.1, and a sample of a chemical-specific worksheet that may be used with it is shown in Appendix B. This is a positive classification scheme in which the categories are not mutually exclusive. It is also a conservative scheme: in some cases a substance is likely to be flagged as belonging to several classes, and then further evaluation may dictate that it does not belong in a particular class or that insufficient data are available to base the criterion on that endpoint. For example, a substance may be initially flagged as belonging to the carcinogen category; subsequent examination of the appropriate studies may indicate that the results were

Table 3.1 Preliminary Categorization of Health Effects Using RTECS

Class	Criteria
A. Carcinogenic	—1. IARC "limited" or "sufficient" evidence in humans or animal models *or*
	2. ACGIH known or suspected human carcinogen *or*
	3. "Clear evidence," "some evidence," or "equivocal evidence" designation in an NCI or NTP bioassay *or*
	4. Listed as a "known" or "reasonably anticipated" carcinogen in the most recent NTP *Annual Report on Carcinogens* *or*
	5. Positive in two or more tests in different species, or the same species by different routes, using acute (single) or chronic exposure protocols
B. Genotoxic	—Positive in one or more assay systems designed to detect mutation (heritable changes in genetic material), DNA damage, chromosomal aberrations, or cell transformation
C. Developmental	—Positive in one or more studies in mammals in which effects on the embryo or fetus, or specific developmental abnormalities or effects on the newborn, were demonstrated
D. Reproductive	—Positive on one or more studies in mammals in which effects on fertility or maternal or paternal effects were demonstrated
E. Systemic Toxicity	—Adverse effects on mammalian systems exclusive of effects in categories A to D above and occurring following acute or chronic exposure by any route
F. Skin/Eye Irritant	—Primary eye or skin irritant in mammals or humans[a]

Note: This categorization should only be considered preliminary since it is very likely that further examination of the secondary data sources will modify it.

[a]The major concern with irritant effects in the context of this methodology is with respect to the respiratory tract; it should be noted that RTECS is of limited value in this context since the irritation data presented usually refer to primary skin and eye irritation test results in animal models, e.g., the Draize procedures. The best source of information on human respiratory and ocular irritant effects is usually TLV documentation and NIOSH criteria documents, if available.

equivocal or the study design deficient, in which case the AALG may be based on another endpoint.

Following a preliminary screening and categorization, which may be done for a large set of chemicals prior to proceeding with the next steps, further information must be obtained based on the categories into which a substance was placed (achievable by review of secondary data sources cited in RTECS and available from other sources) and on the regulatory priorities of the user. A suggested organizational framework for prioritization and application of the methodology is shown in Figure 3.1. For substances placed in multiple classes, it will be necessary to decide whether to (1) set the AALG based upon the endpoint of greatest concern after ascertaining that it definitely belongs in that class or (2) calculate the AALG according to the procedures for each class into which the agent definitely falls and then select the most conservative figure. With respect to the former option, the order of categories in Table 3.1 could be used as presented or with modification as a prioritization tool.

General Issues

The methodology presented in this chapter is based, for the most part, on generally accepted principles of risk assessment; a number of these principles that

Figure 3.1   Prioritization scheme and organizational framework for application of the AALG derivation methodology.

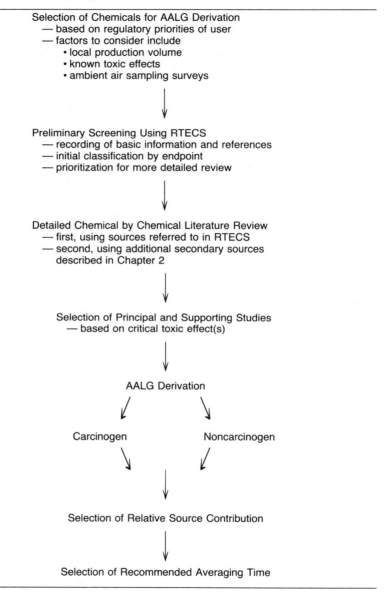

Selection of Chemicals for AALG Derivation
— based on regulatory priorities of user
— factors to consider include
  • local production volume
  • known toxic effects
  • ambient air sampling surveys

↓

Preliminary Screening Using RTECS
— recording of basic information and references
— initial classification by endpoint
— prioritization for more detailed review

↓

Detailed Chemical by Chemical Literature Review
— first, using sources referred to in RTECS
— second, using additional secondary sources
  described in Chapter 2

↓

Selection of Principal and Supporting Studies
— based on critical toxic effect(s)

↓

AALG Derivation

↙   ↘

Carcinogen   Noncarcinogen

↘   ↙

↓

Selection of Relative Source Contribution

↓

Selection of Recommended Averaging Time

are pertinent to AALG derivation for all endpoints will be reviewed in this section. These include the concept of what constitutes an adverse effect and the critical effect, as well as the concept of thresholds, dose scaling, and finally selection of principal and supporting studies.

Adverse Effects and Thresholds

In the previous section the concept that *any AALG must be endpoint based* was described. Two further distinctions are important within the context of this concept. One is the idea of what constitutes an *adverse effect,* and the other is that the AALG should also be based on the *critical effect.* In the context of this methodology, an adverse health effect is considered to be any alteration in structure or function that is clearly deleterious to the organism, or put another way, a degree of injury such that the body's normal compensatory and protective mechanisms are overwhelmed, resulting in irreversible or only partially reversible functional changes. Implicit in this definition is the idea that transitory physiological changes that might be considered adaptive or whose functional significance is uncertain are not considered adverse effects for purposes of AALG derivation. It should be noted that this functional definition of an adverse effect is consistent with current regulatory practice in the United States.[2]

The critical effect has been defined as ''the first adverse effect, or its known precursor, that occurs as the dose rate increases.''[3] For operational purposes, the critical effect for threshold endpoints is usually defined by the most conservative LOAEL. While operational aspects of this are discussed further subsequently, one important point to note here is that the critical effect for a given substance is a function of the available data and frequently may be revised downward as more information becomes available (e.g., studies encompassing further endpoints such as developmental or neurotoxicity) or more sensitive indicators of toxicity are developed and used.[4]

At the present time there two general approaches used in the derivation of numerical health-based guidelines; the approach used is based on the category into which the endpoint falls: *threshold vs. nonthreshold* effects. In the case of nonthreshold effects (i.e., those for which there is theoretically no dose at which there is not some risk of an adverse effect), the general approach has been to use some type of model to estimate either the risk for a given dose (e.g., lifetime exposure to 1 $\mu g/m^3$ of a substance in air) or the dose for a given risk (e.g., one in a million). While both cancer and mutation are generally considered nonthreshold endpoints, in practice numerical estimates of risk, using models such as the linearized multistage and others, are usually derived only for substances identified as carcinogens or associated with tumorigenic outcomes in animal bioassays or human epidemiologic studies, respectively. Most other endpoints, including those associated with developmental and reproductive effects, systemic toxicity, and irritation, are treated operationally as threshold effects, and limits are derived using an uncertainty factor approach. This generally involves modification of a NOAEL or LOAEL from a suitable animal or human study by a series of independent (multiplicative) uncertainty factors. Both the model-based approach and the uncertainty factor approach are described in greater detail in subsequent sections.

While an in-depth discussion of the threshold concept is beyond the scope of this work, the concept is so central to the foregoing distinction between approaches to deriving criteria that a brief review is warranted. In general, nonthreshold, or stochastic, phenomena are considered to be able to arise from damage to a single cell (the target molecule being DNA); threshold phenomena are considered to have a multicellular pathology with a wide variety of possible etiologies involving different physiological processes. For nonthreshold phenomena there is risk of an adverse outcome associated with all exposure levels; the risk of an adverse outcome for threshold phenomena vanishes below a certain dose. In the case of nonthreshold phenomena, it should be noted that at low exposure levels the risk of a given outcome (e.g., tumor induction) may be below the spontaneous or background risk;[2,5] this gives rise to what has been referred to as a "practical threshold." Both the threshold concept and the arguments for and against the existence of thresholds for specific endpoints have been reviewed extensively in a number of articles.[5-10]

In the methodology presented in this chapter, a model-based approach is recommended for AALG derivation for carcinogens in most cases, and an uncertainty factor approach is recommended for endpoints indicative of developmental, reproductive and systemic toxicity, and irritant effects.

Dose Scaling and Conversions

In the course of developing acceptable health-based exposure limits for drinking water or air from animal data, it is necessary to convert the animal dose or exposure to an equipotent value for humans. This process is commonly referred to as *dose scaling*. A *scaling factor* may be defined as the "quantitative estimate of the selected biological characteristic which normalizes the exposure between species."[11] Three general types of scaling techniques will be discussed here, although not all of them were used in this methodology: (1) multiple species regression, (2) functional activity-based techniques (e.g., caloric demand and lifespan), and (3) physical characteristic–based techniques (e.g., body weight and body surface area).

In the *multispecies regression technique*, the dose (mg/kg) of a chemical required to produce a given toxicity endpoint is determined for at least four species. The log of the mean endpoint dose (Y) is regressed on the log of the mean body weight (X) for each test species, e.g., log Y = log a + b log X. This equation is then used to estimate the mean endpoint dose for a 70-kg human, and the test species and the ratio of these two is used as a toxicity scaling factor.[12] Unlike functional activity– and physical characteristic–based techniques, multispecies regression requires multiple data points, and experimental work may be needed to generate the necessary data; these factors may account for the apparent lack of application of this technique in practice. However, it should be noted that the use of multiple data points allows variation in the data to be treated statistically as additive, rather than multiplicative error as is the case when a single data point is used for extrapolation.[11] Further details on this technique are given by Krasovskij.[12,13] While this particular technique has certain attractive aspects, it was not employed in the AALG derivation methodology presented here because the extreme data requirements of the method (studies using multiple species and multiple dose levels) make its application impractical in most cases—necessary data are frequently not available.

Two measures of functional activity that may be used as a basis for dose scaling are caloric demand and lifespan. When using caloric demand, the dose concentra-

Table 3.2 Sequence with Options for Chemicals Flagged in the Carcinogen Class During Preliminary Screening

Utilize EPA weight-of-evidence classification (see Table 3.3)
   Option 1   Use established rating from
           —IRIS
           —Other EPA documentation
   Option 2   Use alternative classification
           —IARC review (see Table 3.4)
   Option 3   Evaluate evidence based on studies in the primary literature using EPA or
           IARC classifications for guidance

Decide on general approach for AALG derivation based on weight-of-evidence
   classification
   —if in Group A, B1, or B2 (or IARC 1 or 2A)
       GO TO [A]
   —if in Group C (or IARC 2B)
       GO TO [B]
   —if in Group D or E (or IARC 3 or 4)
       DERIVE AALG BASED ON ANOTHER ENDPOINT

[A] Group A, B1, B2
     Option 1   Use QRA from IRIS
     Option 2   Use QRA from alternative EPA documentation
     Option 3   Perform QRA using appropriate models and EPA methods
[B] Group C
     Option 1   Use QRA from IRIS if available
     Option 2   Evaluate weight of evidence for genotoxic effects using IARC criteria (see
             Tables 3.5 and 3.6)
             —if weight of evidence is *sufficient*
               QRA is preferred method for AALG derivation
             —if weight of evidence is *limited*
               Uncertainty Factor Approach is preferred
             —if weight of evidence is *inadequate*
               Uncertainty Factor Approach should be used *or* derive AALG based
               on another endpoint

Note: Refer to Table 3.1 for explanation of preliminary categorization. All options are listed in preferred order. Abbreviations in table: IRIS—Integrated Risk Information System; QRA—quantitative risk assessment.

context of the AALG derivation methodology presented here, if any of the five criteria listed in Table 3.1 under subheading *A* are met, then the chemical should be flagged in the preliminary screening stage, and further evaluation, with the possibility of AALG derivation based on carcinogenicity, is necessary.

The steps that should be followed if a substance is flagged for the carcinogen class during the preliminary screening are outlined sequentially with options in preferred order in Table 3.2. The first step is to determine which group within the EPA weight-of-evidence classification that the chemical belongs in; the EPA weight-of-evidence classification is outlined in Table 3.3. Special note should be taken of the use of the terms *sufficient* and *limited*, defined in the table footnotes, since these EPA definitions form the basis of study and data selection for carcinogen AALG derivation. The simplest and most reliable way to make the weight-of-evidence determination is to consult EPA chemical-specific documentation. The best source to consult is IRIS (Integrated Risk Information System) since it is a

Table 3.3 EPA Weight-of-Evidence Classification for Carcinogens and Basis for Suitability of Agents for Quantitative Risk Assessment

Group	Class and Criteria	QRA[a]
A	Carcinogenic to Humans —sufficient[b] evidence from epidemiologic studies to establish a causal association between exposure to agent and cancer	Yes
B	Probably Carcinogenic to Humans —evidence of human carcinogenicity from epidemiologic studies inadequate, class as B1 or B2:	Yes
B1	—limited[c] evidence of carcinogenicity to humans from epidemiologic studies	
B2	—inadequate human data, but sufficient animal study evidence	
C	Possibly Carcinogenic to Humans —limited evidence of carcinogenicity in animals in the absence of human data	Variable[d]
D	Not Classified —inadequate evidence of carcinogenicity	No
E	No Evidence of Carcinogenicity to Humans —no evidence based on both animal data and human epidemiologic evidence	No

[a]Suitability for quantitative risk assessment as determined by EPA.
[b]*Sufficient evidence* means there is an increased incidence of malignant tumors or combined malignant and benign tumors in a well-designed and conducted study. Benign tumors are combined with malignant tumors only if they have the potential to progress to malignancies of the same morphologic type.
[c]*Limited evidence* means that the available evidence is based upon a single species, strain, or experiment or that the experimental design is flawed by inadequate dosage levels, inadequate duration of exposure, inadequate follow-up, poor survival, too few animals, or an increase in benign tumors only.
[d]Suitability for quantitative risk assessment determined on a case-by-case basis.

computerized on-line database and is updated at frequent intervals to reflect new information. Other possible sources are health advisories, health effects assessment documents (HEADs), health assessment documents (HADs), drinking water criteria documents (DWCDs), and ATSDR toxicological profiles. In general, the most recent data source should be consulted.

In the event that the EPA has not evaluated the weight of evidence for carcinogenicity of a specific compound, this must be determined by the user based on a review of the relevant studies using the criteria in Table 3.3. Before proceeding with a detailed review of the relevant studies from the primary literature, it is worthwhile to review secondary sources such as the IARC monographs if these are available since the quality of the primary studies is usually evaluated as well the literature being reviewed. The IARC weight-of-evidence classification (see Table 3.4) is quite similar to that of the EPA and is a reasonable substitute for the EPA evaluation. Again it is important to take careful note of the definitions of *sufficient*, *limited*, and *inadequate* evidence.[32]

If it is necessary to review and evaluate studies from the primary literature, excellent guidance to the evaluation of the quality of studies and criteria to consider in the selection of critical and supporting studies can be found in Appendix B of the IRIS database[33] and the *EPA Guidelines for Carcinogen Risk Assessment*.[25] If specific issues concerning experimental design arise in the course of evaluating a

Table 3.4 IARC Weight-of-Evidence Classification for Carcinogenicity with Definitions for Degree of Evidence

Group 1	The agent is carcinogenic to humans —*sufficient* evidence of carcinogenicity in humans
Group 2A	The agent is probably carcinogenic to humans —*limited* evidence in humans and *sufficient* evidence in experimental animals —only *limited* evidence in humans or only *sufficient* evidence in experimental animals in the presence of other supporting data
Group 2B	The agent is possibly carcinogenic to humans —*limited* evidence in humans in the absence of *sufficient* evidence in experimental animals —*sufficient* evidence of carcinogenicity in experimental animals with *inadequate* evidence or no data in humans —*limited* evidence in experimental animals with other supporting data with *inadequate* or no data in humans
Group 3	The agent is not classifiable as to human carcinogenicity —category used for agents not falling in any other group
Group 4	The agent is probably not carcinogenic to humans —category for agents for which there is *evidence suggesting lack of carcinogenicity* for both humans and experimental animals —occasionally used for agents for which there is *inadequate* evidence or no data in humans and *evidence suggesting lack of carcinogenicity* in experimental animals which is "consistently and strongly supported by a broad range of other relevant data"

Notes: This table was assembled from material by IARC and the definitions below are taken from that reference.[32]

Definitions for degree of evidence used by IARC for human carcinogenicity data:

Sufficient evidence of carcinogenicity—A causal relationship has been established between exposure to the agent and human cancer; i.e., a positive relationship has been observed between exposure to the agent and cancer in studies in which chance, bias, and confounding could be ruled out with reasonable confidence.

Limited evidence of carcinogenicity—A positive association has been observed between exposure to the agent and cancer for which a causal interpretation is considered by the working group to be credible, but chance, bias, or confounding could not be ruled out with reasonable confidence.

Inadequate evidence of carcinogenicity—The available studies are of insufficient quality, consistency, or statistical power to permit a conclusion regarding the presence or absence of a causal association.

Evidence suggesting lack of carcinogenicity—There are several adequate studies covering the full range of doses to which human beings are known to be exposed, which are mutually consistent in not showing a positive association between exposure to the agent and any studied cancer at any observed level of exposure. A conclusion of "evidence suggesting lack of carcinogenicity" is inevitably limited to the cancer sites, circumstances and doses of exposure, and length of observation covered by the available studies. In addition, the possibility of a very small risk at the levels of exposure studied can never be excluded.

Definitions for degree of evidence used by IARC for carcinogenicity in experimental animals:

Sufficient evidence of carcinogenicity—A causal relationship is considered to have been established between an agent and an increased incidence of malignant neoplasms or an appropriate combination of benign and malignant neoplasms in (1) two or more species of animals or (2) in two or more independent studies in one species carried out at different times or in different laboratories or under different protocols. Under exceptional circumstances, a single study in one species might be considered to provide sufficient evidence of carcinogenicity when malignant neoplasms occur to an unusual degree with respect to incidence, site, type of tumor, or age at onset.

Limited evidence of carcinogenicity—The data suggest a carcinogenic effect, but are limited for making a definitive evauation because e.g., (1) the evidence of carcinogenicity is restricted to a single experiment; or (2) there are unresolved questions regarding the adequacy of the design, conduct, or interpretation of the study; or (3) the agent increases the incidence only of benign neoplasms or lesions of uncertain neoplastic potential, or of certain neoplasms that may occur spontaneously in high incidences in certain strains.

Inadequate evidence of carcinogenicity—The studies cannot be interpreted as showing either the presence or absence of a carcinogenic effect because of major qualitative or quantitative limitations.

Evidence suggesting lack of carcinogenicity—Adequate studies involving at least two species are available that show that, within the limits of the tests used, the agent is not carcinogenic. A conclusion of evidence suggesting lack of carcinogenicity is inevitably limited to the species, tumor sites, and doses of exposure studied.

study, then one of the appropriate papers in the bibliography in Appendix E can be consulted. In locating high-quality studies, special note should be taken of the NTP bioassays (described in Chapter 2) and the availability of data from this source ascertained.

The general procedural options for AALG derivation for carcinogens based on the weight-of-evidence classification are outlined in Table 3.2. Based on current theories on the biological basis of cancer as a multistage process consisting of initiation, promotion, and progression[31] and consistency with EPA and NAS methodology, it is proposed that quantitative risk assessment (QRA) procedures, as described in Anderson and EPA,[14] be used to estimate lifetime risk from carcinogen exposure for agents that fall in EPA classes A, B1, or B2, or IARC groups 1 or 2A. A *de minimus* risk level of 10^{-6}, or one in a million, for lifetime cancer risk was selected for carcinogen AALGs in the chemical-specific assessments in Part II. However, depending on the regulatory priorities of the user, a *de minimus* risk level that is lower (10^{-5}) or higher (10^{-7}) may be selected. The best option for deriving AALGs in this manner is to use EPA QRA results since these are extensively peer reviewed prior to final publication and the application of QRA methods to data from the primary literature requires considerable expertise and appropriate computer hardware and software. The best source for EPA QRAs is again IRIS since it is updated most frequently; however, other EPA documents, such as those mentioned previously, may be used. In general, the best source is the most recently published one. In the event that an EPA QRA is not available for a specific compound, it will be necessary for the user either to perform a QRA using EPA methods and the appropriate software or to locate such data from another source. For this latter option, it may be possible to identify a state or local agency that has already performed a QRA for a given compound through use of the yearly NATICH publication "NATICH Data Base Report on State, Local and EPA Air Toxics Activities."[34]

For agents that are classified into either EPA class C or IARC group 2B, it is recommended that the availability of EPA QRAs be ascertained and used as described above if available, particularly if the study from which the data came used the inhalation route. If the inhalation route was not used and a bioassay that used the inhalation route is available, then consideration should be given to either performing a QRA based on that data or using the uncertainty factor approach, depending on the weight-of-evidence determination for genotoxic activity based on short-term tests, discussed in the next section.

If the decision is made to use the uncertainty factor approach to derive an AALG based on carcinogenicity, then the general scheme is the same as that described for systemic and reproductive toxins in the section on the uncertainty factor approach (see Table 3.8). Specific exceptions to the general scheme are (1) use of a 10-fold rather than 5-fold factor is recommended when a LOAEL rather than a NOAEL is used and (2) use of the database factor is usually indicated based on the severity of the effect and depending on the overall weight of evidence for genotoxic effects. The reason for recommending a larger uncertainty factor when using a LOAEL is based on the presumed nonthreshold nature of the carcinogenic response and limitations in the database[35] for this uncertainty factor, described in the section on the basis and use of uncertainty factors. In addition, in their reanalysis of the data, Dourson and Stara pointed out that when the ratios of corresponding LOAELs to NOAELs are compared for both subchronic (27 comparisons) and chronic (25 comparisons) studies, the ratio approached 10 in a few cases, although it was less than 5 in most cases (96%).[36] The issues related to use of the database factor are discussed in the section on the uncertainty factor approach and specific guidelines given in Table 3.13.

Use of Genotoxicity Data

Initial classification of a substance into this category indicates that there is at least some evidence that the agent may cause heritable changes or damage leading to heritable changes in genetic material. Genotoxicity data, usually from a battery of short-term tests, is generally used for one of two purposes: to predict carcino-genicity or to predict potential human germ-cell mutagenicity.[26] Because the de-velopment of batteries of short-term tests predictive for carcinogenicity is an area of continued research and debate,[37] and because complexities and uncertainties are associated with extrapolating results of in vitro short-term tests to in vivo exposure scenarios, genotoxicity data are not used directly to derive AALGs in this meth-odology. Instead genotoxicity data are considered on a case-by-case basis to support decisions on AALG derivation when other toxicologic considerations warrant it, i.e., in the presence of contradictory and inconclusive data on carcinogenic and teratogenic effects. This form of selective consideration of genotoxicity data is also justifiable in the sense that the presumed results of mutational events (i.e., terata and neoplasms) will be accounted for, to some extent, by the consideration of carcinogenic and teratogenic effects.[38]

In general it is recommended that the results of short-term tests indicative of genotoxic activity be summarized or noted from secondary data sources during the initial evaluation of a specific chemical, with in-depth evaluation of studies from the primary literature being reserved for selected chemicals. In the application of this methodology, in-depth evaluation of a specific chemical's genetic activity profile was undertaken if (1) it was classified as an EPA class C carcinogen or IARC group 2B carcinogen and (2) a decision was necessary as to the use of QRA versus the uncertainty factor approach in AALG derivation.

Given the large number of short-term tests that may be used to evaluate genetic activity and the complexity associated with the evaluation of the adequacy of test conditions for all of the relevant genetic endpoints, it was considered important to use an established framework for this task. Due to its comprehensive nature and extensiveness of peer review, the IARC weight-of-evidence approach for evaluating the genetic activity of chemicals in short-term tests was used.[39] This approach is summarized in Table 3.5; it involves scoring results from individual tests as posi-tive, negative, or questionable and entering them in a grid based on endpoint and

Table 3.5 IARC Scheme for the Assessment of Evidence of Activity in Short-Term Tests

Sufficient evidence	at least three positive entries, one of which must involve mammalian cells in vitro or in vivo and which must include at least two of the three endpoints—DNA damage, mutation, and chromosomal effects
Limited evidence	at least two positive entries
Inadequate evidence	only one positive entry or when there are too few data to permit an evaluation of an absence of genetic activity or when there are unexplained, inconsistent findings in different test systems
No evidence	applies when there are only negative entries; these must include entries for at least two endpoints and two levels of biological complexity, one of which must involve mammalian cells in vitro or in vivo

Grid used:

	Genetic activity			Cell transformation
	DNA damage	Mutation	Chromosomal effects	
Prokaryotes				
Fungi/Green plants				
Insects				
Mammalian cells (in vitro)				
Mammals (in vivo)				
Humans (in vivo)				

Source: Material in this table was derived from IARC.[39] Grid is shown exactly as it appears in that source.

level of phylogenetic complexity of the test system. In this system, short-term tests are grouped based on four endpoints:

1. DNA damage
2. mutation
3. chromosomal effects
4. cell transformation

The definitions of these endpoints and examples of outcomes that may be used are given in Table 3.6. The levels of phylogenetic complexity in order of increasing weight are prokaryotes, fungi and green plants, insects, mammalian cells in vitro, mammals in vivo, and humans in vivo.

If a weight-of-evidence determination for genotoxicity is needed, it is recommended that the IARC monographs be consulted to determine if IARC has already evaluated a specific chemical; when the Working Group evaluates a specific chemical, it also considers all of the issues relevant to study design and adequacy of test conditions before making a determination that a given test result was positive, negative, or questionable. It should be noted that IARC has published a listing of short-term test results for the chemicals that it has evaluated.[39] The other source for screened and summarized results of short-term tests for genetic activity is the Gene-Tox database, described in Chapter 2. If it is necessary to evaluate studies

Table 3.6 Endpoints for Short-Term Tests Recognized under the IARC Scheme for the Assessment of Evidence of Activity in Short-Term Tests

DNA damage	tests for covalent binding to DNA, induction of DNA breakage or repair, induction of prophage in bacteria, and differential survival of DNA repair-proficient/-deficient strains of bacteria.
Mutation	tests for the measurement of heritable alterations in phenotype and/or genotype. These include tests for detection of the loss or alteration of a gene product, and change of function through forward or reverse mutation, recombination, and gene conversion; they may involve the nuclear genome, the mitochondrial genome, and resident viral or plasmid genomes.
Chromosomal effects	tests for detection of changes in chromosomal number (aneuploidy), structural chromosomal aberrations, sister-chromatid exchanges, micronuclei and dominant-lethal events. This classification does not imply that some chromosomal effects are not mutational events.
Cell transformation	tests that monitor the production of preneoplastic or neoplastic cells in culture and are of significance because they attempt to simulate essential steps in cellular carcinogenesis. These assays are not grouped with the endpoints listed above (see grid in Table 3.5) because the mechanisms by which chemicals induce cell transformation may not necessarily be the result of genetic change.

Source: Assembled from IARC.[39]

Note: Judgments of individual test results as " + ", " − ", or "?" (which means that there were contradictory results from different laboratories or in different biological systems or that the result was judged to be equivocal) are based on the quality of the data. This includes such factors as purity of the test article, problems of metabolic activation and appropriateness of the test system, and the relative significance of the component tests.

from the primary literature, evaluation of the issues related to adequacy of study design and test conditions will have to be undertaken by the user. Another product of the Gene-Tox program useful in this regard are the evaluations of specific test systems, published as articles in the journal *Mutation Research*. A bibliography of these articles is provided in Appendix F.

The other use of genotoxicity data in this methodology is related to developmental (e.g., terata) and reproductive outcomes (e.g., decreased fertility) that are considered to be a consequence of, or related to, mutations in germ-cell DNA rather than somatic-cell DNA. In general, if there is *sufficient*, as opposed to *suggestive*, evidence of both a chemical's interaction with germ-cell DNA and ability to produce mutations in germ-cell DNA as defined in the EPA *Guidelines for Mutagenicity Risk Assessment*,[26] then consideration should be given to application of the database factor, described in the section on the uncertainty factor approach. The EPA criteria for evaluating the evidence for mutagenicity in germ-cell DNA and chemical interaction with germ-cell DNA is summarized in Table 3.7.

AALG Derivation for Noncarcinogens: The Uncertainty Factor Approach

As indicated previously, the AALG derivation methodology described in this chapter uses the traditional approach of dividing a no observed adverse effect level

Table 3.7 U.S. EPA Classification of Weight of Evidence for Potential Human Germ-Cell Mutagenicity

Evidence for Chemical Interactions in the Mammalian Gonad:

sufficient   demonstration that an agent interacts with germ-cell DNA or other chromatin constituents or that it induces such endpoints as unscheduled DNA synthesis, sister-chromatid exchange, or chromosomal aberrations in germinal cells

suggestive   includes the finding of adverse gonadal effects following acute, subchronic, or chronic toxicity testing or findings of adverse reproductive effects that are consistent with interaction with germ cells

Categories of Weight of Evidence for Potential Human Germ-Cell Mutagenicity (in order of decreasing strength of evidence):

1. Positive data derived from human germ-cell mutagenicity studies, when available, will constitute the highest level of evidence for human mutagenicity.

2. Valid positive results from studies on heritable mutational events (of any kind) in mammalian germ cells.

3. Valid positive results from mammalian germ-cell chromosome aberration studies that do not include an intergeneration test.

4. Sufficient evidence for a chemical's interaction with mammalian germ cells, together with valid positive mutagenicity test results from two assay systems, at least one of which is mammalian (in vitro or in vivo). The positive results may both be for gene mutations or both for chromosome aberrations; if one is for gene mutations and the other for chromosome aberrations, both must be from mammalian systems.

5. Suggestive evidence for a chemical's interaction with mammalian germ cells, together with valid positive mutagenicity evidence from two assay systems as described under 4, above. Alternatively positive mutagenicity evidence of less strength than defined under 4, above, when combined with sufficient evidence for a chemical's interaction with mammalian germ cells.

6. Positive mutagenicity test results of less strength than defined under 4, combined with suggestive evidence for a chemical's interaction with mammalian germ cells.

7. Although definitive proof of nonmutagenicity is not possible, a chemical could be classified operationally as a nonmutagen for human germ cells if it gives valid negative test results for all endpoints of concern.

8. Inadequate evidence bearing on either mutagenicity or chemical interaction with mammalian germ cells.
   It is emphasized in the guidelines that it is not possible to illustrate all possible combinations of weight of evidence and that considerable professional judgment is used in drawing conclusions. It is also noted that "certain responses in tests that do not measure direct mutagenic endpoints (e.g., SCE induction in mammalian germ cells) may provide a basis for raising the weight of evidence from one category to another."

Source: The material in this table was taken directly from the guidelines for mutagenicity risk assessment.
Note: In the AALG derivation methodology presented here, if categories 1, 2, 3, or 4 are met, then consideration should be given to application of the database factor.

(NOAEL), or its surrogate, by an uncertainty factor to derive "acceptable" limits for noncarcinogen pollutant exposure. The general scheme used, with options for AALG derivation for substances identified as developmental, reproductive, or systemic toxicants during preliminary screening, is outlined in Table 3.8.

The uncertainty factor approach has historically been used by both the EPA in deriving exposure limits for pollutants in air and drinking water and the FDA in

Table 3.8 General Scheme with Options for AALG Derivation for Substances Identified as Developmental, Reproductive, or Systemic Toxicants During Preliminary Screening

[A] Check IRIS for availability of RfC
    —if RfC is available, use as AALG with or without RSC (see below) adjustment as appropriate
    —if no RfC is available,
        GO TO [B] and FOLLOW NUMERICAL SEQUENCE

[B] Identify relevant animal and human studies and
    identify occupational limits and their basis
        —based on review of secondary data sources for reference to studies in the primary literature
    1. Evaluate quality of available studies
        —use guidelines in Appendices D.1 and D.2
        —use available EPA health risk assessment guidelines
    2. Identify LOAELs and NOAELs from principal studies
        —convert animal NOAELs or LOAELs to dosimetrically equivalent surrogate human NOAELs or LOAELs, i.e., the $NOAEL_{sh}$ or $LOAEL_{sh}$, as described in Appendix C
        —do not convert NOAELs or LOAELs from human studies or occupational limits
    3. Select most appropriate basis for AALG derivation[a]
        Option 1—use data from human epidemiologic studies or animal species known to most closely resemble human response for the effect in question (based on data from supporting studies)
        Option 2—use data from the most sensitive study, species, and response
    4. Consider application of uncertainty factors
        —use guidelines in Table 3.12
    5. Consider application of database factor
        —use guidelines in Table 3.13
    6. Multiply all uncertainty factors (UFs) to get total UF and divide the NOAEL or other response by the total UF
    7. Apply relative source contribution (RSC) if appropriate
        —refer to the section "Multimedia Exposure and the Relative Source Contribution" and use the guidelines in Table 3.17
    8. Select final AALG

[a]It is entirely likely that multiple endpoints will be identified at this stage that might provide a possible basis for the final AALG. It is recommended that calculations be carried through for all of the endpoints and a final selection be made after reweighing other important factors, including relevancy of the endpoint and animal model response to humans and overall strength of evidence for a particular endpoint.

deriving ADIs for chemicals in foodstuffs.[2,40] Although this approach is widely used, it should noted that it is not without limitations:

1. Focusing solely on the NOAEL ignores potentially valuable information about the shape of the dose-response curve.
2. As measures of physiological alteration become more refined or sensitive, uncertainties regarding their correlation with adverse outcomes further complicates selection of an appropriate NOAEL (this relates to the issue of what constitutes an adverse effect described previously).
3. It is difficult to precisely incorporate factors affecting the differential quality of various studies (e.g., experimental design) into the traditional approach.[4]

It should be noted that EPA has established various working groups to address these issues. Until such time as data are available or guidelines developed to deal effectively with these limitations, the uncertainty factor approach remains a useful one for noncarcinogen criteria derivation.

In this section the concept of the reference dose (RfD) will be introduced, and general criteria for the selection of NOAELs and LOAELs, as well as use of occupational limits as surrogate human NOAELs or LOAELs, will be described. Finally, the scientific basis and guidelines for the use of various uncertainty factors will be discussed.

The Inhalation Reference Concentration* (RfC)

The reference dose is defined by EPA as "an estimate (with uncertainty spanning perhaps an order of magnitude) of a daily exposure to the human population (including sensitive subgroups) that is likely to be without an appreciable risk of deleterious effects during a lifetime." The RfD supplants the earlier concept of the ADI, although it is derived operationally in much the same manner, i.e., division of a NOAEL or other appropriate response (see next section) by a series of independent (multiplicative) uncertainty factors.[4] However, the methodology for the development of RfDs is more rigorously defined and applied, particularly with respect to dosimetry issues. With respect to the inhalation RfD (RfC), which is for continuous lifetime exposure and is expressed in units of mg/m^3, a unique feature is the refined dosimetric adjustments incorporated to adjust an animal inhalation NOAEL or LOAEL to an appropriate human equivalent concentration (HEC). A complete discussion of the RfC is beyond the scope of this document; the interested reader is referred to the brief overview in Appendix G and the in-depth EPA document on the methods for RfC derivation.[3]

With respect to the relationship of AALG derivation to the RfC, it is considered that the RfC should be used as the AALG whenever available with modification by the relative source contribution (RSC), if appropriate. The basis and use of the RSC is described in a later section. The reason for this recommendation is that the RfC represents the culmination of an extensive peer review process, including selection of principal and supporting studies, dosimetric adjustments (determination of the HEC), and appropriate application of uncertainty factors. In addition, the RfCs will be readily available through IRIS, which is updated frequently to reflect new information. Unfortunately, at the time the chemical-specific assessments in Part II were being developed, no RfCs were yet available in IRIS and therefore no examples of their use as a basis for AALG derivation or direct use as the AALG could be included in this document.

Selection of NOAELs and LOAELs

For substances that are classified preliminarily as developmental, reproductive, or systemic toxicants and for which an RfC is unavailable, it is necessary to review secondary literature sources, such as those described in Chapter 2, to identify

---

*As this book was going to publication, EPA released an external review draft in which the term inhalation "reference dose" (RfD$_i$) was changed to inhalation "concentration" (RfC).

Table 3.9 NOAELs and Related Responses Defined

NOAEL	No Observed Adverse Effect Level that dose of a chemical at which there are no statistically or biologically significant increases in frequency or severity of adverse effects seen between the exposed population and its appropriate control. Effects may be produced at this dose, but they are not considered to be adverse.
NOEL	No Observed Effect Level that dose of a chemical at which there are no statistically or biologically significant increases in frequency or severity of effects seen between the exposed population and its appropriate control.
LOAEL	Lowest Observed Adverse Effect Level the lowest dose of a chemical in a study or group of studies which produces statistically or biologically significant increases in frequency or severity of adverse effects between the exposed population and its appropriate control.
FEL	Frank Effect Level that level of exposure which produces a statistically or biologically significant increase in frequency or severity of unmistakable adverse effects, such as irreversible functional impairment or mortality, in an exposed population when compared with its appropriate control.

Source: ATSDR.[41]

appropriate principal and supporting studies. In addition, at this point studies in the primary literature that are referred to in RTECS, but that are not described, reviewed, and critiqued in any of the secondary literature sources, should be obtained. In the evaluation of the quality of data from individual studies, the guidelines described in the section "Selection of Principal and Supporting Studies" should apply. In addition, at this stage, occupational limits (i.e., RELs and TLVs) should be identified and their suitability for use in AALG derivation determined, based on the criteria described in the next section.

A NOAEL or other response identified from an appropriate animal or human study is the central item of data used in the derivation of the AALG. Although throughout this discussion, AALG derivation is discussed in terms of NOAELs, recognition of four types of response is important in the context of this methodology; these responses are the NOAEL, NOEL, LOAEL, and FEL (Table 3.9 contains definitions of each[41]). In addition, general guidelines for the selection and use of NOAELs, and related responses, are provided in Table 3.10. These are the same general guidelines used by EPA in deriving water quality criteria and by the NAS Safe Drinking Water Committees[23] and have been modified to be consistent with the criteria used in the derivation of the RfD.[4]

Once a set of responses (NOAELs, etc.) has been identified for each of the relevant endpoints, it is necessary to convert them to some type of human equivalent dose or exposure level to facilitate interspecies comparisons. In the application of this methodology, the dosimetric adjustments used are described in Appendix C. It should be noted that these conversions are consistent with older EPA methodology[23,42] and are different than those used in calculation of the HEC in the RfD_i methodology.[3]

Since a purely numerical comparison cannot take into account the many factors that contribute to the evaluation of the quality of individual studies, selection of the appropriate response should not be reduced to selection of the most conservative

Table 3.10 Guidelines Used by the U.S. EPA in Selection of Animal Toxicity Data for Deriving Criteria Based on Noncarcinogenic Responses

1. A free-standing FEL (Frank Effect Level)[a] is generally unsuitable for the derivation of criteria.
2. A free-standing NOEL[a] is generally unsuitable for the derivation of criteria. If multiple NOELs[b] are available without additional data on NOAELs or LOAELs, the highest NOEL should be used to derive a criterion.
3. A NOAEL or LOAEL can be suitable for criteria derivation. A well-defined NOAEL, preferably from a lifetime study or at least 13-week (in rodents) study, should be given the greatest weight since its use avoids the uncertainty inherent in extrapolating downward from a LOAEL. If a LOAEL is used, an additional uncertainty factor should be applied.
4. Careful consideration must be given to the severity of the effect to avoid substitution of FELs for LOAELs.
5. If for reasonably closely spaced doses only a NOEL and a LOAEL of equal quality (i.e., from a study or studies equally well-designed and conducted) are available, then the appropriate uncertainty factor is applied to the NOEL.

Source: Adapted from EPA.[4,23]
[a]A free-standing NOEL or FEL refers to a NOEL or FEL from a study in which other responses were not identified, e.g., a study with only two groups, a control and an exposed, in which either no effects were identified in the exposed animals (NOEL) or only severe adverse effects (e.g., high mortality) were observed (FEL). Ideally in a bioassay multiple dose levels are used which enable one to define the shape of the dose-response curve and identify various responses (e.g., NOAEL, LOAEL, and FEL) from a single study. Free-standing NOELs and FELs are not suitable for criteria derivation because they represent a single response level which cannot be placed in the context of the dose-response curve.
[b]Multiple NOELs can refer to either the situation in which a number of free-standing NOELs have been identified in the absence of information on other responses or to a single study in which no effects were identified at any of the exposure levels employed.

figure. Rather this decision must be made in consideration of the quality of the studies and responses with use of the guidelines in Table 3.10. With respect to selection of the most appropriate animal model, the following general guideline was applied in the application of this methodology:

> Presented with data from several animal studies, the risk assessor first seeks to identify the animal model that is most relevant to humans, based on the most defensible biological rationale, for instance using comparative pharmacokinetic data. In the absence of a clearly most relevant species, however, the most sensitive species (i.e., the species showing a toxic effect at the lowest administered dose) is adopted as a matter of scientific policy at EPA, since no assurance exists that humans are not innately more sensitive than any species tested.[4]

This is a conservative assumption used by the EPA in the interests of health protection.[23] Since the process of weighing all the considerations described above can become quite complex, some examples taken from an EPA document[3] are provided in Appendix H. While these examples cannot cover all situations that might be encountered, they are instructive of some of the competing considerations that must be weighed in selection of the most appropriate response to use as a basis for AALG derivation.

Another consideration important to note at this juncture is that it is quite likely that sets of responses may be identified for more than one endpoint. In this event, it is recommended that calculations be carried out to the level of the AALG for all endpoints and then the relative strengths and limitations of the studies and the overall weight of evidence for a given endpoint be considered in selection of the final AALG.

Occupational Limits as Surrogate Human NOAELs and LOAELs

It has been a common practice in recent years for many S/L agencies to derive AALs based on occupational limits; the reasons for, and limitations of, this approach were described in Chapter 1. While OELs are not the ideal basis for AAL derivation, it is our contention that they do have a place in a methodology such as the one described in this chapter because they are a peer-reviewed secondary data source specific for the inhalation route of exposure and because rapid development of scientifically defensible AALGs under resource-limited conditions is a central feature of this methodology. However, OELs are not derived based on a clearly defined and rigorously applied set of criteria (compared to, for example, the RfC), but rather on consensus judgment of a committee (at least in the case of the TLVs), which may implicitly consider other factors such as technical feasibility and detection limits. Therefore, OELs cannot a priori (based on considerations of scientific defensibility) be considered the equivalent of human NOAELs, as is implicitly assumed when they are divided by uniform uncertainty factors without in-depth consideration of their documentation and basis.

In the application of this methodology, the documentation and basis for OELs were considered integrally as part of the overall database for specific compounds, and under certain conditions the OEL was used as the basis for AALG derivation. The general scheme for this is presented in Table 3.11. OSHA PELs were not considered as a basis for AALG derivation because they are legally enforceable standards and as such incorporate nonhealth-based considerations such as detection limits and economic and technical feasibility.[43] NIOSH RELs and ACGIH TLVs, on the other hand, are primarily intended as health-based guidelines, although other considerations are frequently operative. This is the reason why the need for careful review of the background documentation is emphasized in this methodology. In making the decision on whether a specific OEL is a suitable basis for AALG derivation, two criteria need to be met: (1) the endpoint on which the OEL is based should be appropriate for the general population; i.e., there should be evidence that the specific adverse effect that the OEL is intended to prevent is also the critical effect for the population at large, including certain sensitive subgroups such as the very young or very old and (2) there should be evidence that the OEL is a reasonable surrogate human NOAEL or LOAEL. Since the application of considerable judgment may be required in making these two determinations, AALGs derived based on occupational limits should always be considered provisional. This means that these AALGs are associated with a higher degree of uncertainty than AALGs not so designated, and those chemicals that have provisional AALGs should have a higher priority for reevaluation and/or additional, more in-depth literature review.

As mentioned earlier, the documentation and basis for the OELs should be considered as an integral part of the overall database for a given compound. This

Table 3.11 General Considerations and Scheme for AALG Derivation Based on an Occupational Limit

Review background documentation for the OEL(s)
–ACGIH TLV documentation for TLVs
–NIOSH criteria document for RELs

↓

Based on thorough review of the documentation, determine:
(1) Is the endpoint on which the OEL is
    based applicable to the general population?
(2) Is there at least some evidence that the
    OEL is a reasonable surrogate for a human
    NOAEL or LOAEL?

If the answer is unequivocally yes to both (1) and (2), GO TO [A].

If the answer is equivocally yes to both (1) and (2)
or unequivocally yes to one question and
equivocally yes to the other, GO TO [B].

If the answer is equivocally yes to one question
and no to the other, GO TO [C].

If the answer is no to both questions, do not use the
OEL to derive an AALG

[A]   Derive provisional AALG based on OEL
1. Select OEL (TLV or REL) based on (1) and (2) above.
2. Select appropriate uncertainty factors
   • 4.2 for continuous exposure adjustment if the endpoint is any one other than sensory irritation
   • 10 for interindividual variation (sensitive subgroups)
   • 5 if the OEL is judged to be equivalent to a LOAEL
   • 1–10 as appropriate
3. Divide the OEL by the total uncertainty factor
4. Adjust for relative source contribution if appropriate
5. Select the final AALG, based on consideration of the overall quality of other available evidence

[B]   Evaluate quality of available animal and human inhalation studies
   —If the responses from any of these studies are judged to be suitable for AALG derivation, do not use the OEL as a basis for AALG derivation.
   —If the responses from these studies are unsuitable or no animal or human inhalation data are available, GO TO [A].

[C]   Determine the availability of animal and human inhalation data
   —If these data are available, GO TO [B].
   —If these data are unavailable, GO TO [D].

[D]   Evaluate the quality of available animal and human ingestion data
   —If there is evidence to support the conclusion that the same effect might be expected by both the ingestion and inhalation routes, derive a provisional criterion based on this data.
   —If there is no evidence to support the conclusion that the same effect might be expected by both the ingestion and inhalation routes or exposure, reconsider the need for an AALG at this time.

Note: The decision key outlined in this table is intended only as a rough guideline and does not cover all possible contingencies.

means that the quality of the OEL (as defined by the criteria under (1) and (2) above) should be considered in the context of the overall database before making a decision on whether or not to use it as a basis for AALG derivation. For example, unless an OEL is based directly on well-conducted human epidemiologic studies, data from a well-conducted animal inhalation study of at least 90-days duration (particularly one that examined respiratory function endpoints) are generally a better basis for AALG derivation. On the other hand, even if there is some question as to whether an OEL meets the two criteria outlined above, it is still preferable to the use of acute lethality data or results of a subacute (14-day) animal ingestion study as a basis for AALG derivation. It is apparent from these two extreme examples that considerable judgment is necessary in data quality evaluation and that it is not feasible to cover all possible combinations of data quality and availability. However, review of the chemical-specific assessments in Part II for which AALGs were derived based on OELs should be helpful in providing guidance on some of the competing factors that must be weighed.

The Basis and Use of Uncertainty Factors

Uncertainty factors are independent (multiplicative) factors used operationally in deriving criteria for threshold endpoints (e.g., the RfD). Although the terms *safety factor* and *uncertainty factor* have the same meaning and are used interchangeably, uncertainty factor is the preferred term. The reason, as stated in Appendix A to the IRIS database,[4] is that:

> the term "safety factor" suggests, perhaps inadvertently, the notion of absolute safety, i.e., absence of risk. While there is a conceptual basis for believing in the existence of a threshold and "absolute safety" associated with certain chemicals, in the majority of cases a firm experimental basis for this notion does not exist.

The conditions for application and magnitude of uncertainty factors used in this methodology are outlined in Table 3.12.

Uncertainty factors, as they are typically used, are intended to account for

1. human interindividual variation, i.e., variation in sensitivity among the human population or extrapolation from normal humans to sensitive subgroups of the human population
2. interspecies variation, i.e., the uncertainty in extrapolating results of animal data to humans
3. the uncertainty in extrapolating from data obtained in a study that is of less-than-lifetime or less-than-chronic duration
4. the uncertainty in using a LOAEL rather than a NOAEL

The historical and experimental basis for all four of these uncertainty factors has been reviewed by Dourson and Stara;[36] Calabrese has reviewed the experimental basis for the uncertainty factor intended to account for interindividual variation.[44] Due to the thoroughness and depth of these reviews further discussion of the basis for these uncertainty factors will be very limited here; the interested reader is referred to the above reviews and the papers cited in Table 3.12.[13,35,45–51]

With respect to the magnitude of these uncertainty factors, the most recently available guidelines from EPA were generally followed. These guidelines specify

Table 3.12 Criteria and Guidelines for the Application of Uncertainty Factors

Type	Magnitude	Comments
Interindividual variation	10	—intended to account for variation in susceptibility among the human population, i.e., high risk groups —this UF is applied in all cases except studies in which the effects of prolonged exposure have been evaluated in the human population including sensitive subgroups —studies cited as basis for this UF in Dourson and Stara[36] were Mantel and Bryan,[44] Weil,[45] and Krasovskii[13]
Interspecies variation	10	—intended to account for uncertainty in extrapolating results obtained in animals to the general human population —applied in all cases where the NOAEL or LOAEL was derived from an animal study —studies cited as basis for this UF in Dourson and Stara[36] were Rall,[46] Evans et al.,[47] Hayes,[48] and Lehman and Fitzhugh[49]
Subchronic to chronic *or* less-than-lifetime to lifetime	5 or 10	—intended to account for the uncertainty in extrapolating from less-than-lifetime to lifetime exposure or subchronic to chronic exposure —UF of 10 applied whenever study duration is 90 days or less in rodent species or approximately 1/10 to 1/5 lifespan in other species —judgment used in determining whether to apply UF of 5 or 10 if study is between 1/5 and 1/2 of species lifespan —judgment used in determining whether to apply UF of 5 if study is greater than 1/2 the species lifespan —studies cited as basis for this UF in Dourson and Stara[36] were McNamara,[50] and Weil and McCollister[35]
LOAEL to NOAEL	5 or 10	—intended to account for uncertainty inherent in extrapolating downward from a LOAEL to a NOAEL —UF of 5 generally used as a default value in the application of this methodology —UF of 10 used if deriving an AALG based on the uncertainty factor approach for a Class C carcinogen —study cited as basis for this UF in Dourson and Stara[36] was Weil and McCollister[35]

an uncertainty factor of 10 for interspecies and interindividual variation and the use of "up to a 10-fold factor" for less-than-lifetime or less-than-chronic extrapolation and use of a LOAEL rather than a NOAEL.[3,4] In the application of this methodology, an uncertainty factor of 10 was always used for interspecies or interindividual variation.

With respect to the uncertainty factor related to length of the study or exposure period versus the length of species lifespan, there appears to be some inconsistency in recommendations among various sources as to when this UF should be applied. For example, Dourson and Stara specify the use of a factor of 10 when extrapolating from "less than chronic to chronic" exposures, as did the EPA workshop draft;[36,42] however, ATSDR uses the phrase "less than lifetime,"[41] and the most recent EPA source available specifies "subchronic to chronic" and indicates that a factor of "up to 10" may be used.[3]

Part of the difficulty in interpreting these recommendations is related to how chronic and subchronic exposures are defined. In the most recent edition of a standard toxicology text, *subchronic* is defined as exposure of 1 to 3 months duration, and *chronic* as greater than 3 months duration.[52] ATSDR defines chronic exposure as "exposure to a chemical for 365 days or more" and does not have a specific definition for subchronic exposure.[41] The most recently available EPA source that provides definitions defines chronic exposure as "multiple exposures occurring over an extended period of time, or a significant fraction of the animal's or individual's lifetime," and subchronic exposure is defined as "multiple or continuous exposure occurring over about 10% of an experimental species lifetime, usually over 3 months."[3] Based on the the information presented here, it is apparent that there is considerable variation in when an exposure is considered subchronic versus chronic. Therefore, since there is clearly a need for a stronger database to provide a solid scientific basis for allocating uncertainty factors of exact magnitude to studies of various specific exposure durations, and in the interests of keeping the methodology in harmony with current regulatory practice, an UF of up to 10 was used in the application of this methodology. An UF of 5 or 10 was applied based on a combination of exposure duration as a function of species lifespan according to the guidelines in Table 3.12. The specific recommendations were based on a review of how criteria were derived for substances for which there are oral RfDs in the IRIS database. It should be noted that what was considered a normal lifespan was based on Table C.3 in Appendix C, and in cases where there was less than lifetime exposure with lifetime observation, use of a lower UF is often justified.

With respect to the LOAEL-to-NOAEL UF, as previously indicated, use of the NOAEL is preferred because it avoids the uncertainty inherent in extrapolating downward from a LOAEL to a NOAEL. Dourson and Stara indicated that the available experimental data support the use of an UF of between 1 and 10 "depending on the sensitivity of the adverse effect";[36] the most recent EPA source recommends a magnitude of "up to 10."[3] As previously mentioned, in their reanalysis of the Weil and McCollister data,[35] considered to provide the experimental support for the LOAEL-to-NOAEL UF, Dourson and Stara noted that if the ratios of corresponding LOAELs to NOAELs are compared for both subchronic (27 pairs) and chronic (25 pairs) exposures that the ratio is less than 10 in all cases and less than 5 in most cases (96%).* Therefore in this methodology an UF of 5 was generally used except in the specific instance where an AALG was derived for a Class C carcinogen using the uncertainty factor approach. This latter distinction was made on the basis of the severity of the endpoint and limitations of the database, as described in the section "AALG Derivation for Carcinogens."

It is apparent from much of the discussion in various sections of this chapter

---

*The experimental support for this uncertainty factor is very limited and is, in fact, based on a single study.[35] While a diverse spectrum of endpoints was evaluated (including mortality, food intake, body and relative organ weights, gross pathology and histopathology of a variety of organs, hematology, clinical chemistry, etc.), the studies were conducted using mainly rats (some dogs also), and only a limited number (33) of chemicals were evaluated. In addition, due to limited reporting of methodology and results, thorough evaluation of study design and data quality is not possible.

that the process of criteria derivation is limited by the data available for any given compound and that professional judgment is frequently necessary because of deficiencies in the overall database that are not dealt with within the context of the uncertainty factors already described. The U.S. EPA has attempted to deal with this problem by developing the use of an UF specifically to account for this type of uncertainty. It is termed a *modifying factor* and is described as follows:

> Use of professional judgment to determine another uncertainty factor (MF) which is greater than zero and less than or equal to 10. The magnitude of the MF depends on the professional assessment of scientific uncertainties of the study and database not explicitly treated above; e.g., the completeness of the overall data base and the number of species tested. The default value for the MF is 1.[4]

The use of the MF is a relatively new phenomenon. Examples of situations in which it has been used can be found in the IRIS database; however, no specific guidelines have been published for its application. In the recently released EPA publication dealing with the development and application of the RfD$_i$, (now termed RfC) an additional similar uncertainty factor described as "incomplete to complete data base" (D) has been proposed, and the wording in the quote above in regard to the MF ("e.g., the completeness of the overall database") has been removed in the new publication. A magnitude of up to 10 has been proposed for this UF, and it is "intended to account for the inability of any single animal study to adequately address all possible adverse outcomes in humans." No examples of the use of the "D" factor were given nor were more specific guidelines for its use recommended. In addition, it was indicated that use of this UF is currently under discussion within EPA.[3]

Since it is apparent that an UF specifically designed to deal with uncertainties and issues not addressed within the current framework is useful to clearly indicate and compensate for deficiencies in the principal study and overall database for a given chemical, use of a factor somewhat analogous to the MF and D factors proposed by the EPA was adopted for this methodology. This UF is termed the *database factor* and is defined as follows: an uncertainty factor with magnitude 1 to 10 that is designed to account for limitations in the overall database and principal study that are not accounted for by other uncertainty factors. It must be acknowledged that considerable professional judgment is necessary both in determining when to apply this UF and its appropriate magnitude. Based on the extent to which professional judgment is necessary in applying an UF of this type, it was considered a limitation of the analogous factors proposed by the EPA that minimal or no guidelines were provided, or indications given in the documentation of when and how they should be applied. For this reason, guidelines are provided in Table 3.13 as to situations in which the database factor might be applied.[53] However, owing to the complexity of data evaluation in any given case, it is not possible to provide a comprehensive set of guidelines that would anticipate all possible situations, thereby emphasizing the need for professional judgment.

Table 3.13 Guidelines for Application of the Database Factor

1. Route-to-route extrapolation
   Application of the database factor is appropriate whenever a noninhalation study is used to derive a criterion for the inhalation route.
2. Known specific chemical interactions[a]
   Application of the database factor is appropriate whenever there is evidence for a synergistic interaction between the chemical for which the AALG is being derived and other chemicals that are widespread in occurrence or use, e.g., alcohol, tobacco, automobile exhaust components.
3. Severity of effect in the case of UF approach for AALGs for carcinogens
   Application of the database factor, usually a factor of 10, is generally appropriate when the uncertainty factor approach is used to derive an AALG for a Class C carcinogen, particularly where there is evidence for genotoxic effect based on short-term test results.
4. Other significant effects on which the AALG is not based
   Application of the database factor is appropriate when there are other data strongly suggestive of specific adverse effects for endpoints other than the endpoint on which the AALG is based, i.e., data on genotoxicity in the absence of information on carcinogenicity or data indicative of interaction of the chemical with, or effects on, mammalian germ-cell components in the absence of specific data on developmental or reproductive outcomes.
5. Developmental uncertainty factor
   Application of the database factor—which may also be referred to as a developmental uncertainty factor under these circumstances—is appropriate in cases where there is clear evidence of developmental toxicity below levels of exposure that result in maternal toxicity or at levels of exposure that result in developmental toxicity when the maternal animals are minimally effected.

[a]Recently, the NAS in a discussion of the issue of synergism in the risk assessment of chemical mixtures in drinking water, suggested the possibility of incorporating a UF to account for synergisms when developing and applying a hazard index. It was suggested that the UF could vary from 1 (which assumes simple additivity) to up to 10 or 100, depending on the amount of information available and the concentrations of the contaminants. For example, lower concentrations and more information known on synergistic effects would favor a lower UF, and higher concentrations and lack of information would favor a higher uncertainty factor.[52]

Issues Specific to Certain Noncarcinogenic Endpoints

Developmental Toxicity

Developmental toxicity is defined, for purposes of this document, in accordance with the EPA *Guidelines for Health Assessment of Suspect Developmental Toxicants*,[27] as the induction of

> adverse effects on the developing organism that may result from exposure prior to conception (either parent), during prenatal development, or postnatally to the time of sexual maturation. Adverse developmental effects may be detected at any point in the life span of the organism.

Both in the guidelines cited above and their subsequent proposed revisions,[30] as well as in other sources,[54] four general forms of developmental toxicity are recognized: (1) death of the conceptus, (2) structural abnormality, (3) altered growth, and (4) functional deficiency. Although these and other related terms are defined in detail in Table 3.14, it should be noted that functional deficiency, despite its obvious importance, is not routinely evaluated in developmental toxicity studies for

Table 3.14 Definitions of the Major Endpoints of Developmental Toxicity and Other Important Terms

altered growth	any alteration in offspring organ or body weight or size. Changes in body weight may or may not be accompanied by a change in crown-rump length and/or skeletal ossification. Altered growth can be induced at any stage of development, may be reversible, or may result in permanent change.
malformation	any permanent structural abnormality or change that may adversely affect survival, development or function[a]
variation	any structural divergence beyond the usual range of structural constitution that may not adversely affect survival or health[b]
functional deficiency	alteration or delays in functional competence of the organism or organ systems following exposure to an agent during critical periods of development pre- and/or postnatally. Functional deficits are often subtle and may be difficult to detect using standard testing protocols.
embryotoxicity and fetotoxicity	any adverse effect on the conceptus as a result of prenatal exposure. The distinguishing feature between the two terms is the stage of development during which the injury occurred. It should be noted that as these terms are used in EPA[3,27] and in this methodology, they can include malformations, variations, altered growth, and in utero death.

Note: These definitions were taken from EPA.[29,30] The major manifestations of developmental toxicity are (1) death of the conceptus, (2) structural abnormality, (3) altered growth, and (4) functional deficiency.
[a]In the EPA guidelines, [29,30] the term *terata* is used synonymously with *malformation,* and a teratogen is an agent which induces malformations.
[b]Examples of variations include wavy ribs[29] and supernumerary ribs in rats (G.A. de S. Wickramararatne. 1988. "The Post-Natal Fate of Supernumerary Ribs in Rat Teratogenicity Studies." *J. Appl. Toxicol.* 8:91–94).

the following reasons: alterations are frequently subtle and difficult to detect, and the longer time period required for evidence of functional deficits to develop, together with lack of defined testing criteria, often makes their evaluation prohibitively time and cost intensive.[27,55]

Developmental toxicity studies typically differ from conventional subchronic and chronic toxicity studies in that exposure is typically limited to maternal animals during the entire gestation period or during the period of major organogenesis (e.g., days 6 to 15 in rats and mice and days 6 to 18 in rabbits). The maternal animals may be either killed at term pregnancy (e.g., FDA segment II teratology studies) or allowed to deliver and rear the young for a sufficient period (e.g., 21 days for rats and mice) to evaluate viability, weaning, and growth indices (e.g., FDA segment III studies).[56] It should be noted that these latter indices may also be evaluated in multigeneration studies of the type described in the next section (e.g., FDA segment I studies). In terms of specific criteria governing the conduct and evaluation of these studies, the best sources to consult are various EPA guidelines[27,30,57,58] and a summary of FDA test procedures by Manson, Zenick, and Costlow.[56]

A major area of concern unique to developmental toxicity studies is the issue of the influence of maternal toxicity and how this should be dealt with in the context of risk assessment and criteria derivation. One useful way to look at this issue is in terms of the scheme set forth by Schardein,[59] as shown in Figure 3.2. Clearly agents that cause adverse developmental outcomes at exposure levels not producing maternal toxicity present the greatest developmental toxicity hazard. It is also apparent

Figure 3.2   Schematic diagram of possible outcomes of chemical exposure in developmental toxicity studies. *Source:* Schardein.[59]

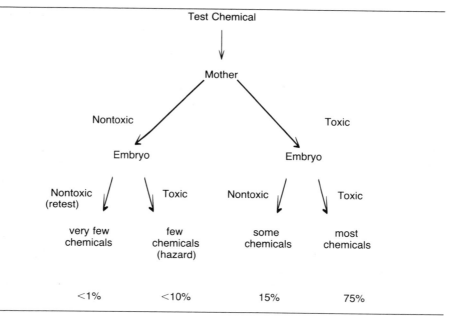

that any agent that causes adverse effects in the developing offspring must be viewed as having the potential to cause developmental toxicity in humans. However, at exposure levels that produce maternal toxicity in animal studies, the majority of chemicals (75% identified by Schardein) also cause developmental toxicity; it has been pointed out that under these circumstances it is extremely difficult to determine the influence that maternal toxicity had on the adverse outcomes observed in the offspring. In a recent review of this area, Chernoff, Rogers, and Kavlock have pointed out that the interpretation of developmental toxicity in the presence of maternal toxicity is further complicated by the fact that the current database does not really allow for an adequate evaluation of underlying trends and mechanisms and, hence, does not provide a basis for extrapolation of such results to humans.[60] The reasons they cite for this problem include

1. the relatively crude measures of maternal toxicity typically used in most testing protocols (e.g., mortality, body weight changes, food and water consumption)
2. possible bias in the types of studies reported
3. incomplete reporting of both maternal and developmental effects
4. frequently, the lack of application of standardized criteria in the evaluation of both maternal and developmental toxicity

In an effort to deal with the problem of interpretation of developmental toxicity in the presence of maternal toxicity, a number of investigators have sought to define this relationship in more quantitative terms. Two major ways that have been suggested are the "relative teratogenic index" (RTI) and the A/D ratio.[59] The RTI is defined as the ratio of the minimum lethal dose (adult LD_{01}) and the dose inducing 5% malformations in live offspring (tD_{05}). This ratio is based on probit analysis of dose-response relationships for both maternal toxicity and teratogenic potency in mice and was developed by Brown et al.[61] and Fabro, Shull, and Brown.[62] The RTI has been criticized on several grounds:

1. It only deals with terata or malformations rather than other endpoints of developmental toxicity.
2. Its use has not been well validated.
3. The index must be generated from single points on the dose-response curves since the slopes of the curves for induction of terata and adult lethality are not the same.[59]

The A/D ratio differs from the RTI in that it uses information that is typically generated in most developmental toxicity studies. It is defined as the ratio of the dose producing "overt toxicity in the the the mother (A) and the dose causing significant developmental toxicity (D) in the embryo." This index was first proposed by Johnson (1980) and subsequently further refined and developed.[63-67] It has been emphasized by Johnson and others that these ratios are likely to be very imprecise considering the crudeness of the maternal endpoints typically used, biologic variations and how these are interpreted by various investigators, and the often limited reporting of studies in the published literature.[59,65] At present the most appropriate use of the A/D ratio appears to be as a tool in priority hazard assessment.[59]

It is apparent from the foregoing discussion that some framework is needed to deal with the influence of maternal toxicity on developmental toxicity in the context of criteria derivation and that the quantitative measures that have been developed suffer from certain limitations. It was concluded that in the context of the AALG derivation methodology described here, the most appropriate mechanism of adjusting the AALG for the effects of agents identified as primary or unique developmental hazards, i.e., agents that produce developmental toxicity at levels below those causing maternal toxicity or for which the developmental effects do not appear to be specifically related to the maternal toxicity observed, was by the application of a higher uncertainty factor—more specifically, by selective application of the database factor described in the previous section based on the guidelines described in Table 3.13. Use of a higher uncertainty factor is supported by both the recommendation of the NAS Safe Drinking Water Committee that a larger uncertainty factor be used in such cases and their conclusion that "humans should be considered at least 50 times more sensitive than animals to agents causing well-defined developmental toxicity in animal studies."[68]

Reproductive Toxicity
For purposes of this document, the definitions of male and female reproductive toxicity provided in the *Proposed Guidelines for Assessing Male Reproductive Risk*

and the *Proposed Guidelines for Assessing Female Reproductive Risk* were accepted; these guidelines, although not yet formally adopted, were generally followed in evaluating studies concerning reproductive endpoints.[28,29] In the document cited above, male reproductive toxicity is defined as:

> the occurrence of adverse effects on the male reproductive system that may result from exposure to environmental agents. The toxicity may be expressed as alterations to the male reproductive organs and/or the related endocrine system. The manifestation of such toxicity may include alteration in sexual behavior, fertility, pregnancy outcomes or modifications in other functions that are dependent on the integrity of the male reproductive system.[28]

Similarly female reproductive toxicity is defined as:

> adverse effects observed in the female reproductive system that may result from exposure to chemical or physical agents. Female reproductive toxicity includes, but is not limited to, adverse effects in sexual behavior, onset of puberty, fertility, gestation, parturition, lactation or premature reproductive senescence.[29]

Although the general scheme for AALG derivation for substances identified as reproductive toxins was outlined in Table 3.8, certain characteristics of this endpoint require further elaboration. One problem that may be encountered is distinguishing between reproductive and developmental toxicity, since certain types of studies (e.g., multigenerational studies), if properly conducted, allow evaluation of both reproductive and developmental outcomes. Another problem that may contribute to confusion of reproductive and developmental outcomes is the overlap in the endpoints that may be considered as related to developmental and reproductive toxicity in the proposed EPA guidelines for assessing male and female reproductive risk and the EPA guidelines for evaluation of suspect developmental toxins.[27-29] For example, depending on the design of the study, certain pregnancy outcomes may be indicative of either male or female reproductive toxicity or developmental toxicity. Some of these outcomes include indices related to litter size, pre- and postimplantation loss, number of live/dead pups, sex ratios, malformations, birth and postnatal weights, and survival. In general, it is only possible to attribute an outcome to male reproductive toxicity with certainty if the outcome is observed in the offspring of unexposed females mated with exposed males; the reverse applies to attributing such an outcome to female reproductive toxicity with certainty. With respect to attributing an outcome to female reproductive toxicity versus developmental toxicity, the important factors to consider are time period of exposure (i.e., prior to gestation, early gestation [preimplantation], late gestation [postimplantation], throughout the entire gestation period, or before and during gestation) and the possible influence on the offspring of any form of maternal systemic toxicity observed. Further guidance on this issue can be found in the EPA health risk assessment guidelines cited above and in documents which specifically discuss criteria to consider in the evaluation of multigeneration studies.[27-29,57,58] In the context of this methodology, it is important to note that while these distinctions may not operationally effect AALG derivation, they are important in the hazard evaluation process and may have implications for risk management.

Another important consideration in the evaluation of male or female reproductive toxicity that may be overlooked is that a number of very different study types may provide an appropriate basis for AALG derivation for this endpoint. This is an important consideration since these different study designs often provide complementary information and the factors that must be weighed in evaluating the quality of the studies are often unique and quite specific. These studies include, in addition to single and multigeneration reproduction studies, conventional chronic and subchronic toxicity studies as well as the dominant lethal assay. A number of examples of outcomes indicative of possible male or female reproductive toxicity that may be either routinely evaluated or included as supplemental evaluations in some of these studies are provided in Table 3.15. With respect to the dominant lethal assay, it should be noted that this test is specific to male reproductive toxicity and is designed to detect mutagenic events in the spermatogenic process that are lethal to the embryo or fetus; a review has been published as part of the EPA Gene-Tox program addressing some of the criteria that should be considered in the design and conduct of this assay.[69]

Sensory Irritation

Sensory irritants have been defined as:

chemicals which when inhaled via the nose will stimulate trigeminal nerve endings, evoke a burning sensation of the nasal passages, and inhibit respiration. Also, most will induce coughing from laryngeal stimulation. . . . These chemicals are also capable of stimulating trigeminal nerve endings of the cornea and induce tearing. At high concentrations, particularly on moist facial skin, they are capable of inducing a burning sensation. Some have odorant and/or gustatory qualities. Most will induce bronchoconstriction, usually at concentrations in the air higher than required for stimulation of nerve endings in the nasal passages.[70]

The general scheme, with options for AALG derivation, for chemicals judged to be sensory irritants is presented in Table 3.16.

Several features of sensory irritation as an endpoint differentiate it from other endpoints discussed in this chapter and have significant implications for criteria derivation. One such characteristic is the localized nature of the response; i.e., it occurs at the portal of entry as a result of direct contact rather than being due to absorption into the systemic circulation and distribution to a distant site of action. Given this characteristic, it follows that the onset of the response is relatively rapid following initialization of exposure and tends to cease quickly when the offending agent is removed. Therefore, sensory irritant responses to airborne chemicals tend to be concentration dependent rather than time dependent, and the injury or effect is generally noncumulative in nature. The implication of this characteristic for AALG derivation is that adjustment for continuous exposure is not warranted. In practice this means that a continuous exposure adjustment is not applied when making dosimetric conversions; for example, if a TLV is used as the basis for

Table 3.15 Examples of Indices and Endpoints Indicative of Male or Female Reproductive Toxicity Which Might Be Evaluated in Various Study Types

A. Subchronic and Chronic Toxicity Studies
  1. Male Reproductive Toxicity Endpoints
    a) organ weights (relative and absolute)
      —testes, epididymides, seminal vesicles, prostate, pituitary
    b) histopathology
      —same organs
  2. Female Reproductive Toxicity Endpoints
    a) organ weights (relative and absolute)
      —ovary, uterus, pituitary
    b) histopathology
      —same organs
B. Single and Multigeneration Reproduction Studies[a,b]
  1. General Types of Endpoints
    a) effects on mating, conception, and fertility
    b) alterations in gestation or parturition (females only)
    c) offspring survival (mainly female, possibly male)
  2. Examples of Specific Indices Used[c]
    a) male mating index (a)
    b) male fertility index (a)
    c) female mating index (a)
    d) female fertility index (a)
    e) female fecundity index (a)
    f) parturition index (b)
    g) gestation index (b)
    h) live litter size (c)
    i) live birth index (c)
    j) viability index (c)
    k) lactation index (c)
    l) weaning index (c)
    m) preweaning index (c)
C. Supplemental Endpoints (not routinely evaluated)
  1. Female Reproductive Alterations
    a) changes in onset of puberty
    b) alterations in reproductive cycle
    c) oocyte toxicity
    d) premature reproductive senescence
  2. Male Reproductive Alterations
    a) spermatogenic endpoints
      —including count, morphology, and motility
    b) endocrine evaluations
      —e.g., hormone levels (also applicable to females)
    c) sexual behavior

Note: The material in this table was assembled from EPA (1988a &b). The examples provided are not exhaustive, and a number of endpoints more commonly associated with developmental toxicity were not included.

[a] In single generation reproduction studies, only parental reproductive capacity is evaluated; in two and three generation studies, postnatal maturation and reproductive ability of the offspring are evaluated and cumulative generational effects may be detected.

[b] Two useful review articles dealing specifically with the design and evaluation of the multigeneration reproduction study are M.S. Christian, "A Critical Review of Multigeneration Studies," *J. Am. Coll. Toxicol.* 5:161–180 and E.M. Johnson, "The Scientific Basis for Multigeneration Safety Evaluations," *J. Am. Coll. Toxicol.* 5:197–201.

[c] Refer to EPA (1988b) for the specific definitions of these indices.

Table 3.16 Sequence of Steps for the Evaluation of Compounds Identified as Sensory Irritants During Preliminary Screening with Options for AALG Derivation

Ascertain as far as possible that irritation is the critical effect before proceeding
—data from occupation sources are preferred

Identify source of data, excluding studies using noninhalation routes of exposure
—if human data, including OELs, GO TO [A]
—if animal data, GO TO [B]

[A] Human sensory irritation data, apply the following quidelines:
  (1) Examine either the background documentation for the TLV or REL or the original study to determine whether the given exposure level was actually a NOAEL or LOAEL. If a LOAEL, apply an additional uncertainty factor of 5.
  (2) Apply an uncertainty factor of 10 for human interindividual variation.
  (3) Consider application of database factor if appropriate.
    —see Table 3.13
  (4) Divide the value (OEL, NOAEL, or LOAEL) by the total uncertainty factor to derive the final AALG. Do not apply a continuous exposure adjustment or relative source contribution.

[B] Animal sensory irritation data only, no usable human data available:[a]
  Determine type of data, NOAEL vs. mouse RD_{50} (defined on p. 71)
    —if mouse RD_{50}, GO TO [C]
    —if NOAEL or LOAEL, GO TO [D]
    —if both are available, mouse RD_{50} is preferred

[C] Calculation of an AALG based on mouse RD_{50}:[b]
  (1) Multiply mouse RD_{50} by 0.001.
  (2) Consider application of database factor, if appropriate.
  (3) Do not use dosimetric conversions in Appendix C or apply the relative source contribution.

[D] Calculation of an AALG based on an animal NOAEL or LOAEL for sensory irritation:
  (1) Determine total uncertainty factor.
    —if LOAEL, use UF of 5
    —apply UF of 10 for interindividual variation
    —apply UF of 10 for interspecies variation
    —consider application of database factor, if appropriate
  (2) Divide LOAEL or NOAEL by total UF.
  (3) Do not use dosimetric conversion described in Appendix C or apply the relative source contribution.

Note: In the application of this methodology, when sensory irritation was judged to be the critical effect, occupational limits were almost always found to be sufficiently supported by other experimental data to warrant their use in AALG derivation.
[a]In the application of this methodology, it was found that RD_{50}s were available for very few chemicals, thereby necessitating the use of animal NOAELs or LOAELs in the absence of human data.
[b]All AALGs based on occupational limits, mouse RD_{50}s, or animal NOAELs or LOAELs should be designated as provisional.

AALG derivation, it is not divided by 4.2 (40-hour workweek/168 hours in one week) nor is a continuous exposure adjustment (see Appendix C) applied when an animal or human NOAEL or LOAEL is used as the basis for AALG derivation.

An additional consequence of the localized, nonsystemic nature of the sensory irritant response is that since it is specific to the respiratory tract, it is not reasonable to set an AALG based on irritation using noninhalation data, as is sometimes done for other endpoints if the overall database for a specific chemical is very limited and there is at least some evidence to support the notion that similar effects might be expected under conditions of inhalation exposure. The nonsystemic nature of the response is also the reason dosimetric adjustments of the type described in Appendix C, which involve conversion to mg/kg/day when calculating a surrogate human NOAEL (NOAEL$_{sh}$) from an animal NOAEL, are not applied when sensory irritation is the endpoint on which the AALG is based.

Another unique aspect of sensory irritation as an endpoint is the subjective nature of the response (particularly at the lower concentration range) and how this relates to animal extrapolation. Since irritant effects often manifest themselves initially as subjective sensations (e.g., tickling, itching, or burning) before proceeding to more readily observed (e.g., lacrimation, rubbing of the eyes, coughing) and measured (e.g., decrease in respiratory rate) responses, determination of thresholds for irritation from animal studies is often difficult compared to other endpoints. It is for this reason that human data, even from studies that were limited, was almost always given precedence over animal data (even from well-designed and conducted studies) in the application of this methodology. It is also the reason that use of an animal RD$_{50}$ (as described in the next paragraphs), if available, was given precedence over other animal data in the absence of reliable human data.

In an effort to quantify and compare the relative potencies of various chemicals as sensory irritants, Alarie developed the RD$_{50}$ index in mice.[69] It was found that exposure of mice to a variety of sensory irritants resulted in a readily measurable, concentration-dependent decrease in respiratory rate; the RD$_{50}$, the concentration that results in a 50% decrease in respiratory rate, is easily determined from concentration-response curves.[71] Kane et al. and Alarie have suggested that exposure guidelines can be developed for occupational settings using some fraction of the RD$_{50}$, and Kane et al. have presented evidence based on literature review to support the idea that it might be expected that 10, 1, 0.1, 0.01, and 0.001 × the RD$_{50}$ would correspond to lethal, toxic, "effective," "ineffective," and "acceptable" exposure levels.[71,72] Buckley et al. have conducted studies in groups of mice exposed to 10 representative sensory irritants at the RD$_{50}$ concentration (exposed 6 hours per day for 5 days) and confirmed the prediction that histopathologic lesions of varying severity would occur as a result of longer exposures at the RD$_{50}$ concentration.[73] Given the quantitative nature of this response and the data supporting its utility for interim criteria derivation, as well as consideration of the special problems related to animal extrapolation due to the subjective nature of the earliest responses to sensory irritants, use of the RD$_{50}$ as a basis for AALG derivation is recommended over the use of animal NOAELs or LOAELs for sensory irritation when only animal data are available. Careful consideration of the overall database is extremely important to determine that sensory irritation is, in fact, the critical effect, and any AALG for sensory irritation based on animal data should be considered provisional.

Options for AALG Derivation When Only Acute Lethality Data Are Available

In the application of this methodology to a given set of compounds, there may be cases where review of RTECS and checking of secondary data sources reveal that the only data available for a given chemical are acute lethality indices such as the LD_{50} or LC_{50}. Under these conditions, it is necessary to weigh various factors including those related to exposure potential (i.e., production volume and environmental release information) in order to decide whether to (1) invest time and resources to perform an extensive primary literature search in hopes of locating studies that might provide a more reliable basis for criteria derivation or (2) use acute lethality data to derive a surrogate NOAEL, which could then be used as a starting point for the derivation of a provisional AALG.

The numerical relationship between acute lethality data (LD_{50}) and chronic NOAELs for the ingestion route of exposure has been evaluated by a number of authors.[51,74-76] On the basis of statistical analysis of this relationship, Venman and Flaga recommended an acute-to-chronic application factor of 0.0001;[75] Layton et al. recommended a factor ranging from 0.0005 to 0.001.[76] More specifically, Venman and Flaga recommend multiplication of the rat oral LD_{50} for a specific chemical by 0.0001 (equivalent to dividing by 10,000) to produce a surrogate chronic rat NOAEL; Layton et al. recommend multiplication of a small mammal oral LD_{50} by 0.0005 to 0.001 (equivalent to dividing by 2000 to 1000) to derive a surrogate chronic animal NOAEL. Following the derivation of a surrogate chronic animal NOAEL by either method, division by a total uncertainty factor of 100 (10 for interindividual variation $\times$ 10 for interspecies variation) can be used to derive a provisional criterion. Alternatively, a provisional criterion may be derived from an oral LD_{50} in a single step by multiplying the LD_{50} by 1×10^{-6} based on the analysis of Venman and Flaga,[75] or by a factor between 5×10^{-6} and 1.0×10^{-5} based on Layton et al.[76]

It is apparent that there are a number of significant uncertainties and limitations in approaches such as the ones described above. A major limitation is that they are purely statistical approaches and it is not possible to know to what extent using a factor derived from such an analysis can predict one member of an unknown pair (unknown chronic oral NOAEL from a given LD_{50}) when complex biological processes are involved. Another limitation that increases the uncertainty of such comparisons is that since the responses are so different (i.e., lethality versus a no-effect level for an adverse response), the mechanisms by which they are produced are also likely to be quite different. Related to this concern is the fact that the pharmacokinetics and pharmacodynamics will be quite different for an acute high-level exposure scenario compared to a chronic low-level exposure scenario. It is due to these limitations that both Venman and Flaga and Layton et al. emphasize that limits derived using these approaches are provisional and are no substitute for appropriate toxicity testing.

It is our recommendation that unless there are compelling reasons to derive a numerical criterion, the credibility of most industrial entities and S/L agencies is not well served by deriving a criterion using methods associated with such a high degree of uncertainty. If a numerical criterion is derived using such methods, it should be clearly identified as provisional, and the methods used and their limitations described in the accompanying documentation. In addition, such substances should have the highest priority for primary literature review and reevaluation, and depending on the outcome of this activity and an evaluation of the exposure potential, these agents should also have a high priority for testing in biological systems.

Multimedia Exposure and the Relative Source Contribution

One issue not addressed in most air toxics AAL derivation methodologies is the contribution of other sources of exposure and how this should be factored into the final AAL derived. In general, this is best dealt with on a chemical-specific basis. (EPA has published final guidelines for exposure assessment in the *Federal Register* 51(185):34042–54, 9/24/86.) However, some generic approaches have been utilized at the federal level. For example, EPA developed two possible ADIs for noncarcinogenic organics in drinking water. One assumed 20% and the other 1% of the total ADI was derived from drinking water and the remainder from other sources. EPA has used a 20:80 ratio in some cases and a 2:98 ratio in other cases in the derivation of approval limits for noncarcinogenic chlorinated hydrocarbon insecticides. The basis for these figures appears highly subjective.[2]

The NAS Safe Drinking Water Committee, in calculating suggested no-adverse response levels (SNARLs), routinely allows 20% of the total exposure to come from drinking water. The rationale for this assumption, admitted to be arbitrary, is to provide a basis for calculation and is based on the lack of information on relative contributions of other sources of exposure. When such information is available, it is used to modify the SNARL calculation.[77] The assumption of a RSC of 20% is also used by the U.S. EPA Office of Drinking Water in the calculation of HAs for organic compounds in drinking water, while an RSC of 10% is assumed for inorganic compounds. Like the NAS SDWC, the EPA ODW also uses existing exposure data to determine the RSC when appropriate information is available rather than relying on the above assumptions.[78]

It would be most desirable to base the contribution of ambient air exposure on chemical-specific exposure assessments; however, this option is generally not feasible, unless the data already exist, because of resource and time limitations. An additional problem with this approach is that since multimedia exposure scenarios may vary considerably for a given chemical depending on the population of interest, it is difficult to come up with a single figure that is applicable to all situations. Therefore, in the absence of specific exposure information, it is proposed that 80, 50, or 20% of the contribution to total exposure be allowed from air in the calculation of the final AALG. The criteria that were used to determine the percentage relative source contribution for the AALGs calculated in Part II are given in Table 3.17. In the absence of data on the considerations outlined in Table 3.17, 50% of the contribution to total exposure was assumed to come from air for the purpose of calculating the final AALG. It is acknowledged that the exact figures chosen are arbitrary and some judgment is involved; however, as the NAS SDWC has pointed out in the case of SNARLs, the final level derived can easily be adjusted if more definitive exposure information should become available.

One final, but significant, point is that use of a relative source contribution adjustment to calculate the final AALG is not appropriate in all cases. Specifically, two circumstances were identified in the application of this methodology in which

Table 3.17 Guidelines for Selection of the Relative Source Contribution Allowed from Ambient Air in the AALG Derivation Process

Relative Source Contribution from Ambient Air	Criteria
80%	Compounds that are highly volatile or in a gaseous state in the typical environmental temperature range
	Specific data in the form of ambient air sampling surveys that indicate that the compound is widespread in the air of a number of different geographic locations, but not generally occurring in food or drinking water
	Use patterns are such that exposure via air is likely, i.e., widespread releases of unconsumed chemical to ambient air (for example as components of fossil fuel combustion)
50%	Evidence that the compound is ubiquitous in the environment based on sampling data from ambient air, water, and foodstuffs
	Chemical characteristics that favor widespread environmental distribution, i.e., intermediate volatility and water solubility
	Use and environmental release patterns that favor both air and water distribution
20%	Specific evidence in the form of sampling surveys indicating that the compound is widespread in foodstuffs and water, but rarely found in ambient air
	Compound physical characteristics such that exposure from air would not be favored, e.g., low volatility, nondusty, highly water soluble
	Specific evidence that compound tends to preferentially accumulate in food plants or animals, i.e., a strong tendency to bioaccumulate

Note: If data on the environmental fate, exposure potential, and physical characteristics of a chemical are absent or very limited, 50% of the RSC is allowed from air.

use of the RSC adjustment was deemed inappropriate: (1) cases where the AALG is based on a noncumulative adverse effect or endpoint and (2) cases where there is strong evidence that the adverse effect is specific to the inhalation route of exposure and would not be expected to occur via other routes of exposure. The reason for the first case listed above is that noncumulative endpoints, specifically irritation, are concentration-dependent rather than time-dependent, and use of an RSC modification to the AALG implicitly assumes time-dependent effects. In the second case, it is apparent that the contribution of other sources of exposure is not relevant when the critical effect is specific to the inhalation route of exposure. However, it should be noted that extreme caution is necessary when making this determination, and such a conclusion must be backed by rigorously designed and conducted studies using multiple species and noninhalation routes of exposure. In the application of this methodology, only two chemicals were found that met this criteria: cadmium and chromium. Both are considered to be carcinogens by the inhalation route, but not by the ingestion route (based on currently available data). Thus, in both cases the U.S. EPA (in the IRIS database) did not calculate unit risk estimates for the ingestion route.[79,80]

Guidelines for the Selection of Averaging Times

The guiding principle in selection of an appropriate averaging time is that it should correspond to the expected duration of exposure required for the manifestation of the endpoint of concern. Thus, longer averaging times should be used when chronic low-level exposure is of greatest concern, and shorter averaging times when short-term effects are most relevant.[34] The principle is applied by both ACGIH and NIOSH in the recommendation of averaging times for TLVs and RELs, respectively; time-weighted averages are usually recommended for effects that are the result of cumulative damage occurring over a period of time, whereas ceiling limits are often recommended for effects that are expected to develop after only a short period of exposure (e.g., irritation). For purposes of this methodology, annual averaging times are recommended for AALGs based on carcinogenicity, and 24-hour averaging times for AALGs based on other endpoints, with the exception of AALGs based on OELs. In the case of AALGs based on occupational limits, the recommended averaging time is the same as that for the OEL on which the AALG is based. This usually means that the recommended averaging time for these AALGs is either 8 hours or is a ceiling limit. It should be noted that as long as the general principle stated above is observed, the actual time period selected for averaging is largely a discretionary matter that can often be selected based on practical considerations of ambient air monitoring.

References

1. National Research Council. 1983. *Risk Assessment in the Federal Government: Managing the Process* (Washington, DC: National Academy Press).
2. Calabrese, E. J. 1978. *Methodological Approaches to Deriving Environmental and Occupational Health Standards* (New York: John Wiley and Sons).
3. U.S. EPA. 1989. "Interim Methods for Development of Inhalation Reference Doses." EPA/600/8–88/066F.
4. U.S. EPA. 1987. "Reference Dose (RfD): Description and Use in Health Risk Assessments." Integrated Risk Information System (IRIS). Appendix A: on line.
5. Brent, R. L. 1986. "Definition of a Teratogen and the Relationship of Teratogenicity to Carcinogenicity." *Teratology* 34:359–60.
6. Hatch, T. F. 1971. "Thresholds: Do They Exist?" *Arch. Environ. Health* 22:687–89.
7. Hermann, E. R. 1971. "Thresholds in Biophysical Systems." *Arch. Environ. Health* 22:699–706.
8. Stockinger, H. E. 1972. "Concepts of Thresholds in Standards Setting." *Arch. Environ. Health* 25:153–57.
9. Aldridge, W. N. 1986. "The Biological Basis and Measurement of Thresholds." *Ann. Rev. Pharmacol. Toxicol.* 26:39–58.
10. Upton, A. C. 1988. "Are There Thresholds for Carcinogenesis?—The Thorny Problem of Low-Level Exposure." *Ann. NY Acad. Sci.* 534:863–84.
11. Hill, T. A., R. C. Wand, and R. W. Leukroth, Eds. 1986. *Biological Bases for Interspecies Extrapolation of Carcinogenicity Data.* Prepared for Center for Food Safety and Applied Nutrition (FDA) by Life Sciences Research Office, Federation of American Societies for Experimental Biology, Bethesda, MD.
12. Krasovskij, G. N. 1975. "Species and Sex Differences in Sensitivity to Toxic Substances." In *Methods Used in the USSR for Establishing Biologically Safe Levels of Toxic Substances* (Geneva, Switzerland: WHO), pp. 109–125.
13. Krasovskij, G. N. 1976. "Extrapolation of Experimental Data from Animals to Man." *Environ. Health Perspect.* 13:51–58.
14. Anderson, E. L., and Carcinogen Assessment Group of the U.S. EPA. 1983. "Quantitative Approaches in Use to Assess Cancer Risk." *Risk Anal.* 3:277–95.

15. Mantel, N., and M. A. Schneiderman. 1975. "Estimating 'Safe' Levels, a Hazardous Undertaking." *Cancer Res.* 35:1379–86.
16. Boxenbaum, H. 1982. "Interspecies Scaling, Allometry, Physiological Time and the Ground Plan of Pharmacokinetics." *J. Pharmacokin. Biopharm.* 10:201–27.
17. Freireich, E. J., E. A. Gehan, D. P. Rall, L. H. Schmidt, and H. E. Skipper. 1966. "Quantitative Comparison of Toxicity of Anticancer Agents in Mouse, Rat, Hamster, Dog, Monkey and Man." *Cancer Chemother. Rep.* 50:219–44.
18. Calabrese, E. J. 1983. *Principles of Animal Extrapolation* (New York: John Wiley and Sons).
19. Kleiber, M. 1961. *The Fire of Life* (New York: John Wiley and Sons), pp. 177–216.
20. Ramsey, J. C., and P. J. Gehring. 1980. "Application of Pharmacokinetic Principles in Practice." *Fed. Proc.* 39:60–65.
21. Pepelko, W. E., and J. R. Withey. 1985. "Methods for Route-to-Route Extrapolation of Dose." *Toxicol. Ind. Health* 1:153–70.
22. Travis, C. C. 1987. "Interspecies and Dose-Route Extrapolations." In Safe Drinking Water Committee (NRC), *Pharmacokinetics in Risk Assessment, Drinking Water and Health, Vol. 8* (Washington, DC: National Academy Press), pp. 208–20.
23. U.S. EPA. 1980. "Appendix D. Guidelines and Methodology Used in Preparation of Health Effects Assessment Chapters of the Consent Decree Water Quality Criteria." *Fed. Reg.* 45:79347–57.
24. National Academy of Sciences, Safe Drinking Water Committee. 1987. *Drinking Water and Health, Vol. 8* (Washington, DC: National Academy Press).
25. U.S. EPA. 1986. "Guidelines for Carcinogen Risk Assessment." *Fed. Reg.* 51:33992–34004.
26. U.S. EPA. 1986. "Guidelines for Mutagenicity Risk Assessment." *Fed. Reg.* 51:34006–12.
27. U.S. EPA. 1986. "Guidelines for the Health Assessment of Suspect Developmental Toxicants." *Fed. Reg.* 51:34028–40.
28. U.S. EPA. 1988. "Proposed Guidelines for Assessing Male Reproductive Risk and Request for Comments." *Fed. Reg.* 53:24850–69.
29. U.S. EPA. 1988. "Proposed Guidelines for Assessing Female Reproductive Risk and Request for Comments." *Fed. Reg.* 53:24834–47.
30. U.S. EPA. 1989. "Proposed Amendments to the Guidelines for the Health Assessment of Suspect Developmental Toxicants; Request for Comments; Notice." 54:9386–403.
31. Williams, G. M., and J. H. Weisburger. 1986. "Chemical Carcinogens." In C. D. Klaassen, M. O. Amdur, and J. Doull, Eds., *Casarett and Doull's Toxicology—the Basic Science of Poisons* (New York: Macmillan Publishing Company), pp. 99–173.
32. IARC. 1987. "Overall Evaluation of Carcinogenicity: An Updating of IARC Monographs Volumes 1 to 42." *IARC Monog.* suppl. 7.
33. U.S. EPA. 1987. "EPA Approach for Assessing the Risk Associated with Exposure to Environmental Carcinogens." Integration Risk Information System (IRIS). Appendix B: on line.
34. U.S. EPA. 1989. "NATICH Data Base Report on State, Local and EPA Air Toxics Activities." EPA-450/3–89–29.
35. Weil, C. S., and D. D. McCollister. 1963. "Relationship Between Short-Term and Long-Term Feeding Studies in Designing an Effective Toxicity Test." *Agric. Food Chem.* 11:486–91.
36. Dourson, M. L., and J. F. Stara. 1983. "Regulatory History and Experimental Support of Uncertainty (Safety) Factors." *Reg. Toxicol. Pharmacol.* 3:224–38.

37. Ashby, J. 1986. "The Prospects for a Simplified and Internationally Harmonized Approach to the Detection of Possible Human Carcinogens and Mutagens." *Mutagenesis* 1:3–16.
38. Brusick, D. 1986. *Principles of Genetic Toxicology, 2nd ed.* (New York: Plenum Press).
39. IARC. 1987. "Genetic and Related Effects: An Updating of Selected IARC Monographs from Volumes 1 to 42." *IARC Monog.* suppl. 6.
40. Bigwood, E. J. 1973. "The Acceptable Daily Intake of Food Additives." *CRC Crit. Rev. Toxicol.* 2:41–93.
41. ATSDR. 1989. "Toxicological Profile for Phenol." ATSDR/TP-89/20.
42. U.S. EPA. 1987. "Interim Methods for Development of Inhalation Reference Doses." Workshop draft, ECAO-CIN-537, September 1987.
43. Rowan, C. A., W. M. Connolly, and H. S. Brown. 1984. "Evaluating the Use of Occupational Standards for Controlling Toxic Air Pollutants." *J. Environ. Sci. Health* B19:618–48.
44. Calabrese, E. J. 1985. "Uncertainty Factors and Interindividual Variation." *Reg. Toxicol. Pharmacol.* 5:190–96.
45. Mantel, N., and W. R. Bryan. 1961. "Safety Testing of Carcinogenic Agents." *J. Natl. Cancer Inst.* 27:455–70.
46. Weil, C. S. 1972. "Statistics Versus Safety Factors and Scientific Judgement in the Evaluation of Safety for Man." *Toxicol. Appl. Pharmacol.* 21:454–63.
47. Rall, D. P. 1969. "Difficulties in Extrapolating the Results of Toxicity Studies in Laboratory Animals to Man." *Environ. Res.* 2:360–67.
48. Evans, R. D., R. S. Harris, and J. W. M. Bunker. 1944. "Radium Metabolism in Rats, and the Production of Osteogenic Sarcoma by Experimental Radium Poisoning." *Amer. J. Roentgenol.* 52:353–73.
49. Hayes, W. J. 1967. "Toxicity of Pesticides to Man: Risks from Present Levels." *Proc. Royal Soc. London* 167(1007):101–27.
50. Lehman, A. J., and O. G. Fitzhugh. 1954. "100-Fold Margin of Safety." *Assoc. Food Drug Off. U.S. Quart. Bull.* 18:33–35.
51. McNamara, B. P. 1976. "Concepts in Health Evaluation of Commercial and Industrial Chemicals." In M. A. Mehlman, R. E. Shapiro, and H. Blumenthal, Eds., *New Concepts in Safety Evaluation* (Washington, DC: Hemisphere Publishing), pp. 61–140.
52. Klaassen, C. D. 1986. "Principles of Toxicology." In C. D. Klaassen, M. O. Amdur, and J. Doull, Eds., *Casarett and Doull's Toxicology—the Basic Science of Poisons* (New York: Macmillan Publishing Company), pp. 11–32.
53. National Academy of Sciences, Safe Drinking Water Committee. 1989. *Drinking Water and Health, Vol. 9* (Washington, DC: National Academy Press), pp. 121–33.
54. Wilson, J. G. 1973. *Environment and Birth Defects* (New York: Academic Press).
55. Newman, L. M., and E. M. Johnson. 1986. "Teratogen-Induced Decrements of Postnatal Functional Capacity." *J. Am. Coll. Toxicol.* 5:517–24.
56. Manson, J. M., H. Zenick, and R. D. Costlow. 1982. "Teratology Test Methods for Laboratory Animals." In A. W. Hayes, Ed., *Principles and Methods of Toxicology* (New York: Raven Press), pp. 141–84.
57. U.S. EPA. 1982. "Pesticides Assessment Guidelines, Subdivision F. Hazard Evaluation: Human and Domestic Animals." EPA-540/9–82–025.
58. U.S. EPA. 1985. "Toxic Substances Control Act Test Guidelines; Final Rules." *Fed. Reg.* 50:39426–434.
59. Schardein, J. L. 1987. "Approaches to Defining the Relationship of Maternal and Developmental Toxicity." *Teratogen. Carcinogen. Mutagen.* 7:255–71.

60. Chernoff, N., J. M. Rogers, and R. J. Kavlock. 1989. "An Overview of Maternal Toxicity and Prenatal Development: Considerations for Developmental Toxicity Hazard Assessments." *Toxicology* 59:111–25.
61. Brown, N. A., G. Shull, J. Kao, E. H. Goulding, and S. Fabro. 1982. "Teratogenicity and Lethality of Hydantoin Derivatives in the Mouse: Structure Toxicity Relationships." *Toxicol. Appl. Pharmacol.* 64:271–88.
62. Fabro, S., G. Shull, and N. A. Brown. 1982. "The Relative Teratogenic Index and Teratogenic Potency. Proposed Components of Developmental Toxicity Risk Assessment." *Teratogen. Carcinogen. Mutagen.* 2:61–76.
63. Johnson, E. M. 1980. "A Subvertebrate System for Rapid Determination of Potential Teratogenic Hazards." *J. Environ. Pathol. Toxicol.* 4:153–56.
64. Johnson, E. M. 1981. "Screening for Teratogenic Hazards: Nature of the Problems." *Ann. Rev. Pharmacol. Toxicol.* 21:417–29.
65. Johnson, E. M. 1984. "A Prioritization and Biological Decision Tree for Developmental Toxicity Safety Evaluation." *J. Am. Coll. Toxicol.* 3:141–47.
66. Johnson, E. M., and B. E. G. Gabel. 1982. "Applications of the Hydra Assay for Rapid Detection of Developmental Hazards." *J. Am. Coll. Toxicol.* 1:57–71.
67. Johnson, E. M., and B. E. G. Gabel. 1983. "An Artificial 'Embryo' for Detection of Abnormal Developmental Biology." *Fund. Appl. Toxicol.* 3:243–49.
68. National Academy of Sciences, Safe Drinking Water Committee. 1986. "Developmental Effects of Chemical Contaminants." In R. D. Thomas, Ed., *Drinking Water and Health, Vol. 6* (Washington, DC: National Academy Press), pp. 11–34.
69. Green, S., A. Auletta, R. Fabricant, M. Kapp, C. Sheu, J. Springer, and B. Whitfield. 1985. "Current Status of Bioassays in Genetic Toxicology—the Dominant Lethal Test." *Mutat. Res.* 154:49–67.
70. Alarie, Y. 1973. "Sensory Irritation by Airborne Chemicals." *CRC Crit. Rev. Toxicol.* 2:299–363.
71. Alarie, Y. 1981. "Bioassay for Evaluating the Potency of Airborne Sensory Irritants and Predicting Acceptable Levels of Exposure in Man." *Food Cosmet. Toxicol.* 19:623–26.
72. Kane, L.E., C. S. Barrow, and Y. Alarie. 1979. "A Short-Term Test to Predict Acceptable Levels of Exposure to Airborne Sensory Irritants." *J. Am. Ind. Hyg. Assoc.* 40:207–29.
73. Buckley, L. A., X. Z. Jiang, R. A. James, K. T. Morgan, and C. S. Barrow. 1984. "Respiratory Tract Lesions Induced by Sensory Irritants at the RD_{50} Concentration." *Toxicol. Appl. Pharmacol.* 74:417–29.
74. Weil, C. S., M. D. Woodside, J. R. Bernard, and C. P. Carpenter. 1969. "Relationship Between Single Peroral, One-Week and Ninety-Day Rat Feeding Studies." *Toxicol. Appl. Pharmacol.* 14:426–31.
75. Venman, B. C., and C. Flaga. 1985. "Development of an Acceptable Factor to Estimate Chronic End Points from Acute Toxicity Data." *Toxicol. Ind. Health* 1:261–69.
76. Layton, D. W., B. J. Mallon, D. H. Rosenblatt, and M. J. Small. 1987. "Deriving Allowable Daily Intakes for Systemic Toxicants Lacking Chronic Toxicity Data." *Reg. Toxicol. Pharmacol.* 7:96–112.
77. National Academy of Sciences, Safe Drinking Water Committee. 1983. *Drinking Water and Health, Vol. 5* (Washington, DC: National Academy Press).
78. Ohanian, E. V. 1989. "Office of Drinking Water's Health Advisory Program." In E. J. Calabrese, C. E. Gilbert, and H. Pastides, Eds., *Safe Drinking Water Act—Amendments, Regulations and Standards* (Chelsea, MI: Lewis Publishers), pp. 85–103.
79. U.S. EPA. 1989. "Cadmium; CASRN 7440–43–9." IRIS (1/1/89).
80. U.S. EPA. 1989. "Chromium (VI); CASRN 7440–47–3." IRIS (3/1/89).

PART II

Chemical-Specific Assessments

ACETIC ACID

Synonyms: ethanoic acid

CAS Registry Number: 64–19–17

Molecular Weight: 60.06

Molecular Formula: CH_3COOH

AALG: irritation (TLV/10)—1.0 ppm (2.45 mg/m^3) 8-hour TWA

Occupational Limits:
- ACGIH TLV: 10 ppm (25 mg/m^3) TWA; 15 ppm (37 mg/m^3) STEL
- NIOSH REL: none
- OSHA PEL: 10 ppm (25 mg/m^3) TWA

Basis for Occupational Limits: The TLVs for acetic acid are based on irritation.[1,2]

Drinking Water Limits: No drinking water limits were found for this substance.

Toxicity Profile

Carcinogenicity: No data were found implicating acetic acid as a carcinogen.

Mutagenicity: Acetic acid is listed as negative in the Ames assay as summarized in RTECS. (No indication was given as to whether a metabolic activation system was used or not.) The following results are listed for acetic acid in RTECS: mutation in *E. coli* without the use of S9 fraction; sex chromosome loss and nondisjunction in *D. melanogaster* exposed via inhalation to 1000 ppm acetic acid for 24 hours or 1000 ppm oral dose; and unscheduled DNA synthesis in mouse skin.[3]

Developmental Toxicity: No data were found implicating acetic acid as a developmental toxin.

Reproductive Toxicity: No data were found implicating acetic acid as a reproductive toxin.

Systemic Toxicity: Data on the systemic toxicity of acetic acid are limited to acute and subacute exposures. Based on reported acute toxicity indices in laboratory animals, acetic acid appears to have a low order of toxicity (e.g., rat oral LD_{50} of 3530 mg/kg and mouse LC_{50} of 5620 ppm [one hour]).[3]

In a study designed to evaluate the toxicity of the acetic acid component of lead acetate, Barrett and Livesey[4] exposed the dams (n = 5/group) of suckling Wistar rats to water, 5.2×10^{-3} M acetic acid, or 2.6×10^{-3} M lead acetate in drinking water from parturition until the pups were 18 days old. When the weanling rats were tested, mean ambulation rates (a measure of open field activity) were significantly decreased in the acetic acid–exposed rat pups compared to both control and lead acetate–exposed pups at 44 days of age, but not at 20 or 28 days of age.

Irritation: Acetic acid is irritating to the eyes, skin, and mucous membranes. It has been reported that concentrations between 800 and 1200 ppm are intolerable after 3 minutes. Based largely on industrial experience, the TLV committee considered that 10 ppm was relatively nonirritating for longer-term exposures.[1] A nasal irritation threshold of 160 ppm is listed by one source for acetic acid.[5] However, Baldi[6] (cited in ACGIH[1] and OSHA[2]) reported mild changes in respiratory activity of guinea pigs exposed to 5 ppm acetic acid.

Vigliani and Zurlo[7] (cited in ACGIH[1] and OSHA[2]) reported respiratory, gastrointestinal, and dermal irritation in workers exposed to 60 ppm acetic acid during the work shift plus one hour daily at concentrations of 100 to 260 ppm. In addition, Parmeggiani and Sassi[8] (cited in ACGIH[1] and OSHA[2]) reported conjunctivitis, bronchitis, pharyngitis, and erosion of exposed teeth in these same workers.

Acetic acid has a characteristic pungent odor.[1] Odor thresholds of 0.48 ppm (calculated geometric mean of several literature citations) and 0.16 ppm (value given in a literature citation that also listed an irritation threshold) are listed by Amoore and Hautala.[5] These authors calculated an "odor safety factor" (TLV-TWA/odor threshold [geometric mean of all literature citations found omitting extreme points and duplicate quotations]) for a number of compounds and developed an odor safety classification using letter designations A to E, in which class A compounds "provide the strongest odorous warning of their presence at threshold limit value concentrations," whereas class E substances are "practically odorless at the TLV concentration." Under this system, acetic acid was placed in class C, i.e., "less than 50% of distracted persons perceive warning of TLV."

Basis for the AALG: Acetic acid is soluble in water and has a vapor pressure of approximately 15.2 mm Hg at 25°C.[5] Acetic acid is used in the manufacture of a variety of chemicals; production of plastics, pharmaceuticals, dyes, insecticides, photographic chemicals, and natural latex coagulant; and textile printing. In addition, it occurs naturally as vinegar.[1] However, since the AALG for acetic acid is based on irritation, no relative source contribution from inhalation was factored into the calculation. The reason is that allocation of a proportion of exposure to a given source is not relevant when the effects of exposure are not cumulative, as is the case with irritant effects.

Since irritation is the primary effect associated with exposure to acetic acid in the absence of information on other toxicity endpoints, the AALG is based on the

TLV/10. The TLV was chosen rather than the irritation threshold given in the literature because this exposure level is much higher than other levels reported to produce irritant effects, and the TLV is backed by many years of industrial experience. However, it should be noted that there have been some isolated reports of respiratory effects below 10 ppm in both animals and humans;[6,9] therefore, the AALG should be considered provisional.

AALG: • irritation (TLV/10)—1.0 ppm (2.45 mg/m^3) 8-hour TWA

References

1. ACGIH. 1986. "Documentation of the Threshold Limit Values and Biological Exposure Indices." 5:4.
2. OSHA. 1989. "Air Contaminants; Final Rule." *Fed. Reg.* 54:2232–959.
3. NIOSH. 1988. AF1225000. "Acetic Acid." RTECS, on line.
4. Barrett, J. and P. J. Livesey. 1982. "The Acetic Acid Component of Lead Acetate: Its Effect on Rat Weight and Activity." *Neurobehav. Tox. Terat.* 4:105–08.
5. Amoore, J. E. and E. Hautala. 1983. "Odor as an Aid to Chemical Safety: Odor Thresholds Compared with Threshold Limit Values for 214 Industrial Chemicals in Air and Water Dilution." *J. Appl. Tox.* 3:272–90.
6. Baldi, G. 1963. *Med. Iavoro* 44:403. Abstract in *Arch. Ind. Hyg. Occ. Med.* 9:349 (1954).
7. Vigliani, E. C. and N. Zurlo. 1955. *Arch. Gewerbepath. Gewerbehyg.* 13:525–28. Abstract in *Arch. Ind. Health* 13:403 (1956).
8. Parmeggiani, L. and C. Sassi. 1954. *Med. Iavoro* 45:319. Cited in: F. A. Patty, Ed. 1963. *Industrial Hygiene and Toxicology,* 2nd ed. (New York: Wiley-Interscience), p. 1779.
9. Amdur, M. 1961. *Am Ind. Hyg. Assoc. J.* 22:1.

ACETONE

Synonyms: 2-propanone

CAS Registry Number: 77–64–1

Molecular Weight: 58.08

Molecular Formula: CH_3COCH_3

AALG: irritation (TLV/50)—15 ppm (36 mg/m^3) 8-hour TWA

Occupational Limits:
- ACGIH TLV: 750 ppm (1780 mg/m^3) TWA; 1000 ppm (2375 mg/m^3) STEL
- NIOSH REL: 250 ppm (590 mg/m^3) TWA
- OSHA PEL: 750 ppm (1780 mg/m^3) TWA; 1000 ppm (2375 mg/m^3) STEL

Basis for Occupational Limits: The TLV is based on respiratory and eye irritation.[1] The REL is similarly based on irritation of the eyes, nose, throat, and respiratory tract. It is noted in the criteria document that "occupational exposure to acetone may lead to its accumulation in the body."[2] OSHA recently adopted the ACGIH TLV-TWA and TLV-STEL as its final rule limits.[3]

Drinking Water Limits: The state of Massachusetts has set an interim drinking water guideline of 0.25 mg/L for acetone, the lowest of several calculated limits. This limit was derived from a LOEL from a study of behavioral effects in baboons assuming 20% of the exposure from drinking water.[4] It should be noted that this guideline is currently under revision.

Toxicity Profile

Carcinogenicity: No data were found implicating acetone as a carcinogen. In the NTP bioassay program, prechronic studies (rats, mice: drinking water) have been completed and the chemical is under review for further testing.[5]

Mutagenicity: Acetone is listed as negative in the following assays in the Gene-Tox database as summarized in RTECS: SHE clonal assay; cell transformation in mouse and F344 rat embryos; Ames assay without metabolic activation; and in vitro cytogenetics and sister chromatid exchange in nonhuman cells. Acetone was reported to be positive in in vitro cytogenetics assays in *S. cerevisiae* and cultured hamster fibroblasts.[6]

Developmental Toxicity: Acetone was one of 5 solvents mentioned in a case report on sacral agenesis in offspring of 5 women exposed during pregnancy[7] (cited in Reproductive Toxicology Center[8]). However, it should be noted that other reviewers have concluded that this study did not provide evidence of acetone toxicity in humans or allow conclusions to be drawn on the developmental toxicity of acetone[9] (cited in Reproductive Toxicology Center[8]).

Reproductive Toxicity: No data were found implicating acetone as a reproductive toxin.

Systemic Toxicity: Data on the subchronic and chronic inhalation toxicity of acetone in experimental animals are limited. Bruckner and Peterson[10] (cited in U.S. EPA[11]) exposed rats to 0 or 19,000 ppm acetone for 3 hours per day, 5 days a week for 8 weeks with serial sacrifices at 2 weeks, 4 weeks, 8 weeks, and 2 weeks postexposure. No deaths were reported, but narcotic effects were noted in exposed animals. No gross or microscopic lesions or changes in clinical chemistry parameters were observed in these animals. Slight but not significant decreases in organ and body weights were reported during exposure, but not at 2 weeks postexposure.

It is of interest to note that the oral RfD for acetone is based on a 90-day gavage study in rats in which a NOAEL of 100 mg/kg/day was identified. At higher doses, i.e., 500 and 2500 mg/kg, effects were observed including increased kidney and liver weights and nephrotoxicity, which appeared histologically in the form of renal tubular degeneration and hyaline droplet accumulation.[12]

Acetone is also capable of producing narcotic effects, e.g., dizziness and weakness, in humans, but the concentrations required have generally been reported to be in excess of 12,000 ppm (at least for short-term exposures).[2]

A long-term study of workers (n = 800) exposed occupationally to acetone vapors (600–2150 ppm) revealed no statistically significant differences in physical condition or hematological or urological parameters in exposed compared to unexposed workers. The findings of this study are strengthened by the fact that the workers were examined yearly for 18 years[13] (cited in U.S. EPA[11]).

Irritation: Raleigh and McGee[14] (cited in NIOSH[2]) reported slight to moderate eye irritation in workers exposed to 800–1000 ppm for 8 hours. The relevancy of this finding is strengthened since breath samples were collected from workers before, during, and after exposure and analyzed by gas chromatography.

Matsushita et al.[15] (cited in NIOSH[2]) studied the effects of acetone on healthy male student volunteers exposed at concentrations of 100, 250, 500, or 1000 ppm for 6 hours. At 500 and 1000 ppm, most subjects reported irritation to the eyes, nose, throat and trachea; the reported severity of the irritation was dose-related and none of the subjects reported any complaints at 100 ppm. This study is limited by the fact that sampling and analytical procedures were not described.

Acetone is reported to have an "aromatic odor" with a reported odor threshold of between 200 and 400 ppm.[1]

Basis for the AALG: Although acetone is highly volatile (vapor pressure = 180 mm Hg at 20°C), it is also soluble in all proportions in water and is widely used as a solvent in many applications;[1] however, since the AALG is based on irritation, no relative source contribution from inhalation is factored into the calculation. The reason is that irritant effects are noncumulative in nature and tend to be concentration-dependent rather than time-dependent.

Although systemic toxicity is associated with oral exposure to acetone, it is clear from the available evidence that the critical toxic effect for the inhalation route of exposure in humans is irritation. This conclusion is supported by the lack of positive findings in the Oglesby et al. study[13] (cited in NIOSH[2]), which involved evaluation of the effects of long-term exposure to concentrations of 600 to 2150 ppm and the fact that irritant effects have been reported at concentrations as low as or lower than 500 ppm in multiple studies.[2] In this connection it should be noted that none of these studies is ideal as a basis for AALG derivation due to limitations in or lack of reporting of the analytical methods. It is for this reason that the AALG is based on the long-standing ACGIH TLV-TWA. However, since there is reasonable evidence that irritation may occur at levels below the TLV, it is treated as a LOAEL, and a total uncertainty factor of 50 (10 for interindividual variation × 5 for a LOAEL) is used.

AALG: • irritation (TLV/50)—15 ppm (36 mg/m^3) 8-hour TWA

References

1. ACGIH. 1986. "Documentation of the Threshold Limit Values and Biological Exposure Indices." 5:6–7.
2. NIOSH. 1978. "Criteria for a Recommended Standard . . . Occupational Exposure to Ketones." DHEW (NIOSH) 78–173.
3. OSHA. 1989. "Air Contaminants; Final Rule." *Fed. Reg.* 54:2332–959.
4. Anderson, P. D. 1986. "Drinking Water Guideline for Acetone." Massachusetts Department of Environmental Quality Engineering, Office of Research and Standards, Boston.
5. NTP. 1989. "Chemical Status Report" (Research Triangle Park, NC: National Toxicology Program).
6. NIOSH. 1987. AL3150000. "Acetone." RTECS, on line.
7. Kucera, J. 1968. "Exposure to Fat Solvents: A Possible Cause of Sacral Agenesis in Man." *J. Pediatr.* 72:857–59.
8. Reproductive Toxicology Center. 1987. "Reproductive Toxicity Review—Acetone." REPROTOX, on line (Washington, DC: Columbia Hospital for Women Medical Center).

9. Barlow, S. M. and F. M. Sullivan. *Reproductive Hazards of Industrial Chemicals* (New York: Academic Press), p. 597.

10. Bruckner, J. V. and R. G. Peterson. 1981. "Evaluation of Toluene and Acetone Inhalant Abuse. II. Model Development and Toxicology." *Tox. Appl. Pharm.*, in press.

11. U.S. EPA. 1984. "Health Effects Assessment for Acetone." EPA/540/1–86/016.

12. U.S. EPA. 1988. "Acetone; CASRN 67–64–1." Integrated Risk Information System (revised 3/1/89).

13. Oglesby, F. L., J. E. Williams, D. W. Fassett, and J. H. Sterner. 1949. Eastman Kodak Co., Rochester, NY. Paper presented at the Annual Meeting of the American Industrial Hygiene Association, Detroit.

14. Raleigh, R. L. and W. A. McGee. 1972. "Effects of Short, High-Concentration Exposures to Acetone as Determined by Observation in the Work Area." *J. Occ. Med.* 14:607–10.

15. Matsushita, T., A. Yoshimune, T. Inoue, S. Yamaka, and H. Suzuki. 1969. "Experimental Studies for Determining the Maximum Permissible Concentrations of Acetone—1. Biological Reactions in One-Day Exposure to Acetone" (Japanese). *Jpn. J. Ind. Health* 11:477–85.

ACETONE CYANOHYDRIN

Synonyms: 2-methyl-lactonitrile

CAS Registry Number: 75–86–5

Molecular Weight: 85.12

Molecular Formula: $(CH_3)_2C(OH)CN$

AALG: systemic toxicity (REL-based)—2.4 ppb (8.3 $\mu g/m^3$) 15-minute ceiling

Occupational Limits:
- ACGIH TLV: none
- NIOSH REL: 1 ppm (4 mg/m^3) 15-minute ceiling
- OSHA PEL: none

Basis for Occupational Limits: The ceiling REL is based on comparative toxicity with acetonitrile. The calculated minimum lethal concentration for rats exposed to acetonitrile by inhalation is approximately 18.3 times higher relative to acetone cyanohydrin. Based on this and the fact that it dissociates readily to release HCN, the REL for acetone cyanohydrin is roughly 1/20 that of acetonitrile.[1]

Drinking Water Limits: No drinking water limits were found for this substance.

Toxicity Profile

Carcinogenicity: No data were found implicating acetone cyanohydrin as a carcinogen.

Mutagenicity: No data were found implicating acetone cyanohydrin as a mutagen.

Developmental Toxicity: No data were found implicating acetone cyanohydrin as a developmental toxin.

Reproductive Toxicity: No data were found implicating acetone cyanohydrin as a reproductive toxin.

Systemic Toxicity: Based on reported LD_{50}s (parenteral, route not specified) of 2.9 mg/kg, 9 mg/kg, 13.3 mg/kg, and 13.5 mg/kg in mice, guinea pigs, rats, and rabbits, respectively, it appears that acetone cyanohydrin has a relatively high acute toxicity in laboratory animals.[1]

Motoc et al.[2] (cited in NIOSH[1]) studied the effects of oral and inhalation exposure to acetone cyanohydrin in albino rats. The animals (n = 50/group) received 5 mg of acetone cyanohydrin orally twice per week or 1 mL in 84 L of air (10.2 g/m^3) by inhalation twice a week for 3, 5, or 8 months (hours per day of exposure not specified). The primary effects in animals exposed orally included gastric ulceration and liver necrosis characterized by compacted off-center nuclei, abnormal fat deposits, patchy thinning of cytoplasm, and absence of cytoplasmic granules. Inhalation of acetone cyanohydrin caused lesions in the lungs with desquamation of bronchial epithelium progressing to ulcerations associated with inflammatory infiltrates. Renal and hepatic necrosis were also observed.

The only reports dealing with human exposure cited in the criteria document involved industrial accidents in which the exposures were mainly dermal and the concentrations unknown. The onset of toxicity was reported to be related to the time required for dissociation to produce free hydrogen cyanide.[1]

Irritation: No data were found implicating acetone cyanohydrin as an irritant.

Basis for the AALG: Acetone cyanohydrin is considered to be a relatively nonvolatile solvent and in occupational settings the principal route of exposure is thought to be dermal.[2] Based on lack of data on its environmental fate and distribution and extent of human exposure from various media, 50% of the contribution to total exposure will be allowed from air.

Data on the inhalation toxicity of acetone cyanohydrin in laboratory animals and humans are very limited, and no suitable studies were available on which to base the AALG. Therefore, the AALG is based on the NIOSH REL. Since the original REL was based on analogy with acetonitrile, it was considered prudent to apply a database factor of 5 to account for the uncertainty associated with an analogy-based REL. Thus, a total uncertainty factor of 210 (4.2 for continuous exposure adjustment × 10 for interindividual variation × 5 for the database factor) was applied. This AALG should be considered provisional. It is recommended that a thorough review of the literature be conducted to identify a more suitable basis for the AALG and as an aid in planning appropriate animal studies and evaluation of the potential for human exposure.

AALG: • systemic toxicity (REL-based)—2.4 ppb (8.3 μg/m^3) 15-minute ceiling

References

1. NIOSH. 1978. "Criteria for a Recommended Standard . . . Occupational Exposure to Nitriles." DHEW (NIOSH) 78–212.
2. Motoc, F., S. Constantinescu, G. Filipescu, M. Dobre, E. Bichir, and G. Pambuccian. 1971. "Noxious Effects of Certain Substances Used in the Plastics Industry (Acetone Cyanohydrin, Methyl Methacrylate, Azo(bis)isobutyronitrile and Anthracene Oil)—Relation Between the Aggressor Agent and its Effects." *Arch. Mal. Prof. Med. Trav. Secur. Soc.* 32:653–58.

ACETONITRILE

Synonyms: methyl cyanide

CAS Registry Number: 75–05–8

Molecular Weight: 41.06

Molecular Formula: $H_3C\text{-}C\equiv N$

AALG: systemic toxicity—0.041 ppm (0.068 mg/m^3) 24-hour TWA

Occupational Limits:
- ACGIH TLV: 40 ppm (70 mg/m^3) TWA; 60 ppm (105 mg/m^3) STEL; skin notation
- NIOSH REL: 20 ppm (34 mg/m^3) TWA
- OSHA PEL: 40 ppm (70 mg/m^3) TWA; 60 ppm (105 mg/m^3) STEL; skin notation

Basis for Occupational Limits: The ACGIH limits are based on prevention of cyanide poisoning and injury to the respiratory tract.[1] NIOSH recommends a lower limit based on (1) an account of an incident of acute effects in workers exposed to high concentrations in which it was commented that after exposure levels were lowered to 17 ppm no further complaints were received and (2) the fact that an effect was noted at 40 ppm (feeling of tightness in the chest) in a controlled human exposure study.[2] OSHA recently adopted the ACGIH STEL for acetonitrile as part of its final rule limits.[3]

Drinking Water Limits: In its 1982 evaluation of acetonitrile the NAS Safe Drinking Water Committee concluded that there were no adequate data from which to calculate a 24-hour, 7-day, or chronic SNARL.[4]

Toxicity Profile

Carcinogenicity: No data were found implicating acetonitrile as a carcinogen. Acetonitrile is presently being tested in the NTP bioassay program (2-year inhalation studies in rats and mice).[5]

Mutagenicity: Positive results were reported for a chromosomal nondisjunction assay in the yeast *Saccharomyces cerevisiae* in the RTECS database.[6]

Developmental Toxicity: Willhite[7] studied the developmental effects of acetonitrile in hamsters exposed via inhalation for 60 minutes on the 8th day of gestation to 0, 1800, 3800, 5000, or 8000 ppm. There was a significant dose-related increase in malformations (exencephaly, encephalocoele, fused ribs) at the two highest exposure levels as well as a significant decrease in fetal weight at the highest exposure level. It is of interest to note that in this study maternal toxicity was observed at levels below those causing fetotoxicity and malformations. At 3800 ppm, 5000 ppm, and 8000 ppm 1 of 6, 1 of 6, and 4 of 12 animals, respectively, showed signs of intoxication (tremors, dyspnea, hypersalivation, ataxia, and hypothermia) and died. At 5000 ppm and 8000 ppm all animals showed signs of irritation. Studies on the in vivo metabolism of acetonitrile indicated that the liberation of cyanide was probably responsible for the production of terata. While this is generally a strong study, the sample sizes were somewhat small and the incidence of malformations was reported in aggregate rather than per litter.

Reproductive Toxicity: No studies were found implicating acetonitrile as a reproductive toxin.

Systemic Toxicity: Acetonitrile is acutely toxic to laboratory animals exposed via inhalation, with reported LC_{50}s of 7551 ppm in rats (8-hour) and 2693 ppm in mice (1-hour).[6]

NTP conducted a subchronic inhalation study in male and female B6C3F$_1$ mice and F344 rats (n = 10/sex/species/group) exposed for 6 hours per day, 5 days per week for 13 weeks to 25, 50, 100, 200, or 400 ppm. The only statistically significant finding in rats was a decreased leukocyte count in females in the highest exposure group and in males at exposure levels of 100 ppm and higher. Cytoplasmic vacuolization of the hepatocytes was observed in controls and all exposure groups but was slightly greater in female rats in the 400-ppm exposure group. In mice, there was a statistically significant decrease in blood urea nitrogen (BUN), erythrocyte counts, and hematocrits in female mice in the 200- and 400-ppm exposure groups. Other observations (not statistically significant) included greater intensity of hepatic lesions (vacuolization and hypertrophy) in both male and female mice exposed to 200 or 400 ppm; elevated relative liver weights; and a dose-related decrease in leukocyte count and serum IgG which did not appear to be related to a specific immune system dysfunction. It was concluded in the IRIS summary that mice were the more sensitive species with respect to acetonitrile toxicity and it is relevant to note that the mouse NOAEL was used to derive the *oral* reference dose (RfD) for acetonitrile. A NOAEL of 100 ppm in mice is suggested by this study with a LOAEL of 200 ppm for hepatic lesions and hematologic changes[8] (cited in U.S. EPA[9]).

Pozzani et al.[10] (cited in NIOSH[2]) studied the effects of inhalation of acetonitrile in humans exposed for 4 hours to 40 ppm (n = 3), 80 ppm (n = 2), and 160 ppm (n = 1). One subject exposed to 40 ppm reported tightness in the chest after but not during exposure that persisted for 24 hours; a slight increase in urinary thiocyanate excretion was also noted. Subjects were also able to detect the odor at

40 ppm for 2–3 hours and then experienced some "olfactory fatigue." Subjects exposed to 80 ppm reported no adverse symptoms, and the subject exposed at 160 ppm reported flushing of the face 2 hours after exposure and a feeling of bronchial tightness.

Irritation: Acetonitrile is reported to have an "aromatic" odor[1] and an odor threshold of 170 ppm (geometric mean of reported literature values) has been reported.[11] At sufficiently high concentrations, acetonitrile is irritating to the respiratory tract.[2]

Basis for the AALG: Acetonitrile is soluble in water and has a vapor pressure of 73 mm Hg at 20°C. Its primary uses are as a specialty solvent and in organic synthesis.[1] Based on lack of information on its environmental fate, distribution, and occurrence, and on the expected contribution of various media to human exposure, 50% of the contribution to total exposure will be allowed from air.

Both developmental and systemic toxicity are relevant endpoints for AALG derivation based on available data. However, it was considered inappropriate to calculate a developmental AALG based on the only available inhalation study, i.e., that of Willhite,[7] since severe maternal toxicity (lethality) occurred at levels below those producing fetotoxic or teratogenic effects and because other studies revealed that the mouse is a more sensitive species, at least for systemic toxicity endpoints.

Both the NIOSH REL and the NOAEL of 100 ppm in mice in the NTP inhalation study are suitable for AALG derivation based on systemic toxicity. Based on available data, the NIOSH REL would be expected to represent a reasonable surrogate human NOAEL; thus, the REL/42 was used. The mouse NOAEL of 100 ppm from the NTP study was adjusted for continuous exposure (100 ppm × 6/24 × 5/7) and converted to a human equivalent inhalation exposure dose according to the procedures described in Appendix C. A total uncertainty factor of 1000 was applied (10 for interindividual variation × 10 for interspecies variation × 10 for less-than-chronic exposure). Since it is more conservative, the AALG based on the mouse NOAEL is recommended; however, this AALG should be considered provisional pending availability of results from the 2-year inhalation NTP bioassay.

AALG: • systemic toxicity (REL/42)—0.24 ppm (0.40 mg/m^3) 8-hour TWA
         • systemic toxicity (animal NOAEL–based)—0.041 ppm (0.068 mg/m^3) 24-hour TWA

References

1. ACGIH. 1986. "Documentation of the Threshold Limit Values and Biological Exposure Indices." 5:8–9.
2. NIOSH. 1978. "Criteria for a Recommended Standard . . . Occupational Exposure to Nitriles." DHEW (NIOSH) 78–212.
3. OSHA. 1989. "Air Contaminants; Proposed Rule." *Fed. Reg.* 54:2332–959.

4. NAS Safe Drinking Water Committee. 1982. *Drinking Water and Health,* Vol. 4 (Washington, DC: National Academy Press).

5. NTP. 1989. "Chemical Status Report" (Research Triangle Park, NC: National Toxicology Program).

6. NIOSH. 1988. AL7700000. "Acetonitrile." RTECS, on line.

7. Willhite, C. C. 1983. "Developmental Toxicology of Acetonitrile in Syrian Golden Hamster." *Teratology* 27:313–25.

8. Hazelton Laboratory. 1983. "90-Day Subchronic Toxicity Study of Acetonitrile in Rats and Mice." Reports prepared for the NTP program on June 27 and July 8, 1983.

9. U.S. EPA. 1988. "Acetonitrile; CASRN 75–05–8." Integrated Risk Information System (6/30).

10. Pozzani, U. C., C. P. Carpenter, P. E. Palm, C. S. Weil, and J. H. Nair. 1959. "An Investigation of the Mammalian Toxicity of Acetonitrile." *J. Occ. Med.* 1:634–42.

11. Amoore, J. E. and E. Hautala. 1983. "Odor as an Aid to Chemical Safety: Odor Thresholds Compared with Threshold Limit Values and Volatilities for 214 Industrial Chemicals in Air and Water Dilution." *J. Appl. Tox.* 3:272–90.

ACETYLENE

Synonyms: ethine, ethyne, narcylen

CAS Registry Number: 74–86–2

Molecular Weight: 26.04

Molecular Formula: HC≡CH

AALG: 2500 ppm (2665 mg/m^3) ceiling limit

Occupational Limits:
- ACGIH TLV: none
- NIOSH REL: 2500 ppm ceiling (absolute maximum)
- OSHA PEL: 2500 ppm ceiling

Basis for Occupational Limits: The basis for the occupational limits is the acetylene environmental standard, which is 10% of the lower explosive limit and was set at this level for protection against the flammable and explosive properties of acetylene. This was deemed appropriate because there were no reports of acetylene toxicity at concentrations below 100,000 ppm.[1] Although there is no TLV for acetylene, it is listed in Appendix E of the TLV booklet as a simple asphyxiant.[2] A simple asphyxiant is a substance that causes death by the exclusion of oxygen.

Drinking Water Limits: No drinking water limits were found for this substance.

Toxicity Profile

Carcinogenicity: No data were found implicating acetylene as a carcinogen.

Mutagenicity: No data were found implicating acetylene as a mutagen.

Developmental Toxicity: No data were found implicating acetylene as a developmental toxin.

Reproductive Toxicity: No data were found implicating acetylene as a reproductive toxin.

Systemic Toxicity: It has been reported that acetylene will produce varying degrees of temporary and reversible narcosis when administered with oxygen as an anesthetic at concentrations of 100,000 ppm or greater. The criteria document for acetylene states that a literature review revealed no reports in which evidence of histopathologic effects attributable to acetylene were established. It is of interest to note that the few reports of fatalities associated with acetylene were attributed to contamination with arsine or phosgine.[1]

Irritation: No data were found implicating acetylene as an irritant.

Basis for the AALG: In view of the lack of toxicity of acetylene and the fact that it presents a substantial explosion and flammability hazard at concentrations well below those associated with narcotic effects, it is not reasonable to establish a health-based AALG. Therefore, it is recommended that the NIOSH limit be used as the AALG.

AALG: • 2500 ppm (2665 mg/m^3) ceiling limit

References

1. NIOSH. 1976. "Criteria for a Recommended Standard . . . Occupational Exposure to Acetylene." DHEW (NIOSH) 76–195.
2. ACGIH. 1988. "Threshold Limit Values and Biological Exposure Indices for 1988–89" (Cincinnati, OH: ACGIH).

ACROLEIN

Synonyms: 2-propenal

CAS Registry Number: 107–02–8

Molecular Weight: 56.06

Molecular Formula: $CH_2 = CHCHO$

AALG: systemic toxicity—0.016 ppb (0.037 $\mu g/m^3$) 24-hour TWA

Occupational Limits:
- ACGIH TLV: 0.1 ppm (0.25 mg/m^3) TWA; 0.3 ppm (0.8 mg/m^3) STEL
- NIOSH REL: none
- OSHA PEL: 0.1 ppm (0.25 mg/m^3) TWA; 0.3 ppm (0.8 mg/m^3) STEL

Basis for Occupational Limits: The occupational limits are based on irritation and it is stated in the TLV documentation[1] that the limits are "sufficiently low to minimize, but not entirely prevent, irritation to all exposed individuals." OSHA recently adopted the ACGIH STEL as part of its final rule limits.[2]

Drinking Water Limits: No drinking water limits were found for this substance. However, EPA derived an ambient water quality criterion for acrolein based on a 90-day study in rats receiving acrolein in their drinking water in which there were no apparent adverse effects at 200 mg/L (the highest level used). Assuming 100% gastrointestinal absorption with a safety factor of 1000, a limit of 0.321 $\mu g/L$ was derived.[3]

Toxicity Profile

Carcinogenicity: Skin application studies in mice[4] (cited in IARC[5]) and inhalation and intratracheal instillation studies in hamsters[6] (cited in IARC[5]) produced negative results. However these studies suffer from design limitations and small sample size and were considered inadequate to determine carcinogenicity of acrolein in laboratory animals.[5]

It was considered relevant by IARC that positive results have been reported for a metabolite of acrolein, glycidaldehyde, in mice exposed via skin painting and in mice and rats exposed by subcutaneous injection.[4] However, at the time of the

IARC evaluation no relevant epidemiologic data were available for either acrolein or glycidaldehyde.

Based on the structural similarity of acrolein to substances possibly carcinogenic to humans, the carcinogenic potential of one of its metabolites, and lack of epidemiologic data, acrolein has been tentatively placed in Group C, possible human carcinogen, under the EPA weight-of-evidence classification.[7]

Mutagenicity: Acrolein has been reported to be negative in the following assays: *E. coli* polA$^+$/polA$^-$ assay without metabolic activation, SCE in Chinese hamster ovary cells, and *E. coli* WP2 uvrA assay. Positive and negative results have been reported in the Ames assay, depending on the strains used and study conditions. Acrolein was reported to be negative in mutagenicity assays using *Aspergillus nidulans* and *S. cerevisiae*, does not induce crosslinks or breakage in *S. cerevisiae* DNA, and was negative in the dominant lethal assay in male mice with a single intraperitoneal injection of 1.5 or 2.2 mg/kg. IARC considered the evidence of genetic activity based on short-term tests to be limited.[7,8]

Developmental Toxicity: Both positive and negative results have been reported in the chick embryo assay, depending on the conditions of the study.[5]

On the 9th day of gestation, Claussen et al.[9] injected rabbits intravenously with 3, 4.5, or 6 mg/kg acrolein and also used the technique of direct injection into the yolk sac of rabbit embryos of 10, 20, or 40 μL/blastoderm of 0.84% acrolein solution. In the former case, dose-dependent embryolethal effects were found and in the latter case dose-dependent increases in resorptions and malformations were observed.

It should be noted that (1) acrolein is a metabolite of cyclophosphamide, a known animal teratogen and (2) reproductive effects have been reported in humans exposed therapeutically to this drug.[4]

Reproductive Toxicity: See comments under "Developmental Toxicity."

Systemic Toxicity: Acrolein is acutely toxic to laboratory animals by the inhalation route, with reported LC$_{50}$s of 130 ppm (30-minute exposure) in rats, 66 ppm (6-hour exposure) in mice, and 25 ppm (4-hour exposure) in hamsters.[5]

The only available inhalation studies in animal models for acrolein were of subacute or subchronic duration. Lyon et al.[10] (cited in IARC[5]) exposed groups of beagle dogs, squirrel monkeys, guinea pigs, and rats to 0.7 or 3.7 ppm acrolein vapor for 8 hours per day, five days per week for 6 weeks. Squamous metaplasia and basal cell hyperplasia were observed in the trachea of both dogs and monkeys, along with squamous cell metaplasia in the lungs of 7 of 9 monkeys.

Feron et al.[11] exposed male and female Syrian golden hamsters (n = 20/ group), Wistar rats (n = 12/group), and Dutch rabbits (n = 4/group) to 0, 0.4, 1.4, or 4.9 ppm acrolein vapor for 6 hours per day, 5 days per week for 13 weeks (90 days). Treatment-related mortality occurred only in rats (6 of 12) at the highest exposure level. At the highest exposure level all species showed signs of irritation including salivation, nasal discharge, sneezing, and occasional difficulty in breathing, and body weight gain was significantly depressed. Variable signs of irritation and symptomatology were observed in the medium exposure group with no signs of irritation in any of the low exposure groups. The only hematologic alterations were seen in hamsters in the highest exposure group. These consisted of significant increases in RBCs, packed cell volume, and hemoglobin content and increased numbers of lymphocytes with decreased numbers of PMN leukocytes; no mechanism for these effects was suggested by the authors. In the high exposure groups in all species there was an increase in amorphous material in the urinary sediment; serum enzymes did not differ from controls in any of the exposure groups. Significant increases in relative organ weights occurred in rats (lung, heart, kidney, adrenals), hamsters (lung, heart, kidneys), and rabbits (lung) in the highest exposure groups. The only gross findings at necropsy were patchy consolidation, hemorrhages, and collapsed dark reddish-purple areas in the lungs of the high-exposure rats. Histopathologic changes attributable to acrolein exposure were found only in the respiratory tract. Squamous cell metaplasia with keratinization and neutrophilic infiltration of the mucosa occurred in all species at the highest exposure level. Such changes were also observed in rats at the intermediate exposure level and in one male rat at the lowest exposure level. Hyperplastic and metaplastic changes were observed in the tracheal and laryngeal epithelium in both rats and hamsters at the highest exposure level, but the lesions were more severe in rats. Bronchopulmonary lesions (metaplasia and hyperplasia of the bronchiolar epithelium, increased mucus production, accumulations of alveolar macrophages, focal interstitial pneumonitis) were observed in rats and rabbits in the highest exposure groups, but were more severe in the rats. A LOAEL of 0.4 ppm in rats and a NOAEL of 0.4 ppm in hamsters and rabbits for alterations in the nasal epithelium and other respiratory tract changes are suggested by this study. The decision was made to treat the 0.4-ppm exposure level as a LOAEL in rats because although the effect may not have been statistically significant, it should be considered biologically significant, since there is evidence that acrolein has potential carcinogenic activity, and the metaplastic changes in rats occurred in a dose-dependent manner.

It is of interest to note that significant reductions in liver and lung mixed-function oxidase activity were reported in rats injected intraperitoneally (twice) with 5 mg/kg acrolein[12] (cited in IARC[5]).

Irritation: Acrolein is a potent eye and respiratory irritant in humans and laboratory animals.[1,5] The eye irritation, odor, and respiratory response thresholds in humans have been reported to be 0.2, 0.33–0.40, and 0.62 ppm, respectively.[7]

Basis for the AALG: Acrolein is soluble in water (20.6 wt% at 20°C) and volatile (vapor pressure = 220 mm Hg at 20°C). It is used as a chemical intermediate, aquatic herbicide, microbicide (in oil wells), and slimicide (in manufacture of paper and paperboard). Acrolein occurs in nature and has also been identified in urban and suburban air, municipal effluents, and various food products and is a constituent of marijuana and cigarette smoke.[5] Based on these considerations, 50% of the contribution to total exposure will be allowed from air.

The AALG for acrolein is based on the LOAEL of 0.4 ppm for respiratory tract changes in rats (the most sensitive species tested) in the Feron et al. study.[11] The LOAEL was adjusted for continuous exposure (0.4 ppm $\times$ 6/24 $\times$ 5/7) and converted to a human equivalent LOAEL according to the methods described in Appendix C. A total uncertainty factor of 5000 was applied (10 for interindividual variation $\times$ 10 for interspecies variation $\times$ 10 for less-than-chronic duration of study $\times$ 5 for a LOAEL). This AALG should be considered provisional pending availability of more complete data on the potential carcinogenic and developmental effects of acrolein by the inhalation route.

AALG: • systemic toxicity—0.016 ppb (0.037 $\mu g/m^3$) 24-hour TWA

References

1. ACGIH. 1986. "Documentation of the Threshold Limit Values and Biological Exposure Indices." 5:11.
2. OSHA. 1989. "Air Contaminants; Final Rule." *Fed. Reg.* 54:2332–959.
3. U.S. EPA. 1980. "Ambient Water Quality Criteria for Acrolein." EPA 440/5–80–016.
4. Salaman, M. H. and F. J. C. Roe. 1956. "Further Test for Tumour-Initiating Activity: N,N-Di-(2-chloroethyl)-p-aminophenyl-butyric Acid (CB1348) as an Initiator of Skin Tumour Formation in the Mouse." *Br. J. Can.* 10:363–78.
5. IARC. 1985. "Acrolein." *IARC Monog.* 36:133–61.
6. Feron, V. J. and A. Kruysse. 1977. "Effects of Exposure to Acrolein Vapor in Hamsters Simultaneously Treated with Benzo(a)pyrene or Diethylnitrosamine." *J. Tox. Env. Health* 3:379–94.
7. U.S. EPA. 1986. "Health Assessment Document for Acrolein." EPA-600/8–86–014A (review draft).
8. IARC. 1987. "Acrolein." *IARC Monog.* suppl. 6:24–26.
9. Claussen, U., W. Hellmann, and G. Pache. 1980. "The Embryotoxicity of the Cyclophosphamide Metabolite Acrolein in Rabbits, Tested In Vivo by I.V. Injection and by the Yolk-Sac Method." *Arzneim.-Forsch.* 30:2080–83.
10. Lyon, J. P., L. J. Jenkins, R. A. Jones, R. A. Coon, and J. Siegel. 1970. "Repeated and Continuous Exposure of Laboratory Animals to Acrolein." *Tox. Appl. Pharm.* 17:726–32.
11. Feron, V. J., A. Kruysse, H. P. Til, and H. R. Immel. 1978. "Repeated Exposure to Acrolein Vapor: Subacute Studies in Hamsters, Rats and Rabbits." *Toxicology* 9:47–57.
12. Patel, J. M. and K. C. Leibman. 1979. "Biochemical Effects of Acrolein on Rat Liver and Lung, as Influenced by Various Treatments." Abstract 1657. *Fed. Proc.* 38:865.

ACRYLAMIDE

Synonyms: propenamide, acrylamide monomer

CAS Registry Number: 79–06–1

Molecular Weight: 71.08

Molecular Formula: $CH_2 = CHC(O)NH_2$

AALG: 10^{-6} 95% UCL—6.5×10^{-5} ppb (1.9×10^{-4} $\mu g/m^3$) annual TWA

Occupational Limits:
- ACGIH TLV: 0.01 ppm (0.03 mg/m^3) TWA; A2 suspect human carcinogen; skin notation
- NIOSH REL: 0.1 ppm (0.3 mg/m^3) TWA
- OSHA PEL: 0.01 ppm (0.03 mg/m^3) TWA; skin notation

Basis for Occupational Limits: The TLV is based on carcinogenicity[1] and the NIOSH limit is based on the prevention of neurological effects, particularly delayed neuropathy.[2] It should be noted that the TLV was previously the same as the REL, but an additional safety factor of 10 was incorporated to account for potential carcinogenic effects.[1] OSHA recently adopted the ACGIH-TLV as part of its final rule limits.[3]

Drinking Water Limits: An MCLG of 0 mg/L has been proposed for acrylamide based on carcinogenic effects.[4] In the draft drinking water document for acrylamide,[5] a 90-day rat NOAEL (0.2 mg/kg/day) was used to derive an AADI of 0.007 mg/L. In the health advisory for acrylamide, 1- and 10-day HAs of 1.5 mg/L and 0.3 mg/L, respectively, were recommended for the 10-kg child. Longer-term HAs of 0.02 mg/L and 0.07 mg/L were derived for the 10-kg child and the 70-kg adult, respectively. No lifetime HA was recommended due to the B2 classification of acrylamide under the EPA weight-of-evidence classification.[6]

Toxicity Profile

Carcinogenicity: Acrylamide has been shown to be a skin tumor initiator in female Sencar mice and to induce lung adenomas in male and female A/J mice by oral and intraperitoneal administration.[7]

Johnson et al.[8] studied the carcinogenic response of male and female F344 rats (n = 90/sex/group) exposed to 0, 0.01, 0.1, 0.5, and 2.0 mg/kg/day acrylamide

(96–99% purity; major impurity was water) in drinking water for two years. Serial sacrifice of 10 rats of each sex and treatment level was performed at 6, 12, and 18 months to detect early tumor development. Body weight gain was decreased at the highest exposure level and in male rats in the 0.5-mg/kg/day exposure group; there was also an increase in mortality in the highest exposure group during the last 4 months of the study. The major nonneoplastic lesion observed was peripheral neuropathy at the highest exposure level. No statistically significant increases in tumors occurred at the 0.01- or 0.1-mg/kg/day exposure levels and the only statistically significant increase in tumor type at 0.5 mg/kg/day was for scrotal mesotheliomas. However, at 2.0 mg/kg/day statistically significant increases in mammary gland, central nervous system, thyroid follicular epithelium, oral tissue, uterine, and clitoral gland tumors were observed in females and in central nervous system (relative to historical controls), thyroid follicular epithelial, and scrotal mesothelial tumors in males.

Acrylamide has been classified in group B2, probable human carcinogen, under the EPA weight-of-evidence classification.[6]

Mutagenicity: The following Gene-Tox results were summarized in RTECS: positive for cytogenetic effects in male germ cells, negative for in vivo cytogenetic effects in nonhuman bone marrow cells, and inconclusive for in vivo SCE in nonhuman cells.[9] Consistently negative results have been reported in the Ames assay using multiple strains, with and without metabolic activation.[10] In a 1988 review, it was concluded: "The available evidence suggests that acrylamide does not produce detectable gene mutations, but that the major concern for its genotoxicity is its clastogenic activity. This clastogenic activity has been observed in germinal tissues which suggests the possible heritability of acrylamide-induced DNA alterations."[11]

Developmental Toxicity: It has been demonstrated that acrylamide readily crosses the placenta in several species[12] (cited in Reproductive Toxicology Center[13]) and that acrylamide in fetal blood crosses the blood-brain barrier in dogs and pigs[14] (cited in Reproductive Toxicology Center[13]). In one study[15] (cited in Reproductive Toxicology Center[13]) acrylamide was reported to alter the activity of several intestinal enzymes in rat pups exposed in utero and via the milk from treated dams.

Agrawal and Squibb[16] (cited in IARC[10] and Reproductive Toxicology Center[13]) studied the effects of acrylamide on the developing dopamine system in rat pups whose dams had been exposed to 0 or 20 mg/kg acrylamide by gavage from day 7 to day 17 of gestation. In two-week-old pups (but not three-week-old) there was significantly decreased ³H-spiroperidol binding in striatal tissue in all groups (except treated dams with control pups) relative to control dams with control pups; this finding suggests that the effect is reversible.

In a recent review of the nonneurotoxic effects of acrylamide, Dearfield et al.[11] concluded that there was evidence indicating that acrylamide may produce neurotoxic effects in neonatal rats at exposure levels not causing overt toxicity in the dams. It should be noted that no developmental toxicity studies using the inhalation route were cited in this review.

Reproductive Toxicity: In addition to its other effects, acrylamide is also a reproductive toxin, as evidenced by positive results in a number of dominant lethal assays and its capability of producing testicular degeneration and spermhead abnormalities in animals exposed via drinking water or by gavage. A number of these studies are reviewed in Dearfield et al.[11]

Zenick et al.[17] evaluated the effects of acrylamide on copulatory behavior, semen, and (for controls and 100-ppm group only) fertility and fetal outcomes in male Long-Evans rats exposed to 0, 50, 100, or 200 ppm in drinking water (n = 15/group). Statistically significant increases in number of mounts and intromissions occurred in the 100- and 200-ppm groups prior to the development of neurologic effects, and sperm counts were significantly reduced in the 100-ppm group. In addition, significantly fewer females were successfully impregnated by the 100-ppm males, and postimplantation loss was significantly increased in those females that were pregnant.

Zenick et al.[17] also studied the effects of acrylamide on mating performance, pregnancy rates, litter size, and pup growth and survival in female Long-Evans rats (n = 15/group) exposed to 0, 25, 50, or 100 ppm in drinking water from 2 weeks prior to mating through gestation and lactation. No significant differences were found in mating performance, pregnancy rates, litter size, or survival for any group. Pup body weight at birth was significantly decreased in the 100-ppm group and pup weight gain during lactation through weaning was significantly decreased in the 50- and 100-ppm groups. Vaginal patency was also delayed in the 100-ppm dams. However, because a cross-fostering design was not used it is not possible to clearly separate the influence of decreases in weight gain and food consumption by the dams as well as other signs of maternal toxicity.

It should be noted that neurological signs, i.e., ataxia and hind limb splaying, were noted in the 100- and 200-ppm males and 50- and 100-ppm females. These signs occurred earlier in the higher exposure groups.

Systemic Toxicity: Subchronic and/or chronic exposures to acrylamide in the diet or drinking water of rats, monkeys, cats, and dogs produces neurotoxic effects, including muscular weakness, ataxia, tremors, and paralysis. Histologically, moderate to severe degenerative lesions of the peripheral nerves are seen.[2]

In humans exposed occupationally (dermal exposure thought to be most important) similar clinical signs have been observed as well as sensory loss in the extremities. The neuropathy observed is classified as delayed-type "dying back" neuropathy. The effects appear to be reversible with the recovery period dependent on the severity of poisoning.[1,2]

Irritation: Mild skin (abraded) and conjunctival irritation with no permanent corneal injury have been reported in rabbits.[2] No mention of an odor or odor threshold for acrylamide was found in the literature reviewed.

Basis for the AALG: Acrylamide is very soluble in water (215 g/100 mL at 30°C) and has a very low vapor pressure (0.007 mm Hg at 25°C). It is used in the production of polyacrylamides, which are used in water and wastewater treatment, crude oil production processes, paper and pulp processing, mineral and concrete processing, coating applications, etc. Acrylamide has been detected in effluents, tap water, and surface waters.[6] Based on its physical characteristics and available use and occurrence information, 20% of the contribution to total exposure will be allowed from air for acrylamide.

Based on sufficient evidence of carcinogenicity in animals, the AALG for acrylamide is based on this endpoint using the potency estimate (q_1*) recommended by the EPA Carcinogen Assessment Group. In its quantitative risk assessment, EPA-CAG derived several q_1*'s using different dose-response data sets from the Johnson et al. study.[8] The recommended human q_1* for acrylamide, 3.7 (mg/kg/day)$^{-1}$, was derived from the combined incidence of mammary gland, thyroid, and uterine tumors in female rats.[6] If the oral slope factor is used to derive a unit risk for inhalation, assuming a 70-kg human breathing 20 m^3 air per day and 100% absorption, the resultant unit risk is 1.06×10^{-3}, which corresponds to an inhalation q_1* of 1.06×10^{-3} (μg/m^3)$^{-1}$. Applying the relation $E = q_1* \times d$ (where E is the specified level of extra risk and d is the dose) and factoring in the 20% relative source contribution yields the estimates shown below. This AALG should be considered provisional, since it was derived from a noninhalation study; there is a high degree of uncertainty associated with this type of extrapolation.

AALG: • 10^{-6} 95% UCL—5.3×10^{-6} ppb (1.34×10^{-5} μg/m^3)
        • 10^{-5} 95% UCL—5.3×10^{-5} ppb (1.34×10^{-4} μg/m^3)—annual TWAs

References

1. ACGIH. 1986. "Documentation of the Threshold Limit Values and Biological Exposure Indices." 5:12–13.
2. NIOSH. 1976. "Criteria for a Recommended Standard . . . Occupational Exposure to Acrylamide." DHEW (NIOSH) 77–112.
3. OSHA. 1989. "Air Contaminants; Final Rule." *Fed. Reg.* 54:2332–959.
4. U.S. EPA. 1985. "National Primary Drinking Water Regulations; Synthetic Organic Chemicals, Inorganic Chemicals and Microorganisms; Proposed Rule." *Fed. Reg.* 50: 46936–7025.
5. U.S. EPA. 1985. "Drinking Water Criteria Document for Acrylamide (Final Draft)." NTIS PB86–117744.
6. U.S. EPA. 1987. "Health Advisory for Acrylamide." U.S. EPA, Office of Drinking Water.
7. Bull, R. J., M. Robinson, R. D. Laurie, G. D. Stoner, E. Greisiger, J. R. Meier, and J. Stober. 1984. "Carcinogenic Effects of Acrylamide in Sencar and A/J Mice." *Cancer Res.* 44:107–11.
8. Johnson, K. A., S. J. Gorzinski, K. M. Bodner, R. A. Campbell, C. H. Wolf, M. A. Friedman, and R. W. Mast. 1986. "Chronic Toxicity and Oncogenicity Study on Acryl-

amide Incorporated in the Drinking Water of Fisher 344 Rats.'' *Tox. App. Pharm.* 85:154–68.

9. NIOSH. 1987. AS3325000. ''Acrylamide.'' RTECS, on line.

10. IARC. 1986. ''Acrylamide.'' *IARC Monog.* 39:41–66.

11. Dearfield, K. L., C. O. Abernathy, M. S. Ottley, J. H. Branter, and P. F. Hayes. 1988. ''Acrylamide: Its Metabolism, Developmental and Reproductive Effects, Genotoxicity and Carcinogenicity.'' *Mut. Res.* 195:45–77.

12. Ikeda, G. J., E. Miller, P. P. Sapienza, T. C. Michel, M. T. King, V. A. Turner, H. Blumenthal, W. E. Jackson, and S. Levin. 1983. ''Distribution of ^{14}C-labelled Acrylamide and Betaine in Foetuses of Rats, Rabbits, Beagle Dogs and Miniature Pigs.'' *Fd. Chem. Tox.* 21:49–58.

13. Reproductive Toxicology Center. 1987. ''Reproductive Toxicity Review: Acrylamide.'' REPROTOX, on line (Washington, DC: Columbia Hospital for Women Medical Center).

14. Ikeda, G. J. et al. 1985. ''Maternal-Foetal Distribution Studies in Late Pregnancy. II. Distribution of [1-^{14}C]acrylamide in Tissues of Beagle Dogs and Miniature Pigs.'' *Fd. Chem. Tox.* 23:757–61.

15. Walden, R. et al. 1981. ''Effects of Prenatal and Lactational Exposure to Acrylamide on the Development of Intestinal Enzymes in the Rats.'' *Tox. App. Pharm.* 58:363–69.

16. Agrawal, A. K. and R. E. Squibb. 1981. ''Effects of Acrylamide Given During Gestation on Dopamine Receptor Binding in Rat Pups.'' *Tox. Let.* 7:233–38.

17. Zenick, H., E. Hope, and M. K. Smith. 1986. ''Reproductive Toxicity Associated with Acrylamide Treatment in Male and Female Rats.'' *J. Tox. Environ. Health* 17:457–72.

ACRYLIC ACID

Synonyms: acroleic acid, ethylene carboxylic acid, 2-propenoic acid, vinyl formic acid

CAS Registry Number: 79–10–7

Molecular Formula: $CH_2 = CHCOOH$

Molecular Weight: 72.07

AALG: systemic toxicity—1.2 $\mu g/m^3$ (0.4 ppb) 24-hour TWA

Occupational Limits:
- ACGIH TLV: 2 ppm (6 mg/m^3) TWA; skin notation
- NIOSH REL: none
- OSHA PEL: 10 ppm (30 mg/m^3) TWA; skin notation

Basis for Occupational Limits: The TLV-TWA for acrylic acid was formerly 10 ppm and based on prevention of nasal and ocular irritation by analogy with glacial acetic acid. However, the "Documentation of the Threshold Limit Values" published in 1986 stated: "This limit seems surprising, in view of the much greater toxicity of acrylates and acrolein, etc., in comparison with the corresponding acetic compounds." At this time acrylic acid was placed under review by the TLV committee.[1] In the TLV booklet for 1988–89, a new TWA limit of 2 ppm was listed under the "notice of intended changes."[2] OSHA, which previously had no limit for acrylic acid, has recently adopted the former ACGIH TLV.[3]

Drinking Water Limits: No drinking water limits were found for this substance.

Toxicity Profile

Carcinogenicity: Cote et al.[4] (cited in Segal et al.[5]) reported that squamous cell carcinomas of the skin were observed in 2 of 30 ICR/Ha mice treated topically with 1 mg (14 μmol) acrylic acid in acetone three times a week for 1.5 years. However, DePass et al.[6] (cited in Segal et al.[5]) observed no tumors at the site of application in 40 male C3H/HeJ mice treated topically with 0.2 mg (3 μmol) acrylic acid in acetone three times a week for their entire lifetime.

Segal et al.[5] injected female ICR/Ha mice (n = 30/group) with either 0.05 μL trioctanoin or 20 μmol acrylic acid in 0.05 μL trioctanoin once a week for 52 weeks and observed them for an additional 13 weeks; a group of 100 untreated controls was also maintained. Two of 30 mice treated with acrylic acid developed injection site sarcomas, but none were observed in the untreated or vehicle controls.

Based on available data, acrylic acid has some potential for tumor induction at the site of application and further study is warranted. At this time, acrylic acid is most appropriately placed in group C (limited animal data in the absence of human data) bordering on group D (unclassified) under the EPA weight-of-evidence classification.

Mutagenicity: Acrylic acid was reported to be nonmutagenic to *S. typhimurium* in a series of studies in which multiple tester strains were used with and without metabolic activation in both plate and liquid suspension assays[7] (cited in Segal et al.[5]).

Developmental Toxicity: Singh et al.[8] studied the developmental toxicity of methacrylate esters and acrylic acid in Sprague-Dawley rats administered the compounds by intraperitoneal injection. Groups of five rats were untreated or injected with distilled water, normal saline, cottonseed oil (controls), or 0.0023, 0.0045, or 0.0075 mL/kg acrylic acid on gestation days 5, 10, and 15. The doses used were 1/10, 1/5, and 1/3 of the previously determined intraperitoneal LD_{50} for this compound. The percentage of dead fetuses was increased only in the highest dose group compared to all control groups. It should be noted that the process of injection itself apparently resulted in decreased fetal body weights and increased resorptions. There were significant increases in gross malformations (hemangiomas) in the two highest dose groups and skeletal malformations (elongation and fusion of sternebrae and ribs) in the highest dose group.

While the above study is of interest and points to a need for more extensive evaluation of the developmental toxicity of acrylic acid, it is unsuitable for criteria derivation due to the route of exposure. In addition, various limitations of the study design, i.e., small group sizes, lack of historical or concurrent positive controls, and no evaluation of maternal toxicity, limit the conclusions that may be drawn from this study.

Reproductive Toxicity: DePass et al.[9] conducted a one-generation reproduction study in F344 rats exposed to acrylic acid in drinking water. Male (n = 10/group) and female (n = 20/group) rats received dose levels of 0, 83, 250, or 750 mg/kg/day acrylic acid in their drinking water for 90 days prior to mating and throughout gestation and lactation. Food and water consumption and hence weight gain were reduced in a dose-related manner, particularly in the two highest dose groups, and the authors concluded that a number of adverse effects observed were secondary to the reduction in food and water consumption. Although interpretation of the reproductive effects was confounded by these latter effects at the highest dose level, comparison with both concurrent and historical controls indicated no adverse reproductive effects at the two lower exposure levels.

Systemic Toxicity: Data on the systemic toxicity of acrylic acid in humans are lacking. In lethality studies in animals, oral LD_{50}s in rats are reported to be variable with figures of 2590 mg/kg, 1250 mg/kg, and 360 mg/kg having been reported[10-12] (cited in ACGIH[1]). Oral LD_{50}s of 830 mg/kg and 250 mg/kg have been reported in mice and rabbits, respectively[11] (cited in ACGIH[1]). A two-hour LC_{50} of 1797 ppm has been reported in mice.[13] Carpenter et al.[12] reported that exposure to 2000 ppm acrylic acid for 4 hours caused no mortality in rats.

Limited data are available on the effects of acrylic acid inhalation in experimental animals. Majka et al.[14] (cited in IARC[15]) reported that exposure of rats for 4 hours per day for 5 weeks to 240 ppm acrylic acid resulted in reduced body weight gain, increased reticulocyte count, and profound irritant effects with irreversible skin and eye changes. Gage[16] conducted subacute inhalation toxicity studies on a number of compounds using Aderley Park specific pathogen-free rats. Four male and four female rats per group were exposed to 0, 80, or 300 ppm acrylic acid for 6 hours per day, 5 days a week for 4 weeks and weighed and observed daily with urine, blood, and histologic examination at terminal sacrifice. Rats exposed to 300 ppm acrylic acid showed signs of nasal irritation, were lethargic, and had decreased body weight gain. However, no gross or microscopic lesions were apparent at necropsy. Rats exposed to 80 ppm acrylic acid exhibited no signs of toxicity and their organs were normal at necropsy; a NOAEL of 80 ppm is suggested by this study. Gage also reported that rats given four 6-hour exposures to 1500 ppm acrylic acid had nasal discharge, were lethargic, and lost weight, and congestion of the kidneys was apparent at necropsy.

Miller et al.[17] exposed male and female $B6C3F_1$ mice and F344 rats (n = 15/sex/species/group) to 0, 5, 25, or 75 ppm acrylic acid for 6 hours per day, 5 days a week for 13 weeks. Body weight gain was not significantly affected in rats or male mice by exposure to acrylic acid. However, female mice in the 25- and 75-ppm exposure groups had significantly decreased body weight gain at week 12 relative to controls. No gross pathologic lesions were observed in rats or mice and the only histopathologic finding of significance was lesions of the olfactory nasal epithelium in both rats and mice. In rats, lesions consisting of focal degeneration of the olfactory epithelium were observed in 7 of 10 males and 10 of 10 females in the highest exposure group, but were not observed in the 5- or 25-ppm exposure groups. In mice, such lesions were found in all exposure groups, but not in controls. The incidence in male and female mice, respectively, was as follows: 5 ppm, 1 of 10 and 4 of 10; 25 ppm, 10 of 10 and 9 of 10; and 75 ppm, 10 of 10 and 10 of 10. A dose-response relationship was apparent in these lesions with respect to size of the area affected and severity of lesions within the affected area. A NOAEL for effects on the nasal epithelium in rats of 25 ppm and a LOAEL of 5 ppm for the same effect in mice are suggested by this study.

Concurrent with their reproductive toxicity study, DePass et al.[9] also conducted a subchronic study (90 days) of rats (n = 15/sex/group) exposed to acrylic acid via the drinking water at dose levels of 0, 83, 250, or 750 mg/kg/day. At the middle and high dose levels there was a dose-related decrease in body weight gain and alterations in organ weights, which were considered secondary to decreased food and water consumption. No treatment-related histologic or gross lesions were observed. There was a slight decrease in water consumption at the lowest dose and a NOAEL of 83 mg/kg/day is suggested by this study. This study served as the basis for calculation of the oral RfD for acrylic acid by the U.S. EPA.[18]

It should be noted that acrylic acid was recently accepted for general toxicology study by the NTP program.[19]

Irritation: Acrylic acid vapors have been reported to cause nasal and eye irritation in workers (no concentrations given) and contact with the liquid may produce skin burns and blindness.[1] It has a "distinctive, acrid" odor[1] and an odor threshold of 0.094 ppm (geometric mean of several reported literature values).[20]

Basis for the AALG: Acrylic acid is miscible with water and has a vapor pressure of 31 mm Hg at 25°C.[1] In the United States, most acrylic acid is used captively in surface coatings and in paints, polishes, and adhesives.[13] Based on limited information on its environmental fate, distribution, and occurrence and the expected contribution of various media to human exposure, 50% of the contribution to total exposure from acrylic acid is allowed from air.

The AALG for acrylic acid is based on the LOAEL of 5 ppm in mice for degeneration of the olfactory nasal epithelium reported in the study of Miller et al.,[17] since mice were the most sensitive species. The LOAEL was adjusted for continuous exposure (5 ppm x 5/7 $\times$ 6/24) and converted to a human equivalent LOAEL according to the procedures described in Appendix C. A total uncertainty factor of 5000 (10 for interindividual variation $\times$ 10 for interspecies variation $\times$ 10 for less-than-lifetime exposure $\times$ 5 for a LOAEL) was applied. This AALG should be considered provisional, since it is based on a subchronic study and because there is suggestive evidence for potential carcinogenic and developmental effects.

AALG: • systemic toxicity—1.2 μg/m^3 (0.4 ppb) 24-hour TWA

References

1. ACGIH. 1986. "Documentation of the Threshold Limit Values and Biological Exposure Indices." 5:14, plus updated TLV documentation for acrylic acid, 1987–88.
2. ACGIH. 1988. "Threshold Limit Values and Biological Exposure Indices for 1988–89" (Cincinnati, OH: ACGIH).
3. OSHA. 1989. "Air Contaminants; Final Rule." *Fed. Reg.* 54:2332–959.
4. Cote, I. L., A. Hochwalt, I. Seidman, G. Budzilovich, J. J. Solomon, and A. Segal. 1986. "Acrylic Acid: Skin Carcinogenesis in ICR/Ha Mice." *The Toxicologist* 6:235.
5. Segal, A., J. Fekyk, S. Melchionne, and I. Seidman. 1987. "The Isolation and Characterization of 2-Carboxyethyl Adducts Following In Vitro Reaction of Acrylic Acid with Calf Thymus DNA and Bioassay of Acrylic Acid in Female Hsd:(ICR)Br Mice." *Chem. Biol. Interact.* 61:189–97.
6. DePass, L. R., E. H. Fowler, D. R. Meckley, and C. S. Weil. 1984. "Dermal Onco-

genicity Bioassays of Acrylic Acid, Ethyl Acrylate and Butyl Acrylate.'' *J. Tox. Env. Health* 14:115–20.

7. Lijinsky, L. and A. W. Andrews. 1980. ''Mutagenicity of Vinyl Compounds in *Salmonella typhimurium. Teratogen. Carcinogen. Mutagen.* 1:259–67.

8. Singh, A. R., W. H. Lawrence, and J. Autian. 1972. ''Embryonic-Fetal Toxicity and Teratogenic Effects of a Group of Methacrylate Esters in Rats.'' *J. Dent. Res.* 51: 1632–38.

9. DePass, L. R., M. D. Woodside, R. H. Garman, and C. S. Weil. 1983. ''Subchronic and Reproductive Toxicology Studies on Acrylic Acid in the Drinking Water of the Rat.'' *Drug Chem. Tox.* 6:1–20.

10. Smyth, H. F., et al. 1962. *Am. Ind. Hyg. Assoc. J.* 23:95.

11. Klimkina, N. V. et al. 1969. ''Prom. Zagryazeniya Vodoemov'' (Russian). 9:171. Chem. Abstract No. 71:1284105 (1969).

12. Carpenter, C. P., C. S. Weil, and H. F. Smyth. 1974. ''Range Finding Toxicity Data: List VIII.'' *Tox. Appl. Pharm.* 28:313–19.

13. NIOSH. 1988. AS4375000. ''Acrylic Acid.'' RTECS, on line.

14. Majka, J., K. Knobloch, and J. Stetkiewicz. 1974. ''Evaluation of Acute and Subacute Toxicity of Acrylic Acid'' (Polish). *Med. Pr.* 25:427–35.

15. IARC. 1979. ''Acrylic Acid, Methyl Acrylate, Ethyl Acrylate and Polyacrylic Acid.'' *IARC Monog.* 19:47–71.

16. Gage, J. C. 1970. ''The Subacute Inhalation Toxicity of 109 Industrial Chemicals.'' *Brit. J. Ind. Med.* 27:1–18.

17. Miller, R. R., J. A. Ayres, G. C. Jersey, and M. J. McKenna. 1981. ''Inhalation Toxicity of Acrylic Acid.'' *Fund. Appl. Tox.* 1:271–77.

18. U.S. EPA. 1988. ''Acrylic Acid.'' IRIS (3/1/88).

19. NTP. 1989. ''Chemical Status Report'' (Research Triangle Park, NC: National Toxicology Program).

20. Amoore, J. E. and E. Hautala. 1983. ''Odor as an Aid to Chemical Safety: Odor Thresholds Compared with Threshold Limit Values and Volatilities for 214 Industrial Chemicals in Air and Water Dilution.'' *J. Appl. Tox.* 3:272–90.

ACRYLONITRILE

Synonyms: vinyl cyanide, propenenitrile

CAS Registry Number: 107–13–1

Molecular Weight: 53.05

Molecular Formula: $CH_2 = CHC \equiv N$

AALG: carcinogenicity (10^{-6})—3.37×10^{-3} ppb $(7.3 \times 10^{-3}$ $\mu g/m^3)$ annual TWA

Occupational Limits:
- ACGIH TLV: 2 ppm $(4.5$ mg/m$^3)$ TWA; Appendix A2 suspect human carcinogen; skin notation
- NIOSH REL: 1 ppm $(2.25$ mg/m$^3)$ TWA; 10 ppm $(22.5$ mg/m$^3)$ 15-minute ceiling; skin notation
- OSHA PEL: 1 ppm $(4.5$ mg/m$^3)$ TWA; 10 ppm $(22.5$ mg/m$^3)$ 15-minute ceiling; skin notation

Basis for Occupational Limits: All of the occupational limits are based on carcinogenic effects, particularly cancer of the brain, lung, and bowel.[1,2]

Drinking Water Limits: No drinking water limits were found for this substance, but it is one of the synthetic organic chemicals scheduled for regulation by 1989 under the 1986 amendments to the Safe Drinking Water Act.[3]

The ambient water quality criterion of 0 mg/L was recommended based on carcinogenic effects. Since this may not be attainable in practice, criteria were estimated for risk levels of 10^{-7}, 10^{-6}, and 10^{-5}; these were 0.006, 0.058, and 0.58 $\mu g/L$, respectively. These levels were calculated based on the consumption of 2 L of water and 6.5 g of fish/shellfish per day.[4]

Toxicity Profile

Carcinogenicity: The Health Assessment Document [HAD] for Acrylonitrile[5] cites 10 epidemiologic studies, four of which were sufficient to address the issue of lung cancer risk. Statistically significant increases in the risk of lung cancer were reported in all four studies[6–9] (cited in U.S. EPA[5]); increased risk of lymph system cancer was reported in one[8] (cited in U.S. EPA[5]) and increased risk of stomach cancer in another[9] (cited in U.S. EPA[5]).

The HAD also cites seven cancer bioassays conducted in rats; exposure was via drinking water in four of these studies, via inhalation in two, and via gastric intubation in one. Data from three of the rat drinking water studies were used in the calculation of a human potency factor, q_1^* (geometric mean of 3 q_1^* values). Although different strains of rats received lifetime exposure to various levels of acrylonitrile in these studies, the tumor types reported were consistent among studies and included tumors in the CNS, Zymbal gland, stomach, tongue, and small intestine in both male and female rats and mammary gland tumors in female rats.[5]

Quast et al.[10] (cited in U.S. EPA[5]) exposed male and female Sprague-Dawley rats (n = 100/sex/group) to 0, 20, or 80 ppm acrylonitrile for 6 hours per day, 5 days a week for 2 years. Statistically significant increases in tumors of the central nervous system (glial cell), Zymbal gland, tongue, small intestine, and mammary gland were found at 80 ppm except for glial cell tumors in female rats, which were statistically significant at 20 ppm.

It is relevant to note that acrylonitrile is thought to be an indirect-acting carcinogen, i.e., the proximate carcinogen is a metabolite (possibly 2-cyanoethylene oxide) of acrylonitrile. In addition, there appears to be a definite interspecies difference in the tumorigenic response to acrylonitrile in that no lung tumors have been reported in animal models and no brain tumors associated with acrylonitrile have been observed in humans.[5] Acrylonitrile has been placed in group B1, probable human carcinogen, under the EPA weight-of-evidence classification.[11]

Mutagenicity: Acrylonitrile is listed as positive in the following assays by the Gene-Tox program as summarized in RTECS:[12] cell transformation, Ames assay, TRP reversion, and dose-response TRP reversion; it is listed as inconclusive in the *D. melanogaster* sex-linked recessive lethal assay. A recent IARC review[13] summarized the genetic activity of acrylonitrile as follows: "In animals treated in vivo, acrylonitrile did not induce dominant lethal mutations, chromosomal aberrations (in bone-marrow cells or spermatogonia) or micronuclei in mice, or chromosomal aberrations in rat bone-marrow cells. It bound covalently to rat liver DNA in vivo and induced unscheduled DNA synthesis in rat liver but not brain. It induced sister chromatid exchanges, mutation and unscheduled DNA synthesis, but not chromosomal aberrations in human cells in vitro. Acrylonitrile induced cell transformation in several test systems and inhibited intercellular communication in Chinese hamster V79 cells. It did not induce aneuploidy but induced chromosomal aberrations, micronuclei and sister chromatid exchanges in Chinese hamster cells; in one study, it did not induce chromosomal aberrations or sister chromatid exchanges in rat cells in vitro. It induced mutation and DNA stand breaks in rodent cells in vitro. It induced somatic mutation in *Drosophila* and was weakly mutagenic in plants. It induced aneuploidy, mutation, mitotic crossing-over and gene conversion in fungi. Acrylonitrile was mutagenic to bacteria. Urine from treated mice and rats, but not bile from rats, was mutagenic to bacteria. It bound covalently to isolated DNA."

It is our judgement that there is sufficient evidence for genetic activity of acrylonitrile in short-term tests based on the IARC weight-of-evidence classification.

Developmental Toxicity: Acrylonitrile has been shown to be teratogenic (statistically significant increase in total malformations) in the pups of rats exposed orally on days 6–15 of gestation at a dose of 65 mg/kg; however, this dose was also toxic to the dam.[5] In the same study no statistically significant effects on reproductive success or fetal development were noted at exposures of 40 or 80 ppm for 6 hours per day on days 6–15 of gestation[14] (cited in U.S. EPA[5]).

Acrylonitrile is also teratogenic in hamsters. Willhite et al.[15] found that intraperitoneal injection of dams on day 8 of gestation with 80–120 mg/kg acrylonitrile induced exencephaly, encephalocoeles, and rib fusions and bifurcations in the offspring. However, these doses were also toxic to the dams.

It is stated in the HAD that limitations of the available data on reproductive and developmental toxicity do not allow for a full assessment of these effects.[5]

Reproductive Toxicity: See comments above under "Developmental Toxicity."

Systemic Toxicity: Acrylonitrile is highly toxic to laboratory animals regardless of the route of administration, but there is considerable interspecies variation in susceptibility to intoxication. Chronic and subchronic exposures in laboratory animals have been associated with both peripheral and central nervous system effects and damage to the adrenal glands, as well as disturbances in liver and kidney function.[5] It has been concluded that the systemic toxicity observed in laboratory animals is not entirely due to cyanide.[1]

In humans mild intoxication is associated with headaches, nausea, feelings of apprehension, and nervous irritability, while more severe intoxication is associated with low-grade anemia and jaundice.[5]

Irritation: Workers unintentionally exposed to 16 to 100 ppm acrylonitrile for 20–45 minutes reported nasal irritation and an "oppressive feeling" in the upper respiratory tract.[5]

An odor threshold of 21.6 ppm has been reported for acrylonitrile. However, it is commonly believed (but not confirmed) that concentrations, as low as 1 ppm are detectable by olfaction.

Basis for the AALG: Acrylonitrile is volatile (vapor pressure = 100 mm Hg at 23°C) and soluble in water (7.3% by weight).[1,16] Although it has been detected in drinking water and ambient air near production facilities, it was concluded in the HAD that insufficient data were available to determine the human intake through food and drinking water. Based on these considerations, 50% of the contribution to total exposure will be allowed from air.

The AALG for acrylonitrile is based on carcinogenicity, using data from both human occupational and animal inhalation studies. The HAD provides quantitative risk assessments based on several studies. The estimates generated using the data from the Quast et al. study[10] are presented here due to the greater relevance of the inhalation route of exposure. These estimates are based on the combined tumor incidence in four target organs (Zymbal gland, small intestine, brain, and/or spinal cord) in female rats. Female rat data were used in preference to male rat data, since the former provide more conservative estimates of the slope coefficients. The raw tumor incidence data were as follows: at 0 ppm, 0 of 100; at 20 ppm, 9 of 100; and at 80 ppm, 31 of 100. The exposure duration was 6 hours per day, 5 days per week for 2 years. Using the linearized multistage model and assuming a 70-kg human breathing 20 m^3 air per day, the estimates for q_1 and q_1* were 2.61×10^{-2} $(ppm)^{-1}$ and 3.35×10^{-2} $(ppm)^{-1}$, respectively. Applying the relationship $E = q_1* \times d$ (where E is the specified level of extra risk and d is the dose) and allowing a 50% relative source contribution from air gives the estimates shown below.

In the HAD[5] and in the IRIS review,[11] data from an occupational epidemiologic study by O'Berg[6] (cited in U.S. EPA[5,11]) were used to calculate a potency factor or unit risk. A relative risk model adjusted for smoking and based on a continuous lifetime equivalent of occupational exposure was used and the potency factor derived was 6.85×10^{-5} $(\mu g/m^3)^{-1}$. The estimates for exposures corresponding to the 10^{-5} and 10^{-6} risk levels allowing a 50% relative source contribution from air are shown below. The recommended AALG is for the 10^{-6} risk level using the relative risk model with the O'Berg study, since it is based on human studies and is also the most conservative figure.

AALG:
- 10^{-6} 95% UCL—1.49×10^{-2} ppb $(3.23 \times 10^{-2}$ $\mu g/m^3)$
  10^{-5} 95% UCL—1.49×10^{-1} ppb $(3.23 \times 10^{-1}$ $\mu g/m^3)$

  The above estimates are annual TWAs based on animal inhalation data.
- 10^{-6}—3.37×10^{-3} ppb $(7.3 \times 10^{-3}$ $\mu g/m^3)$
  10^{-5}—3.37×10^{-2} ppb $(7.3 \times 10^{-2}$ $\mu g/m^3)$

  The above estimates are annual TWAs.
- *Note:* The parameters q_1 and q_1* are the maximum likelihood coefficients of the dose and the upper 95% confidence limit on the dose. For a given level of extra risk, E, the dose corresponding to that level of risk is given by E/q. See Anderson et al.[17] or U.S. EPA pp. 13–140 to 13–162[5] for more detailed discussions of quantitative risk assessment.

References

1. ACGIH. 1986. "Documentation of the Threshold Limit Values and Biological Exposure Indices." 5:15–6.
2. NIOSH. 1978. "Criteria for a Recommended Standard . . . Occupational Exposure to Acrylonitrile." DHEW (NIOSH) 78–116.
3. Ohanian, E. V. 1989. "National Primary Drinking Water Regulations for Additional Contaminants to be Regulated by 1989." In Calabrese, E. J., C. E. Gilbert, and H. Pastides, Eds. *Safe Drinking Water Act: Amendments, Regulations and Standards* (Chelsea, MI: Lewis Publishers, Inc.), pp. 71–82.
4. U.S. EPA. 1980. "Ambient Water Quality Criteria for Acrylonitrile." EPA 440/5–80–017.
5. U.S. EPA. 1983. "Health Assessment Document for Acrylonitrile." EPA 600/8–82–007F.

6. O'Berg, M. 1980. "Epidemiologic Study of Workers Exposed to Acrylonitrile." *J. Occ. Med.* 22:245–52.

7. Delzell, E. and R. R. Monson. 1982. "Mortality Among Rubber Workers. VI. Men with Exposure to Acrylonitrile." *J. Occ. Med.* 24:767–69.

8. Theiss, A. M., R. Frentzel-Beyme, R. Link, and H. Wild. 1980. "Mortalitatsstudie bei Chemiefacharbeitern Verschiedener Produktionsbetriebe mit Exposition auch Gegenuber Acrylonitrile Zentralbl." *Arbeitsmed.* 30:359–67.

9. Werner, J. B. and J. T. Carter. 1981. "Mortality of United Kingdom Acrylonitrile Polymerization Workers." *Brit. J. Ind. Med.* 38:247–53.

10. Quast, J. F., D. J. Schuetz, M. F. Balmer, T. S. Gushow, C. N. Park, and M. J. McKenna. 1980. "A Two-Year Toxicity and Oncogenicity Study with Acrylonitrile Following Inhalation Exposure of Rats." Prepared by the Toxicology Research Laboratory, Health and Environmental Sciences, Dow Chemical, USA, Midland, MI, for the Chemical Manufacturing Association, Washington, DC.

11. U.S. EPA. 1989. "Acrylonitrile; CASRN 107–13–1." Integrated Risk Information System (2/1/89).

12. NIOSH. 1987. AT5250000. "Acrylonitrile." RTECS, on line.

13. IARC. 1987. "Acrylonitrile." *IARC Monog.* suppl. 6:27–31.

14. Murray, F. J., B. A. Schwetz, K. D. Nitschke, J. A. John, J. M. Norris, and P. J. Gehring. 1978. "Teratogenicity of Acrylonitrile Given to Rats by Gavage or by Inhalation." *Fd. Cosmet. Tox.* 16:547–52.

15. Willhite, C. C., V. H. Ferm, and R. P. Smith. 1981. "Teratogenic Effects of Aliphatic Nitriles." *Teratology* 23:317–23.

16. IARC. 1979. "Acrylonitrile, Acrylic Acid and Modacrylic Fibres and Acrylonitrile-Butadiene-Styrene and Styrene-Acrylonitrile Copolymers." *IARC Monog.* 19:17–113. (See also *IARC Monog.* Suppl. 7:79–80, 1988.)

17. Anderson, E. L. and the Carcinogen Assessment Group of the U.S. EPA. 1983. "Quantitative Approaches in Use to Assess Cancer Risk." *Risk Anal.* 3:277–295.

ALLYL CHLORIDE

Synonyms: 3-chloro-1-propene, 3-chloropropene

CAS Registry Number: 107–05–1

Molecular Weight: 76.53

Molecular Formula: $H_2C = CHCH_2Cl$

AALG: systemic toxicity (TLV-based)—3.8 ppb (12 $\mu g/m^3$) 8-hour TWA

Occupational Limits:
- ACGIH TLV: 1 ppm (3 mg/m^3) TWA; 2 ppm (6 mg/m^3) STEL
- NIOSH REL: 1 ppm (3 mg/m^3) TWA; 3 ppm (9 mg/m^3) 15-minute ceiling
- OSHA PEL: 1 ppm (3 mg/m^3) TWA; 2 ppm (6 mg/m^3) STEL

Basis for Occupational Limits: The TLVs were set to "protect against the possible injuries suggested by animal studies." The studies in question[1-3] (cited in ACGIH[4]) are reviewed under systemic toxicity. The effects observed include irritation and liver and kidney injury. The NIOSH recommended limits have a similar basis, although more weight was given to an epidemiologic study in which abnormal liver function tests were reported following exposure to allyl chloride in the range of 1 to 113 ppm.[5] OSHA recently adopted the TLV STEL as part of its final rule limits.[6]

Drinking Water Limits: No drinking water limits were found for this substance.

Toxicity Profile

Carcinogenicity: In an NCI cancer bioassay,[7] male and female B6C3F$_1$ mice (n = 50/sex/group) received 172 or 199 and 129 or 258 mg/kg allyl chloride (purity 98%), respectively, in corn oil by gavage 5 days a week for 78 weeks and were retained until 92 weeks. Groups of 20 animals of each sex were maintained as vehicle and untreated controls. Excessive mortality occurred in high-dose male mice; all animals in that group had died or been euthanized by week 56. Survival at the end of the study was 8 of 20, 14 of 20, and 23 of 50 in untreated controls, vehicle controls, and low-dose male mice, respectively. Survival was not adversely affected in female mice; 70–90% were still alive at the conclusion of the study. Treatment-related lesions were observed in the forestomach of both male and female mice and included a dose-dependent increase in acanthosis (increased thickness) and hyperkeratosis; no such lesions were observed in vehicle or untreated controls. Squamous cell carcinomas were observed in 2 of 46 low-dose males and 2 of 47 low-dose females. The NCI bioassay report stated: ''The occurrence of these neoplasms at the incidences observed in this allyl chloride bioassay was statistically and significantly higher than in the historical incidences. The proliferative non-neoplastic stomach lesions, squamous-cell papillomas and squamous-cell carcinomas, may all represent progressive stages in a neoplastic process. When the probable pathogenesis of this tumor is coupled with the known chemical reactivity of the compound and the statistical evidence for the rare occurrence of this tumor, the results are strongly suggestive of the carcinogenic action of allyl chloride in mice.''

In the same bioassay described above, male and female Osborne-Mendel rats (n = 50/sex/group) received TWA doses of 57 or 77 and 55 or 73 mg/kg allyl chloride, respectively, in corn oil by gavage for 78 weeks. Control groups of rats were maintained analogous to those of mice as described above. Excess mortality in the high-dose groups precluded evaluation of carcinogenicity and no treatment-related increases in tumors occurred in the low-dose group.

Two skin painting studies have been reported in mice. Negative results were reported in one study in which mice were exposed to allyl chloride alone[8] (cited in IARC[9]), but in a two-stage mouse skin painting assay in which a single application of allyl chloride was followed by treatment with 12-O-tetradecanoyl phorbol 13-acetate, there was some evidence that allyl chloride acted as an initiator[8] (cited in IARC[9]).

In a mouse lung adenoma bioassay, Theiss et al.[10] injected male and female Strain A mice (n = 10/sex/group) with tricaprylin alone or 1.2, 2.9, and 5.9 g/kg total dose of allyl chloride in tricaprylin by the intraperitoneal route three times a week for 8 weeks with terminal sacrifice 24 weeks after the first injection. There was a slight but statistically significant increase in lung adenomas in high-dose animals compared to controls (both sexes combined), but not in the low- or medium-dose groups.

IARC concluded that there was inadequate evidence of carcinogenicity for allyl chloride in experimental animals with no data in humans.[9] However, it is the judgment here that allyl choride is most appropriately placed in group C, possible human carcinogen, under the EPA weight-of-evidence classification based on the following evidence: (1) the interpretation of the forestomach lesions in mice as preneoplastic and the low incidence of forestomach carcinoma as significant by the NCI, (2) evidence of initiator activity in a two-stage mouse skin painting study, (3) suggestive evidence in a mouse lung adenoma bioassay, and (4) positive results in bacterial mutagenicity and DNA damage assays (see next section).

Mutagenicity: Under adequate test conditions (prevention of loss due to volatilization) allyl chloride is mutagenic to *S. typhimurium* strains TA100 and TA1535, but not TA1538, and its mutagenic activity is greatly decreased by the addition of an exogenous metabolic activation system. Allyl chloride has been reported to be mutagenic to *Streptomyces coelicoler* but not *A. nidulans*. It also induces gene conversions in *S. cerevisiae* and is positive in the *E. coli* polA$^+$/polA$^-$ DNA repair assay.[8] It has also been reported to induce unscheduled DNA synthesis in HeLa cells, a human-origin cell line.[11]

Developmental Toxicity: John et al.[12] (cited in IARC[9]) exposed Sprague-Dawley rats (n = 25 to 39/group) and New Zealand white rabbits (n = 20 to 25/group) to 0, 30, or 300 ppm allyl chloride (98.6% purity) for 7 hours per day on gestation days 6 to 15 and 6 to 18, respectively. Developmental effects in the form of delayed ossification of fetal sternebrae and vertebral centra occurred only at the highest dose level, at which maternal toxicity in the form of decreased body weight gain was also observed.

In another study, intraperitoneal injection of 80 mg/kg allyl chloride in corn oil on days 1 to 15 of gestation in Sprague-Dawley rats resulted in maternal toxicity in the form of increased organ weights with no histological changes and fetotoxicity in the form of edematous fetuses and fetuses with shortened snouts and protruding tongues[13] (cited in IARC[9]).

Reproductive Toxicity: No data were found implicating allyl chloride as a reproductive toxin.

Systemic Toxicity: The acute toxicity of allyl chloride is relatively greater by the oral route compared to the inhalation route in laboratory animals. Oral LD$_{50}$s of 425 mg/kg and 460 mg/kg have been reported for mice and rats, respectively (purity of allyl chloride >99%). Two-hour LC$_{50}$s of 11,500, 11,400, and 5800 mg/m^3 have been reported for mice, rats, and guinea pigs, respectively, using a static exposure system.[9]

Both animal and human studies have demonstrated that extended exposure to allyl chloride may result in liver injury. In an epidemiologic study of workers exposed to allyl chloride at concentrations in the range of 1 to 113 ppm over a period of 16 months, Hausler and Lenich[14] (cited in NIOSH[5]) reported that altered liver function tests, e.g., serum enzyme changes, were indicative of early stages of liver damage. It was further reported that liver function tests had returned to normal within 6 months after exposure ceased. While the criteria used to judge abnormal results were reported, this study was limited because of the absence of preexposure control values and the fact that no results were presented to support the conclusion that liver function tests had returned to normal after 6 months.

Torkelson et al.[3] (cited in NIOSH[5]) exposed 5 male and 5 female rats, 4 male guinea pigs and 1 female rabbit to 8 ppm (range 7.9 to 10 ppm) allyl chloride 7 hours per day, 5 days a week for 7 weeks. Matched controls were exposed to room air under similar conditions. No differences in gross appearance, mortality, or weight gain were noted between exposed and control animals. However, histologic examination of tissues (lungs, heart, liver, kidney, spleen, testes) revealed definite liver and kidney damage in the form of dilation of the sinusoids; cloudy swelling and focal necrosis in the liver; and glomerular changes, necrosis of the convoluted tubular epithelium, and proliferation of the interstitial tissues. In further studies,

Torkelson et al.[3] (cited in NIOSH[5]) exposed rats, rabbits, guinea pigs, and dogs (n = 48, 6, 18, and 2, respectively with equal numbers of each sex) to 3 ppm (range 1.8 to 3.9 ppm) allyl chloride 7 hours per day, 5 days a week for 127 to 134 exposures. Both air-exposed and unexposed control groups were maintained. Histologic examination of tissues revealed no abnormalities in exposed animals with the exception of slight centrilobular degeneration in the livers of female rats. These changes were judged to be reversible, since they were not observed in female rats allowed to recover for 2 months.

Irritation: Allyl chloride is irritating to the eyes, skin, and mucous membranes of laboratory animals and humans and in sufficiently high concentrations is capable of producing severe injury to pulmonary tissues. In reports cited in the NIOSH criteria document,[5] nasal irritation and "pulmonary discomfort" have been reported in humans at concentrations of less than 25 ppm. The odor of allyl chloride has been described as "unpleasant, pungent," and "garlic-like,"[1,8] and it has been reported that at concentrations between 3 and 6 ppm, 50% of human subjects are able to detect the odor.[5]

Basis for the AALG: Allyl chloride is slightly soluble in water (0.36 grams in 100 mL) and is volatile with a vapor pressure of 295 mm Hg at 20°C.[1,8] It is used primarily in organic syntheses, i.e., as a chemical intermediate, and in one survey was found in urban air at a concentration of 64 ng/m^3 in a single city and at less than 16 ng/m^3 in several other cities. No information was found on its occurrence in other media.[8] Based on these considerations, 80% of the contribution to total exposure will be allowed from air.

Two AALGs were calculated for allyl chloride based on carcinogenicity and systemic toxicity. Since allyl chloride was classified as a group C carcinogen under the EPA weight-of-evidence classification, an uncertainty factor approach was used to derive the AALG for this endpoint. The LOAEL for forestomach tumors in female mice of 129 mg/kg was adjusted for continuous exposure (129 × 5/7 × 78/92) and converted to a surrogate human LOAEL for the inhalation route based on the procedures described in Appendix C. A total uncertainty factor of 5000 was applied (10 for interindividual variation × 10 for interspecies variation × 10 for a LOAEL × 10 for the database factor). The decision for application of the database factor was based on the use of the noninhalation route in this study and less-than-lifetime exposure (as distinct from less-than-lifetime duration of experiment) to the agent, as well as the severity of the effect (i.e., carcinogenicity) and the existence of positive mutagenicity data.

That allyl chloride is capable of causing liver injury is supported by both epidemiologic and animal studies. Industrial experience and epidemiologic evidence support the conclusion that the TLV might reasonably correspond to a human LOAEL; thus, the TLV-TWA of 1 ppm was divided by a total uncertainty factor of 210 (10 for interindividual variation × 4.2 for continuous exposure adjustment × 5 for a LOAEL) with an 80% relative source contribution from air factored into the calculation.

Although both AALGs should be considered provisional, the AALG based on the TLV is recommended for use since it is more conservative.

AALG: • carcinogenicity—7 ppb (22 μg/m³) annual TWA; systemic toxicity— 3.8 ppb (12 μg/m³) 8-hour TWA

References

1. Adams, E. M., H. C. Spencer, and D. D. Irish. 1940. *J. Ind. Hyg. Tox.* 22:79.
2. Silverman, M. and B. E. Abreu. 1938. *Univ. Cal. Publ. Pharm.* 1:119.
3. Torkelson, T. R., M. A. Wolf, F. Oyen, and V. K. Rowe. 1959. *Am. Ind. Hyg. Assoc. J.* 20:217.
4. ACGIH. 1986. "Documentation of the Threshold Limit Values and Biological Exposure Indices." 5:19.
5. NIOSH. 1976. "Criteria for a Recommended Standard . . . Occupational Exposure to Allyl Chloride." DHEW (NIOSH) 76–204.
6. OSHA. 1989. "Air Contaminants; Final Rule." *Fed. Reg.* 54:2332–959.
7. NCI. 1977. "Bioassay of Allyl Chloride for Possible Carcinogenicity." DHEW (NIH) 78–1323.
8. Van Duuren, B. L., B. M. Goldschmidt, G. Loewengart, A. C. Smith, S. Melchionne, I. Seldman, and D. Roth. 1979. "Carcinogenicity of Halogenated Olefinic and Aliphatic Hydrocarbons in Mice." *J. Natl. Cancer. Inst.* 63:1433–39.
9. IARC. 1985. "Allyl Chloride." *IARC Monog.* 36:39–54.
10. Theiss, J. C., M. B. Shimkin, and L. A. Poirier. 1979. "Induction of Pulmonary Adenomas in Strain A Mice by Substituted Oganohalides." *Cancer Res.* 39:391–95.
11. NIOSH. 1988. UC7350000. "Propene, 3-chloro-." RTECS, on line.
12. John, J. A., T. S. Gushow, J. A. Ayres, T. R. Hanley, J. F. Quast, and K. S. Rao. 1983. "Teratologic Evaluation of Inhaled Epichlorohydrin and Allyl Chloride in Rats and Rabbits." *Fund. Appl. Tox.* 3:437–42.
13. Hardin, B. D., G. P. Bond, M. R. Sikov, F. D. Andrew, R. P. Belilies, and R. W. Niemeier. 1981. "Testing of Selected Workplace Chemicals for Teratogenic Potential." *Scand. J. Work Environ. Health* 7:66–75.
14. Hausler, M. and R. Lenich. 1968. "Zur Wirkung von Allylchlorid bei Chronischer Gewerblicher Exposition" (German). *Archiv fur Toxikologie* 23:209–14.

AMMONIA

Synonyms: ammonia gas

CAS Registry Number: 7664–41–7

Molecular Weight: 17.03

Molecular Formula: NH_3

AALG: irritation (TLV/10)—2.5 ppm (1.8 mg/m^3) 8-hour TWA

Occupational Limits:
- ACGIH TLV: 25 ppm (35 mg/m^3) TWA; 35 ppm (27 mg/m^3) STEL
- NIOSH REL: 50 ppm (35 mg/m^3) 5-minute ceiling
- OSHA PEL: 50 ppm (35 mg/m^3) TWA (transitional limit); 35 ppm (27 mg/m^3) STEL (final rule limit)

Basis for Occupational Limits: All of the occupational limits are based on the prevention of eye and respiratory tract irritation.[1-3] OSHA recently adopted the ACGIH-STEL and plans to phase out the present OSHA TWA limit.[4]

Drinking Water Limits: No drinking water limits were found for this substance.

Toxicity Profile

Carcinogenicity: No data were found implicating ammonia as a carcinogen.

Mutagenicity: Inconclusive results are listed for ammonia in the Gene-Tox database as summarized in RTECS for the *Drosophilia melanogaster* sex-linked recessive lethal assay.[5]

Developmental Toxicity: No data were found implicating ammonia as a developmental toxin.

Reproductive Toxicity: No data were found implicating ammonia as a reproductive toxin.

Systemic Toxicity: Kapeghian et al.[6] reported a 1-hour LC_{50} of 4230 ppm for ammonia in male albino ICR mice.

Irritation: The criteria document for ammonia states: "Other than sensory effects—either irritation or annoyance—there is no evidence of acute or chronic adverse effects of ammonia exposure except after accidental exposure at extremely high concentrations, estimated in one fatal exposure to have been 10,000 ppm." It is also noted that concentrations as low as 50 ppm are moderately irritating, but that workers generally become accustomed to these levels.[2]

The principal effects of ammonia exposure in laboratory animals are also irritant in nature. Coon et al.[7] conducted studies in which groups of male and female Sprague-Dawley and Long-Evans rats were exposed continuously to 0, 40 ± 2, 127 ± 8, 262 ± 10, 455 ± 23, or 470 ± 16 mg/m³ ammonia for 90 days (114 days for the 40-mg/m³ group); 90-day continuous exposure studies were also run using male and female Princeton-derived guinea pigs, male New Zealand white rabbits, male squirrel monkeys, and male purebred beagle dogs at exposure levels of 0, 40, and 470 mg/m³ ammonia. At the lowest exposure level, no findings of clinical significance were reported. At 127 and 262 mg/m³, nonspecific inflammatory changes in the lungs and kidneys were reported, but their toxicologic significance was limited due to the fact that similar changes were observed in controls. At 262 mg/m³ a number of rats exhibited signs of irritation in the form of mild nasal discharge. At the highest exposure levels (455 and/or 470 mg/m³), signs of irritation were present in all species and were accompanied by high mortality in the rats, rabbits, and guinea pigs.

The odor threshold for ammonia is estimated at between 1 and 5 ppm. Industrial experience with workers exposed to ammonia from blueprinting and copy machines seems to indicate a maximum acceptable concentration of 20 to 25 ppm without severe complaint.[1]

Basis for the AALG: Ammonia occurs naturally as a degradation product of protein metabolism and is used in a variety of industrial applications including in fertilizers and as a chemical intermediate. Ammonia is very soluble in water and is a gas at room temperature and pressure.[1] Since the AALG for ammonia is based on irritation, no relative source contribution is factored into the calculations.

Since long-term low-level exposures are of most concern in the general population, a TWA occupational limit is the most appropriate choice for AALG derivation. The AALG is based on the TLV/10 for irritant effects, since available evidence indicates that the TLV is a reasonable surrogate human NOAEL.

AALG: • irritation (TLV/10)—2.5 ppm (1.8 mg/m³) 8-hour TWA

References

1. ACGIH. 1986. "Documentation of the Threshold Limit Values and Biological Exposure Indices." 5:27.
2. NIOSH. 1974. "Criteria for a Recommended Standard . . . Occupational Exposure to Ammonia." DHEW (NIOSH) 74–136.

3. OSHA. 1988. "Air Contaminants; Proposed Rule." *Fed. Reg.* 53:20959–21550.

4. OSHA. 1989. "Air Contaminants; Final Rule." *Fed. Reg.* 54:2332–959.

5. NIOSH. 1988. BO0875000. "Ammonia." RTECS, on line.

6. Kapeghian, J. C., A. B. Jones, H. H. Mincer, A. J. Verlangieri, and I. W. Waters. 1982. "Acute Toxicity of Ammonia Gas in the Mouse." *Fed. Proc.* 41:1568.

7. Coon, R. A., R. A. Jones, L. J. Jenkins, and J. Siegel. 1970. "Animal Inhalation Studies on Ammonia, Ethylene Glycol, Formaldehyde, Dimethylamine and Ethanol." *Tox. Appl. Pharm.* 16:646–55.

ANTIMONY AND COMPOUNDS

Synonyms: "Antimony and compounds" includes elemental antimony and the tri- and pentoxides, chlorides, and sulfides of antimony, i.e., Sb_2O_3 (CAS 1309–64–4), Sb_2O_5 (CAS 1314–60–9), $SbCl_3$ (CAS 10025–91–9), $SbCl_5$ (CAS 7647–18–9), Sb_2S_3 (CAS 1345–04–6) and Sb_2S_5 (CAS 1315–04–4)

CAS Registry Number: 7440–36–0

Molecular Weight: 121.75

Molecular Formula: Sb

AALG: 10^{-6} 95% UCL—3.0×10^{-5} µg (Sb)/m^3 annual TWA

Occupational Limits:
- ACGIH TLV: 0.5 mg/m^3 TWA as Sb
- NIOSH REL: 0.5 mg/m^3 TWA as Sb
- OSHA PEL: 0.5 mg/m^3 TWA as Sb

Basis for Occupational Limits: The occupational limits have been set to protect workers from dermatitis, mucous membrane irritation, pneumoconiosis, and electrocardiogram (ECG) alterations. Possible carcinogenic effects of antimony and its compounds have not been factored into the occupational limits.[1,2] However, the industrial process of antimony trioxide production is listed in Appendix A2 (suspect human carcinogen) with no TLV.[1]

Drinking Water Limits: The U.S. EPA ambient water quality criterion for antimony is 146 µg/L for a 70-kg human assuming consumption of 2 L of water and 6.5 g fish/shellfish per day. It was derived based on a LOEL of 5 ppm in drinking water for decreased growth and longevity in rats.[3] In the proposed rule for national primary drinking water regulations, it was noted for Sb and a few other inorganics that preliminary analysis indicated the compounds had limited potential for causing a significant risk via drinking water exposure. However, it was further stated that data collection on exposures and health effects was not yet complete and that these compounds would be considered in later phases of the revised regulations.[4] Antimony is listed as one of the drinking water contaminants scheduled for regulation by 1989.[5]

Toxicity Profile

Carcinogenicity: Studies on the carcinogenicity of antimony in animals were prompted by reports of excess lung cancers in British antimony workers, particularly those engaged in smelting and related activities[6] (cited in NIOSH[2]). However, it is noted in the NIOSH criteria document that much key information is not included in these reports, e.g., community and factory population descriptions, biological and environmental monitoring data, and methods of data analysis.[2]

Groth et al.[7] exposed male and female Wistar-derived rats (n = 90/sex/group) by inhalation to filtered air, antimony trioxide (TWA 45 mg/m^3), or antimony ore concentrate (TWA 36 mg/m^3) for 7 hours per day, 5 days per week over periods of up to a year with terminal sacrifice 20 weeks after the final exposure. It should be noted that the rats were 8 months old at the beginning of exposure. No lung neoplasms were found in male rats in any group or in control females; however, incidences of lung tumors (squamous cell carcinomas, bronchioalveolar adenomas, bronchioalveolar carcinomas, scirrhous carcinomas) in female Sb$_2$O$_3$-exposed rats and female Sb ore–exposed rats were 27% and 25%, respectively.

A number of inorganic elements commonly present in Sb ore were analyzed for and found in both the Sb ore concentrate and Sb$_2$O$_3$ used in this study, e.g., Al, Mg, Ti, Fe, Cu, Zn, As, Sn, Pb, etc., with higher concentrations usually present in the Sb ore. The principal component of Sb ore concentrate was antimony trisulfide. It is relevant to note that higher concentrations of arsenic were found in the tissues of female rats relative to males, but that the concentration of Sb was higher in the lungs of males than females. The authors state the following with respect to the influence of arsenic on Sb carcinogenesis: "If As is to be implicated in promoting the carcinogenic effect of Sb, then either it is not related to a local tissue effect in the lung or the As in the Sb$_2$O$_3$ is not readily biologically available. It is possible that the systemic concentration of As is more critical in affecting the carcinogenic response."

It is noteworthy with respect to antimony trioxide that much lower concentrations are apparently capable of inducing lung tumors in female rats. Watt[8] exposed female CDF rats to 0, 1.6, or 4.2 mg/m^3 antimony as antimony trioxide for 6 hours per day, 5 days a week for one year (approximately one-half of their lifespan). Exposure was started when the animals were 18 weeks of age and continued until they were 75 weeks of age, with final sacrifice and examination of surviving animals at 12 to 15 months postexposure. The Sb$_2$O$_3$ was 99.4% pure with the major contaminants being arsenic and lead. There was a significant increase in lung neoplasms of various types in the high exposure group animals examined at the end of the study. These included scirrhous carcinomas of the lung (controls, 0 of 13; low, 0 of 17; and high, 9 of 18), squamous cell carcinomas (control, 0 of 13; low, 0 of 17; and high 2 of 18), and bronchioalveolar adenomas (control, 0 of 13; low, 1 of 17; and high, 3 of 18). It is noteworthy that these animals were approximately 132 weeks of age at final sacrifice and if only those animals alive and examined after 12 months of exposure are included, the incidence of all lung neoplasms was 21 of 34 or 62%.[7] The major limitations of this study are that the numbers of animals used were relatively small and the study was limited to female rats; however, a clear dose-related increase in lung neoplasms did occur at the highest exposure level.

Based on the available evidence, antimony trioxide, and possibly other antimony compounds as well, are most appropriately placed in group B2, probable human carcinogen, under the EPA weight-of-evidence classification.

Mutagenicity: Antimony trioxide and antimony tri- and pentachloride are reported to be mutagenic to *Bacillus subtilis*.[9]

Developmental Toxicity: Developmental and reproductive effects were reported in a Soviet study comparing women working in an antimony metallurgical plant with control workers not exposed to antimony[10] (cited in NIOSH[2]). Higher incidences were reported in the exposed women for late-occurring spontaneous abortions, premature births, and a variety of gynecologic problems including menstrual cycle disorders and pelvic inflammatory disease. It is interesting to note that while the birth weight of offspring of control and exposed mothers did not differ, the weight of offspring of exposed mothers began to lag behind the weight of offspring of control mothers at 3 months of age and was significantly less when the offspring were one year of age. However, it is pointed out in the NIOSH criteria document that a definitive interpretation of the significance of these findings is not possible because of the lack of detail in the reporting of procedural aspects of the study. It was further pointed out that the incidence of gynecologic problems reported in the control population was quite high.

Belyaeva[10] (cited in NIOSH[2] and U.S. EPA[3]) also exposed female rats to 250 mg/m^3 antimony trioxide for 4 hours per day over a period of 1.5 to 2 months encompassing the time prior to mating and throughout gestation. All of the control animals conceived, whereas one-third of the treated animals failed to conceive. Histological examination of uterine and ovarian tissue revealed changes such as lack of ova in follicles, misshapen ova, cysts, and uterine metaplasia.

Three studies are cited in the REPROTOX summary for antimony in which this substance was found not to be teratogenic in the chick embryo assay, rats, and sheep.[11] However, in view of the report cited above, conclusions on the developmental and reproductive toxicity of antimony compounds should be made cautiously and the literature on these effects reviewed periodically.

Reproductive Toxicity: Refer to the discussion above, particularly the findings with respect to antimony trioxide–exposed rats by Belyaeva[10] (cited in NIOSH[2] and U.S. EPA[3]).

Systemic Toxicity: Occupational exposure to antimony compounds has been reported to result in adverse effects on the heart, lungs, and gastrointestinal tract. Following the identification of 8 deaths as attributable to occupational exposure to antimony trisulfide, Brieger et al.[12] (cited in NIOSH[2]) conducted an investigation in which it was found that 37 of 75 workers in the same department had abnormal ECGs. The exposure levels were in the range of 0.58 to 5.5 mg/m^3. The same authors found that definite and consistent ECG changes occurred in rats and rabbits exposed at similar levels for 6 hours per day, 5 days per week over a period of 6 weeks. Other investigators have reported finding abnormal ECGs in workers exposed to both antimony trisulfide and trioxide in a metallurgical plant[13] (cited in NIOSH[2]) where arsenic contamination was also present. However, it is noted in the criteria document that patients treated with arsenic-free antimony drugs for parasitic infections had similar ECG changes.[2]

Occupational exposures to antimony trioxide, antimony metal, and antimony ore have been associated with the development of pneumoconiosis as well as impairment of lung function.[2] Occupational exposure to antimony has also been reported to result in gastrointestinal toxicity that appears to be systemic in nature (as opposed to arising from the irritant properties of antimony). In addition to the ECG abnormalities found by Brieger et al.[12] (cited in NIOSH[2]), a higher incidence of gastrointestinal ulcers was observed in exposed relative to unexposed workers. Other investigators have noted that workers exposed to antimony reported experiencing nausea, vomiting, abdominal pain, and diarrhea and similar effects are noted when antimony drugs are given parenterally.[2]

Irritation: In reports of occupational inhalation exposures, skin and mucous membrane irritation have been observed at concentrations of 0.40 to 70.7 mg/m^3 for antimony trioxide and 73 mg/m^3 for antimony trichloride with exposure times of a few months and up to 8 hours, respectively.[2]

Basis for the AALG: Since food is considered to be the primary exposure route for antimony for the general population,[9] 20% of the contribution to total exposure will be allowed from air.

Two AALGs have been calculated for antimony and compounds (as Sb): one based on the TLV, which is applicable to the chronic effects (TLV/42) and another based on carcinogenicity. The issue of which antimony compounds should be considered carcinogenic is complex. There is sufficient evidence from animal studies and limited evidence from human epidemiologic studies that antimony trioxide is carcinogenic. In addition, Groth et al.[7] reported that exposure to antimony ore concentrate containing large amounts of antimony trisulfide produced lung neoplasms in female Wistar-derived rats. Therefore, at least these two antimony compounds should be treated as potential carcinogens. Another complicating factor in antimony carcinogenesis is the presence—both in occupational settings and in experimental studies in animals—of contaminants that are in and of themselves carcinogenic, e.g., arsenic. The influence of these contaminants on the process of antimony carcinogenesis is difficult to determine given the limitations of presently available data. It is the judgment here that in the interests of health protection it is prudent to consider all airborne antimony compounds as potential carcinogens until further evidence becomes available to allow for a more precise determination. Given these considerations and limitations in the data, the AALGs derived below should be considered provisional.

The AALG for carcinogenicity was based on the application of the linearized multistage model to data for total lung tumors in antimony trioxide–exposed female rats in the Groth et al. study.[7] The major limitation of this study is that only a single exposure level in addition to the control group was used. However, the Watt study[8] suffered from such limitations as small group sizes; in addition, there were problems with lack of fit when the linearized multistage model was applied to the data.

In performing the quantitative risk assessment using the Groth et al.[7] data, several points should be noted: (1) the exposure level used in the study was adjusted for continuous exposure, i.e, 45 mg/m^3 × 7/24 × 5/7 × 1/2; (2) for purposes of the dose conversions, antimony trioxide is considered to be absorbed proportional to the amount of air breathed in (for a complete discussion of this issue see Anderson et al.[14]; and (3) the absorption fraction (r) used in the calculations was 0.5. The resultant q_1^* assuming a 70-kg human breathing 20 m^3 of air per day was 6.69 × 10^{-3} (μg/m^3)$^{-1}$. Applying the relationship $E = q_1^* × d$ (where E is the specified level of extra risk and d is the dose) and allowing a 20% relative source contribution from air gives the estimates shown below.

AALG: • systemic toxicity (TLV/42)—2.4 μg (Sb)/m^3 8-hour TWA
   • 10^{-6} 95% UCL—3.0 × 10^{-5} μg (Sb)/m^3 annual TWA
   • 10^{-5} 95% UCL—3.0 × 10^{-4} μg (Sb)/m^3 annual TWA

References

1. ACGIH. 1986. "Documentation of the Threshold Limit Values and Biological Exposure Indices." 5:32–3.
2. NIOSH. 1978. "Criteria for a Recommended Standard . . . Occupational Exposure to Antimony." DHEW (NIOSH) 78–216.
3. U.S. EPA. 1980. "Ambient Water Quality Criteria for Antimony." NTIS PB81–117319.
4. U.S. EPA. 1985. "National Primary Drinking Water Regulations; Synthetic Organic Chemicals, Inorganic Chemicals and Microorganisms; Proposed Rule." *Fed. Reg.* 50: 46936–7025.
5. Ohanian, E. V. 1989. "National Primary Drinking Water Regulations for Additional Contaminants to be Regulated by 1989." In: Calabrese, E. J., C. E. Gilbert, and H. Pastides, Eds. *Safe Drinking Water Act—Amendments, Regulations and Standards* (Chelsea, MI: Lewis Publishers, Inc.), pp. 71–82.
6. Davies, T. A. L. 1973. "The Health of Workers Engaged in Antimony Oxide Manufacture—A Statement" (London: Employment Medical Advisory Service, Department of Employment).
7. Groth, D. H., L. E. Stettler, J. R. Burg, W. M. Busey, G. C. Grant, and L. Wong. 1986. "Carcinogenic Effects of Antimony Trioxide and Antimony Ore Concentrate in Rats." *J. Tox. Environ. Health* 18:607–26.
8. Watt, W. D. 1983. "Chronic Inhalation Toxicity of Antimony Trioxide: Validation of the Threshold Limit Value." PhD Thesis. Wayne State University, Detroit.
9. Carson, B. L., H. V. Ellis, and J. L. McCann. 1986. *Toxicology and Biological Monitoring of Metals in Humans* (Chelsea, MI: Lewis Publishers, Inc.), pp. 21–26.
10. Belyaeva, A. P. 1967. "The Effect of Antimony on Reproductive Function" (Russian). *Gig. Tr. Prof. Zabol.* 11:32–37.
11. Reproductive Toxicology Center. 1987. "Reproductive Toxicity Review—Antimony." REPROTOX, on line (Washington, DC: Columbia Hospital for Women Medical Center).

12. Brieger, H., C. W. Semisch, J. Stasney, and D. A. Piatnek. 1954. "Industrial Antimony Poisoning." *Ind. Med. Surg.* 23:251–53.
13. Klucik, I. and L. Ulrich. 1960. "Electrocardiographic Examination of Workers in an Antimony Metallurgical Plant" (Czech). *Prac. Lek.* 12:236–43.
14. Anderson, E. L., and the Carcinogen Assessment Group of the U.S. EPA. 1983. "Quantitative Approaches in Use to Assess Cancer Risk." *Risk Anal.* 3:277–295.

ARSENIC AND COMPOUNDS

Synonyms: "Arsenic and compounds" refers to inorganic arsenicals, both tri- and pentavalent, including arsenic trioxide (As_2O_3—CAS 1327–53–3, molecular weight 197.84) and excluding arsine gas.

CAS Registry Number: 7440–38–2

Molecular Weight: 74.92

Molecular Formula: As

AALG: carcinogenicity: 10^{-6}—4.66×10^{-5} $\mu g/m^3$ (as As) as an annual TWA

Occupational Limits:
- ACGIH TLV: 0.2 mg/m^3 (as As) TWA
- NIOSH REL: 0.002 mg/m^3 (as As) 15-minute ceiling
- OSHA PEL: 0.01 mg/m^3 (as As) TWA

Basis for Occupational Limits: The ACGIH limit for arsenic given above includes arsenic and its soluble compounds; arsenic trioxide production is treated separately and is listed in Appendix A2 as a suspect human carcinogen with no TLV. The TLV documentation states: "A search of the world literature reveals no reports of industrial or experimental exposures solely to arsenic compounds which contain both environmental and toxicological criteria from which a TLV can be unequivocally based." However, a report is cited of a pharmaceutical plant with concentrations averaging around 0.2 mg/m^3 where no evidence of toxicity in the workers was found. Another report is cited where smelter workers experienced dermatitis, nasal septum perforation, pharyngitis, and conjuctivitis; exposure was deduced to be about 0.2 mg/m^3 based on urinary excretion of arsenic.[1]

The NIOSH REL is set for inorganic arsenic including arsenic trioxide, and carcinogenic effects were considered. It is stated in the criteria document that the limit set is "intended to achieve the greatest practicable reduction in worker exposure while avoiding spurious sampling results produced by natural background concentrations of inorganic arsenic."[2] The health effects considered in establishing the OSHA PEL were lung and lymphatic cancer and dermatitis.[3]

Drinking Water Limits: The current MCL for arsenic is 0.05 mg/L. This was based on human data that showed no adverse health effects were associated with consumption of water with this level of arsenic. One-day HAs of 0.14 mg/L and 1.0 mg/L have been derived for the 10-kg child and 70-kg adult, respectively.[4] The ambient water quality criterion for arsenic is based on carcinogenicity and is 2.2 ng/L at the 10^{-6} risk level, assuming a person drinking 2 L of water and eating 6.5

g of fish/shellfish per day.[5] It is noteworthy that in establishing the MCLG, EPA considered setting it at 0, but rejected the idea because doing so would ignore the essentiality of arsenic in the diet.[6] It is also relevant to note that the document on quantification of toxicological effects of arsenic stated that there is currently no suitable basis for quantitative risk assessment for excess cancer risk due to arsenic ingestion that is applicable to the United States.[4]

Toxicity Profile

Carcinogenicity: A number of arsenical compounds, particularly trivalent inorganics, have been associated with the development of lung and skin neoplasms in humans ingesting arsenic. These include arsenic trioxide, arsenic sulfides, and other salts of arsenic used therapeutically. Occupational exposure to arsenic at copper smelters is associated with the development of lung cancer.[7] It has been observed that there is an increasing risk of lung cancer mortality with increasing age of initial exposure, independent of time after exposure ceased. This is cited as evidence that arsenic acts as a late-stage carcinogen[8] (cited in U.S. EPA[7]).

Arsenic is unique among carcinogens in that most animal bioassays have produced negative results.[7] IARC has classified the evidence for carcinogenicity in animals as limited, while the evidence in humans was considered sufficient.[9,10] Animal studies pertinent to the evaluation of arsenic carcinogenicity have been reviewed and evaluated by both IARC[9,10] and the U.S. EPA.[4,5,11]

Extensive epidemiologic studies have been conducted in populations exposed to naturally high levels of arsenic in drinking water and in copper smelter workers and populations in the vicinity of copper smelters. These studies are extensively reviewed and the basis for quantitative assessments of cancer risk for different types of arsenic exposure are discussed in the health assessment documents for arsenic.[4,11]

Inorganic arsenic has been placed in group A, known human carcinogen, under the EPA weight-of-evidence classification.[7]

Mutagenicity: The genetic activity of both tri- and pentavalent arsenic was recently reviewed by IARC. It was reported that neither tri- nor pentavalent arsenic induced mutations in bacteria, but both induce gene conversion in yeast and oncogenic transformation of Syrian hamster embryo cells. Trivalent arsenic did not induce mutations in rodent cells in vitro, but did induce chromosomal aberrations and SCEs in human and rodent cells in vitro. In vivo studies in mice, trivalent arsenic did not induce dominant lethal mutations, but a slight increase in the incidence of chromosomal aberrations and micronuclei in bone marrow cells was reported. Pentavalent arsenic has been found to induce chromosomal aberrations in both human and rodent cells in vitro, but evidence for the induction of SCEs was equivocal.[12]

Arsenic compounds have been reported to produce chromosomal aberrations in peripheral lymphocytes of humans exposed therapeutically and occupationally.[7]

Developmental Toxicity: Arsenical compounds, particularly sodium arsenate, have been shown to be teratogenic in hamsters, mice, and rats when injected or ingested. A wide range of developmental abnormalities have been reported, including gross, visceral, and skeletal abnormalities with various neural tube defects being quite common.[5,13] It was noted in the ambient water quality criteria document that although arsenites are generally more toxic than arsenates, their developmental toxicity has seldom been evaluated.[5]

Nagymajtenyi et al.[14] exposed CFLP mice to 0, 0.26 ± 0.01, 2.9 ± 0.04, or 28 ± 0.30 mg/m^3 arsenic trioxide aerosol for 4 hours per day on gestation days 9 through 12. The maternal animals were killed and the fetuses harvested on day 18 of gestation. The fetuses were weighed, examined grossly for retarded growth and malformations, and stained for examination of skeletal malformations, and the fetal liver cells were examined for chromosomal aberrations. There was a statistically significant dose-related decrease in fetal body weight at all exposure levels relative to controls, but retarded fetal growth and skeletal malformations (primarily ossification defects) were significant only at the highest dose. There was also an increased incidence of chromosomal aberrations in the liver cells of fetal mice in the highest exposure group. A LOAEL of 0.26 mg/m^3 for decreased fetal weight is suggested by this study. The major limitations of the study include lack of examination for visceral malformations, no mention of maternal toxicity, and relatively small sample sizes (8 to 11 litters per exposure level).

In the REPROTOX assessment of arsenic it was noted that five cases of arsenic poisoning during pregnancy have been reported in humans in which no adverse effects on the offspring were noted. However, it is further pointed out that all of these exposures occurred during the second trimester; the data are insufficient to evaluate the effects of first-trimester exposure.[11]

Reproductive Toxicity: No data were found implicating arsenic as a reproductive toxin.

Systemic Toxicity: Based predominantly on animal studies, trivalent forms of inorganic arsenic are generally considered to be more toxic than pentavalent forms. Exposure to arsenic compounds in drinking water (average concentration 0.598 mg/L) over periods of up to 15 years has been associated with hyperkeratosis, chronic coryza, abdominal pain, and Raynaud's syndrome. Exposure to 0.01 to 1.82 mg/L arsenic for periods of greater than 45 years has been associated with hyperpigmentation, keratosis, skin cancer, and blackfoot disease.

Studies, particularly of smelter workers, have indicated that chronic exposure to arsenic compounds (especially arsenic trioxide) may result in skin lesions, cirrhosis of the liver, respiratory effects (e.g., perforated nasal septum), and peripheral neuropathy.[1,7] Although Landau et al.[15] (cited in U.S. EPA[7]) reported a direct relationship between length and intensity of smelter workers' exposure to arsenic (especially As$_2$O$_3$) and peripheral neuropathy, existing studies are generally insufficient to characterize the exposure-response relationship for airborne arsenic.[7]

Irritation: Refer to "Basis for Occupational Limits" and "Systemic Toxicity," above.

Basis for the AALG: Since arsenic is considered to be widely distributed in food and water and most exposure is believed to be via the diet,[11] 20% of the contribution to total exposure will be allowed from air.

AALGs were calculated for arsenic based on both developmental toxicity and carcinogenicity. The AALG for developmental toxicity was derived from the LOAEL of 0.26 mg/m^3 for decreased fetal body weight in mice in the Nagymajtenyi et al. study.[14] The NOAEL was adjusted for continuous exposure (0.26 mg/m^3 × 4/24) and converted to a human equivalent exposure level according to the procedures described in Appendix C, and a total uncertainty factor of 2500 was applied (10 for interindividual variation × 10 for interspecies variation × 5 for a LOAEL × 5 for the developmental uncertainty factor).

The AALGs for carcinogenicity are based on the risk assessment performed by the U.S. EPA Carcinogen Assessment Group as presented in the IRIS summary.[16] Unit risk estimates were based on studies of workers at two different smelters[17-22] (cited in U.S. EPA[16]). The final estimate of unit risk—4.29 × 10^{-3} —was based on the geometric mean of unit risks calculated from several different data sets. The extrapolation method used was the absolute-risk linear model. It was assumed that the increase in the age-specific mortality rate of lung cancer was a function only of cumulative exposure. The concentrations in air corresponding to the 10^{-5} and 10^{-6} risk levels based on the above potency factor (4.29 × 10^{-3} (μg/m^3)$^{-1}$) are 2.33 × 10^{-3} μg/m^3 and 2.33 × 10^{-4} μg/m^3, respectively. Applying a 20% relative source contribution gives the estimates shown below. It should be noted that it is considered inappropriate to use the above unit risk if the air concentration exceeds 2 μg/m^3, since the slope factor may differ at concentrations above this level. In addition, this assessment applies to total arsenic, because no distinction was made between pentavalent and trivalent arsenic in the documentation.[16]

The recommended AALG is based on carcinogenicity for the 10^{-6} risk level, since this is the most conservative of the derived limits.

AALG: • developmental toxicity: 0.014 μg/m^3 24-hour TWA
  • carcinogenicity: 10^{-6}, 4.66 × 10^{-5} μg/m^3; 10^{-5}, 4.66 × 10^{-4} μg/m^3— annual TWAs

References

1. ACGIH. 1986. "Documentation of the Threshold Limit Values and Biological Exposure Indices." 5:37–38.
2. NIOSH. 1975. "Criteria for a Recommended Standard . . . Occupational Exposure to Inorganic Arsenic." DHEW (NIOSH) 75–149.
3. NIOSH. 1985. "NIOSH Recommendations for Occupational Safety and Health Standards." *Morb. Mort. Wkly. Rpt.* (MMWR) 34:5s-31s.

4. U.S. EPA. 1984. "Quantification of Toxicological Effects of Arsenic" (draft—scientific review) (Washington, DC: U.S. EPA Office of Drinking Water), PB86–117892.
5. U.S. EPA. 1980. "Ambient Water Quality Criteria for Arsenic." EPA 440/5–80–021.
6. U.S. EPA. 1985. "National Primary Drinking Water Regulations; Synthetic Organic Chemicals, Inorganic Chemicals and Microorganisms; Proposed Rule." *Fed. Reg.* 50: 46936–7022.
7. U.S. EPA. 1984. "Health Effects Assessment for Arsenic." EPA/540/1–86/020.
8. Brown, C. C. and K. C. Chu. 1983. "Implications of the Multistage Theory of Carcinogenesis Applied to Occupational Arsenic Exposure." *J. Natl. Cancer Inst.* 70: 455–63.
9. IARC. 1980. "Arsenic." *IARC Monog.* 23:38–149.
10. IARC. 1987. "Arsenic." *IARC Monog.* suppl. 7:100–06.
11. U.S. EPA. 1984. "Health Assessment Document for Inorganic Arsenic." EPA-600/8–83–021F.
12. IARC. 1986. "Arsenic." *IARC Monog.* suppl. 6:71–76.
13. Reproductive Toxicology Center. 1987. "Reproductive Toxicity Review—Arsenic." REPROTOX, on line (Washington, DC: Columbia Hospital for Women Medical Center).
14. Nagymajtenyi, L., A. Selypes, and G. Berencsi. 1985. "Chromosomal Aberrations and Fetotoxic Effects of Atmospheric Arsenic in Mice." *J. Appl. Tox.* 5:61–3.
15. Landau, E. et al. 1977. "Selected Noncarcinogenic Effects of Industrial Exposure to Inorganic Arsenic." EPA/569/6–77–018.
16. U.S. EPA. 1988. "Arsenic, Inorganic; CASRN 7440–38–2." Integrated Risk Information System summary (12/1/88).
17. Brown, C. C. and K. C. Chu. 1983. "Approaches to Epidemiologic Analysis of Prospective and Retrospective Studies: Example of Lung Cancer and Exposure to Arsenic." In: Risk Assessment: Proceedings of the SIMS Conference on Environmental Epidemiology. June 28–July 2, 1982, Alta, UT. SIAM Publication.
18. Brown, C. C. and K. C. Chu. 1983. "Implications of the Multistage Theory of Carcinogenesis Applied to Occupational Arsenic Exposure." *J. Natl. Cancer Inst.* 70: 455–63.
19. Brown, C. C. and K. C. Chu. 1983. "A New Method for the Analysis of Cohort Studies; Implications of the Multistage Theory of Carcinogenesis Applied to Occupational Arsenic Exposure." *Env. Health Perspect.* 50:293–308.
20. Lee-Feldstein, A. 1983. "Arsenic and Respiratory Cancer in Man: Follow-up of an Occupational Study." In: *Arsenic: Industrial, Biomedical and Environmental Perspectives,* W. Lederer and R. Fensterheim, Eds. (New York: Van Nostrand Reinhold Company).
21. Higgins, I. 1982. "Arsenic and Respiratory Cancer Among a Sample of Anaconda Smelter Workers." Report submitted to the Occupational Safety and Health Administration in the comments of the Kennecott Minerals Company on the inorganic arsenic rulemaking (Exhibit 203–5).
22. Enterline, P. E. and G. M. Marsh. 1982. "Cancer Among Workers Exposed to Arsenic and Other Substances in a Copper Smelter." *Am. J. Epidemiol.* 116:895–911.

ASBESTOS

Definition: The term asbestos refers to a variety of hydrated silicates containing metal cations such as sodium, magnesium, calcium, and iron. Although several types of asbestos exist, only chrysotile (CAS 12001–29–5), amosite (CAS 12172–73–5), and crocidolite (CAS 12001–28–4) will be discussed here, since they constitute over 99% of the asbestos produced in the United States.[1] It should be noted that of these three types, chrysotile constitutes 95% of that used.[2]

CAS Registry Number: 1332–21–4

AALG: 10^{-6} 95% UCL–4 $\times$ 10^{-6} fibers/mL annual TWA

Occupational Limits:
- ACGIH TLV: amosite, 0.5 fibers >5 μm/cm^3 TWA; chrysotile, 2 fibers >5 μm/cm^3 TWA; crocidolite, 0.2 fibers >5 μm/cm^3 TWA; other forms, 2 fibers >5 μm/cm^3 TWA; Appendix A1a—recognized human carcinogen
- NIOSH REL: 0.1 fibers >5 μm/cm^3 TWA; 0.5 fibers >5 μm/cm^3 peak concentration in a 15-minute sampling period
- OSHA PEL: 0.2 fibers >5 μm/cm^3 TWA; 1 fiber >5 μm/cm^3 STEL (over a 30-minute sampling period)

Basis for Occupational Limits: The health effects considered in setting all of the limits were asbestosis, lung cancer, and mesothelioma.[2,3] The ACGIH limits are unique in that limits are set based on fiber type. With respect to the NIOSH limit, it is stated in the criteria document that the limit is set to "(1) protect against the noncarcinogenic effects of asbestos, (2) materially reduce the risk of asbestos-induced cancer (only a ban can assure protection against carcinogenic effects of asbestos) and (3) be measured by techniques that are valid, reproducible, and available to industry and official agencies."[3] On September 14, 1988, OSHA issued a STEL to go with the TWA PEL.[4]

Drinking Water Limits: The MCLG for asbestos (medium and long fibers) is 7.1 million fibers/L.[5] The ambient water quality criterion for asbestos fibers of all sizes is 30,000 fibers/L at the 10^{-6} risk level assuming consumption of 2 L of water per day.[6]

Toxicity Profile

Carcinogenicity: The relationship between asbestos exposure and cancer is summarized as follows in the ''Airborne Asbestos Health Assessment Update'' (AAHAU):

> Lung cancer and mesothelioma are the most important asbestos-related causes of death among exposed individuals. Gastrointestinal cancers are also increased in most studies of occupationally exposed workers. Cancer at other sites (larynx, kidney, ovary) has also been shown to be associated with asbestos exposure in some studies, but the degree of excess risk and the strength of the association are less for these and the gastrointestinal cancers than for lung cancer or mesothelioma. The International Agency for Research on Cancer lists asbestos as a group 1 carcinogen, meaning that exposure to asbestos is carcinogenic to humans. EPA's proposed guidelines would categorize asbestos as Group A, human carcinogen.[7]

The role of animal data is also well-summarized in the AAHAU:

> Animal studies confirm the human epidemiological results. All major asbestos varieties produce lung cancer and mesothelioma with only limited differences in carcinogenic potency. Implantation and injection studies show that fiber dimensionality, not chemistry, is the most important factor in fiber-induced carcinogenicity. Long (>4 μm) and thin (<1 μm) fibers are the most carcinogenic at a cancer-inducible site. However, the size dependence of the deposition and migration of fibers also affects their carcinogenic action in humans.[7]

Mutagenicity: In general, most studies seem to indicate that asbestos probably does not cause gene mutations and/or chromosomal breakage. However, there is some evidence that asbestos causes aneuploidy.[7]

Developmental Toxicity: Teratogenic effects were not observed in a study in which mice were exposed to doses between 4 and 400 mg/kg asbestos in drinking water on gestation days 1 through 15.[8]

Reproductive Toxicity: In one study, a dose of 20 mg/kg/day of asbestos did not cause genotoxic effects on mouse sperm.[8]

Systemic Toxicity: Asbestosis is the chronic noncarcinogenic form of lung disease caused by asbestos exposure and is characterized by diffuse interstitial fibrosis, often associated with pleural fibrosis (thickening) and calcification.[3] Fibrosis results in the lungs being less compliant, and there is decreased diffusion of gases through the alveolar walls, resulting in breathlessness.[2]

Irritation: No data were found implicating asbestos as an irritant.

Basis for the AALG: Risk calculations for asbestos exposure are presented and explained in detail in the "Airborne Asbestos Health Assessment Update" document. Extensive calculations are provided due to a number of factors: (1) human occupational data from a variety of studies are used to calculate risks for the general population; (2) parameters such as age at onset of exposure and length of exposure are factored into the model; (3) gender-specific and smoking status–specific rates are also given; and (4) risks for mesothelioma versus lung cancer differ and are thus also provided.

The figures given below were stated to be the best estimates for risk to the U.S. general population and the following conditions and assumptions are operative: (1) exposure is continuous, beginning at birth and continuing for the entire lifetime; (2) U.S. general population death rates were used and smoking habits were not considered; and (3) the exposure concentration was set at 0.0001 fibers/cm^3 (3 ng/m^3) which is stated to be typical of urban air.

> *mesothelioma risks*: females—2.8 deaths/100,000 population
> males—1.9 deaths/100,000 population
> *lung cancer risks:* females—0.5 deaths/100,000 population
> males—1.7 deaths/100,000 population

In the IRIS database,[9] estimates were provided for the composite risk of lung cancer and mesothelioma, not accounting for smoking, in both males and females. Additive risk was assumed, and a relative risk model was used for lung cancer and an absolute risk model for mesothelioma; it should be noted that the epidemiologic data indicate that the interaction of cigarette smoking with asbestos exposure is synergistic for lung cancer but not for mesothelioma. The final unit risk figure, 2.3 $\times$ 10^{-1} (fibers/mL)$^{-1}$ was based on data from a number of studies referenced and described briefly in the IRIS summary and calculated as a weighted geometric mean. It is noted that the unit risk is based on fiber counts made by phase contrast microscopy and should not be applied directly to measurements made using other analytical methods. No relative source contribution from air was factored into these calculations due to the fact that the inhalation route is unique to the endpoint on which the AALG is based.

AALG:
- 10^{-6} 95% UCL—4 $\times$ 10^{-6} fibers/mL annual TWA
- 10^{-5} 95% UCL—4 $\times$ 10^{-5} fibers/mL annual TWA

Note: Federal emission standards apply to milling, manufacturing, and fabrication sources as well as demolition, renovation, and waste disposal. In general, standards allow compliance alternatives, i.e., (1) no visible emissions or (2) employment of specified control techniques. The standards do not include any mass or fiber count emission limitations.[7]

References

1. U.S. EPA. 1984. "Health Effects Assessment for Asbestos." NTIS PB86–134608.
2. ACGIH. 1986. "Documentation of the Threshold Limit Values and Biological Exposure Indices." 5:40–42.

3. NIOSH. 1976. "Criteria for a Recommended Standard . . . Revised Recommended Asbestos Standard." DHEW (NIOSH) 77–169.

4. OSHA. 1988. "Occupational Exposure to Asbestos, Tremolite, Anthophyllite and Actinolite; Final Rule, Amendment." *Fed. Reg.* 53:35610.

5. U.S. EPA. 1985. "National Primary Drinking Water Regulations; Synthetic Organic Chemicals, Inorganic Chemicals and Microorganisms; Proposed Rule." *Fed. Reg.* 50: 46936–47022.

6. U.S. EPA. 1980. "Ambient Water Quality Criteria for Asbestos." EPA 440/5–80–022.

7. U.S. EPA. 1986. "Airborne Asbestos Health Assessment Update." EPA 600/8–84/003F.

8. Reproductive Toxicology Center. 1987. "Reproductive Toxicity Review: Asbestos." REPROTOX, on line (Washington, DC: Columbia Hospital for Women Medical Center).

9. U.S. EPA. 1989. "Asbestos; CASRN 1332–21–4." IRIS (5/1/89).

BENZENE

CAS Registry Number: 71–43–2

Molecular Weight: 78.11

Molecular Formula:

AALG: 10^{-6}—3.02×10^{-2} ppb (9.6×10^{-2} µg/m³) annual TWA

Occupational Limits:
- ACGIH TLV: 10 ppm (30 mg/m³) TWA; Appendix A2 suspect human carcinogen
- NIOSH REL: 1 ppm (3 mg/m³) as a 60-minute ceiling
- OSHA PEL: 10 ppm (30 mg/m³) TWA; 25 ppm (80 mg/m³) acceptable ceiling; 50 ppm (160 mg/m³) acceptable maximum peak (maximum duration, 10 minutes) above the acceptable ceiling concentration for an 8-hour shift

Basis for Occupational Limits: Although benzene is listed as a suspect human carcinogen, the TLV appears to have been based on a noncancer endpoint. The TLV documentation states: "There is little evidence that exposure to benzene at concentrations below 25 ppm causes blood dyscrasias of any kind. Setting the TLV at 10 ppm, as a time-weighted average, provides an added margin of safety."[1] For the NIOSH and OSHA limits the health effects considered were blood changes, including leukemia.[2,3]

Drinking Water Limits: The final MCLG for benzene is 0 µg/L and the MCL is 5 µg/L in consideration of carcinogenic effects.[4] The ambient water quality criterion for benzene is 0.66 µg/L at the 10^{-6} risk level assuming a person drinking 2 L of water and eating 6.5 g of fish/shellfish per day.[5] Both the National Academy of Sciences[6] and the U.S. EPA Office of Drinking Water[7] concluded that there were insufficient data to calculate a one-day SNARL or HA, respectively. However, a 10-day HA of 235 µg/L was calculated assuming a 10-kg child consuming 1 L of water per day. This was considered sufficiently protective for a one-day exposure. Longer-term and lifetime HAs were not calculated for benzene due to its carcinogenic potency.[7]

Toxicity Profile

Carcinogenicity: A number of case studies and epidemiologic investigations have established the causal association between inhalation exposure to benzene and leukemia in humans. These studies have been extensively reviewed by IARC[8] and in the ATSDR (Agency for Toxic Substances and Disease Registry) toxicological profile.[9] However, with respect to the ingestion carcinogenicity of benzene in humans two sources have reported that no epidemiologic studies are available pertaining to benzene carcinogenicity in drinking water.[9,10] The evidence for carcinogenicity of benzene in humans is designated as sufficient by IARC.[8]

The evidence for benzene carcinogenicity is also sufficient in experimental animals based on IARC criteria.[11] With respect to the role of animal data, it is stated in the Health Effects Assessment Document (HEAD)[10] that "animal bioassays which demonstrate increased incidence of Zymbal and mammary gland carcinoma in orally exposed rats and suggest increased incidence of hematopoietic tumors in C57B1 mice exposed via inhalation, may be considered corroborative data supportive of a carcinogenic role for benzene."

Benzene has been placed in group A, known human carcinogen, under the EPA weight-of-evidence classification.[9]

Mutagenicity: Benzene is nonmutagenic to *S. cerevisiae*, *E. coli*, and *S. typhimurium* with and without metabolic activation and is negative in the *Drosophila* sex-linked recessive lethal assay. However, it is positive in the mouse micronucleus test and induces chromosomal aberrations in bone marrow cells of mice, rats, and rabbits. Benzene also induces chromosomal aberrations in human bone marrow cells and peripheral lymphocytes, persisting years after exposure ceases in some cases.[10]

In an updated review of benzene genotoxicity, it is noted that benzene has been reported to cause mutations in vitro as well as DNA damage in both human and rodent cells. Both positive and negative results have been reported in cell transformation assays, depending on the test system used.[12]

Developmental Toxicity: In most inhalation studies in laboratory animals, benzene has been shown to be fetotoxic at doses that are also maternally toxic. These studies have been extensively reviewed.[5,8-10]

Murray et al.[13] (cited in U.S. EPA[10]) exposed CF-1 mice (n = 36/group) and New Zealand white rabbits (n = 20/group) to 0 or 500 ppm benzene for 7 hours per day on days 6 through 15 and 6 through 18 of gestation, respectively. Maternal toxicity was not seen in either species and no treatment-related effects were seen in the rabbit offspring. However, a statistically significant increase in minor skeletal variants and decreased fetal body weight were seen in mice, indicating delayed development; there were no increases in major malformations.

Kuna and Kapp[14] exposed Sprague-Dawley rats to 0 (n = 17), 10 (n = 18), 50 (n = 20), or 500 (n = 20) ppm benzene via inhalation for 7 hours per day on days 6 through 15 of gestation. Hematologic parameters were unaffected in dams in all groups, but reduced weight gain was noted on days 6 through 15 in the 50- and 500-ppm dams. No developmental effects were seen in the control and 10-ppm offspring; mean fetal body weight was decreased and delayed ossification occurred in the 50- and 500-ppm groups. At the highest dose, mean crown-rump length was significantly reduced and several malformations with a very low incidence in historical controls occurred together. NOAELs of 10 ppm for both maternal and fetal toxicity in rats are suggested by this study.

Reproductive Toxicity: The available information on reproductive toxicity of benzene is very limited and has been reviewed by the ATSDR.[9] One noteworthy study among the several reviewed is that of Ward et al.[15] (cited in ATSDR[9]) in which male and female CD-1 mice were exposed to 1, 10, 30, or 300 ppm benzene vapor for 6 hours per day, 5 days a week for 13 weeks. At the highest exposure level, histological changes were observed in both the ovaries (bilateral cysts) and testes (atrophy/degeneration, decreased spermatozoa, increase in morphological abnormalities in sperm) and the lesions were considered more severe in the male mice. The general conclusions reached in the ATSDR review[9] were summarized as follows: "... evidence of an effect of benzene exposure on human reproduction is not sufficient to demonstrate a definite association. Exposure to benzene occurs along with exposure to many other chemicals, so no conclusion can be drawn relative to any single agent. There are insufficient animal data to propose NOAELs and LOAELs."

Systemic Toxicity: The major systemic effect associated with chronic benzene exposure in humans is depression of bone marrow resulting in pancytopenia, which is a decrease in numbers of erythrocytes, leukocytes, and thrombocytes, sometimes progressing to aplastic anemia. In early chronic benzene poisoning, numbers and morphology of only a single blood cell type may be altered and the effects are usually reversible if exposure is discontinued.[1,2,10] According to the HEAD,[10] the lower limit of exposure that will result in hematologic effects in humans is not well defined, but is probably less than 100 ppm; several apparently well-documented instances of blood dyscrasias (40–50 ppm) and even fatalities (60 ppm) at lower concentrations exist.

In addition to hematopoietic effects, there is also evidence that chronic benzene exposure may impair the immune system. This evidence comes from studies in which decreased serum complement, IgG, and IgA levels were found in workers exposed to, but not seriously intoxicated by, benzene; IgM levels were not decreased, but were slightly higher[16,17] (cited in IARC[8]).

Irritation: Benzene is reported to have a "strong, rather pleasant" odor[2] and an odor threshold of approximately 1.5 ppm (4.9 mg/m^3) has been reported.[18]

Basis for the AALG: Although benzene is widely distributed in the environment and somewhat soluble in water (720 to 820 mg/L at 22°C), it is volatile (75 mm Hg at 20°C) and the major source of human exposure is thought to be the respiratory route.[1,5,9] Therefore, 80% of the contribution to total exposure will be allowed from air.

Two AALGs may be calculated for benzene: one for carcinogenicity (leukemia) based on the epidemiologic studies cited in the IRIS summary[18] and another based on the developmental toxicity study of Kuna and Kapp.[14]

The AALG for carcinogenicity is based on human leukemias in occupationally exposed individuals using the analysis presented in the IRIS summary,[18] which gives an inhalation unit risk of 8.3×10^{-6} corresponding to a q_1* of 8.3×10^{-6} $(\mu g/m^3)^{-1}$. The extrapolation method used was the one-hit model with pooled data and the basis for the unit risk estimate is described as follows: "The unit risk estimate is the geometric mean of four ML (maximum likelihood) point estimates using pooled data from the Rinsky et al.[19] (cited in U.S. EPA[18]) and Ott et al.[20] (cited in U.S. EPA[18]) studies, which was then adjusted for the results of the Wong et al.[21] (cited in U.S. EPA[18]) study. The Rinsky data used were from an updated tape which reports one more case of leukemia than was published in 1981. Equal weight was given to cumulative dose and weighted cumulative dose exposure categories as well as to relative and absolute risk model forms. The results of the Wong et al. study were incorporated by assuming that the ratio of the Rinsky-Ott-Wong studies to the Rinsky-Ott studies for the relative risk cumulative dose model was the same as for other model-exposure category combinations and multiplying this ratio by the Rinsky-Ott geometric mean. The age-specific U.S. death rates for 1978 (the most current year available) were used for background leukemia and total death rate." Applying the relationship $E = q_1* \times d$ (where E is the specified risk level and d is the dose) and using an 80% relative source contribution from air gives the estimates shown below.

The developmental AALG is based on the NOAEL of 10 ppm for fetotoxicity in rats reported in the study by Kuna and Kapp.[14] The AALG is calculated based on a 225-g rat (average weight at the beginning of the study) and assuming a weight of 60 kg for the human. The NOAEL was adjusted for continuous exposure (10 ppm $\times$ 7/24) and converted to a human equivalent inhalation exposure level as described in Appendix C, and a total uncertainty factor of 100 (10 for interindividual variation $\times$ 10 for interspecies variation) was applied. Application of the developmental uncertainty factor was considered unwarranted, since relatively minor fetotoxicity occurred only at doses where maternal toxicity was also manifest. The recommended AALG is based on carcinogenicity, since it is more conservative.

AALG:
- 10^{-6}—3.02×10^{-2} ppb (9.6×10^{-2} $\mu g/m^3$) annual TWA
- 10^{-5}—3.02×10^{-1} ppb (9.6×10^{-1} $\mu g/m^3$) annual TWA
- developmental toxicity—0.05 ppm (0.16 mg/m^3) 24-hour TWA

References

1. ACGIH. 1986. "Documentation of the Threshold Limit Values and Biological Exposure Indices." 5:50–52.

2. NIOSH. 1974. "Criteria for a Recommended Standard . . . Occupational Exposure to Benzene." DHEW (NIOSH) 74–137.

3. OSHA. 1989. "Air Contaminants; Final Rule." *Fed. Reg.* 54:2332–959.

4. U.S. EPA. 1987. "National Primary Drinking Water Regulations; Public Notification; Final Rule." *Fed. Reg.* 52:41533–50.

5. U.S. EPA. 1980. "Ambient Water Quality Criteria for Benzene," EPA 440/5–80–018.

6. Safe Drinking Water Committee. 1982. *Drinking Water and Health,* Vol. 4 (Washington, DC: National Academy Press).

7. U.S. EPA. 1987. "Health Advisory for Benzene." Office of Drinking Water.

8. IARC. 1982. "Benzene." *IARC Monog.* 29:93–148.

9. ATSDR. 1987. "Toxicological Profile for Benzene" (draft). Published by Oak Ridge National Laboratory under DOE Interagency Agreement no. 1425–1425-A1 (contract no. DW89932147–01–3).

10. U.S. EPA. 1984. "Health Effects Assessment for Benzene." NTIS PB86–134483.

11. IARC. 1987. "Benzene." *IARC Monog.* suppl. 7:120–22.

12. IARC. 1986. "Benzene." *IARC Monog.* suppl. 6:91–95.

13. Murray, F. J., J. A. John, L. W. Rampy, R. A. Kuna, and B. A. Schwetz. 1979. "Embryotoxicity of Inhaled Benzene in Mice and Rabbits." *Am. Ind. Hyg. Assoc. J.* 40:993–98.

14. Kuna, R. A. and R. W. Kapp. 1981. "The Embryotoxic/Teratogenic Potential of Benzene Vapor in Rats." *Tox. App. Pharm.* 57:1–7.

15. Ward, C. O., R. A. Kuna, N. K. Snyder, R. D. Alsaker, W. B. Coate, and P. H. Craig. 1985. "Subchronic Inhalation Toxicity of Benzene in Rats and Mice." *Am J. Ind. Med.* 7:457–73.

16. Lange, A., R. Smolik, W. Zatonski, and J. Szymanska. 1973. "Serum Immunoglobulin Levels in Workers Exposed to Benzene, Toluene and Xylene." *Int. Arch. Arbeitsmed.* 31:37–44.

17. Smolik, R., K. Grzybek-Hryncewicz, A. Lange, and W. Zatonski. 1973. "Serum Complement Level in Workers Exposed to Benzene, Toluene and Xylene." *Int. Arch. Arbeitsmed.* 31:243–47.

18. U.S. EPA. 1988. "Benzene; CASRN 71–43–2." IRIS update (12/1/88).

19. Rinsky, R. A., R. J. Young, and A. B. Smith. 1981. "Leukemia in Benzene Workers." *Am. J. Ind. Med.* 2:217–45.

20. Ott, M. G., J. C. Townsend, W. A. Fishbeck, and R. A. Langner. 1978. "Mortality Among Individuals Occupationally Exposed to Benzene." *Arch. Environ. Health* 33:3–9.

21. Wong, O., R. W. Morgan, and M. D. Whorton. 1983. "Comments on the NIOSH Study of Leukemia in Benzene Workers." Technical report submitted to Gulf Canada, Ltd., by Environmental Health Associates.

BENZOYL PEROXIDE

Synonyms: benzoperoxide, benzoyl superoxide, dibenzoyl peroxide

CAS Registry Number: 94–36–0

Molecular Weight: 242.22

Molecular Formula:

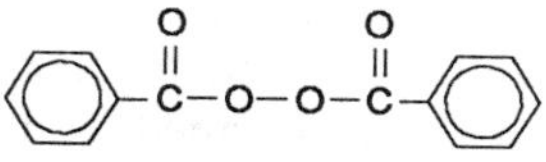

$$\text{C}_6\text{H}_5-\overset{\displaystyle O}{\overset{\|}{C}}-O-O-\overset{\displaystyle O}{\overset{\|}{C}}-\text{C}_6\text{H}_5$$

AALG: systemic toxicity—2.0 μg/m^3 (0.20 ppb) 24-hour TWA

Occupational Limits:
- ACGIH TLV: 5 mg/m^3 TWA
- NIOSH REL: 5 mg/m^3 TWA
- OSHA PEL: 5 mg/m^3 TWA

Basis for Occupational Limits: The occupational limits for benzoyl peroxide are based on the prevention of irritation.[1,2]

Drinking Water Limits: No drinking water limits were found for this substance.

Toxicity Profile

Carcinogenicity: The carcinogenicity of benzoyl peroxide has been evaluated in mice and rats exposed via the diet[3] (cited in IARC[4]) and by subcutaneous/intramuscular injection[3,5,6] (cited in IARC[4]). All of these studies provided no evidence for carcinogenic effects; however, IARC noted major limitations in the parenteral exposure studies. Benzoyl peroxide has also been evaluated in several mouse skin painting studies[3,7-10] (cited in IARC[4]). While three studies were inadequate to evaluate complete carcinogenic activity, two studies[9,10] (cited in IARC[4]) indicated that benzoyl peroxide acts as a promoter in the mouse skin model.

When benzoyl peroxide became recognized as a promoter, its use as an ingredient in over-the-counter (OTC) acne medications was evaluated by organizations concerned with OTC drugs in the United States and Canada (Non-Prescription Drug Manufacturers Association of Canada and Proprietary Association of the United States). It was concluded that while benzoyl peroxide is a promoter, there is no evidence of complete carcinogenic activity, and it does not pose an increased risk for patients using acne medications containing it. It is worthy of note that unlike phorbol esters, benzoyl peroxide does not act as a tumor promoter after initiation by ultraviolet light.[11]

Mutagenicity: Benzoyl peroxide was not mutagenic in *S. typhimurium* strains TA1537, TA1535, TA92, TA94, TA98, and TA100 with and without metabolic activation, nor did it induce polyploidy or chromosomal aberrations in Chinese hamster ovary cells. It was also negative in a mouse dominant lethal assay in which intraperitoneal doses of 54 and 62 mg/kg were used. However, benzoyl peroxide did induce a dose-dependent increase in sister chromatid exchanges in CHO cells with but not without the addition of S9 fraction. In addition, it inhibited intracellular communication between Chinese hamster V79 cells at noncytostatic concentrations (metabolic cooperation assay); it also inhibited intracellular communication between human epidermal keratinocytes as measured by H^3-uridine transfer.[3] It is noteworthy that inhibition of intracellular communication is consistent with a compound having promotor activity. In the Gene-Tox database as summarized in RTECS,[12] inconclusive results were reported in the *E. coli* polA$^+$/polA$^-$ assay without S9 fraction (a primary DNA damage assay).

Developmental Toxicity: No data were found implicating benzoyl peroxide as a developmental toxin.

Reproductive Toxicity: No data were found implicating benzoyl peroxide as a reproductive toxin.

Systemic Toxicity: Data on systemic toxicity of benzoyl peroxide in laboratory animals following inhalation exposure are lacking. It is stated in the TLV documentation[1] that "systemic toxicity in man has not been reported for benzoyl peroxide." However, based on case reports, benzoyl peroxide can induce sensitization in humans following dermal contact.[1,2,4]

Sharratt et al.[3] exposed male and female "albino" rats and mice (n = 25/sex/group/species, age at start of exposure not specified) to diets containing varying concentrations of an 18% benzoyl peroxide–containing powder (used in bleaching flour) for 80 weeks (mice) or 120 weeks (rats). IARC calculated the concentrations of benzoyl peroxide in the diet to be 28, 280, and 2800 mg/kg of diet for animals in the exposed groups.[4] For both mice and rats, overall tumor incidence did not differ significantly between treated and control animals. With respect to mice, no indications were given that weight gain was decreased or treatment-related lesions were observed. However, survival in all groups toward the end of the study was low. In rats in the medium and high exposure groups there was a slight but statistically significant decrease in body weight gain and in the highest exposure group there was an increased incidence of testicular atrophy. The authors attributed this latter finding to vitamin E deficiency based on the idea that benzoyl peroxide destroyed tocopherols in the diet; however, similar findings were not observed in mice. A NOAEL of 28 ppm benzoyl peroxide in the diet of rats was suggested by this study.

Irritation: Based on eye irritation tests in rabbits and dermal irritancy tests in rabbits and guinea pigs, benzoyl peroxide has been classified as a low-grade irritant.[2]

On the basis of an inspector's report in which two investigators reported eye, nose, and throat irritation in a plant where airborne concentrations of benzoyl peroxide were in the range of 1.34 to 82.5 mg/m³,[2] the TLV committee concluded that there were "no objectionable subjective symptoms" when exposures were in the range of 1.34 to 5.25 mg/m³ benzoyl peroxide dust, but that concentrations of 12.2 mg/m³ and higher caused "pronounced irritation."[1] NIOSH interpreted this report somewhat differently and considered that the lack of detail in the original report precluded verification of the TLV committee interpretation. It was further stated in the NIOSH criteria document[2] that "the inspection report did not specifically state that benzoyl peroxide was the cause of the discomfort or whether potassium aluminum sulfate or magnesium carbonate in the dust caused or contributed to the irritating effects. The methods of analysis were not described. Since there is no validated method of sampling and analysis for benzoyl peroxide, the method used to analyze the collected samples was probably not specific for this compound. The possible toxic effects of airborne benzoyl peroxide in humans cannot be accurately assessed because the report lacks essential data."

Basis for the AALG: Benzoyl peroxide is considered "very sparingly" soluble in water[4] and has a vapor pressure of less than 0.1 mm Hg at 20°C.[1] It is a highly reactive oxidizing agent and is used as a catalyst in plastics manufacturing, as an oxidizing agent in bleaching oils and flours,[1] and in OTC acne medications.[4] Due to its explosive properties it is usually handled diluted in an unreactive medium.[1] Based on these considerations, 20% of the contribution to total exposure will be allowed from air.

Since the numerical basis of the TLV is questionable and in view of the known promoter effects and systemic toxicity of benzoyl peroxide demonstrated in animal studies, the decision was made to base the AALG on the NOAEL of 28 ppm benzoyl peroxide in the diet of rats from the study of Sharratt et al.[3] The NOAEL of 28 ppm in the diet was converted to a human equivalent NOAEL for the inhalation route as described in Appendix C and a total uncertainty factor of 500 was applied (10 for interindividual variation × 10 for interspecies variation × 5 for the database factor). Application of the database factor was based on the use of a noninhalation study as the basis for the AALG and on uncertainty about the known promoter effects of benzoyl peroxide and their implications for inhalation exposure. Since this AALG was based on a noninhalation study, it should be considered provisional.

AALG: • systemic toxicity—2.0 μg/m³ (0.20 ppb) 24-hour TWA

References

1. ACGIH. 1986. "Documentation of the Threshold Limit Values." 5:54.
2. NIOSH. 1977. "Criteria for a Recommended Standard . . . Occupational Exposure to Benzoyl Peroxide." DHEW (NIOSH) 77–166.
3. Sharratt, M., A. C. Frazer, and O. C. Forbes. 1964. "Study of the Biological Effects of Benzoyl Peroxide." *Food Cosmet. Tox.* 2:527–38.
4. IARC. 1985. "Benzoyl Peroxide." *IARC Monog.* 36:267–83.
5. Poirier, L. A., J. A. Miller, E. C. Miller, and K. Sato. 1967. "N-benzoyloxy-N-methyl-4-aminoazobenzene: Its Carcinogenic Activity in the Rat and its Reactions with Proteins and Nucleic Acids and Their Constituents In Vitro." *Cancer Res.* 27:1600–13.
6. Hueper, W. C. 1964. "Cancer Induction by Polyurethane and Polysilicon Plastics." *J. Natl. Cancer Inst.* 33:1005–27.
7. Van Duuren, B. L., N. Nelson, L. Orris, E. D. Palmes, and F. L. Schmitt. 1963. "Carcinogenicity of Epoxides, Lactones and Peroxy Compounds." *J. Natl. Cancer Inst.* 31:41–45.
8. Saffiotti, U. and P. Shubik. 1963. "Studies on Promoting Action in Skin Carcinogenesis." *Natl. Cancer Inst. Monog.* 10:489–507.
9. Slaga, T. J., A. J. P. Klein-Szanto, L. L. Triplett, and L. P. Yotti. 1981. "Skin Tumor–Promoting Activity of Benzoyl Peroxide, a Widely Used Free Radical–Generating Compound." *Science* 213:1023–25.
10. Reiners, J. J., S. Nesnow, and T. J. Slaga. 1984. "Murine Susceptibility to Two-Stage Skin Carcinogenesis is Influenced by the Agent Used for Promotion." *Carcinogenesis* 5:301–7.
11. Zbinden, G. 1988. "Scientific Opinion on the Carcinogenic Risk due to Topical Administration of Benzoyl Peroxide for the Treatment of Acne Vulgaris." *Pharm. Toxicol.* 63:307–9.
12. NIOSH. 1988. DM8575000. "Benzoyl Peroxide." RTECS, on line.

BENZYL ALCOHOL

Synonyms: phenyl carbinol

CAS Registry Number: 100–51–6

Molecular Weight: 108.15

Molecular Formula: $\langle\bigcirc\rangle$—CH$_2$OH

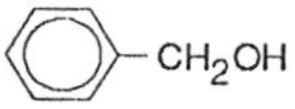

AALG: irritation—0.02 ppm (0.09 mg/m^3) 24-hour TWA

Occupational Limits:
- ACGIH TLV: none
- NIOSH REL: none
- OSHA PEL: none

Drinking Water Limits: No drinking water limits were found for this substance.

Toxicity Profile

Carcinogenicity: In an NTP bioassay, male and female B6C3F$_1$ mice and F344/N rats (n = 50/sex/group) were gavaged with 0, 100, or 200 mg/kg and 0, 200, or 400 mg/kg, respectively, of benzyl alcohol in corn oil 5 days a week for 103 weeks. Benzyl alcohol administration did not significantly affect body weight gain in either rats or mice, nor were there adverse effects on survival in mice or male rats. However, survival was reduced by 50% in female rats; this was considered to be due primarily to gavage-related deaths. It was reported that there were apparently no compound-related neoplastic or nonneoplastic lesions and it was concluded that under the conditions of this bioassay there was no evidence of carcinogenic activity in B6C3F$_1$ mice or F344/N rats.[1]

Mutagenicity: Benzyl alcohol is listed as negative in the *E. coli* polA$^+$/polA$^-$ assay without metabolic activation in the Gene-Tox database as summarized in RTECS.[2]

In conjunction with the NTP carcinogenicity bioassay, the genetic activity of benzyl alcohol was evaluated in several short-term tests. Using a preincubation protocol, benzyl alcohol did not induce mutations in *S. typhimurium* strains TA98, TA100, TA1535, or TA1537, with or without metabolic activation. It induced forward mutations in mouse L5178Y/TK$^{+/-}$ lymphoma cells without (but not with) metabolic activation and chromosomal aberrations in Chinese hamster ovary cells with (but not without) metabolic activation. In addition, benzyl alcohol caused an equivocal increase in sister chromatid exchanges with and without the addition of a metabolic activation system.[1]

Developmental Toxicity: No data were found implicating benzyl alcohol as a developmental toxin.

Reproductive Toxicity: No data were found implicating benzyl alcohol as a reproductive toxin.

Systemic Toxicity: Since benzyl alcohol is used primarily as a preservative or bacteriostatic agent in a number of medicinal preparations intended for parenteral use, most studies have centered on this route.

Kimura et al.[3] reported that 1 mL/kg of 0.9% benzyl alcohol injected intravenously had no effect on blood pressure, heart rate, respiration, ECG, or hematologic parameters in anesthetized and unanesthetized dogs and anesthetized monkeys. Intracarotid and intrarenal injections also produced no effects on these parameters as well as the EEG of anesthetized dogs. Rapid intravenous injection of 480 mg/kg was nonlethal in mice.

Based on reports of aseptic meningitis following intrathecal injection of radiopharmaceuticals containing benzyl alcohol as a preservative, DeLand[4] studied the effects of intrathecal injection of this compound in dogs. It was found that concentrations as high as 10 times those normally used did not produce meningitis in adult or immature dogs.

Opdike,[5] in summarizing the effects of benzyl alcohol, noted that vapors of benzyl alcohol can penetrate intact skin and that concentrations of approximately 100 ppm can produce systemic effects and deaths in laboratory animals. It is further noted that benzyl alcohol may demonstrate narcotic effects and cause death due to respiratory paralysis at "high" concentrations.

Irritation: Jones[6] (cited in Opdike[5]) suggested a tentative occupational exposure level of 1 ppm for workroom air and stated that exposures above 1 ppm would not be tolerated for prolonged periods due to lacrimatory effects.

Benzyl alcohol is reported to have a faint aromatic odor,[7] but no specific odor threshold was identified.

Basis for the AALG: Benzyl alcohol is soluble in water to the extent of 4% by weight and is used in the manufacture of other benzyl compounds, as a solvent, in perfuming and flavoring, and in microscopy as an embedding material.[7] Since the AALG for benzyl alcohol is based on irritant effects, no relative source contribution from air is factored into the calculation. The reason is that allocation of a proportion of exposure from a specific medium is not relevant when the effects of exposure are noncumulative, as is the case with irritation.

Based on available animal and human data, it appears that the critical effect for benzyl alcohol in humans is irritation. This conclusion is based on studies indicating low toxicity by the parenteral route in animals and the observation that relatively high concentrations are required to produce narcosis, together with the observation of lacrimatory effects at concentrations in excess of 1 ppm. In the absence of appropriate inhalation studies in humans or animals, the concentration of 1 ppm benzyl alcohol given by Jones[6] (cited in Opdike[5]) as the level at which lacrimatory effects might be expected is treated as a LOAEL in humans. A total uncertainty factor of 50 (10 for interindividual variation × 5 for a LOAEL) was applied. Given the limited information available on the effects of inhalation exposure to benzyl alcohol in humans and animal models, the AALG should be considered provisional.

AALG: • irritation—0.02 ppm (0.09 mg/m^3) 24-hour TWA

References

1. NTP. 1987. "National Toxicology Program Draft Report Abstracts for Four Chemical Carcinogenesis Bioassays." *Chemical Regulation Reporter* 11/13/87, pp. 1270–74.
2. NIOSH. 1987. DN3150000. "Benzyl Alcohol." RTECS, on line.
3. Kimura, E. T., T. D. Darby, R. A. Krause, and H. D. Brondyk. 1971. "Parenteral Toxicity Studies with Benzyl Alcohol." *Tox. Appl. Pharm.* 18:60–68.
4. DeLand, F. H. 1973. "Intrathecal Toxicity Studies with Benzyl Alcohol." *Tox. Appl. Pharm.* 25:153–56.
5. Opdike, D. L. J. 1973. "Monographs on Fragrance Raw Materials." *Fd. Cosmet. Tox.* 11:1011–81.
6. Jones, W. H. 1967. "Toxicity and Health Hazard Summary." Laboratory of Industrial Medicine, Eastman Kodak Company, Kodak Park, NC.
7. "Benzyl Alcohol." *The Merck Index.* 1138. p. 148.

BENZYL CHLORIDE

Synonyms: alpha-chlorotoluene, chloromethylbenzene

CAS Registry Number: 100–44–7

Molecular Weight: 126.59

Molecular Formula: ⬡–CH$_2$Cl

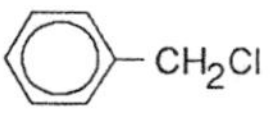

AALG: 10^{-6} 95% UCL—1.2×10^{-3} ppb (6.4×10^{-3} μg/m^3) annual TWA

Occupational Limits:
- ACGIH TLV: 1 ppm (5 mg/m^3) TWA
- NIOSH REL: 1 ppm (5 mg/m^3) 15-minute ceiling
- OSHA PEL: 1 ppm (5 mg/m^3) TWA

Basis for Occupational Limits: The occupational limits are all based on the prevention of irritant effects.[1,2] It is relevant to note that at the time the criteria document was published (1978), NIOSH concluded that the health risk from chronic low-level occupational exposure to benzyl chloride was probably negligible.[2]

Drinking Water Limits: No drinking water limits were found for this substance.

Toxicity Profile

Carcinogenicity: Benzyl chloride has been tested in skin application and intraperitoneal injection experiments in mice and subcutaneous injection experiments in rats[3-5] (cited in IARC[6]). Skin tumors were reported in some mice, but their incidence was not statistically significant.[4] In mice receiving benzyl chloride in tricaprylin by intraperitoneal injection (three groups with total doses of 0.6, 1.5, and 2 g/kg), lung tumors occurred in some animals, but again their incidence was not statistically different from controls receiving tricaprylin alone.[5] In rats receiving weekly subcutaneous injections of 40 or 80 mg/kg benzyl chloride in arachis oil for a year, local sarcomas with lung metastases were observed in the high-dose group, whereas the low-dose group developed sarcomas only at the injection site, with an average induction time of 500 days.[6]

In an NCI cancer bioassay,[7] male and female F344 rats and B6C3F$_1$ mice received benzyl chloride in corn oil by gavage three times per week for two years at doses of 15 and 30 mg/kg and 50 and 100 mg/kg, respectively. There were statistically significant increases in C-cell carcinomas of the thyroid gland in female

159

rats at the highest dose level and hepatocellular carcinomas in male mice at the lowest (but not highest) dose level. In mice of both sexes there was a statistically significant increase in carcinomas and papillomas of the forestomach at the highest dose. No increase in forestomach tumors was observed in rats in this study, but the authors note that the true maximally tolerated dose may not have been achieved in rats and also concluded that "the experiment in rats was inadequate to properly evaluate the carcinogenicity of benzyl chloride in that species."

It is of interest to note that excess respiratory cancers have been reported in benzoyl chloride manufacturing workers who were also potentially exposed to benzyl chloride.[6]

It is our judgement that benzyl chloride is most appropriately placed in group B2, probable human carcinogen, under the EPA weight-of-evidence classification.

Mutagenicity: Benzyl chloride is a direct-acting bacterial mutagen and alkylating agent[6] and is listed as positive in the EPA Gene-Tox database as summarized in RTECS in the SHE clonal assay, *B. subtilis* rec assay, *E. coli* polA$^+$/polA$^-$ assay with and without metabolic activation, Ames assay, and induction of unscheduled DNA synthesis in human fibroblasts.[8] It is also reported to increase mitotic recombination in *S. cerevisiae* and oncogenic transformation in Syrian hamster embryo cells.[6] The genotoxic activity of benzyl chloride was recently reviewed by IARC[9] and it is our judgment that there is sufficient evidence of genotoxic activity in short-term tests based on IARC criteria.

Developmental Toxicity: In a Russian study[10] (cited in IARC[6]), it was reported that increases in embryolethality occurred in Wistar rats exposed to 208 and 0.006 mg/kg (but not 0.0006 or 0.00006 mg/kg) benzyl chloride orally on days 1 to 19 of gestation. With respect to postnatal effects, the offspring of dams receiving 208 mg/kg benzyl chloride showed delayed development, decreased resistance to anoxia, and impaired fertility, while no adverse postnatal effects were seen in offspring of dams receiving 0.00006 mg/kg benzyl chloride. Although detailed reporting of this study was lacking, the wide dose range used is noteworthy.

Reproductive Toxicity: No data were found implicating benzyl chloride as a reproductive toxin.

Systemic Toxicity: For 2-hour exposures, the LC_{16} and LC_{50} values for mice and rats have been reported as 230 and 390 mg/m^3 and 440 and 740 mg/m^3, respectively.[6]

In the subchronic study conducted adjunct to the NCI cancer bioassay using F344 rats and B6C3F$_1$ mice, lethality at the higher dose levels was mainly due to severe acute and chronic gastritis of the forestomach.[7]

Irritation: Benzyl chloride is reported to be a potent lacrimator and to be irritating to the eyes, nose, and throat.[2] Exposure to 6–8 mg/m^3 for 5 minutes can cause conjunctivitis, and the eye and nasal irritation thresholds have been reported as 41 mg/m^3 for 10 seconds and 180 mg/m^3 for a single breath, respectively.[2]

Odor thresholds of 0.21 mg/m^3 and 0.24 mg/m^3 have been reported for benzyl chloride. The odor has been described as pungent[1] and "benzene-like."[2]

Basis for the AALG: Benzyl chloride has a vapor pressure of 1.0 mm Hg at 22°C and is considered insoluble in water.[1] It is used as a chemical intermediate in the manufacture of a variety of dyes, plasticizers, lubricants, gasoline additives, pharmaceuticals, tanning agents, and quaternary ammonium compounds.[1,6] Benzyl chloride has been detected in surface waters. Based on lack of information on its environmental fate and distribution and the expected contribution of various media to human exposure, 50% of the contribution to total exposure will be allowed from air.

The AALG for benzyl chloride is based on carcinogenicity for the following reasons: (1) there was a statistically significant increase in tumors in the NCI bioassay at sites in addition to the portal of entry, and (2) there is sufficient evidence of genotoxic activity for benzyl chloride in short-term tests. The linearized multistage model was applied to three data sets from the NCI bioassay: (1) combined forestomach carcinomas and papillomas in female mice: control, 0 of 52; low-dose, 5 of 50; and high-dose, 19 of 51; (2) combined forestomach carcinomas and papillomas in male mice: control, 0 of 51; low-dose, 4 of 52; and high-dose, 32 of 52; and (3) forestomach carcinomas only in male mice: control, 0 of 51; low-dose, 2 of 52; and high-dose, 8 of 52. Due to poor fit of the model to the data for combined forestomach carcinomas and papillomas in male mice and the fact that this data set was a partial duplication of the data for forestomach carcinomas only in male mice, this data set was dropped from consideration and the geometric mean of the q_1*s for the first (2.85 $\times$ 10^{-2} (mg/kg/day)$^{-1}$) and third (1.5 $\times$ 10^{-2} (mg/kg/day)$^{-1}$) data sets described above were used. The animal q_1* of 2.07 $\times$ 10^{-2} (mg/kg/day)$^{-1}$ was converted to a human q_1* of 2.75 $\times$ 10^{-1} (mg/kg/day)$^{-1}$, which yields an inhalation unit risk of 7.9 $\times$ 10^{-5} (or inhalation q_1* of 7.9 $\times$ 10^{-5} (μg/m^3)$^{-1}$), assuming 100% absorption and a 70-kg human breathing 20 m^3 of air per day. Applying the relationship E = q_1* $\times$ d (where E is the specified level of extra risk and d is the dose) and allowing 50% of the contribution to total exposure from air yields the figures shown below.

This AALG should be considered provisional, since it is based on non-inhalation data, and a high degree of uncertainty is associated with this type of extrapolation.

AALG:
- 10^{-6} 95% UCL—1.2 $\times$ 10^{-3} ppb (6.4 $\times$ 10^{-3} μg/m^3) annual TWA
- 10^{-5} 95% UCL—1.2 $\times$ 10^{-2} ppb (6.4 $\times$ 10^{-2} μg/m^3) annual TWA

References

1. ACGIH. 1986. "Documentation of the Threshold Limit Values and Biological Exposure Indices." 5:55.

2. NIOSH. 1978. "Criteria for a Recommended Standard . . . Occupational Exposure to Benzyl Chloride." DHEW (NIOSH) 78–182.
3. Fukuda, K., H. Matsushita, H. Sakabe, and K. Takemoto. 1981. "Carcinogenicity of Benzyl Chloride, Benzal Chloride, Benzotrichloride and Benzoyl Chloride in Mice by Skin Application." *Gann* 72:655–64.
4. Poirier, L. A., G. D. Stoner, and M. B. Shimkin. 1975. "Bioassay of Alkyl Halides and Nucleotide Base Analogs by Pulmonary Tumor Response in Strain A Mice." *Cancer Res.* 35:1411–15.
5. Druckrey, H., H. Kruse, R. Preussmann, S. Ivankovic, and C. Landschutz. 1970. "Cancerogenic Alkylating Substances. III. Alkylhalogenides, Sulphonates and Strained Heterocyclic Compounds" (German). *Z. Krebsforsch.* 74:241–73.
6. IARC. 1982. "Benzyl Chloride." *IARC Monog.* 29:49–63.
7. Lijinsky, W. 1986. "Chronic Bioassay of Benzyl Chloride in F344 Rats and (C57BL/6J × BALB/c)F₁ Mice." *J. Natl. Cancer Inst.* 76:1231–36.
8. NIOSH. 1988. XS8925000. "Toluene, Alpha-chloro." RTECS, on line.
9. IARC. 1986. "Benzyl Chloride." *IARC Monog.* suppl. 6:105–09.
10. Leonskaya, G. I. 1980. "Evaluation of the Embryotoxic and Teratogenic Effect of Butyl and Benzyl Chlorides to Establish Hygienic Standards for Reservoir Water" (Russian). *Gig. Naselen. Mest.* 19:40–43.

BIS(2-CHLOROETHYL) ETHER

Synonyms: bis(β-chloroethyl) ether, dichloroethyl ether, BCEE, 2,2'-dichloroethyl ether

CAS Registry Number: 111–44–4

Molecular Weight: 143.02

Molecular Formula:

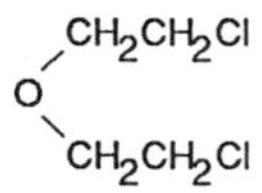

$$O \begin{array}{l} \diagup CH_2CH_2Cl \\ \diagdown CH_2CH_2Cl \end{array}$$

AALG: 10^{-6} 95% UCL—2.6 × 10^{-4} ppb (1.5 × 10^{-3} μg/m^3) annual TWA

Occupational Limits:
- ACGIH TLV: 5 ppm (30 mg/m^3) TWA; 10 ppm (60 mg/m^3) STEL; skin notation
- NIOSH REL: none
- OSHA PEL: 5 ppm (30 mg/m^3) TWA; 10 ppm (60 mg/m^3) STEL; skin notation

Basis for Occupational Limits: The TLVs were set to "prevent eye and upper respiratory tract irritation, as well as lung injury."[1] The numerical basis for the TLV appears to be the fact that based on acute toxicity tests, Carpenter et al.[2] (cited in ACGIH[1]) considered BCEE to have an acute toxicity "comparable to that of substances with TLVs of 5 ppm or less." OSHA recently adopted the TLVs as part of its final rule limits.[3]

Drinking Water Limits: The ambient water quality criterion for BCEE at the 10^{-6} risk level is 0.030 μg/L, assuming a person consuming 2 L of water and 6.5 g fish/shellfish per day.[4]

Toxicity Profile

Carcinogenicity: BCEE has been reported to produce hepatomas in mice following oral administration and to induce sarcomas at the injection site when administered subcutaneously.[5] Innes et al.[6] administered BCEE to male and female (C57BL/6 × C3H/Anf)F$_1$ and (C57BL/6 × AKR)F$_1$ mice, in water by intubation daily from 7 to 28 days of age and in the diet thereafter for 80 weeks. The single dose level used—100 mg/kg (300 ppm)—had been determined to be the maximally tolerated dose in previous studies. Excessive numbers of hepatomas were found in male mice

163

(14 of 16 and 9 of 17 in the (C57BL/6 × C3H/Anf)F_1 and (C57BL/6 × AKR)F_1 strains, respectively). The data of Innes et al.[6] were used in the calculation of the quantitative risk estimates presented in both the ambient water quality criteria document[3] and in the IRIS summary.[7]

Norpoth et al.[8] subcutaneously injected male and female Sprague-Dawley rats (n = 50/sex/group) with 4.36 or 13.1 μmol of BCEE in 0.25 mL DMSO weekly for two years. No statistically significant increases in tumors were found in treated animals compared to vehicle or untreated controls. Similarly, Ulland et al.[9] found no statistically significant increases in tumors in CD rats exposed to BCEE (ingestion) at the MTD for 18 months and then observed for an additional 6 months.

BCEE has been placed in group B2, probable human carcinogen, in the EPA weight-of-evidence classification based on positive results in two strains of mice and evidence of mutagenicity.[7]

Mutagenicity: BCEE has been reported to produce basepair substitution mutations in *E. coli*, *S. typhimurium*, and *B. subtilis* without metabolic activation. It has also been reported to be weakly positive in inducing frameshift mutations in some strains of *S. typhimurium* exposed to the vapor phase. BCEE was also found to induce mutations in *S. cerevisiae* but was negative in a heritable translocation test in mice.[7]

Developmental Toxicity: The Reproductive Toxicology Center indicated that no reports of developmental toxicity for BCEE have been identified.[10]

Reproductive Toxicity: No data were found implicating BCEE as a reproductive toxin.

Systemic Toxicity: LC_{50} values of 330 mg/m^3 in 4 hours and 650 mg/m^3 in 2 hours have been reported for rats and mice, respectively.[11]

In a study[12] (cited in ACGIH[1]) summarized in the TLV documentation, it was reported that exposure of rats and guinea pigs to 69 ppm BCEE for 7 hours per day, 5 days a week for 130 days (93 exposures) resulted in growth depression but no "serious injury" or microscopic lesions.

Irritation: BCEE is irritating to the eyes, skin, and respiratory tract of both humans and animal models.[1,3]

Schrenk et al.[13] (cited in ACGIH[1]) reported the following findings in human volunteers exposed to various concentrations of BCEE: (1) above 500 ppm—intolerable irritation of the eyes and nasal passages with coughing, nausea, and retching; (2) 260 to 100 ppm—irritation, but not intolerable; and (3) 35 ppm—no irritation, but "nauseating" odor still detectable.

The odor of BCEE has been described as "nauseating" and is reported to be similar to ethylene dichloride.[1] Amoore and Hautala[14] reported an odor threshold of 0.049 ppm for BCEE (geometric mean of reported literature values).

Basis for the AALG: BCEE has a vapor pressure of 0.4 mm Hg at 20°C and is considered insoluble in water.[1,5] It has been found in drinking water in New Orleans, Louisiana. It has been postulated that BCEE may be formed from hydrocarbons present in water and chlorine added for disinfection purposes.[5] BCEE is used as a solvent for certain lacquers, resins, and oils; as a dewaxing agent for lubricating oils; as an intermediate in chemical syntheses; as a wetting agent and penetrant; and as an insecticidal soil fumigant.[1] Based on lack of information on its environmental fate and distribution and the expected contribution of various media to human exposure, 50% of the contribution to total exposure will be allowed from air.

BCEE is somewhat unusual in that it is treated as a carcinogen by EPA[7] but is not listed as such by ACGIH.[1] There is not a strong basis for calculation of the AALG because (1) the TLV is based on irritation, but the reasons for selecting this particular exposure level are not clear from the documentation, and (2) the only available study for calculation of quantitative risk estimates did not use the inhalation route and had small numbers of animals, a single dose level, and a low IRIS confidence rating for the quantitative risk estimates. Based on these considerations, the AALGs calculated must be considered tentative, and periodic review of the literature for a more appropriate basis for the AALG is of greater than usual importance.

Data from the study by Innes et al.[6] provide the basis for the quantitative risk estimates presented here. These estimates are based on the incidence of hepatomas in male mice. The adjusted lifetime doses were 0 and 39 mg/kg/day; the corresponding incidences of hepatoma were 8 of 79 and 14 of 16. Based on the application of the linearized multistage model to these data, a q_1* (or carcinogenic potency factor) for humans of 1.14 $(mg/kg/day)^{-1}$ was calculated. In the IRIS summary an inhalation equivalent q_1* of 3.3×10^{-4} $(\mu g/m^3)^{-1}$ was calculated for BCEE based on this same ingestion data. Applying the relationship $E = q_1^* \times d$ (where E is the specified level of extra risk and d is the dose) and allowing a 50% relative source contribution from air gives the estimates shown below. The AALG should be considered provisional based on the limitations of the original study and the fact that ingestion data were used to derive an inhalation criterion.

AALG: • 10^{-6} 95% UCL—2.6×10^{-4} ppb (1.5×10^{-3} $\mu g/m^3$) annual TWA
   • 10^{-5} 95% UCL—2.6×10^{-3} ppb (1.5×10^{-2} $\mu g/m^3$) annual TWA

References

1. ACGIH. 1986. "Documentation of the Threshold Limit Values and Biological Exposure Indices." 5:186.
2. Carpenter, C. P., H. F. Smyth, and U. C. Pozzani. 1949. *J. Ind. Hyg. Tox.* 31:343.
3. OSHA. 1989. "Air Contaminants; Final Rule." *Fed. Reg.* 54:2332–959.
4. U.S. EPA. 1980. "Ambient Water Quality Criteria for Chloroalkyl Ethers." EPA 440/5–80–030.
5. IARC. 1975. "Bis(2-chloroethyl) Ether." *IARC Monog.* 9:117–23.

6. Innes, J. R. M., B. M. Ulland, M. G. Valerio, L. Petrucelli, L. Fishbein, E. R. Hart, A. J. Pallotta, R. R. Bates, H. L. Falk, J. J. Gart, M. Klein, I. Mitchell, and J. Peters. 1969. "Bioassay of Pesticides and Industrial Chemicals for Tumorigenicity in Mice: A Preliminary Note." *J. Natl. Cancer Inst.* 42:1101–14.

7. U.S. EPA. 1987. "Bis(chloroethyl) Ether; CASRN 111–44–4." IRIS summary.

8. Norpoth, K., M. Heger, G. Muller, E. Mohtashamipar, A. Kemena, and C. Witting. 1986. "Investigations on Metabolism, Genotoxic Effects and Carcinogenicity of 2,2'-Dichloroethylether." *J. Cancer Res. Clin. Oncol.* 112:125–30.

9. Ulland, B., E. K. Weissburger, and J. H. Weissburger. 1973. "Chronic Toxicity and Carcinogenicity of Industrial Chemicals and Pesticides." *Tox. Appl. Pharm.* 25:446 (abstract).

10. Reproductive Toxicology Center. 7/5/87. REPROTOX Substance Request: Dichloroethyl Ether (Washington, DC: Columbia Hospital for Women Medical Center). Personal communication.

11. NIOSH. 1988. KN0875000. "Ether, Bis(2-chloroethyl)." RTECS, on line.

12. Kosyan, S. A. 1967. *Tr. Erevan. Gos. Inst. Usoversh. Vrachel.* 3:617. In: *Chem. Abstr.* 71:37213p, 1969.

13. Schrenk, H. H., F. A. Patty, and W. P. Yant. 1933. "Acute Response of Guinea Pigs to Vapors of Some New Commercial Organic Compounds." *Publ. Health Rep.* 48:1389–97.

14. Amoore, J. E. and E. Hautala. 1983. "Odor as an Aid to Chemical Safety: Odor Thresholds Compared with Threshold Limit Values and Volatilities for 214 Industrial Chemicals in Air and Water Dilution." *J. Appl. Tox.* 3:272–90.

BIS(CHLOROMETHYL) ETHER

Synonyms: bisCME, BCME

CAS Registry Number: 542–88–1

Molecular Weight: 114.96

Molecular Formula: $ClCH_2OCH_2Cl$

AALG: 10^{-6} 95% UCL—5.83×10^{-8} ppb (2.74×10^{-7} $\mu g/m^3$) annual TWA

Occupational Limits:
- ACGIH TLV: 1 ppb (5 $\mu g/m^3$) TWA; Appendix A1 recognized human carcinogen
- NIOSH REL: none
- OSHA PEL: carcinogen; no numerical criterion

Basis for Occupational Limits: The TLV has been set to provide a "satisfactory margin of safety from the development of nasal tumors, at least in the nonsmoking worker."[1] OSHA has not promulgated a numerical criterion for BCME. However, it is considered by OSHA to be a carcinogen and regulations pertaining to workplace practices are given in 29 *Code of Federal Regulations* (CFR) 1910.1008.[2]

Drinking Water Limits: The ambient water quality criterion for BCME is 3.76×10^{-6} $\mu g/L$ at the 10^{-6} risk level based on the consumption of 2 L of water and 6.5 g of fish/shellfish per day.[3]

Toxicity Profile

Carcinogenicity: BCME is an alkylating agent and a potent cutaneous carcinogen in mice. It has also been demonstrated to be carcinogenic to adult mice and rats by inhalation and subcutaneous injection and to newborn mice following a single subcutaneous injection.[4]

Thiess et al.[5] (cited in IARC[4]) conducted a retrospective study of a small group of laboratory workers exposed to BCME for 6–9 years. Six cases of lung cancer, primarily oat-celled carcinomas, were found among 18 men (5 of the 6 cases occurred in smokers) with a latency period ranging from 8 to 16 years. BCME may occur as an impurity in chloromethyl methyl ether (CMME); in a study of chemical plant workers exposed to CMME, 14 cases of lung cancer (primarily oat-celled carcinoma) were found[6] (cited in IARC[4]).

Leong et al.[7] conducted a 6-month inhalation study (exposure period 6 hours/

day, 5 days/week—lifetime observation) using male Sprague-Dawley–derived rats (120 per group) and male Ha/ICR mice (144–157 per group) exposed to 0, 1, 10, or 100 ppb for the purpose of establishing an animal NOEL. No significant tumorigenic response was observed in mice or rats exposed to 1 or 10 ppb. Rats exposed to 100 ppb had a significant increase in nasal tumors (esthesioneuroepithelioma—96 of 111) whereas mice exposed to this same concentration had a significant increase in pulmonary adenomas. It is relevant to note that at the end of the 6-month exposure period no abnormalities were observed in hematologic parameters, exfoliative cytology of lung washes, or cytogenetic characteristics of bone marrow cells in exposed rats.

In another study, male Sprague-Dawley rats were exposed to 100 ppb BCME for 10, 20, 40, 60, 80, or 100 6-hour periods. Dose-related esthesioneuroepitheliomas and squamous cell carcinomas of the lung occurred.[8]

BCME is most appropriately placed in group A, known human carcinogen, under the EPA weight-of-evidence classification.

Mutagenicity: BCME has been reported to induce mutations in *S. typhimurium* (Ames assay) with metabolic activation, unscheduled DNA synthesis in human fibroblasts and mouse skin cells, inhibition of DNA synthesis in mouse skin, and oncogenic transformation in hamster kidney cells.[9,10] In addition, it was reported in a recent IARC review that a slight increase in chromosomal aberrations was found in peripheral lymphocytes of workers exposed to BCME in the manufacture of ion exchange resins.[10]

Developmental Toxicity: No data were found implicating BCME as a developmental toxin.

Reproductive Toxicity: No data were found implicating BCME as a reproductive toxin.

Systemic Toxicity: BCME LC_{50}s for mice, rats, and hamsters are reported to be between 5 and 7 ppm for 6- or 7-hour exposures.[1]

Chronic occupational exposure to CMME contaminated with BCME has been reported to cause bronchitis in workers.[3]

Irritation: BCME is reported to have a "suffocating" odor,[1] but no specific odor threshold was found. Irritant properties were not mentioned in association with BCME in the references consulted.

Basis for the AALG: BCME is produced inadvertently in the manufacture of CMME and is used as a monitoring indicator for CMME, since it is more stable in air. BCME is no longer used in industrial chloromethylation processes but is used as an intermediate in certain anion exchange resins.[1] Since BCME decomposes in the presence of water to form hydrochloric acid and formaldehyde, has a high vapor pressure, and is stable for several hours in moist air,[1,4] 80% of the contribution to total exposure is allowed from air.

Based on positive results in animal studies and evidence in humans linking BCME exposure to oat-celled carcinoma of the lung, the AALG is based on carcinogenicity. Although the study by Leong et al.[7] is stronger, it was judged to be better to use the Kuschner et al. study[8] due to problems with the fit of the model with the Leong et al. data. It should be noted that the quantitative estimates presented here are the same as those given in the ambient water quality criteria document.

The lifetime adjusted doses (expressed as mg $\times$ 10^{-4}/kg/day) and the corresponding incidence of respiratory cancers in male rats are as follows: 0.0, 0 of 240; 0.35, 1 of 41; 0.70, 3 of 46; 1.4, 4 of 18; 2.1, 4 of 18; 2.8, 15 of 34; and 3.5, 12 of 20. The estimates of q_1 and q_1^* for humans are 0.41 $\times$ 10^{-4} $(mg/kg/day)^{-1}$ and 1.05 $\times$ 10^{-4} $(mg/kg/day)^{-1}$, respectively. The maximum likelihood estimates for the dose and the upper 95% confidence limit of the dose at the 10^{-6} risk level are 2.44 x 10^{-10} mg/kg/day and 3.43 $\times$ 10^{-10} mg/kg/day, respectively. Converting back to the inhalation dose and allowing 80% of the contribution to total exposure from air gives the figures shown below.

AALG:
- 10^{-6} 95% UCL—5.83 $\times$ 10^{-8} ppb (2.74 $\times$ 10^{-7} $\mu g/m^3$) annual TWA
- 10^{-5} 95% UCL—5.83 $\times$ 10^{-7} ppb (2.74 $\times$ 10^{-6} $\mu g/m^3$) annual TWA

References

1. ACGIH. 1986. "Documentation of the Threshold Limit Values and Biological Exposure Indices." 5:131.
2. OSHA. 1988. 29 *Code of Federal Regulations* 1910.1008.
3. U.S. EPA. 1980. "Ambient Water Quality Criteria for Chloroalkyl Ethers," EPA 440/5–80–030.
4. IARC. 1974. "Bis(chloromethyl) Ether." *IARC Monog.* 4:231–58.
5. Thiess, A. M., W. Hey, and H. Zeller. 1973. "Zur Toxikologie von Dichlordimethyl-ather—Verdacht auf Kanzerogene Wirkung auch beim Menschen." *Zbl. Arbeitsmed.* 23:97.
6. Figueroa, W. G., R. Raszkowski, and W. Weiss. 1973. "Lung Cancer in Chloromethyl Methyl Ether Workers." *New Engl. J. Med.* 288:1096.
7. Leong, B. K. J., R. J. Kociba, and G. C. Jersey. 1981. "A Lifetime Study of Rats and Mice Exposed to Vapors of Bis(chloromethyl) Ether." *Tox. Appl. Pharm.* 58:269–81.
8. Kuschner, M., S. Laskin, R. T. Drew, V. Cappiello, and N. Nelson. 1975. "Inhalation Carcinogenicity of Alpha Halo Ethers. III. Lifetime and Limited Period Inhalation Studies with Bis(chloromethyl) Ether at 0.1 ppm." *Arch. Env. Health* 30:73–77.
9. NIOSH. 1987. KN1575000. "Ether, Bis(chloromethyl)." RTECS, on line.
10. IARC. 1987. "Bis(chloromethyl) Ether and Chloromethyl Methyl Ether." *IARC Monog.* suppl. 7:131–33.

BORON TRIFLUORIDE

Synonyms: trifluoroborane

CAS Registry Number: 7637–07–2

Molecular Weight: 67.81

Molecular Formula: BF_3

AALG: irritation (TLV/50)—0.02 ppm (0.06 mg/m^3) ceiling

Occupational Limits:
- ACGIH TLV: 1 ppm (3 mg/m^3) ceiling
- NIOSH REL: none recommended
- OSHA PEL: 1 ppm (3 mg/m^3) ceiling

Basis for Occupational Limits: The ACGIH limit is based on the prevention of pulmonary irritation.[1] NIOSH recommended no limit due to the absence of a reliable monitoring method and relevant human data.[2]

Drinking Water Limits: No drinking water limits were found for this substance.

Toxicity Profile

Carcinogenicity: No data were found implicating boron trifluoride as a carcinogen.

Mutagenicity: No data were found implicating boron trifluoride as a mutagen.

Developmental Toxicity: No data were found implicating boron trifluoride as a developmental toxin.

Reproductive Toxicity: No data were found implicating boron trifluoride a reproductive toxin.

Systemic Toxicity: Torkelson et al.[3] studied the effects of inhalation of boron trifluoride on guinea pigs, rats, and rabbits exposed for 7 hours per day, 4–5 days per week, for periods of 2 to 6 months at concentrations of 12.8 ppm (nominal) and 1.5 and 3–4 ppm (analyzed). At 1.5 ppm, no differences from controls were noted in any of the species except for rats, which had a slightly higher incidence of pneumonitis with cellular infiltration and congestion. At exposures of 3–4 ppm for 4 months, no significant structural changes were noted in rats, but guinea pigs developed subacute tracheitis and bronchitis, which had disappeared 30 days after the last test dose. At a concentration of 12.9 ppm for 2 months, rats and guinea pigs developed a pneumonitis suggestive of chemical injury in the hilar and alveolar areas of the lung. Dental fluorosis was observed in some animals and was attributed to excessive licking of the fur. The authors concluded that the primary hazard of overexposure to BF_3 is irritation of the respiratory tract.

Irritation: Boron trifluoride is considered to be a severe pulmonary irritant in both animals and humans[1] and repeated exposure may lead to the development of chemical pneumonia in animal models.[1,3]

Boron trifluoride has been reported to have a "pungent, suffocating" odor,[1] but no specific odor threshold was mentioned in the literature reviewed.

Basis for the AALG: Boron trifluoride is a colorless nonflammable gas that hydrolyzes in moist air (e.g., in the environment of the respiratory tract) to form boric acid, hydrogen fluoride, fluoroboric acid, and related chemical species. Boron trifluoride is used as a catalyst in organic syntheses, in soldering fluxes, and for neutron measurement.[1] Since the AALG is based on irritation, no relative source contribution was allocated from air. The reason is that the contribution of various media to total exposure is not relevant when the effects of exposure are noncumulative, as is the case with irritation.

Based on available data it appears that the critical effect for inhalation exposure to boron trifluoride is irritation. The decision was made to treat the TLV ceiling limit as a human LOAEL, based on marginal evidence of pneumonitis in rats exposed to 1.5 ppm BF_3 in the Torkelson et al. study[3] and the statement in the TLV documentation[1] that "the original basis for the threshold limit value of 1 ppm appears to be in some question, providing little margin of safety." Therefore, the AALG is based on the TLV/50 (10 for interindividual variation × 5 for a LOAEL).

It should be noted that the potential for inhalation exposure of the general population to boron trifluoride is probably quite low due to its rapid expected degradation in ambient air.

AALG: • irritation (TLV/50)—0.02 ppm (0.06 mg/m^3) ceiling

References

1. ACGIH. 1986. "Documentation of the Threshold Limit Values and Biological Exposure Indices." 5:63.
2. NIOSH. 1976. "Criteria for a Recommended Standard . . . Occupational Exposure to Boron Trifluoride." DHEW (NIOSH) 77–122.
3. Torkelson, T. R., S. E. Sadek, and V. K. Rowe. 1961. "The Toxicity of Boron Trifluoride when Inhaled by Laboratory Animals." *Am. Ind. Hyg. Assoc. J.* 22:263–70.

BROMINE

Definition: This assessment applies only to the diatomic form of bromine and not to inorganic or organic bromides.

Molecular Weight: 159.82

CAS Registry Number: 7726–95–6

Molecular Formula: Br_2

AALG: irritation (TLV/10)—0.01 ppm (0.07 mg/m^3) 8-hour TWA

Occupational Limits:
- ACGIH TLV: 0.1 ppm (0.7 mg/m^3) TWA; 0.3 ppm (2.0 mg/m^3) STEL
- NIOSH REL: none
- OSHA PEL: 0.1 ppm (0.7 mg/m^3) TWA; 0.3 ppm (2.0 mg/m^3) STEL

Basis for Occupational Limits: The ACGIH limits are set to prevent irritation of the respiratory tract and "injury" to the lungs.[1] OSHA recently adopted the ACGIH TLV STEL as part of their final rule limits.[2]

Drinking Water Limits: No drinking water limits were found for bromine (Br_2), but a chronic SNARL of 2.3 mg/L was calculated for bromide (Br⁻). It is relevant to note that bromine has been used as a disinfectant for swimming pool waters, but its use for this purpose in drinking water is not recommended due to its cumulative neurotoxicity.[3]

Toxicity Profile

Carcinogenicity: No data were found implicating bromine as a carcinogen.

Mutagenicity: No data were found implicating bromine as a mutagen.

Developmental Toxicity: No data were found implicating bromine as a developmental toxin.

Reproductive Toxicity: No data were found implicating bromine as a reproductive toxin.

Systemic Toxicity: An LC_{50} of 750 ppm (9 minutes) has been reported for bromine in mice.[4]

Various inorganic bromides were formerly used in medicine for their sedative effects. At toxic doses neurological signs ranging from neuroses and psychoses through severe ataxia have been reported.[3] No reports of similar effects in workers exposed to bromine were noted in the TLV documentation.[1]

The toxicology of the bromide ion was recently the subject of an in-depth review by van Leeuwen and Sangster.[5]

Irritation: Bromine vapors are irritating to the lungs and Henderson and Haggard[6] (cited in ACGIH[1]) have reported the following conclusions on the effects of inhalation exposure to bromine in humans: (1) maximum concentrations allowable for prolonged (time period not specified) and short (30 minutes–1 hour) exposure are 0.1 to 0.15 and 4 ppm, respectively; (2) dangerous concentration for short-term exposure is 40 to 60 ppm; (3) fatal short-term exposure level is 1000 ppm; and (4) bromine acts as a respiratory irritant leading to lung edema.

An odor threshold of 0.05 ppm (geometric mean of several reported literature values) has been reported for bromine.[7]

Basis for the AALG: Bromine is considered to be very slightly soluble in water and has a vapor pressure of 175 mm Hg at 20°C. It is used in the synthesis of a variety of bromine-containing compounds as well as in bleaching, water purification, and a variety of other applications.[1] Van Leeuwen and Sangster[5] considered that the diet was the major source of bromine exposure in humans and that concentrations in ambient air are due to ocean-derived aerosols and volcanic activity, with some contributions from automobile exhaust and use of bromine-containing compounds. They concluded: "Although the occurrence of bromine-containing compounds in the atmosphere is not of toxicologic concern, it poses an indirect threat to human health since bromine, like chlorine, can reduce ozone, and therefore is a risk to the protective ozone layer of the earth." Since the AALG for bromine is based on irritant effects, no relative source contribution from air was factored into the calculation.

Based on available information, it appears that the critical effect for inhalation exposure to bromine in humans is irritation and that the TLV constitutes a reasonable surrogate human NOAEL. Therefore, the AALG is based on the TLV/10.

AALG: • irritation (TLV/10)—0.01 ppm (0.07 mg/m^3) 8-hour TWA

References

1. ACGIH. 1986. "Documentation of the Threshold Limit Values and Biological Exposure Indices." 5:65.
2. OSHA. 1989. "Air Contaminants; Final Rule." *Fed. Reg.* 54:2332–959.

3. Safe Drinking Water Committee. 1980. *Drinking Water and Health,* Vol. 3 (Washington, DC: National Academy Press).

4. NIOSH. 1988. EF9100000. ''Bromine.'' RTECS, on line.

5. Van Leeuwen, F. X. R. and B. Sangster. 1987. ''The Toxicology of Bromide Ion.'' *CRC Crit. Rev. Tox.* 18:189–213.

6. Henderson, Y. and H. W. Haggard. 1943. *Noxious Gases* (New York: Reinhold Publishing Co.), p. 133.

7. Amoore, J. E. and E. Hautala. 1983. ''Odor as an Aid to Chemical Safety: Odor Thresholds Compared with Threshold Limit Values and Volatilities for 214 Industrial Chemicals in Air and Water Dilution.'' *J. Appl. Tox.* 3:272–90.

1,3-BUTADIENE

Synonyms: butadiene, biethylene, erythene, divinyl

CAS Registry Number: 106–99–0

Molecular Weight: 54.09

Molecular Formula: $CH_2 = CHCH = CH_2$

AALG: 10^{-6} 95% UCL—1.29×10^{-3} ppb (2.86×10^{-3} $\mu g/m^3$) annual TWA

Occupational Limits:
- ACGIH TLV: 10 ppm (22 mg/m^3) TWA; Appendix A2 suspect carcinogen
- NIOSH REL: reduce to lowest feasible level
- OSHA PEL: 1000 ppm (2200 mg/m^3) TWA

Basis for Occupational Limits: Butadiene is listed in Appendix A2 on the basis of its carcinogenicity in rats and mice, and this was considered in setting the TLV; however, the 10 ppm figure was "recommended based upon data that indicate average workplace exposures are presently controlled below 10 ppm."[1] The health effects considered in establishing the NIOSH recommendation were cancer, teratogenicity, and reproductive effects.[2] The OSHA PEL listed above has been designated as a transitional limit in the recent final rule on air contaminants and 1,3-butadiene is one of several chemicals currently undergoing 6(b) rulemaking.[3]

Drinking Water Limits: No drinking water limits were found for this substance.

Toxicity Profile

Carcinogenicity: An NTP bioassay[4,5] (cited in U.S. EPA[6] and IARC[7]) was conducted in male and female B6C3F$_1$ mice (n = 50/sex/group) exposed to 0, 625, or 1250 ppm 1,3-butadiene via inhalation for 6 hours per day, 5 days per week; the study was terminated at week 60–61 due to excessive tumor-related deaths among the treated mice. Hemangiosarcoma of the heart, malignant lymphomas, and bronchioalveolar adenomas and carcinomas occurred in a statistically significant dose-related manner for both male and female mice. Statistically significant increases in acinar (mammary) cell carcinomas, granulosa cell tumor or carcinoma (ovary), forestomach papillomas and carcinomas, and hepatocellular carcinomas and/or adenomas also occurred in female mice. It is noteworthy that lymphomas developed quite early in this study, i.e., 22 and 20 weeks in the high-dose group and 24 and

29 weeks in the low-dose group for male and female mice, respectively. Testicular atrophy without tumor development was observed in both the high- and low-dose male mice; ovarian atrophy was observed in the female mice.

In a study sponsored by Hazelton labs[8] (cited in U.S. EPA[6]) male and female CD rats (n = 110/sex/group) were exposed to 0, 1000, or 8000 (not established as the maximally tolerated dose) ppm 1,3-butadiene for 6 hours per day, 5 days per week for 111 (males) or 105 (females) weeks. Statistically significant increases in mammary and thyroid follicular tumors were observed in females and Leydig's cell tumors and pancreatic exocrine adenomas in males. While this study provides qualitative evidence of a carcinogenic effect, its quantitative assessment value was judged to be limited by incomplete reporting of the data and lack of independent data quality evaluation.[6]

Several occupational epidemiologic studies have been conducted in humans and are reviewed in the mutagenicity and carcinogenicity assessment of 1,3-butadiene.[6] However, it was concluded that none of the studies were adequate to determine a causal association on account of inconsistency in results and concurrent exposures to styrene and solvents.[6]

1,3-Butadiene has been placed in group B2, probable human carcinogen, under the EPA weight-of-evidence classification.[6,9]

Mutagenicity: In a recent IARC review of the genetic activity of 1,3-butadiene in short-term tests, it was reported to induce forward mutations in *S. typhimurium* strains TA1530 and TA1535 with, but not without, metabolic activation.[10] In mice exposed to 100 ppm 1,3-butadiene for 6 hours on two consecutive days, micronuclei and sister chromatid exchanges were observed in bone marrow cells.[10,11] In addition, 1,3-butadiene is reported to induce mutations in mouse lymphocytes in culture,[11] and there is evidence that metabolites of 1,3-butadiene alkylate DNA.[6]

Developmental Toxicity: In a study conducted at Hazelton Laboratories Europe, Ltd.[12] (cited in U.S. EPA[6]) pregnant CD rats were exposed to 1,3-butadiene via inhalation at 0, 200, 1000, or 8000 ppm (40, 24, 24, and 24 rats/group, respectively) for 6 hours per day on days 6 through 15 of gestation. There was a statistically significant increase in the percentage of fetuses with ''minor'' skeletal defects at the 200-ppm level, but not at the higher levels, as well as a statistically significant increase in the percentage of fetuses with ''major'' skeletal defects at the 1000- and 8000-ppm dose levels. However, it is noted[6] that some investigators might not consider these defects—marked or severe wavy ribs—as major. Maternal toxicity as evidenced by reduced weight gain was observed in dams at the highest exposure level.

Another investigator[13] (cited in U.S. EPA[6]) has reported decreases in litter size of rats exposed to 2300 and 6700, but not 600 ppm butadiene; however, the reporting of this study was considered incomplete.[6] It was concluded in the EPA mutagenicity and carcinogenicity assessment of 1,3-butadiene that these studies provide only suggestive evidence of adverse developmental effects in rats.

Reproductive Toxicity: Although no multigeneration reproduction studies were found for 1,3-butadiene, Carpenter et al.[13] (cited in IARC[7]) reported that mating was unimpaired in rats, guinea pigs, and rabbits exposed to concentrations of up to 6700 ppm 1,3-butadiene via inhalation for 7.5 hours per day, 6 days a week for eight months.

In the NTP bioassay, testicular atrophy without tumor development was reported in male mice exposed to 625 or 1250 ppm, and ovarian atrophy and tumor development were reported in female mice at both exposure levels.[5]

Systemic Toxicity: The acute inhalation toxicity of 1,3-butadiene is relatively low in rats and mice, with reported LC_{50}s of 129,000 ppm (4-hour) and 122,170 ppm (2-hour), respectively.

In the two-year inhalation study in rats described earlier[8] (cited in U.S. EPA[6]), nephropathy occurred in 27% of the male rats in the high-dose group and increased alveolar metaplasia was also observed. In addition, dose-related increases in liver weights, possibly indicative of enzyme induction, were reported in animals killed halfway through and at the termination of the study.

In a series of Russian studies summarized in the assessment document,[6] toxic effects were reported in rats at concentrations of 1,3-butadiene much lower than those used in other studies. Rats exposed continuously (length of time not stated) to 13.5 ppm butadiene experienced elevated blood cholinesterase, hypotension, and decreased motor activity. Histological examination of tissues from rats exposed to 0.45 ppm revealed congestion of the spleen and hyperemia and leukocyte infiltration in cardiac tissue.

Industrial use of 1,3-butadiene has not generally been associated with serious injury.[1] However, in several Russian studies, hematological disorders, liver enlargement, renal malfunction, a variety of skin disorders, and neurasthenic symptoms have been reported; it must be noted, however, that such reports have not been accompanied by details of exposure concentrations or duration, concurrent exposure to other substances, or information on appropriate control groups.[6]

Irritation: Carpenter et al.[13] (cited in ACGIH[1]) exposed human volunteers to 2000, 4000, and 8000 ppm butadiene for 6 to 8 hours. Subjects reported smarting of the eyes, difficulty in focusing on instrument scales, and that the odor was objectionable. Butadiene is reported to have a "mild aromatic" odor[1] and an odor threshold of 1.3 ppm (geometric mean of reported literature values).[14]

Basis for the AALG: Although 1,3-butadiene is a gas at room temperature and pressure,[6] it is slightly soluble in water and has been detected in drinking water in the United States.[7] In the IRIS summary,[9] no quantitative estimate of carcinogenic risk was calculated for ingestion of 1,3-butadiene because oral exposure was considered unlikely, since it is a gas at room temperature. Therefore, 80% of the contribution to total exposure will be allowed from air.

Because there is sufficient evidence of 1,3-butadiene carcinogenicity in laboratory animals, the AALG is based on this endpoint. In the assessment for 1,3-

butadiene,[6,9] the mouse study[5] was considered to be the primary study for quantitative risk assessment for reasons explained in the toxicity profile section above. Since, as noted in the IRIS summary,[9] pharmacokinetic analysis revealed that the "effective doses" were the same for both treatment groups in the rat study,[8] the quantitative estimates presented here are based on this assessment.

The recommended q_1* (based on the inhalation unit risk) of 2.8×10^{-4} $(\mu g/m^3)^{-1}$ given in the IRIS summary was originally derived from the geometric mean of the q_1*s for male mice data sets (control, 2 of 50; low, 43 of 49; and high, 40 of 45) and female mice data sets (control, 4 of 48; low, 31 of 48; and high, 45 of 49); the numbers of mice were based on those with at least one of the statistically significant increased tumor types or tumors considered unusual at the time of sacrifice. The extrapolation method used was the linearized multistage model with extra risk. Animals dying prior to the onset of the first tumors at 20 weeks were excluded from the analysis and adjustment was made for early sacrifice in the q_1* calculation. The concentration of ppm in air was assumed to be equivalent for experimental animals and humans and an absorption rate of 54% was assumed at low exposure concentrations. It is noted in the IRIS summary that the "unit risk should not be used if the air concentration exceeds 40 $\mu g/m^3$ since above this concentration the slope factor may differ from that stated."

Applying the relationship $E = q_1$* $\times$ d, where E is the specified level of extra risk and d is the dose and allowing an 80% relative source contribution from air gives the figures shown below.

AALG: • 10^{-6} 95% UCL—1.29×10^{-3} ppb (2.86×10^{-3} $\mu g/m^3$)
   • 10^{-5} 95% UCL—1.29×10^{-2} ppb (2.86×10^{-2} $\mu g/m^3$) annual TWA

References

1. ACGIH. 1986. "Documentation of the Threshold Limit Values and Biological Exposure Indices." 5:68–69.
2. NIOSH. 1985. "NIOSH Recommendations for Occupational Safety and Health Standards." *Morb. Mort. Wkly. Rpt.* 34:5s-31s.
3. OSHA. 1989. "Air Contaminants; Final Rule." *Fed. Reg.* 54:2332–959.
4. Huff, J. E., R. L. Melnick, H. A. Solleveld, J. K. Haseman, M. Powers, and R. A. Miller. 1985. "Multiple Organ Carcinogenicity of 1,3-Butadiene in B6C3F$_1$ Mice After 60 Weeks of Inhalation Exposure." *Science* 277:548–49.
5. NTP. 1984. "Toxicology and Carcinogenesis Studies of 1,3-Butadiene (CAS No. 106–99–0) in B6C3F$_1$ Mice (Inhalation Studies)" (Research Triangle Park, NC: National Toxicology Program), TR-288.
6. U.S. EPA. 1985. "Mutagenicity and Carcinogenicity Assessment of 1,3-Butadiene." EPA/600/8–85/004F.
7. IARC. 1986. "1,3-Butadiene." *IARC Monog.* 39:165–79.
8. Hazelton Laboratories Europe, Ltd. 1981. "The Toxicity and Carcinogenicity of Butadiene Gas Administered to Rats by Inhalation for Approximately 24 Months." Prepared

for the International Institute of Synthetic Rubber Producers, New York, (unpublished) (discussed in reference no. 6).

9. U.S. EPA. 1988. "1,3-Butadiene; CASRN 106–99–0." IRIS (3/1/88).
10. IARC. 1986. "1,3-Butadiene." *IARC Monog.* suppl. 6:126–28.
11. NIOSH. 1988. EI9275000. "1,3-Butadiene." RTECS, on line.
12. Hazelton Laboratories Europe, Ltd. 1981. "1,3-Butadiene Inhalation Teratogenicity Study No. 2788 522/3." Prepared for the International Institute of Synthetic Rubber Producers, New York, NY (unpublished) (discussed in reference no. 6).
13. Carpenter, C. P., C. B. Shaffer, C. S. Weil, and H. F. Smyth. 1944. "Studies on the Inhalation of 1,3-Butadiene; with a Comparison of Its Narcotic Effect with Benzol, Toluol, and Styrene, and a Note on the Elimination of Styrene by the Human." *J. Ind. Hyg. Tox.* 26:69–78.
14. Amoore, J. E. and E. Hautala. 1983. "Odor as an Aid to Chemical Safety: Odor Thresholds Compared with Threshold Limit Values and Volatilities for 214 Industrial Chemicals in Air and Water Dilution." *J. Appl. Tox.* 3:272–90.

BUTYL ACRYLATE

Synonyms: *n*-butyl acrylate, butyl 2-propenoate

CAS Registry Number: 141–32–2

Molecular Weight: 128.17

Molecular Formula:

$$CH_2 = CH\text{-}\overset{\displaystyle O}{\overset{\displaystyle \|}{C}}O(CH_2)_3CH_3$$

AALG: developmental toxicity—0.06 ppm (0.31 mg/m^3) 24-hour TWA

Occupational Limits:
- ACGIH TLV: 10 ppm (55 mg/m^3) TWA
- NIOSH REL: none
- OSHA PEL: 10 ppm (55 mg/m^3) TWA

Basis for Occupational Limits: The TLV is based on the "close similarity in toxic response by inhalation, skin and eye to methyl acrylate, which has a TLV of 10 ppm."[1] OSHA recently adopted the TLV-TWA as part of its final rule limits.[2]

Drinking Water Limits: No drinking water limits were found for this substance.

Toxicity Profile

Carcinogenicity: Two studies are cited in the IARC review of *n*-butyl acrylate. In one, male C3H/HeJ mice were exposed to *n*-butyl acrylate by skin application (0.2 mg/mouse/application) three times a week for life. No treatment-related tumors were observed in the mice on histological examination of back skin; IARC noted that no mention was made of controlling for loss of the parent compound due to volatilization or polymerization. In another study, negative results were obtained when Sprague-Dawley rats were exposed to 0, 15, 45, or 135 ppm *n*-butyl acrylate by inhalation for 6 hours per day, 5 days per week for 24–30 months; however, IARC noted difficulty in evaluating this study due to incomplete reporting of design and results, since the study was available only as an abstract.[3]

Butyl acrylate is most appropriately placed in group D, unclassified substance, under the EPA weight-of-evidence classification.

Mutagenicity: No data were found implicating *n*-butyl acrylate as a mutagen. Negative results have been consistently obtained in the Ames assay using various strains with and without metabolic activation. Negative results were also reported in cytogenetic analysis of bone marrow cells of Chinese hamsters and Sprague-Dawley rats exposed to 820 ppm n-butyl acrylate via inhalation for 5–6 hours per day for 4 days. IARC judged the degree of evidence for genetic activity of n-butyl acrylate in short-term tests to be inadequate.[3]

Developmental Toxicity: Merkle and Klimisch[4] exposed Sprague-Dawley rats to 0, 25, 135, or 250 ppm n-butyl acrylate via inhalation for 6 hours per day on days 7 to 16 of gestation. Although no gross external, visceral, or skeletal abnormalities were detected, there was a statistically significant increase in postimplantation loss at the two highest concentrations. However, the two highest concentrations also produced maternal toxicity, i.e., maternal body weight gain was reduced and signs of irritation were observed.

Reproductive Toxicity: No data were found implicating n-butyl acrylate as a reproductive toxin.

Systemic Toxicity: Data on n-butyl acrylate's inhalation toxicity is limited mainly to acute studies.[1,3] LC_{50} values have been reported as 1490 ppm (2 hours) in mice and 2730 ppm (4 hours) in rats.[5] It is noted in the TLV documentation[1] that the LC_{50} for rats and the dermal LD_{50} in rabbits were lower for methyl acrylate compared to butyl acrylate.

Irritation: Butyl acrylate is reported to be moderately irritating to rabbit skin and to cause corneal necrosis in unwashed rabbit eyes. It is noted that methyl acrylate is a more potent irritant than butyl acrylate.[1] An odor threshold of 0.1 ppm is reported for this substance.[3]

In a study available only as an abstract, Klimisch and Reininghaus[6] exposed male and female Sprague-Dawley rats (n = 86/sex/group) to 0, 15, 45, or 135 ppm n-butyl acrylate via inhalation for 6 hours per day, 5 days a week for 2 years. The rats were held for 6 months postexposure. Although no signs of systemic toxicity were reported, histologic examination of the nasal turbinates revealed local changes in the nasal epithelium. These changes were dose-related and were described as follows: ". . . atrophy of the neurogenic portion of the olfactory epithelium with proliferation of the reserve cells to a multilayered epithelium was found. These changes were primarily observed at the anterior portion of the olfactory mucosa at the transition of the respiratory to the olfactory epithelium. A regeneration (replacement of the olfactory epithelium by respiratory epithelium) was observed after the post-exposure period in the BA exposed animals." Unfortunately, insufficient detail was given in the abstract to determine a NOAEL or LOAEL for these effects.

Basis for the AALG: Butyl acrylate is slightly soluble in water (0.14 g/100 mL at 20°C) and has a vapor pressure of 4 mm Hg at 20°C.[3] It is used in paint formulation and in the manufacture of polymers and resins used in textile and leather finishing.[1] No data on the occurrence of n-butyl acrylate in waters or ambient air were reported in the IARC review.[3] Based on the lack of data on its environmental fate and distribution and the expected contribution of various media to human exposure, 50% of the contribution to total exposure will be allowed from air.

There are two possible endpoints on which the AALG for *n*-butyl acrylate may be based—irritation and developmental toxicity. Conceptually, the TLV is not the most desirable basis for the AALG, since it is based on analogy with methyl acrylate. However, due to the general paucity of data on this compound such a calculation is warranted. Given the lack of data on the concentrations at which humans might be expected to experience irritant effects from butyl acrylate exposure and the fact that the TLV was based on analogy with methyl acrylate, it was the judgment here that a database factor of 5 was warranted. Thus, the TLV was divided by 50 (10 for interindividual variation × 5 for the database factor).

The AALG for developmental toxicity is based on the NOAEL of 25 ppm for fetotoxicity in rats in the Merkle and Klimische study.[4] The NOAEL was adjusted for continuous exposure (25 × 6/24) and converted to a human equivalent NOAEL according to the procedures described in Appendix C. A total uncertainty factor of 100 (10 for interindividual variation × 10 for interspecies variation) was applied. Application of the developmental uncertainty factor was considered unwarranted since significant fetotoxicity was only observed at levels that also produced maternal toxicity. The recommended AALG is the one based on developmental toxicity, since it is more conservative.

AALG: • irritation (TLV-based)—0.2 ppm (1 mg/m^3) 8-hour TWA
• developmental toxicity—0.06 ppm (0.31 mg/m^3) 24-hour TWA

References

1. ACGIH. 1986. "Documentation of the Threshold Limit Values and Biological Exposure Indices." 5:75.
2. OSHA. 1989. "Air Contaminants; Final Rule." *Fed. Reg.* 54:2332–2959.
3. IARC. 1986. "*n*-Butyl Acrylate." *IARC Monog.* 39:67–79.
4. Merkle, J., and H. J. Klimisch. 1983. "*n*-Butyl Acrylate: Prenatal Inhalation Toxicity in the Rat." *Fund. App. Tox.* 3:443–47.
5. NIOSH. 1987. UD3150000 "2-Propenoic Acid, Butyl Ester." RTECS, on line.
6. Klimisch, H. J., and W. Reininghaus. 1984. "Carcinogenicity of Acrylates: Long-term Inhalation Studies on Methyl Acrylate (MA) and *n*-Butyl Acrylate (BA) in Rats (Abstract)." *Toxicologist* 4:53.

BUTYL BENZYL PHTHALATE

Synonyms: *n*-butyl benzyl phthalate, BBP

CAS Registry Number: 85–68–7

Molecular Weight: 312.39

Molecular Formula:

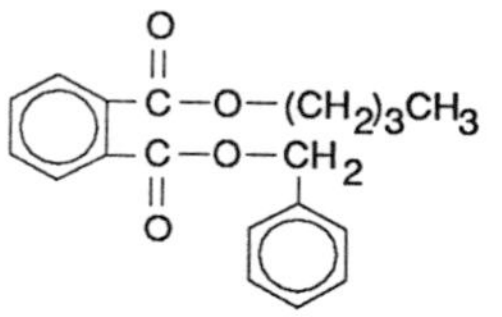

AALG: carcinogenicity—0.42 mg/m^3 (0.032 ppm) annual TWA

Occupational Limits:
- ACGIH TLV: none
- NIOSH REL: none
- OSHA PEL: none

Drinking Water Limits: No criterion was developed for BBP in the "Ambient Water Quality Criteria for Phthalate Esters" document.[1]

Toxicity Profile

Carcinogenicity: In an NTP bioassay[2] male and female B6C3F$_1$ mice and F344 rats (n = 50/sex/group) were fed 0, 6000, or 12,000 ppm BBP (97.2% purity) in the diet for 103 weeks (with the exception of male rats in all groups, which were terminated by 28 weeks due to excess mortality associated with BBP exposure). No signs of toxicity or of non-neoplastic or neoplastic lesions were identified as attributable to BBP exposure in either male or female mice. Early mortality in male rats precluded adequate evaluation of carcinogenicity. In female rats, weight gain and food consumption were consistently lower in dosed vs control animals, but survival was comparable among all groups. There was a statistically significant dose-related trend in the occurrence of mononuclear cell leukemia in treated animals compared to both historical and concurrent controls. The incidence of mononuclear cell leukemia was also significantly higher in the high-dose group compared to the low-dose group and historical and concurrent controls (controls, 7 of 49; low-dose, 7 of 49; and high-dose, 18 of 50). No other treatment-related neoplastic or non-neoplastic lesions were identified. It was concluded that under the conditions of this bioassay, BBP was probably carcinogenic to female F344 rats.

Theiss et al.[3] (cited in NTP[2]) reported that BBP was negative in a Strain A mouse lung adenoma bioassay; however, it should be noted that a negative result in this experimental model does not constitute evidence of noncarcinogenicity.[4] IARC considered the above studies to constitute inadequate evidence of carcinogenicity in laboratory animals, and no epidemiologic data were available. It is the judgement here that the animal data are limited (positive data in a single sex of only one species) and BBP is most appropriately placed in group C ("possible human carcinogen") bordering on group D ("unclassified substance") under the EPA weight-of-evidence classification.

Mutagenicity: BBP has been reported as negative in the following short-term tests: (1) nonmutagenic to *E. coli* and *B. subtilis*, as Santicizer 160 (commercial BBP); and (2) nonmutagenic with and without metabolic activation in *S. typhimurium* strains TA1535, TA1537, TA1538, TA98, or TA100 and *S. cerevisiae* strain D4 at concentrations of 10 μL per plate. In addition, Santicizer 160 did not induce forward mutations at the TK locus in L5178Y mouse lymphoma cells.[4]

Developmental Toxicity: No data were found implicating BBP as a developmental toxin.

Reproductive Toxicity: In 14- and 90-day feeding studies conducted in F344 rats,[2] testicular atrophy was observed in all rats exposed to 50,000 or 100,000 ppm BBP in the diet (14-day studies) and 25,000 ppm BBP in the diet (90-day studies).

Systemic Toxicity: No inhalation studies were found for BBP. Based on reported oral LD_{50} values in F344 rats (2330 mg/kg) and B6C3F$_1$ mice (4170 mg/kg females and 6160 mg/kg males), BBP is of relatively low acute toxicity.[2]

Prior to conducting the cancer bioassay for BBP, 14- and 90-day feeding studies were conducted in male and female B6C3F$_1$ mice and F344 rats. The findings in these studies may be summarized as follows:

- 14-day studies: no compound-related mortality at dietary concentrations of up to 100,000 ppm BBP in rats and 25,000 ppm BBP in mice; decreased body weight gain at 25,000 ppm and higher in male and female rats; thymic atrophy at 100,000 ppm; testicular degeneration at 50,000 ppm and higher in male rats
- 90-day studies: no compound-related mortality at dietary concentrations of up to 25,000 ppm BBP in rats or mice; decreased body weight gain in male but not female rats; testicular degeneration in male rats at 25,000 ppm[2]

Based on available data, rats are clearly the species most sensitive to BBP-induced toxicity. The cause of the early mortality in male rats in the chronic study was apparently internal hemorrhaging, but the cause of this hemorrhaging was not identified and the effect had not been observed at higher doses in earlier studies.[2]

Irritation: Based on available data and analogy with other phthalates, the potential for BBP irritancy is relatively low and it is not generally recognized as an irritant.

Basis for the AALG: BBP has a relatively high boiling point (377°C), hence probably a low vapor pressure, and is soluble in water to the extent of 2.9 mg/L. BBP is used exclusively as a plasticizer and has been identified in drinking water, surface waters, well water, groundwater, and industrial effluents in various areas of the United States.[4] Based on these considerations, 20% of the contribution to total exposure will be allowed from air.

The AALG is based on carcinogenicity using an uncertainty factor approach, since BBP was classified as a group C carcinogen. The NOAEL of 6000 ppm in the diet of female rats for the occurrence of mononuclear cell leukemia was converted to a human equivalent NOAEL for the inhalation route according to the procedures described in Appendix C and assuming a 0.35-kg rat. A total uncertainty factor of 500 was applied (10 for interindividual variation × 10 for interspecies variation × 5 for the database factor). Application of a database factor of 5 was based on use of a noninhalation study as the basis for the AALG and on the severity of the endpoint (i.e., carcinogenicity) in the presence of negative mutagenicity data. This AALG should be considered provisional, since it is based on a noninhalation study and the data on carcinogenicity are limited.

AALG: • carcinogenicity—0.42 mg/m^3 (0.032 ppm) annual TWA

References

1. U.S. EPA. 1980. "Ambient Water Quality Criteria for Phthalate Esters." EPA 440/5–80–067.
2. NTP. 1982. "Carcinogenesis Bioassay of Butyl Benzyl Phthalate in F344/N Rats and B6C3F₁ Mice (Feed Study)." NTP Carcinogen Technical Report No. 213, DHHS (NIH) 82–1769.
3. Theiss, J. C., G. D. Stoner, M. B. Shimkin, and E. K. Weisburger. 1977. "Test for Carcinogenicity of Organic Contaminants of United States Drinking Waters by Pulmonary Tumor Response in Strain A Mice." *Cancer Res.* 37:2717–20.
4. IARC. 1982. "Butyl Benzyl Phthalate." *IARC Monog.* 29:193–201.

n-BUTYL ALCOHOL

Synonyms: *n*-butanol, normal butanol, 1-butanol

CAS Registry Number: 71–36–6

Molecular Weight: 74.12

Molecular Formula: $CH_3(CH_2)_3OH$

AALG: systemic toxicity (TLV-based)—0.12 ppm (0.36 mg/m^3) ceiling limit

Occupational Limits:
- ACGIH TLV: 50 ppm (150 mg/m^3) ceiling; skin notation
- NIOSH REL: none
- OSHA PEL: 50 ppm (150 mg/m^3) ceiling; skin notation

Basis for Occupational Limits: The TLV is set to prevent irritation as well as hearing loss and vestibular injury in workers.[1] OSHA recently adopted the ACGIH ceiling limit of 50 ppm as part of their final rule limits.[2]

Drinking Water Limits: No drinking water limits were found for this substance.

Toxicity Profile

Carcinogenicity: No data were found implicating *n*-butanol as a carcinogen.

Mutagenicity: In the Gene-Tox database as summarized in RTECS, negative results are listed in the Ames assay (no indication of whether a metabolic activation system was used or not) and for in vitro induction of sister chromatid exchanges (cell type not specified).[3] It has been reported that *n*-butanol is capable of causing disturbances in spindle function in cultured Chinese hamster V79 lung cells, resulting in aneuploidization and nondisjunction.[4]

Developmental Toxicity: Nelson et al.[5] conducted studies in which the developmental toxicity of three butanol isomers was evaluated. Sprague-Dawley rats (n = 15–20/group) were exposed to 0, 3500, 6000, or 8000 ppm *n*-butanol (purity > 99%) via inhalation for 7 hours per day on gestation days 1 through 19. *n*-Butyl alcohol was subjectively the least effective of the butanol isomers in producing narcosis in the maternal animals and no evidence of narcosis was observed at the low and middle concentrations. However, 2 of 18 animals in the highest exposure group died and there was a significant reduction in body weight gain in this group (which became nonsignificant when adjusted for multiple comparisons); food consumption was significantly reduced at both 6000 and 8000 ppm. Fetotoxicity in the form of a dose-related decrease in fetal body weight was statistically significant in the two highest exposure groups. In addition, there was a significant decrease in the percentage of normal fetuses (mainly due to rudimentary cervical ribs, a variation) in the highest exposure group compared to the control group. A NOAEL of 3500 ppm for fetotoxicity is suggested by this study. The authors concluded that the butanols were not strongly selective developmental toxins in the rat and that if teratogenic effects were likely to occur they would most likely be in the presence of maternal toxicity.

Reproductive Toxicity: No data were found implicating *n*-butanol as a reproductive toxin.

Systemic Toxicity: Studies on the inhalation toxicology of *n*-butanol in laboratory animals have generally been confined to short-term, high-concentration exposures; a 4-hour LC$_{50}$ of 8000 ppm has been reported in rats.[3,6] Weese[7] (cited in Wimer, Russell, and Kaplan[6]) reported signs of narcosis and reversible fatty changes in the livers of mice exposed to 8000 ppm *n*-butanol for 130 hours over a period of several days. Starrek[8] (cited in Wimer, Russell, and Kaplan[6]) reported that mice exposed to 1650 or 3300 ppm *n*-butanol for 7 hours showed no evidence of intoxication, but those exposed to 6600 ppm started to stagger after 1 hour, tolerated a side position after 1.5 to 2 hours, and were "deeply narcotized" after 3 hours. Smyth and Smyth[9] (cited in Wimer, Russell, and Kaplan[6]) exposed guinea pigs to 100 ppm *n*-butanol for 4 hours per day, 6 days a week for 64 weeks; a variety of pathologic lesions, including hemorrhagic areas in the lungs, early degenerative lesions in the liver, and renal cortical and tubular degeneration as well as hematologic changes, were observed. In the IRIS summary,[10] a 4-month inhalation study is mentioned (the study was apparently available only as an abstract and no reference was given) in which a NOAEL of 0.26 ppm for reversible changes in blood cholinesterase activity and increased thyroid activity in rats was identified.

In an ingestion study sponsored by the U.S. EPA,[10] male and female rats (n = 30/sex/group) were gavaged daily with 0, 30, 125, or 500 mg/kg/day *n*-butanol for 13 weeks. Interim (6-week) sacrifice revealed no biochemical or histopathologic lesions, other than reductions in certain hematologic parameters in the two highest exposure groups that were not evident at final sacrifice. Ataxia and hypoactivity were observed in high-dose rats of both sexes during the latter half of this study; 125 mg/kg/day was identified as a NOAEL for central nervous system effects.

In humans, systemic effects in the form of auditory nerve and vestibular injury have been reported in French and Mexican workers. Several investigators[11-13] (cited in ACGIH[1] and OSHA[2]) have reported that workers exposed to 80 ppm *n*-butanol without hearing protection experienced hearing loss and impairment after 3 to 11 years. Hearing loss was related to length of exposure and was particularly marked in younger workers.

In the United States, it has been reported that exposure to concentrations of greater than 100 ppm *n*-butanol results in narcotic effects. Exposure to 50 ppm or greater has been reported to cause headaches.[2]

Irritation: Irritation of the eyes, nose, and throat has been reported in humans occupationally exposed to *n*-butanol. In controlled exposure studies, Nelson et al.[14] exposed males and females (n = 10) to various concentrations (in random order) of *n*-butanol vapor for periods of 3 to 5 minutes. At 25 ppm *n*-butanol, the majority of human subjects experienced mild nose and throat irritation; at 50 ppm, eye irritation was experienced as well and most subjects termed the effects "objectionable." In addition, several subjects reported experiencing mild headaches.

n-Butanol is reported to have a "mild, vinous" odor; an odor threshold of 10 ppm has been reported.[1]

Basis for the AALG: *n*-Butanol is soluble in water and has a relatively low vapor pressure (4.2 mm Hg at 20°C) compared to other alcohols.[6] It is used as a solvent for fats, waxes, shellac, resins, gums, and varnish, and in the manufacture of lacquers, detergents, rayon, and other butylated compounds. Based on these considerations and the lack of information on the environmental fate, distribution, and occurrence of *n*-butanol, 50% of the contribution to total exposure will be allowed from air.

AALGs were calculated for *n*-butanol based on both systemic and developmental toxicity. The AALG for systemic toxicity is based on the TLV. Although irritant effects were a major consideration in setting this limit, other endpoints, including headaches, narcotic effects, and hearing loss and vestibular injury, were also factors. Thus, inclusion of a factor of 4.2 for continuous exposure adjustment is appropriate. Since the majority of subjects in the study by Nelson et al.[14] experienced eye, nose, and throat irritation as well as mild headaches at levels at or below the level of the TLV, the TLV is treated as a LOAEL and an additional uncertainty factor of 5 is used; the total uncertainty factor applied to the TLV is 210 (4.2 for continuous exposure × 10 for interindividual variation × 5 for a LOAEL).

The AALG for developmental toxicity is based on the NOAEL of 3500 ppm for fetotoxicity in the Nelson et al. study.[5] The NOAEL was adjusted for continuous exposure (3500 ppm × 7/24) and converted to a human equivalent NOAEL according to the procedures described in Appendix C (assuming a 250-g rat, the midpoint of the range of weights at the time of mating), and a total uncertainty factor of 100 (10 for interindividual variation × 10 for interspecies variation) was applied. Application of the developmental uncertainty factor was considered unwarranted because fetotoxicity was observed only at maternally toxic doses.

Based on these observations, the recommended AALG is the one based on systemic toxicity. However, this AALG should be considered provisional since more study is needed on the neurological and vestibular effects of *n*-butyl alcohol.

AALG: • systemic toxicity (TLV-based)—0.12 ppm (0.36 mg/m^3) ceiling limit
  • developmental toxicity—10.9 ppm (33.1 mg/m^3) 24-hour TWA

References

1. ACGIH. 1986. "Documentation of the Threshold Limit Values and Biological Exposure Indices." 5:76.
2. OSHA. 1989. "Air Contaminants; Final Rule." *Fed. Reg.* 54:2332–959.
3. NIOSH. 1988. EO1400000. "Butyl Alcohol." RTECS, on line.
4. Onfelt, A. 1987. "Spindle Disturbances in Mammalian Cells. III. Toxicity, c-Mitosis, and Aneuploidy with 22 Different Compounds. Specific and Unspecific Mechanisms." *Mut. Res.* 182:135–54.
5. Nelson, B. K., W. S. Brightwell, A. Khan, J. R. Burg, and P. T. Goad. 1989. "Lack of Selective Developmental Toxicity of Three Butanol Isomers Administered by Inhalation to Rats." *Fund. Appl. Tox.* 12:469–79.
6. Wimer, W. W., J. A. Russell, and H. L. Kaplan. 1983. *Alcohols Toxicology* (Park Ridge, NJ: Noyes Data Corporation), pp. 56–64.
7. Weese, H. 1928. "Comparative Studies of the Efficacy and Toxicity of the Vapors of Lower Aliphatic Alcohols." *Arch. Exp. Pathol. Pharm.* 135:118.
8. Starrek, E. 1938. "The Effect of Some Alcohols, Glycols, and Esters." Doctoral dissertation. Julius Maxmillian University, Wurtzburg, Germany.
9. Smyth, H. F. and H. F. Smyth, Jr. 1928. "Inhalation Experiments with Certain Lacquer Solvents." *J. Ind. Hyg.* 10:261.
10. U.S. EPA. 1988. "*n*-Butanol." IRIS (3/31).
11. Seitz, B. 1972. *Soc. Med. Hyg. du Travail* (4/10).
12. Velasquez, J. 1964. *Med. del Trabajo* 1:43.
13. Velasquez, J., et al. 1969. Presented at the XVI International Congress on Occupational Health, Tokyo (Sept. 21–28, 1969).
14. Nelson, K. W., J. F. Ege, M. Ross, L. E. Woodman, and L. Silverman. 1943. "Sensory Response to Certain Industrial Solvent Vapors." *J. Ind. Hyg. Tox.* 25:282–85

s-BUTYL ALCOHOL

Synonyms: *sec*-butyl alcohol, secondary butanol, 2-butanol, *s*-butanol

CAS Registry Number: 78–92–2

Molecular Weight: 74.12

Molecular Formula: $CH_3CH_2CH(OH)CH_3$

AALG: systemic toxicity (TLV/42)—1.2 ppm (3.6 mg/m^3) 8-hour TWA

Occupational Limits:
- ACGIH TLV: 100 ppm (305 mg/m^3) TWA
- NIOSH REL: none
- OSHA PEL: 100 ppm (305 mg/m^3) TWA

Basis for Occupational Limits: The occupational limits for 2-butanol are based on prevention of irritation and narcotic effects.[1] In a communication to the TLV committee an individual reported that "many years of industrial experience with exposures approximating 100 ppm 'have resulted in no difficulties.'"

Drinking Water Limits: No drinking water limits were found for this substance.

Toxicity Profile

Carcinogenicity: No data were found implicating 2-butanol as a carcinogen.

Mutagenicity: No data were found implicating 2-butanol as a mutagen.

Developmental Toxicity: No data were found implicating 2-butanol as a developmental toxin.

Reproductive Toxicity: No data were found implicating 2-butanol as a reproductive toxin.

Systemic Toxicity: Based on the limited available data on 2-butanol toxicity in experimental animals, the major toxic effect appears to be narcosis.[2] Starrek[3] (cited in Wimer, Russell, and Kaplan[2]) reported the following observations in mice:

1. at 10 mg/L—staggering after 1 to 1.5 hours, toleration of side position after 2 to 3 hours, deep narcosis after 5 hours and rapid recovery without effects after cessation of exposure
2. 19,800 ppm for 40 minutes—no deaths
3. 1650 ppm for 420 minutes—no signs of intoxication

Weese[4] (cited in Wimer, Russell, and Kaplan[2]) also reported similar findings in mice:

1. 5300 ppm for 117 hours—narcosis apparent with recovery upon cessation of exposure
2. 10,670 ppm for 225 minutes—lethal
3. 16,000 ppm for 160 minutes—lethal

Secondary butanol has been reported to have relatively greater narcotic effects compared to normal butanol and normal propanol. However, this is most likely linked to its higher vapor pressure, which would facilitate its penetration into the blood and central nervous system.[2]

Irritation: In standard skin and eye irritancy tests in rabbits, liquid 2-butanol has been reported to be mildly irritating to the skin and moderately irritating to the eyes.[5] However, it is less irritating to the eyes than *n*-butanol, and there are no data on the eye irritancy of *s*-butanol vapor.[1] Although *s*-butanol is apparently considered irritating to the respiratory tract, there are no data indicating the concentrations at which this effect might be expected.

Secondary butanol has been described as having a "strong, vinous" odor[1] and an odor threshold (geometric mean of reported literature values) of 2.6 ppm has been reported.[6]

Basis for the AALG: Secondary butanol is moderately soluble in water and has a vapor pressure of 13 mm Hg at 20°C. It is used in the synthesis of flotation agents, flavors, perfumes, dyes, and wetting agents, and in industrial paint removers and cleaners.[1] Given these properties and the fact that no data were available on the environmental fate, distribution, and occurrence of 2-butanol, 50% of the contribution to total exposure will be allowed from air.

Given the limited database on 2-butanol and the fact that the TLV is long-standing and that there is some industrial experience indicating that it approximates a human NOAEL, the AALG is based on the TLV/42 with the 50% relative source contribution factored into the calculation. Because of the limited database for this compound, the AALG should be considered provisional.

AALG: • systemic toxicity (TLV/42)—1.2 ppm (3.6 mg/m³) 8-hour TWA

References

1. ACGIH. 1986. "Documentation of the Threshold Limit Values and Biological Exposure Indices." 5:77.

2. Wimer, W. W., J. A. Russell, and H. L. Kaplan. 1983. *Alcohols Toxicology* (Park Ridge, NJ: Noyes Data Corporation).
3. Starrek, E. 1938. "The Effect of Some Alcohols, Glycols and Ethers." Doctoral Dissertation, Julius Maximillian University, Wurzburg, West Germany.
4. Weese, H. 1928. "Comparative Studies on the Efficacy and Toxicity of the Vapors of Lower Aliphatic Alcohols." *Arch. Exp. Pathol. Pharm.* 135:118.
5. NIOSH. 1988. EO1750000. "*sec*-Butyl Alcohol." RTECS, on line.
6. Amoore, J. E. and E. Hautala. 1983. "Odor as an Aid to Chemical Safety: Odor Thresholds Compared with Threshold Limit Values and Volatilities for 214 Industrial Chemicals in Air and Water Dilution." *J. Appl. Tox.* 3:272–90.

t-BUTYL ALCOHOL

Synonyms: *t*-butanol, 2-methyl-2-propanol

CAS Registry Number: 75–65–0

Molecular Weight: 87.14

Molecular Formula: $(CH_3)_3COH$

AALG: systemic toxicity (TLV-based)—1.2 ppm (4.3 mg/m^3) 8-hour TWA

Occupational Limits:
- ACGIH TLV: 100 ppm (300 mg/m^3) TWA; 150 ppm (450 mg/m^3) STEL
- NIOSH REL: none
- OSHA PEL: 100 ppm (300 mg/m^3) TWA; 150 ppm (450 mg/m^3) STEL

Basis for Occupational Limits: The TLV is based on prevention of narcosis.[1] OSHA recently adopted the TLV-STEL as part of its final rule limits.[2]

Drinking Water Limits: No drinking water limits were found for this substance.

Toxicity Profile

Carcinogenicity: No data were found implicating *t*-butyl alcohol as a carcinogen. *t*-Butyl alcohol is currently on test in the NTP bioassay program with histopathology from two-year studies in progress (drinking water, rats and mice).[3]

Mutagenicity: *t*-Butyl alcohol is listed as negative in the *Neurospora crassa* reversion test in the Gene-Tox database as summarized in RTECS, and it is listed as positive for cytogenetic effects in *S. cerevisiae* in another reference cited in RTECS.[4]

Developmental Toxicity: Daniel and Evans[5] fed pregnant mice liquid diets containing 3.6% ethanol or 0.50, 0.75, and 1.00% *t*-butyl alcohol from days 6 to 20 of gestation. A dose-dependent developmental delay was found for several physiological and psychomotor performance tests. It was noted that "significant postnatal maternal nutritional and behavioral factors affecting lactation and/or nesting behavior were also evident at the higher concentrations of alcohol''; a decrease in maternal weight gain relative to the control diet was also noted at the two highest concentrations of *t*-butyl alcohol.

Nelson et al.[6] conducted studies in which the developmental toxicity of three butanol isomers was evaluated. Sprague-Dawley rats (n = 15/sex/group) were exposed to 0, 2000, 3500, or 5000 ppm *t*-butanol (purity > 99%) via inhalation (7 hr/day) on gestation days 1 through 19. *t*-Butyl alcohol was subjectively more effective in producing narcosis in the maternal animals than other butanol isomers; locomotor activity was definitely impaired in the middle and high exposure groups, and some unsteadiness of gait was reported in the lowest exposure group at the end of each exposure period. In addition, maternal body weight gain and food consumption were significantly reduced at the highest exposure level. Fetotoxicity in the form of a dose-related decrease in fetal body weight was statistically significant at all exposure levels compared to controls. In addition, there was a significant increase in skeletal variants (rudimentary cervical ribs) in the middle and high exposure groups. A LOAEL of 2000 ppm *t*-butanol for fetotoxicity is suggested by this study. The authors concluded that the butanols were not strongly selective developmental toxins in the rat and that if teratogenic effects were likely to occur, it would most likely be in the presence of maternal toxicity.

Reproductive Toxicity: No data were found implicating *t*-butyl alcohol as a reproductive toxin.

Systemic Toxicity: It is reported (no concentrations given) that ingestion of *t*-butyl alcohol may cause drowsiness and irritation of the skin and that inhalation of the vapors may produce headache, dizziness, and dry skin. In addition, *t*-butyl alcohol is noted to have a stronger narcotic action in mice than other butanol isomers.[7]

The TLV documentation notes that "eighteen repeated daily narcotic doses were not fatal to animals and no injurious effects resulted from a long-continued, easily tolerated dosage."[1]

Irritation: See reference to skin irritation above. *t*-Butyl alcohol is reported to have a "camphor-like" odor;[1] an odor threshold of 42 ppm (geometric mean of reported literature values) has been reported.[8]

Basis for the AALG: *t*-Butyl alcohol is soluble in water and has a vapor pressure of 13 mm Hg at 20°C. It is used as a solvent in the manufacture of flavors, perfumes, and flotation agents, as a denaturant for ethanol, and as an octane booster in gasoline.[1] Based on a lack of data on its environmental fate and distribution and the expected contribution of various media to human exposure, 50% of the contribution to total exposure will be allowed from air.

The AALG might reasonably be based on either developmental toxicity or systemic toxicity endpoints. The AALG for systemic toxicity is based on the TLV/42, which is long-standing and set to prevent narcotic effects, since available evidence indicates that it is a reasonable surrogate human NOAEL.

The AALG for developmental toxicity is based on the LOAEL for fetotoxicity in rats in the Nelson et al. study.[6] The LOAEL of 2000 ppm was adjusted for continuous exposure (2000 × 7/24) and converted to a human equivalent LOAEL (assuming a 250-g rat, the midpoint of the range of weights at the time of mating given in the original paper) according to the procedures described in Appendix C, and a total uncertainty factor of 500 (10 for interindividual variation × 10 for interspecies variation × 5 for a LOAEL) was applied. Application of the developmental uncertainty factor was considered unwarranted since mild fetotoxicity was observed at levels that produced maternal toxicity.

Based on these observations the recommended AALG is the one based on systemic toxicity. The AALG should be considered provisional pending availability of results from the NTP bioassay.

AALG: • systemic toxicity (TLV-based)—1.2 ppm (4.3 mg/m^3) 8-hour TWA
• developmental toxicity—1.25 ppm (4.45 mg/m^3) 24-hour TWA

References

1. ACGIH. 1986. "Documentation of the Threshold Limit Values and Biological Exposure Indices." 5:78.
2. OSHA. 1989. "Air Contaminants; Final Rule." *Fed. Reg.* 54:2332–2959.
3. NTP. 1989. "Chemical Status Report." National Toxicology Program, Research Triangle Park, NC (2/7).
4. NIOSH. 1988. EO1925000. "*tert*-Butyl Alcohol." RTECS, on line.
5. Daniel, M. A. and M. A. Evans. 1982. "Quantitative Comparison of Maternal Ethanol and Maternal Tertiary Butanol Diet on Postnatal Development." *J. Pharm. Exp. Ther.* 222:294–300.
6. Nelson, B. K., W. S. Brightwell, A. Khan, J. R. Burg, and P. T. Goad. 1989. "Lack of Selective Developmental Toxicity of Three Butanol Isomers Administered by Inhalation to Rats." *Fund. Appl. Tox.* 12:469–79.
7. Wimer, W. W., J. A. Russell, and H. L. Kaplan. 1983. *Alcohols Toxicology* (Park Ridge, NJ: Noyes Data Corporation), pp. 63–64.
8. Amoore, J. E. and E. Hautala. 1983. "Odor as an Aid to Chemical Safety: Odor Thresholds Compared with Threshold Limit Values and Volatilities for 214 Industrial Chemicals in Air and Water Dilution." *J. Appl. Tox.* 3:272–90.

CADMIUM AND COMPOUNDS

Synonyms: "Cadmium and compounds" includes Cd dust, salts, and cadmium oxide (CAS 1306–19–0)

CAS Registry Number: 7440–43–9

Molecular Weight: 112.40

Molecular Formula: Cd

AALG: 10^{-6} point estimate—1.1×10^{-4} $\mu g/m^3$ annual TWA

Occupational Limits:
- ACGIH TLV: 0.01 mg (Cd)/m^3 TWA (for Cd and compounds); Appendix A2, suspected human carcinogen
- NIOSH REL: reduce to lowest feasible level
- OSHA PEL: fume—0.1 mg/m^3 TWA; 0.3 mg/m^3 ceiling; dust—0.2 mg/m^3 TWA; 0.6 mg/m^3 ceiling

Basis for Occupational Limits: The TLV is based on the prevention of preclinical kidney dysfunction (as evidenced by B_2 microglobulin excretion of greater than 290 $\mu g/L$) and increased risk of lung cancer.[1] The intention of lowering the TLV-TWA from 0.05 mg/m^3 to 0.01 mg/m^3 for cadmium and compounds (including CdO) and the addition of cadmium to appendix A2 was published in the intended changes for 1987–88.[2] The NIOSH recommendation and OSHA limits are set in consideration of lung cancer in occupationally exposed workers.[3] However, the OSHA limits were recently designated as transitional in the final rule on air contaminants and cadmium fume and dust are currently undergoing 6(b) rulemaking.[4]

Drinking Water Limits: The 1-day health advisory for the 10-kg child, 43 $\mu g/L$, is based on a human NOAEL and is also recommended as the 10-day HA. The longer-term HAs for the 70-kg adult and 10-kg child are 18 $\mu g/L$ and 5 $\mu g/L$, respectively. The lifetime HA of 5 $\mu g/L$ was calculated from an RfD of 0.5 $\mu g/kg/day$ assuming 25% relative source contribution from drinking water.[5] The ambient water quality criterion is the same as the current MCL of 0.01 mg/L and the proposed MCLG is 0.005 mg/L.[6,7] These limits were based on chronic toxicity since there is no evidence that ingested Cd causes cancer.[7]

Toxicity Profile

Carcinogenicity: Although the evidence is limited, cadmium is considered by CAG to present a significant risk of lung cancer. However, it is stated in the health effects assessment document that "evidence that Cd is a potent lung carcinogen is not compelling."[8] The epidemiologic studies that serve as the basis of the quantitative risk estimates will be reviewed here. Many supporting epidemiologic studies are reviewed in the health assessment document for cadmium and the update to the health assessment document.[9,10] It is of interest to note that CAG has concluded that there is no evidence that Cd is carcinogenic via ingestion, which is the major route of human exposure, and that the upper limit of potency via ingestion would be at least 100 times lower than via inhalation.[10]

The epidemiologic study of Thun et al.[11,12] (cited in U.S. EPA[8,10]) provides the basis for the quantitative risk estimates by CAG and is an update and extension of earlier studies by Lemen et al.[13] and Varner[14] (cited in U.S. EPA[9,10]). In line with these earlier studies, Thun et al. reported a 2 times excess risk of lung cancer in smelter workers and extended the analysis to evaluate the effects of smoking and arsenic as confounding variables. It was concluded that neither factor alone or in combination could account for the excess lung cancers observed. However, the extended analysis did negate the significant dose-response relationship observed in the earlier studies. It is stated in the IRIS summary[15] that: "An evaluation by the Carcinogen Assessment Group of these possible confounding factors has indicated that the assumptions and methods used in accounting for them may not be valid. As the SMRs observed were low and there is a lack of clear-cut evidence of a causal relationship for the cadmium exposure only, this study is considered to supply only limited evidence of human carcinogenicity."

With respect to animal models, injected cadmium salts have been found to induce injection site sarcomas and testicular tumors in rats and mice; inhalation exposure to cadmium chloride aerosol causes lung cancer in rats.[10] Takenaka et al.[16] (cited in U.S. EPA[10]) exposed male Wistar rats (n = 40/group) to filtered air or 12.5, 25, or 50 $\mu g/m^3$ $CdCl_2$ aerosol continuously for 18 months with an observation period of 13 months. The animals experienced a statistically significant dose-related increase in lung adenomas and carcinomas (controls, 0 of 38; low, 6 of 39; medium, 20 of 38; and high, 25 of 35) with no statistically significant differences in survival or body weight gain. Additional animal studies relevant to carcinogenicity of Cd in animal models are reviewed in the HAD and its update.[9,10] It is noted in the IRIS summary[15] that no evidence of a carcinogenic response was found in seven studies in which mice and rats were exposed to cadmium salts (acetate, sulfate, chloride) via ingestion.

Based on the EPA weight-of-evidence classification, Cd is most appropriately placed in group B1, probable human carcinogen.[15]

Mutagenicity: The evidence for mutagenicity of cadmium and its salts is best described as mixed and in many cases conflicting; the spectrum of results observed is most thoroughly reviewed in the HAD and its update.[9,10] In addition, the spectrum of genotoxic responses in short-term tests for cadmium was recently reviewed by IARC.[17]

Developmental Toxicity: Cadmium has been demonstrated to be teratogenic and to decrease fertility in animal models following intravenous, intraperitoneal, and subcutaneous injection.[8] Data relevant to the developmental effects of inhaled cadmium were not found in the literature reviewed.

With respect to the effects of ingested Cd, a NOAEL of 10 mg/L for fetal growth retardation was suggested from a study by Ahokas et al.[18] (cited in U.S. EPA[5]) in which rats were exposed to 0, 0.1, 10, or 100 mg (Cd)/L during gestation. In another rat ingestion study, a NOAEL of 1.0 mg/kg/day was identified for decreased total implants and live fetuses, increased resorptions, decreased fetal body weight gain, and delayed ossification. Male and female rats were exposed to 0, 0.1, 1.0, or 10 mg/kg/day $CdCl_2$ for 6 weeks during which they were mated; exposure for pregnant animals continued through gestation[19] (cited in U.S. EPA[5]).

Reproductive Toxicity: Refer to the Sutou et al. study[19] (cited in U.S. EPA[5]) described briefly in ''Developmental Toxicity'' above.

Systemic Toxicity: Chronic exposure to cadmium via ingestion and inhalation is associated with renal toxicity in both humans and animal models once a critical body burden is reached. In humans, this target organ toxicity is initially manifest as proteinuria which may progress, if exposure continues, to more severe renal dysfunction resulting in disturbances in mineral metabolism, kidney stones, and osteomalacia.[8,9]

Acute occupational exposure to freshly generated fumes from heated cadmium results in a syndrome of tracheobronchitis, pneumonitis, and pulmonary edema which may not appear until several hours after exposure. Fatalities have resulted from exposure to 40–50 mg/m³ for one hour and 9 mg/m³ for five hours. Nonfatal pneumonitis has been associated with exposures between 0.5 and 2.5 mg/m³.[1]

Chronic cadmium poisoning resulting from many years of inhalation exposure to cadmium dust, salts, and fume is characterized by a nonhypertrophic emphysema and proteinuria with or without renal tubular injury, depending on the duration and level of exposure.[1] Respiratory effects have generally not been reported at levels below 0.1 mg/m³, while there is some evidence for renal effects (proteinuria) at concentrations below 0.1 mg/m³.[3]

Effects on other systems, including the nervous system, liver, gonads, thyroid, and pancreas, have been reported; the evidence for these effects is reviewed in the NIOSH criteria document and current intelligence bulletin and in the HAD.[3,9,20]

Irritation: Exposure to high concentrations of freshly generated fumes from heated cadmium is reported to be severely irritating.[1]

Basis for the AALG: Since ingestion is thought to be the major route of exposure to cadmium,[5,9,10] 20% of the contribution to total exposure will be allowed from air.

The AALG for cadmium and compounds is based on carcinogenicity. In the update to the HAD, quantitative risk assessment was performed on the epidemiologic data of Thun et al. and the rat inhalation study of Takenaka et al. However, CAG based its potency estimates on the human data, considering it intrinsically more reliable than the animal data in spite of certain limitations in the original epidemiologic data (reviewed in detail in the HAD update). In a departure from its usual practice, CAG used the MLE estimate of the linear parameter q_1, 1.8×10^{-3} $(\mu g/m^3)^{-1}$, rather than the 95% upper bound. It was felt that the latter was unnecessarily conservative since the model tends to inflate the risk estimate if nonlinear components exist or confounding factors are present. It should be noted that the extrapolation method used was the two-stage model (only the first stage is affected by exposure) with extra risk. It is noted in the IRIS summary[15] that: "The unit risk should not be used if the air concentration exceeds 6 $\mu g/m^3$, since above this concentration the slope factor may differ from that stated." Applying the relationship $E = q_1 \times d$, where E is the specified level of extra risk and d is the dose, and allowing a 20% relative source contribution, gives the estimates shown below.

AALG: • 10^{-6} point estimate—1.1×10^{-4} $\mu g/m^3$ annual TWA
           • 10^{-5} point estimate—1.1×10^{-3} $\mu g/m^3$ annual TWA
Note: The ATSDR has recently published a toxicological profile for cadmium.[21]

References

1. ACGIH. 1987. "Documentation of the Threshold Limit Values and Biological Exposure Indices." 87(87)–88(87) (update).
2. ACGIH. 1987. "Threshold Limit Values and Biological Exposure Indices for 1987–88" (Cincinnati, OH: ACGIH).
3. NIOSH. 1984. "Cadmium. Current Intelligence Bulletin #42." DHHS (NIOSH) 84–116.
4. OSHA. 1989. "Air Contaminants; Final Rule." *Fed. Reg.* 54:2332–2959.
5. U.S. EPA. 1987. "Health Advisory Draft—Cadmium." Office of Drinking Water (3/31).
6. U.S. EPA. 1980. "Ambient Water Quality Criteria Document." EPA-440/5–80–025.
7. U.S. EPA. 1985. "National Primary Drinking Water Regulations; Synthetic Organic Chemicals, Inorganic Chemicals, and Microorganisms; Proposed Rule." *Fed. Reg.* 50:46936–47022.
8. U.S. EPA. 1984. "Health Effects Assessment for Cadmium" (Springfield, VA: NTIS), PB86–134491, EPA/540/1–86/038.
9. U.S. EPA. 1981. "Health Assessment Document for Cadmium." EPA/600/8–81/023.
10. U.S. EPA. 1985. "Updated Mutagenicity and Carcinogenicity Assessment of Cadmium." EPA/600/8–83/025F.
11. Thun, M. J., T. M. Schnorr, A. B. Smith, and W. E. Halperin. 1984. "Mortality

Among a Cohort of U.S. Cadmium Production Workers: An Update.'' NIOSH, Washington, DC (unpublished).

12. Thun, M. J., T. M. Schnorr, A. B. Smith, and W. E. Halperin. 1985. ''Mortality Among a Cohort of U.S. Cadmium Production Workers: An Update.'' *J. Natl. Cancer Inst.* 74:325–33.

13. Lemen, R. A., J. S. Lee, J. K. Wagoner, and H. P. Blejer. 1976. ''Cancer Mortality Among Cadmium Production Workers.'' *Ann. NY Acad. Sci.* 271:273.

14. Varner, M. O. 1983. ''Updated Epidemiologic Study of Cadmium Smelter Workers.'' Presented at the 4th International Cadmium Conference, April (unpublished).

15. U.S. EPA. 1989. ''Cadmium; CASRN 7440–43–9.'' IRIS (1/1/89).

16. Takenaka, S., H. Oldiges, H. Konig, O. Hochrainer, and G. Oberdorster. 1983. ''Carcinogenicity of Cadmium Chloride Aerosols in W Rats.'' *J. Natl. Cancer Inst.* 70: 367–73.

17. IARC. 1986. ''Cadmium.'' *IARC Monog.* suppl. 6:132–35.

18. Ahokas, R. A., P. V. Dilts, and E. B. LaHaye. 1980. ''Cadmium-Induced Fetal Growth Retardation: Protective Effect of Excess Dietary Zinc.'' *Am. J. Obstet. Gynecol.* 136: 216–26.

19. Sutou, S., K. Yamamoto, H. Sendota, and M. Sugiyama. 1980. ''Toxicity, Fertility, Teratogenicity, and Dominant Lethal Tests in Rats Administered Cadmium Subchronically. III. Fertility, Teratogenicity, and Dominant Lethal Test.'' *Ecotox. Environ. Safety* 4:51–56.

20. NIOSH. 1976. ''Criteria for a Recommended Standard . . . Occupational Exposure to Cadmium.'' DHEW (NIOSH) 76–192.

21. ATSDR. 1989. ''Toxicological Profile for Cadmium.'' ATSDR/TP-88/08.

CAPROLACTAM

Synonyms: 2-oxohexamethylenimine, ε-caprolactam

CAS Registry Number: 105–60–2

Molecular Weight: 113.16

Molecular Formula:

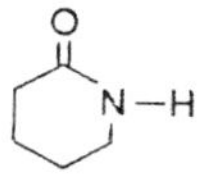

AALG: systemic toxicity and irritation
vapor—0.02 ppm (0.11 mg/m^3) 8-hour TWA
dust—0.005 mg/m^3 8-hour TWA

Occupational Limits:
- ACGIH TLV: vapor—5 ppm (20 mg/m^3) TWA;
  10 ppm (40 mg/m^3) STEL
  dust—1 mg/m^3 TWA;
  3 mg/m^3 STEL
- NIOSH REL: none
- OSHA PEL: vapor—5 ppm (20 mg/m^3) TWA;
  10 ppm (40 mg/m^3) STEL
  dust—1 mg/m^3 TWA;
  3 mg/m^3 STEL

Basis for Occupational Limits: The limits for dust were set to prevent irritation to the skin; when only the vapors are present, the limits were set to prevent early signs of eye and throat irritation in sensitive workers.[1] Caprolactam appeared on the list of intended changes in the TLV booklet in 1987–88; it has been proposed to extend the TWA for vapor to include vapor and aerosols and to eliminate the STELs for both vapor and dust.[2] OSHA recently adopted the old ACGIH limits for caprolactam vapor and dust as part of its final rule limits.[3]

Drinking Water Limits: In an assessment in 1977, NAS concluded that sufficient data were not then available to calculate a SNARL.[4] No drinking water limits were found for this substance.

Toxicity Profile

Carcinogenicity: Negative results were reported for an NTP lifetime cancer bioassay in which B6C3F$_1$ mice and F344 rats of both sexes were exposed to caprolactam in the diet at 7500 or 15,000 ppm, and 3750 or 7500 ppm, respectively.[5]

IARC has classified caprolactam in group 4 of its weight-of-evidence classification (i.e., evidence suggesting lack of carcinogenicity).[6]

Mutagenicity: Caprolactam was tested in a large number and variety of short-term assays in multiple labs as part of the Collaborative Study on Short-Term Tests for Genotoxicity and Carcinogenicity of the International Program on Chemical Safety[7] (cited in IARC[8]). For the most part caprolactam showed no activity, with the exception of consistently positive results in tests for somatic mutations in *D. melanogaster*. In a recent IARC supplement, it was reported that there was some evidence that caprolactam induces chromosomal aberrations in cultured human cells and point mutations in yeast.[6]

Developmental Toxicity: Khadzhieva[9] (cited in IARC[8]) exposed groups of rats via inhalation to 140 or 475 mg/m^3 caprolactam dust for 4 hours per day on gestation days 1–5 or 6–12, or at 13 days post parturition. Increases in pre- and post-implantation loss and decreased fetal body weight were reported at both concentrations.

In Russian studies cited in the IARC monograph,[8] women exposed occupationally to caprolactam were reported to experience an increase in a variety of complications during pregnancy (no further details were given).

Reproductive Toxicity: Gabrielyan et al.[10] (cited in IARC[8]) exposed male rats to 11 or 125 mg/m^3 caprolactam dust for 4 hours per day; after 2.5 months spermatogenesis was reduced in the high-exposure group, but no effect was seen in the low-exposure group. Khadzhieva[11] (cited in IARC[8]) found that female rats exposed to 140 or 475 mg/m^3 caprolactam dust experienced a disruption of the estrous cycle and had reduced fertility.

Serota et al.[12] conducted a three-generation reproduction study in which F344 rats were exposed to levels of 0, 1000, 5000, or 10,000 ppm caprolactam in the diet. No treatment-related effects were observed with respect to reproductive performance, mortality, clinical signs, or gross necropsy findings; there was, however, a decrease in body weight gain and food consumption in both parental rats and pups, exposed in utero through weaning, at 5000 and 10,000 ppm. In addition, "minimal kidney toxicity" occurred in males in the highest exposure group as was manifest by histological findings (nephropathy and the presence of granular casts).

In Russian studies summarized in the IARC monograph,[8] male workers exposed to caprolactam concurrently with cyclohexanone, 1,1'-biphenyl, and 1,1'-oxybis(benzene) were reported to have abnormal sperm production, but no controls were made for possible confounding factors, and semen collection and processing methods were not described.

Systemic Toxicity: In the study by Gabrielyan et al.[10] (cited in IARC[8]), in addition to the reduced spermatogenesis mentioned above, increased excitability, decreased respiratory rate, and decreased excretion of chloride were reported at exposures of 125 mg/m^3, but not at 11 mg/m^3.

It is noted in the IARC monograph[8] that functional disorders of the nervous system, genitourinary tract, and cardiovascular system have been reported among occupationally exposed women in the USSR (no further details were given).

Irritation: Ferguson and Wheeler[13] studied the effects of vapor exposure by examination of plant records and exposure of human volunteers. A difference in response was noted depending on relative humidity conditions; at high relative humidity no discomfort was reported at concentrations up to 14 ppm, while at low relative humidity transient eye and throat irritation occurred in some workers at 10 ppm, but no eye irritation was reported below 7 ppm. In contrast to vapor exposure, in which the effects cease once exposure is terminated, dust exposure can cause skin irritation and sensitization that may continue after exposure stops.[1]

Caprolactam is reported to have an ''unpleasant'' odor,[1] but no specific odor threshold was found.

Basis for the AALG: Caprolactam is highly water soluble (525 g in 100 g of water at 25°C) and has a low vapor pressure (0.001 mm Hg at 20°C). It is used primarily as a monomer in the production of polymeric synthetic fibers (e.g., Nylon-6).[1,8] Caprolactam has been found in finished drinking water in the United States.[8] Based on these considerations, 20% of the contribution to total exposure will be allowed from air.

Because exposure data from the studies of Russian workers are lacking, data from these studies are not usable in AALG derivation. In addition, Russian animal exposure studies should be regarded with caution on account of design limitations and other problems noted in the IARC monographs. Given these considerations, the most appropriate bases for the AALGs are the TLVs that were established to prevent irritation. However, because systemic effects have been reported in occupationally exposed workers in the USSR and in Russian animal inhalation studies, it was considered appropriate to apply a correction for continuous exposure: the TLVs were divided by 42 and a 20% relative source contribution was allowed from air.

The AALGs should be considered provisional because the TLVs are under review by the ACGIH and because reports of reproductive and developmental toxicity in Soviet workers need to be investigated further.

Interestingly, if the NOAEL for spermatogenesis effects from the Gabrielyan et al. study[10] (cited in IARC[8]) is used to derive an AALG, the resultant limit of 0.20 mg/m^3 for caprolactam dust is less conservative than the calculated limit for dust based on the TLV (assuming a 350-g rat and an uncertainty factor of 100).

AALG: • systemic toxicity and irritation
• vapor—0.02 ppm (0.11 mg/m^3) 8-hour TWA
• dust—0.005 mg/m^3 8-hour TWA

References

1. ACGIH. 1986. "Documentation of the Threshold Limit Values and Biological Exposure Indices." 5:95–96.
2. ACGIH. 1987. "Threshold Limit Values and Biological Exposure Indices for 1987–88" (Cincinnati, OH: ACGIH) and communication with ACGIH.
3. OSHA. 1989. "Air Contaminants; Final Rule." *Fed. Reg.* 54:2332–2959.
4. Safe Drinking Water Committee. 1977. *Drinking Water and Health,* Vol. 1 (Washington, DC: National Academy Press).
5. NTP. 1982. "Carcinogenesis Bioassay of Caprolactam in F344 Rats and B6C3F$_1$ Mice" (Research Triangle Park, NC: National Toxicology Program), TR-214.
6. IARC. 1987. "Caprolactam." *IARC Monog.* suppl. 7:390–91.
7. Ashby, J., F. J. DeSerres, M. Draper, M. Ishidate, B. H. Margolin, B. E. Matter, and M. D. Shelby (Eds.). 1985. "Evaluation of Short-Term Tests for Carcinogens." *Progress in Mutation Research,* Vol. 5. Report of the International Program on Chemical Safety's Collaborative Study on In Vitro Assays (Elsevier: Amsterdam).
8. IARC. 1986. "Caprolactam." *IARC Monog.* 39:247–76.
9. Khadzhieva, E. D. 1969. "Effect of Caprolactam on the Reproductive Functions of Albino Rats." *Hyg. Sanit.* 34:28–32.
10. Gabrielyan, N. I., G. E. Kuchukhidze, and E. M. Chirkova. 1975. "Characterization of the General and Gonadotropic Action of Caprolactam" (Russian). *Gig. Tr. Prof. Zabol.* 10:40–42.
11. Khadzhieva, E. D. 1969. "The Effect of Caprolactam on the Sexual Cycle (Experimental Investigation)" (Russian). *Gig. Tr. Prof. Zabol.* 13:22–25.
12. Serota, D. G., A. M. Hoberman, M. A. Friedman, and S. C. Gad. 1988. "Three-Generation Reproduction Study with Caprolactam in Rats." *J. Appl. Tox.* 8:285–93.
13. Ferguson, W. S. and D. D. Wheeler. 1973. "Caprolactam Vapor Exposures." *Am. Ind. Hyg. Assoc. J.* 34:384–89.

CARBON DISULFIDE

Synonyms: carbon bisulfide

CAS Registry Number: 75–15–0

Molecular Weight: 76.12

Molecular Formula: CS_2

AALG: systemic toxicity—0.05 ppm (0.15 mg/m³) 24-hour TWA

Occupational Limits:
- ACGIH TLV: 10 ppm (31 mg/m³) TWA; skin notation
- NIOSH REL: 1 ppm (3 mg/m³) TWA; 10 ppm (31 mg/m³) 15-minute ceiling
- OSHA PEL: 4 ppm (12 mg/m³) TWA; 12 ppm (36 mg/m³) STEL; skin notation

Basis for Occupational Limits: The TLV-TWA was formerly set at 20 ppm to prevent neurological effects; however, the additional finding of cardiovascular effects in workers exposed to "relatively low" concentrations caused the TLV-TWA to be reduced to 10 ppm.[1] The LOAEL of 10 ppm for cardiovascular effects in humans with a safety factor of 10 is the basis for the NIOSH REL.[2] OSHA recently adopted the more stringent limits above as part of its final rule limits.[3]

Drinking Water Limits: No drinking water limits were found for this substance. As of 1977, the NAS Safe Drinking Water Committee, after reviewing the literature, concluded that there were insufficient data on which to base a SNARL.[4]

Toxicity Profile

Carcinogenicity: No data were found implicating carbon disulfide as a carcinogen.

Mutagenicity: Citations in the RTECS database indicate that carbon disulfide has been reported to be mutagenic to *S. typhimurium* (Ames assay) and to induce sister chromatid exchanges in human lymphocytes.[5]

Developmental Toxicity: Petrov[6] (cited in NIOSH[2]) compared the pregnancy experience of viscose rayon workers exposed to approximately 9 ppm carbon disulfide before and during pregnancy with unexposed controls working in the same factory. Workers exposed to carbon disulfide had significantly greater incidences of "threatened'' pregnancy terminations, spontaneous abortions, and premature births. The effects remained significant even after adjusting for age and job longevity.

Several Hungarian investigators have studied the developmental effects of carbon disulfide on rats, particularly neurobehavioral outcomes. Tabacova et al.[7] evaluated the teratogenic effects of CS_2 in Wistar strain rats exposed via inhalation to 0, 50, 100, or 200 mg/m^3 for 8 hours per day throughout the entire period of gestation. Its effect on postnatal development in the F_1 and F_2 generations was also evaluated. Significant increases in pre-implantation lethality were observed at the highest concentration, and significant decreases in fetal weight were observed at the two highest concentrations. Statistically significant increases in gross malformations, particularly hydrocephalus and club foot, were found at the two highest exposure levels, and skeletal abnormalities (retarded ossification and deformation of skull bones) were seen in all exposed groups.

Behavioral alterations were observed in both F_1 and F_2 generations; the authors state that ''reduced exploratory activity and increased emotional activity, more pronounced in the females and persistent up to the 90th day of life, were observed in all test groups.'' In addition, similar types of malformations were found in the F_2 generation, which were the progeny produced by mating F_1 males and F_1 females not exposed to CS_2 before or during gestation. Although the effects reported in this study confirm the effects reported in humans by other workers, evaluation of the study itself is difficult because the control group, exposure chamber design and function, methods of monitoring concentrations in the chamber, and the statistical analysis are either inadequately described or not mentioned. In addition, detailed reporting of the results was lacking.

In another study, Nikiforov and Tabacova,[8] using an exposure scenario like the one described above except with levels of 0.03 and 10 mg/m^3, found that carbon disulfide exposure at the higher level resulted in statistically significant inhibition and delayed development of the mixed function oxidase system, as evidenced by prolongation of hexobarbital sleeping time in the early postnatal period. This study has the same limitations as the one described above.

Lehotzky et al.[9] studied the behavioral effects of prenatal exposure to carbon disulfide in the offspring of Lati:CFY rats exposed to CS_2 vapor at concentrations of 0, <10, 700, or 2000 mg/m^3 for 6 hours per day on days 7 through 15 of gestation. The following effects were reported by the authors:

1. Eye opening and auditory startle response were delayed.
2. Immature gait, motor incoordination, diminished open-field activity, and altered behavioral patterns were manifest in 21- and 36-day-old offspring, but these parameters were nearly age-appropriate in 90-day-old offspring.
3. Although avoidance conditioning performance was nearly age-appropriate, the latency of the response was significantly lengthened at all exposure levels, an indication of disturbed learning ability.

Although this study is more completely reported than those described previously, sufficient detail for complete evaluation is still lacking, particularly with regard to the results.

In a developmental study conducted by Hardin et al.[10] that was also used as the basis for calculation of the oral RfD in the IRIS database,[11] rats (Sprague-Dawley or Wistar) and New Zealand white rabbits (exact numbers per group not specified) were exposed to 0, 20, or 40 ppm carbon disulfide for 6 to 7 hours per day, 34 weeks before breeding and on gestation days 1 to 19 and 1 to 24 in rats and rabbits, respectively. No fetotoxic or teratogenic effects or maternal toxicity were reported at either exposure level for either species. It should be noted that the reporting of this study was very limited.

Reproductive Toxicity: Although adverse reproductive effects such as painful, irregular, and delayed menstruation and decreased fertility have been reported in female viscose rayon workers exposed at low levels (3 ppm), the criteria document states that these effects are questionable because there is a lack of reporting of sampling and analytical methods and because the results have not been confirmed in other studies.[2]

In one study[12] (cited in NIOSH[2]), male viscose rayon workers exposed to 13–26 ppm CS_2 were reported to have significantly decreased sperm counts and a higher frequency of malformed sperm relative to controls. However, in a thorough, well-conducted study of workers exposed to carbon disulfide at levels of up to 10 ppm for at least one year, no differences were found in semen quality relative to controls.[13]

Zenick et al.[14] reported that male rats exposed to CS_2 at 600 ppm for several weeks experienced significant alterations in copulatory behavior and decreased ejaculated sperm counts with no alterations in blood levels of several sex hormones. The authors concluded that CS_2 does not exert a direct effect on the testes but may effect the processes regulating sperm transport and ejaculation.

Systemic Toxicity: Humans occupationally exposed to CS_2 have experienced adverse effects on the eyes and nervous and cardiovascular systems. It should be noted that most occupational exposures to CS_2 also involve exposure to hydrogen sulfide and that reports of apparent synergisms exist. The following ocular effects have been reported in viscose rayon workers:[2]

- vascular encephalopathy at 10 to 48 ppm
- disturbed retinal vascular circulation at CS_2 concentrations of 10 to 30 ppm
- conjunctival inflammation, temporary corneal opacities, and color vision disturbances at concentrations below 3 ppm

In the concentration range at which ocular disorders are manifest, signs of nervous system toxicity are also reported. These findings have included abnormal EMGs and EEGs, polyneuropathy, diminished muscular power, markedly weakened knee and ankle reflexes, slowed motor conduction velocities, and psychological and behavioral abnormalities.[2]

The most extensive and thoroughly documented studies on CS_2 exposure have centered on its cardiovascular effects. In studies conducted by Hernberg et al.[15,16]

(cited in NIOSH[2]), Finnish viscose rayon workers were reported to have significantly elevated rates of coronary heart disease, mortality, angina, hypertension, and disturbances in plasma glucose and creatinine concentrations. In the 10 years previous to the publication of these studies, the carbon disulfide concentrations in the working environment had been between 10 and 30 ppm; in the 10 years prior to that, concentrations were between 20 and 40 ppm. The workers were concurrently exposed to hydrogen sulfide, estimated to make up 10% of the total concentration of the two compounds. In another study a significant dose-response relationship was found between carbon disulfide exposure and ECG patterns indicative of coronary artery disease[17] (cited in NIOSH[2]). Cardiovascular effects have been demonstrated in numerous other studies reviewed and referenced in the NIOSH criteria document.[2]

Irritation: The reported odor threshold for carbon disulfide is 0.1–0.2 ppm. Technical grade CS_2 may have an offensive odor due to contamination with hydrogen sulfide.[2]

Basis for the AALG: Carbon disulfide is slightly water soluble (0.22 weight %) and highly volatile with a vapor pressure of 300 mm Hg at 20°C. It is used as a solvent and in the manufacture of rayon, carbon tetrachloride, electronic vacuum tubes, and soil disinfectants.[1] Although carbon disulfide has been detected in water supplies surveyed by the U.S. EPA,[4] its high volatility makes rapid evaporation from waters and exposure via inhalation more likely. Based on these considerations, 80% of the contribution to total exposure will be allowed from air.

It is stated in the criteria document that the human LOAEL for cardiovascular and neurological effects based on studies using "well-documented, reproducible and accurate environmental monitoring procedures" is 31 mg/m[3] (10 ppm). This LOAEL divided by an uncertainty factor of 210 (10 for interindividual variation × 5 for a LOAEL × 4.2 adjustment for continuous exposure) provides the basis for the AALG.

It was deemed inadvisable at this time to set a limit based on developmental or reproductive effects because of the lack of good correlations between exposure and effect in human studies and the serious limitations in reporting in the available animal studies that precluded a proper evaluation of the studies.

In view of the unresolved questions surrounding the developmental and reproductive effects reported in humans and animal models and the serious nature of the neurobehavioral effects reported in animal studies, the AALG should be considered provisional. It is recommended that ongoing periodic review of the literature should be conducted for studies that could provide an acceptable basis for the derivation of an AALG based on these endpoints.

AALG: • systemic toxicity—0.05 ppm (0.15 mg/m[3]) 24-hour TWA

References

1. ACGIH. 1986. "Documentation of the Threshold Limit Values and Biological Exposure Indices." 5:104–5.
2. NIOSH. 1977. "Criteria for a Recommended Standard . . . Occupational Exposure to Carbon Disulfide." DHEW (NIOSH) 77–156.
3. OSHA. 1989. "Air Contaminants; Final Rule." *Fed. Reg.* 54:2332–2959.
4. Safe Drinking Water Committee. 1977. *Drinking Water and Health,* Vol. 1 (Washington, DC: National Academy of Sciences), pp. 700–703.
5. NIOSH. 1987. FF6650000. "Carbon disulfide." RTECS, on line.
6. Petrov, M.V. 1969. "Some Data on the Course and Termination of Pregnancy in Female Workers of the Viscose Industry." *Pediatr. Akush. Ginekol.* 3:50–52.
7. Tabacova, S., L. Hinkova, and L. Balabaeva. 1978. "Carbon Disulphide Teratogenicity and Postnatal Effects in Rats." *Tox. Let.* 2:129–33.
8. Nikiforov, B., and S. Tabacova. 1980. "Activity of Some Mixed Function Oxidases in Prenatally Carbon Disulfide (CS_2) Treated Rats." *Arch. Toxicol.* suppl. 4:296–8.
9. Lehotzky, K., J.M. Szeberenyi, G. Ungvary, and A. Kiss. 1985. "Behavioural Effects of Prenatal Exposure to Carbon Disulphide and to Aromatol in Rats." *Arch. Toxicol.* suppl. 8:442–6.
10. Hardin, B.D., G.P. Bond, M.R. Sikor, F.D. Andrew, R.P. Beliles, and R.W. Niemer. 1981. "Testing of Selected Work Place Chemicals for Teratogenic Potential." *Scand. J. Work Environ. Health* 7(suppl. 4):66–75.
11. U.S. EPA. 1989. "Carbon disulfide; CASRN 75–15–0." IRIS (2/1/89).
12. Lancranjan, I., H.I. Popescu, and I. Klepsch. 1969. "Changes of the Gonadic Function in Chronic Carbon Disulphide Poisoning" (Russian). *Med. Lav.* 60:566–71.
13. Meyer, C.R. 1981. "Semen Quality in Workers Exposed to Carbon Disulfide Compared to a Control Group from the Same Plant." *J. Occ. Med.* 23:435–9.
14. Zenick, H., K. Blackburn, E. Hope, and D. Baldwin. 1984. "An Evaluation of the Copulatory, Endocrinologic, and Spermatogenic Effects of Carbon Disulfide in the Rat." *Tox. Appl. Pharm.* 73:275–83.
15. Hernberg, S., T. Partanen, C.H. Nordman, and P. Sumari. 1970. "Coronary Heart Disease Among Workers Exposed to Carbon Disulphide." *Br. J. Ind. Med.* 27:313–25.
16. Hernberg, S., C.H. Nordman, T. Partanen, V. Christiansen, and P. Virkola. 1971. "Blood Lipids, Glucose Tolerance, and Plasma Creatinine in Workers Exposed to Carbon Disulphide." *Work. Environ. Health* 8:11–16.
17. Cirla, A.M., A. Villa, and M. Tomasini. 1972. "Investigation of the Incidence of Coronary Disease in Workers Exposed to Carbon Disulfide in a Viscose Rayon Industry" (Italian). *Med. Lav.* 63:431–41.

CARBON TETRACHLORIDE

Synonyms: tetrachloromethane, perchloromethane

CAS Registry Number: 56–23–5

Molecular Weight: 153.84

Molecular Formula: CCl_4

AALG: 10^{-6} 95% UCL—8.4×10^{-3} ppb (5.3×10^{-2} $\mu g/m^3$) as an annual TWA

Occupational Limits:
- ACGIH TLV: 5 ppm (31.5 mg/m^3) TWA; skin notation; Appendix A2, suspected human carcinogen
- NIOSH REL: 2 ppm (12.6 mg/m^3) ceiling (60 minutes)
- OSHA PEL: 2 ppm (12.6 mg/m^3) TWA

Basis for Occupational Limits: The TLV was recommended based on animal data indicating fatty degeneration of the liver at 10 ppm and potentiation of CCl_4 toxicity by ethanol and other common chemicals.[1] The NIOSH limit was based on these endpoints and consideration of hepatocarcinogenicity.[2] OSHA recently adopted the NIOSH limit as a TWA rather than a ceiling limit as part of its final rule on air contaminants.[3]

Drinking Water Limits: The ambient water quality criterion for carbon tetrachloride is 0.40 $\mu g/L$ at the 10^{-6} risk level assuming consumption of 2 L of water and 6.5 g fish/shellfish per day.[4] The following HAs have been calculated for CCl_4:

- 1-day child—4.0 mg/L
- 10-day child—0.16 mg/L
- longer-term child—0.071 mg/L
- longer-term adult—0.25 mg/L

No lifetime HA was recommended due to the B2 (probable human carcinogen) designation of carbon tetrachloride under the EPA weight-of-evidence classification.[5] The MCLG for CCl_4 is 0 mg/L and the final MCL is 0.005 mg/L.[6]

Toxicity Profile

Carcinogenicity: Based primarily on case studies, there is suggestive evidence that liver cancer may develop in humans years after high levels of exposure to CCl_4. However, these data are considered limited or inadequate to infer a causal association between liver cancer and CCl_4 exposure.[7]

There is sufficient evidence to classify CCl_4 as a hepatocarcinogen based on positive results in multiple strains of mice, rats, and hamsters given the compound via ingestion. Evidence that CCl_4 is carcinogenic via inhalation is confined to a single study in rats. These studies are thoroughly reviewed in the HAD,[8] but only those used in the CAG's quantitative risk assessment will be described here.

In all of the studies[9-13] (cited in U.S. EPA[8]) used for quantitative risk assessment, the animals were given the CCl_4 by gavage in either olive oil or corn oil. The details on these studies are as follows:

- Hamsters exposed for 20 weeks and observed until 55 weeks of age were given an average daily dose (ADD) of 0 or 0.95 mg and the incidence of hepatocellular carcinoma was 0 of 80 and 10 of 19, respectively.
- Mice exposed for 17 weeks and observed until 31 weeks of age were given an ADD of 0 or 15 mg and the incidence of hepatocellular carcinoma was 2 of 52 and 34 of 73, respectively.
- Mice exposed for 78 weeks and observed until 92 weeks of age were given an ADD of 0, 21, or 42 mg and the incidence of hepatocellular carcinoma was 6 of 157, 89 of 89, and 90 of 93, respectively.
- Rats exposed for 78 weeks and observed until 110 weeks of age were given an ADD of 0 (males, females), 11 or 21 mg (males), and 18 or 36 mg (females), and the incidence of hepatocellular carcinoma was 0 of 37, 2 of 45, 2 of 47, 4 of 46, and 1 of 30, respectively.

It should be noted that some investigators have suggested that hepatomas occur only subsequent to liver necrosis and fibrosis, but several studies do not support this finding and it must be regarded as controversial.[8]

Carbon tetrachloride would be classed as a B2, probable human carcinogen, based on the EPA weight-of-evidence classification.[7]

Mutagenicity: CCl_4 has been evaluated for its mutagenic potential in a variety of test systems including bacteria, yeast, and a mammalian cell line and for its capacity to damage DNA in rat hepatocytes in vivo. The bacterial assays generally yielded negative results. Negative results were also reported in the in vivo UDS assays. However, CCl_4 was mutagenic in a yeast system, and positive DNA binding studies have been reported with metabolically activated CCl_4. The EPA has concluded that there is insufficient evidence to draw firm conclusions about the genotoxic potential of CCl_4 and that additional study is warranted.[8] The genotoxic potential of CCl_4 has also been recently reviewed and quantified by IARC.[14]

Developmental Toxicity: In two inhalation studies[15,16] (cited in U.S. EPA[8]) in rats described in the HAD, maternal and prenatal toxicity, but not teratogenic effects, were reported. At least one of the offspring effects (subcutaneous edema) could not be associated with maternal toxicity. Regarding the teratogenic and reproductive toxicity potential of CCl_4, it was concluded in the HAD that: "In general, the studies do not provide adequate dose groups for concluding the existence of teratogenic or reproductive effects according to testing criteria such as those currently used for the U.S. EPA Office of Pesticides Programs or Office of Toxic Substances."[8]

Reproductive Toxicity: Relatively high i.p. doses of CCl_4 (lower doses were not tested) have been reported to cause degenerative histologic changes in the testes of male rats that ultimately resulted in cessation of spermatogenesis.[8]

Systemic Toxicity: The C.N.S., liver, and renal toxicity of CCl_4 is well established in both humans and animal models via inhalation, ingestion, and dermal exposure routes. The database for these effects is extensive and includes acute, subchronic, and chronic studies, which are reviewed in the HAD. In this connection, it is of interest to note that short-term exposure to high concentrations is associated with greater hepatotoxicity than longer-term exposure to low concentrations, even when the time × concentration products are equal.[8] It is reported in the TLV documentation[1] that the lowest level of chronic exposure reported to cause hepatic dysfunction was 20 ppm.

Irritation: Carbon tetrachloride is irritating to the skin within a few minutes of direct application, causing erythema and wheal formation.[8] No other data were found implicating CCl_4 as an irritant. It is reported to have a "sweet" odor[1] and an odor threshold of 96 ppm (geometric mean of reported literature values).[17]

Basis for the AALG: Carbon tetrachloride is considered to be "almost insoluble" in water and is volatile with a vapor pressure of 90 mm Hg at 20°C.[1] It is considered to be ubiquitous in the environment and, at "minimum" and "typical" exposure levels, uptake from air appears to be the major source of exposure.[8] Thus, 80% of the contribution to total exposure will be allowed from air.

The AALG is based on carcinogenicity; the quantitative risk assessment data used were based on ingestion (gavage) studies because suitable inhalation studies were not available. Since all of the studies suffered from limitations, the geometric mean of the 95% UCL was used. These limitations included—in some but not all studies—lack of concurrent controls, less than lifetime exposure and observation, infrequent dosing (weekly), small sample sizes, and low survival.

The oral estimates of q_1^* (from the linearized multistage model with extra risk) were converted to $(\mu g/m^3)^{-1}$ assuming a 70-kg man breathing 20 m^3 of air per day and a 40% absorption rate of CCl_4 per m^3 of air. The basis for this absorption rate is discussed in the HAD. The upper limit unit risk for continuous lifetime exposure to 1 $\mu g/m^3$ ranged from 1.2×10^{-6} to 1.4×10^{-4} with a geometric mean of 1.5×10^{-5}. This latter figure would correspond to a q_1^* of 1.5×10^{-5} $(\mu g/m^3)^{-1}$. Applying the relationship $E = q_1^* \times d$, where E is the specified level of extra risk and d is the dose, and allowing 80% of the contribution to total exposure from air, gives the figures shown below. Further details of the quantitative risk assessment are described in the HAD.[8] The basis for the CAG quantitative risk assessment is also reviewed in the IRIS summary[18] and it is noted that "the unit risk should not be used if the air concentration exceeds 700 $\mu g/m^3$, since above this concentration the slope factor may differ from that stated."

AALG:
- 10^{-6} 95% UCL—8.4 × 10^{-3} ppb (5.3 × 10^{-2} $\mu g/m^3$) annual TWA
- 10^{-5} 95% UCL—8.4 × 10^{-2} ppb (5.3 × 10^{-1} $\mu g/m^3$) annual TWA

References

1. ACGIH. 1986. "Documentation of the Threshold Limit Values and Biological Exposure Indices." 5:109–10.
2. NIOSH. 1985. "NIOSH Recommendations for Occupational Safety and Health Standards." *Morb. Mort. Wkly. Rpt.* suppl. 34:1s–34s.
3. OSHA. 1989. "Air Contaminants; Final Rule." *Fed. Reg.* 54:2332–2959.
4. U.S. EPA. 1980. "Ambient Water Quality Criteria for Carbon Tetrachloride." EPA/440/5–80–026.
5. U.S. EPA. 1987. "Health Advisory—Carbon Tetrachloride." Office of Drinking Water (3/31).
6. U.S. EPA. 1987. "Drinking Water Regulations; Public Notification; Final Rule." *Fed. Reg.* 52:41533–550.
7. U.S. EPA. 1984. "Health Effects Assessment for Carbon Tetrachloride" (Springfield, VA: NTIS), PB86–134509, EPA/540/1–86–039.
8. U.S. EPA. 1984. "Health Assessment Document for Carbon Tetrachloride." EPA/600/8–82–001F.
9. Della Porta, G., B. Terracini, and P. Shubik. 1961. "Induction with Carbon Tetrachloride of Liver Cell Carcinomas in Hamsters." *J. Natl. Cancer Inst.* 26:855–63.
10. Edwards, J. and A. Dalton. 1942. "Induction of Cirrhosis of the Liver and Hepatomas in Mice with Carbon Tetrachloride." *J. Natl. Cancer Inst.* 3:19.
11. NCI. 1976. "Report on the Carcinogenesis Bioassay of Chloroform." Carcinogenesis Program, Division of Cancer Cause and Prevention (3/1).
12. NCI. 1976. "Carcinogenesis Bioassay of Trichloroethylene." NCI TR-2 (2/76).
13. NCI. 1977. "Bioassay of 1,1,1-Trichloroethylene for Possible Carcinogenicity." NCI TR-3 (1/77).
14. IARC. 1986. "Carbon Tetrachloride." *IARC Monog.* suppl. 6:136–38.
15. Schwetz, B. A., B. J. K. Leong, and P. J. Gehring. 1974. "Embryo- and Fetotoxicity of Inhaled Carbon Tetrachloride, 1,1-Dichloroethane, and Methyl Ethyl Ketone in Rats." *Tox. Appl. Pharm.* 28:452–64.
16. Gilman, M. R. 1971. "A Preliminary Study of the Teratogenic Effects of Inhaled Carbon Tetrachloride and Ethyl Alcohol Consumption in the Rat." Dissertation. Drexel University.
17. Amoore, J. E. and E. Hautala. 1983. "Odor as an Aid to Chemical Safety: Odor Thresholds Compared with Threshold Limit Values and Volatilities for 214 Industrial Chemicals in Air and Water Dilution." *J. Appl. Tox.* 3:272–90.
18. U.S. EPA. 1988. "Carbon Tetrachloride; CASRN 56–23–5." IRIS (6/30/88).

CHLORINE

Synonyms: molecular chlorine

Note: This assessment focuses primarily on the effects of inhalation exposure to gaseous Cl.

CAS Registry Number: 7782–50–5

Molecular Weight: 70.90

Molecular Formula: Cl_2

AALG: irritation—6 ppb (17.4 $\mu g/m^3$) 24-hour TWA

Occupational Limits:
- ACGIH TLV: 0.5 ppm (1.5 mg/m^3) TWA; 1 ppm (3 mg/m^3) STEL
- NIOSH REL: 0.5 ppm ceiling (15 minutes)
- OSHA PEL: 0.5 ppm (1.5 mg/m^3) TWA; 1 ppm (3 mg/m^3) STEL

Basis for Occupational Limits: The TLVs for chlorine were set to "minimize chronic changes in the lungs, accelerated aging and erosion of the teeth."[1] NIOSH based its recommendation on reported ocular and respiratory irritation at concentrations of 0.5 ppm in controlled human exposure studies.[2] OSHA recently adopted the ACGIH TLVs as part of its final rule limits.[3] The TLV-TWA and TLV-STEL were lowered from 1.0 ppm and 3.0 ppm to 0.5 ppm and 1.0 ppm, respectively (as noted above) in 1987–88 on the basis of new, well-designed studies in human subjects clearly indicating that 1.0-ppm chlorine levels may result in symptoms of irritation of the nose, throat, and conjunctiva in exposed individuals.[4] These studies are reviewed in the section on irritation.

Drinking Water Limits: Chlorine (usually in the form of hypochlorite or hypochlorous acid in the United States) is used for water disinfection and treatment of sewage effluent.[5]

Toxicity Profile

Carcinogenicity: No data were found implicating chlorine as a carcinogen.

Mutagenicity: Chlorine has been reported to induce chromosomal aberrations in cultured human lymphocytes at concentrations of 20 ppm or higher, but not at concentrations below this level.[6] Chlorine (in the form of hypochlorous acid or hypochlorite) has been reported to be weakly mutagenic to bacteria, and oral exposure of B6C3F$_1$ mice to chlorine at concentrations of 4 or 8 mg/kg/day (pH 8.5, hypochlorite ion is the predominant form) resulted in a significantly increased incidence of spermhead abnormalities.[7]

Developmental Toxicity: No data were found implicating chlorine as a developmental toxin.

Reproductive Toxicity: Druckrey[8] (cited in International Program on Chemical Safety[9]) reported no adverse effects on growth, lifespan, fertility, or hematologic parameters in rats exposed to chlorine at 100 ppm in the drinking water for their entire lifespan over seven generations. No evidence of increased tumor incidence or other histologic lesions were found.

Systemic Toxicity: Chronic lung disease, electrocardiographic changes, and death have occurred in humans exposed to high concentrations of chlorine as a consequence of industrial accidents. However, no data on exposure levels were available in any of these incidents.[2]

Several retrospective epidemiologic studies have been conducted on workers repeatedly exposed to chlorine; these are reviewed in the NIOSH criteria document. Ferris et al.[10] (cited in NIOSH[2]) reported no specific adverse effects in workers exposed to 0–64 ppm over an average of 20.4 years. Patil et al.[11] (cited in NIOSH[2]) conducted a prevalence study of 332 workers exposed to chlorine for 2–14 years. Although this study is stronger than most because sufficient exposure information was available to calculate TWAs, no dose-response relationship was established for exposure to chlorine at levels of up to 1.42 ppm (the highest levels found in this group of workers), and most workers experienced TWA concentrations of less than 0.44 ppm.

Mechanistic studies in animals have indicated that chlorine reacts with water to produce hypochlorous acid (under physiologic conditions), which is capable of penetrating the cell membrane and altering permeability as well as reacting with sulfhydryl groups in the amino acid cysteine. In addition, chlorine and chlorides are thought to induce toxic effects via the production of oxygen radicals.[9]

Chlorine is fairly toxic to laboratory animals following acute inhalation exposure, with one-hour LC$_{50}$s of 137 ppm and 293 ppm reported in mice and rats, respectively.[12] Studies on the effects of subacute and subchronic inhalation exposure to chlorine in laboratory animals are well reviewed in reference 9 and only a few more recent studies will be discussed here. It should be noted that a confounding factor in the interpretation of many older and some newer studies is concurrent exposure to chloramines formed from ammonia derived from animal urine and feces.[13]

Barrow et al.[14] exposed male and female F344 rats (n = 10/sex/group) to 0,

1, 3, or 9 ppm chlorine gas via inhalation for 6 hours per day, 5 days a week for 6 weeks. Decreased body weight gain occurred in male rats at the two highest exposure levels and in female rats at all exposure levels; some mortality also occurred at the highest exposure levels in female rats. Clinical signs of irritation, including lacrimation, hyperemia of the conjunctiva, and nasal discharge, were severe to moderately severe at the two highest exposure levels and slight at 1 ppm. Histological changes noted in the respiratory tract of animals exposed to 1 or 3 ppm chlorine included focal mucopurulent inflammation of the nasal turbinates, submucosal inflammation of the tracheal epithelium, slight to moderate inflammatory reaction around the respiratory bronchioles and alveolar ducts, increased numbers of alveolar macrophages within the alveoli, and isolated areas of atelectasis. Histological changes in the respiratory tract of animals exposed to 9 ppm chlorine were more severe and extensive and included focal to multifocal mucopurulent inflammation of the nasal turbinates with necrotic erosions of the mucosal epithelium; inflammation and epithelial hyperplasia in the trachea and bronchiolar areas; epithelial hyperplasia and hypertrophy of the respiratory bronchioles and alveolar ducts; increased numbers of alveolar macrophages and amounts of secretory material in the alveoli; and necrosis of the alveolar epithelium with areas of atelectasis and interstitial inflammation. Other pathologic lesions observed included slight degenerative changes in the renal tubules at the highest exposure level with elevated urine specific gravity and blood urea nitrogen, and slight degenerative changes in hepatocytes at the two highest exposure levels with elevated levels of various serum enzymes, indicative of liver damage. The authors note that concurrent exposure to chloramines could not be ruled out in this study.

Some studies have indicated that exposure to chlorine via inhalation may alter disease resistance in laboratory animals. Bell and Elmes[15] (cited in International Program on Chemical Safety[9]) reported that rats with spontaneous pulmonary disease experienced higher mortality, a more severe inflammatory reaction, and higher incidences of emphysema and pneumonia than conventional rats when both groups were exposed to 90 ppm chlorine for 3 hours per day for 20 days or 104 ppm for 3 hours per day for 6 days. Arloing et al.[16] (cited in International Program on Chemical Safety[9]) reported that development of tuberculosis was accelerated in guinea pigs exposed to 1.69 ppm chlorine for 5 hours per day for 47 days either before or after bacterial challenge with a virulent strain of *Mycobacterium tuberculosis*.

Irritation: Studies in both humans and laboratory animals have shown that chlorine gas is a potent primary irritant to the eyes and both the upper respiratory tract and deeper structures of the lung.[1,2,9]

Based on the results of several controlled human exposure studies from the older literature[17-19] (cited in NIOSH[2]), the following exposure-effect correlations can be developed.

1. 0.014 to 0.054 ppm (mean 0.027 ppm): tickling of the nose
2. 0.04 to 0.097 ppm (mean 0.058 ppm): tickling of the throat

3. 0.06 to 0.3 ppm (mean 0.19 ppm): itching of the nose and cough, stinging, or dryness in the nose and throat in some subjects
4. 0.35 to 0.72 ppm (mean 0.452 ppm): burning of the conjunctiva and pain after 15 minutes
5. above 1.0 ppm: discomfort with reactions ranging from ocular and respiratory irritation to coughing, shortness of breath, and headaches, reported by nearly all exposed persons

Concentrations above 3.5 to 4.0 ppm are virtually intolerable due to immediate burning of the eyes and nasal congestion.

In a more recent study, Anglen[20] (cited in ACGIH[1]) studied the responses of 29 subjects, both males and females, to 0, 0.5, 1.0, and 2.0 ppm chlorine for 4- and 8-hour periods. Subjective sensation (e.g., smell, taste, itching, burning) and symptoms were recorded by the subjects at 1-hour intervals during exposure; severity of sensations and symptoms were rated. In addition, pulmonary function tests and ophthalmologic examinations were conducted before, during, and following exposure. Exposure at 1.0 ppm for 8 hours produced statistically significant increases in subjective irritation as well as significant changes in pulmonary function, i.e., measures of forced vital capacity (FVC) and forced expiratory volume at one second (FEV$_1$). In addition, medical examination of 6 of 14 subjects at this exposure level revealed increased mucous secretion from the nose and increased mucous in the hypopharynx. No significant changes in pulmonary function and less severe subjective irritation resulted from 8-hour exposure at 0.5 ppm.

Another recent controlled human exposure study by Rotman et al.[21] (cited in ACGIH[1]) has confirmed that exposure to 1.0 ppm chlorine for 8 hours has a significant, although transient, effect on pulmonary function as measured by a number of parameters, whereas exposure to 0.5 ppm chlorine for 8 hours does not result in significant pulmonary function changes, but does produce signs of irritation.

Chlorine is reported to have a "suffocating" odor[1] and an odor threshold (geometric mean of several reported literature values) of 0.31 ppm.[22] Amoore and Hautala[22] calculated an "odor safety factor" (TLV-TWA/odor threshold [geometric mean of all literature citations found, omitting extreme points and duplicate quotations]) for a number of compounds and developed an odor safety classification using letter designations A to E, in which class A compounds "provide the strongest odorous warning of their presence at threshold limit value concentrations," whereas class E substances are "practically odorless at the TLV concentration." Under this system, chlorine was placed in odor safety class C ("less than 50% of distracted persons perceive warning of TLV").

Basis for the AALG: Chlorine is a gas at room temperature but is also freely soluble and present in water by virtue of its extensive use in drinking water disinfection. Data on atmospheric chlorine levels are limited because most studies measure "gaseous chloride" and do not distinguish between Cl, HCl, or other chloride ion species. Some investigators have also questioned the presence of chlorine in the atmosphere based on its high reactivity and hence limited solubility. It is thought that atmospheric reactions involving sodium chloride aerosols are the major source of "gaseous chlorides."[9] However, since the AALG for chlorine is based on irritation, no relative source contribution from inhalation was factored into the calculation. The reason is that allocation of a proportion of exposure to a given source is not relevant when the effects of exposure are not cumulative, as is the case with irritant effects.

The AALG for chlorine is based on irritant effects. Although effects such as "tickling of the nose and throat" have been reported at concentrations ranging from approximately 0.015 ppm to 0.1 ppm, the validity of these sensations may be questionable, since the subjects were aware that they were being exposed to increasing concentrations of chlorine, rather than levels in a randomly selected order. However, based on several reports, distinct signs of irritation (ocular, nasal, respiratory) seem to begin in the range of 0.3 to 0.35 ppm. Therefore, the AALG was calculated by treating 0.3 ppm as a human LOAEL for irritant effects with an uncertainty factor of 50 (10 for interindividual variation × 5 for a LOAEL).

It should be noted that the occupational limits were considered unsuitable as a basis for AALG derivation because, although they are supported by recent well-conducted studies, 0.5 ppm was still a LOAEL for signs of subjective irritation and lower levels of exposure were not tested in these studies. Similarly, data from the study of Barrow et al.[14] were considered unsuitable for AALG derivation because concurrent exposure to chloramines could not be ruled out.

For comparative purposes only, an AALG was also calculated on the basis of the RD_{50} (concentration causing a 50% decrease in respiratory rate) in mice multiplied by 10^{-3} as suggested by Kane et al.[23] to derive an exposure level at which no effects would be expected. Using the RD_{50} of 9.3 ppm determined by Buckley et al.,[24] this yields a figure of 9.3 ppb.

AALG: • irritation—6 ppb (17.4 $\mu g/m^3$) 24-hour TWA

Note: The Task Group of the International Program on Chemical Safety (sponsored in part by the WHO) recommended a limit of 0.034 ppm (0.1 mg/m^3) for ambient chlorine levels to protect the general population from irritant effects. However, how this figure was derived was not explained in the documentation.[9]

References

1. ACGIH. 1986. "Documentation of the Threshold Limit Values and Biological Exposure Indices." 5:117.
2. NIOSH. 1976. "Criteria for a Recommended Standard . . . Occupational Exposure to Chlorine." DHEW (NIOSH) 76–170.
3. OSHA. 1989. "Air Contaminants; Final Rule." *Fed. Reg.* 54:2332–959.
4. ACGIH. 1988. "Threshold Limit Values and Biological Exposure Indices for 1988–89" (Cincinnati, OH: ACGIH).

5. Safe Drinking Water Committee. 1986. *Drinking Water and Health,* Vol. 7 (Washington, DC: National Academy Press).

6. Bishun, N. P., D. C. Williams, J. Mills, N. Lloyd, R. W. Raven, and D. V. Parke. 1973. "Chromosome Damage Induced by Chemicals." *Chem. Biol. Interact.* 6:375–92.

7. Meier, J. R., R. J. Bull, J. A. Stober, and M. C. Cimino. 1985. "Evaluation of Chemicals Used for Drinking Water Disinfection for Production of Chromosomal Damage and Spermhead Abnormalities in Mice." *Env. Mut.* 7:201–11.

8. Druckrey, H. 1968. "Chlorinated Water, Toxicity Tests, Involving Seven Generations of Rats." *Food Cosmet. Tox.* 6:147–54.

9. International Program on Chemical Safety. 1982. "Chlorine and Hydrogen Chloride." Environmental Health Criteria 21 (Geneva: World Health Organization).

10. Ferris, B. G., W. A. Burgess, and J. Worcester. 1967. "Prevalence of Chronic Respiratory Disease in a Pulp and Paper Mill in the United States." *Br. J. Ind. Med.* 24:26–37.

11. Patil, L. R. S., R. G. Smith, A. J. Vorwald, and T. F. Mooney. 1970. "The Health of Diaphragm Cell Workers Exposed to Chlorine." *Am. Ind. Hyg. Assoc. J.* 312:678–86.

12. NIOSH. 1988. FO2100000. "Chlorine." RTECS, on line.

13. Barrow, C. S., and D. E. Dodd. 1979. "Ammonia Production in Inhalation Chambers and its Relevance to Chlorine Inhalation Studies." *Tox. Appl. Pharm.* 49:89–95.

14. Barrow, C. S., R. J. Kociba, L. W. Rampy, D. G. Keyes, and R. R. Albee. 1979. "An Inhalation Toxicity Study of Chlorine in Fischer 344 Rats Following 30 Days of Exposure." *Tox. Appl. Pharm.* 49:77–88.

15. Bell, D. P., and P. C. Elmes. 1965. "The Effect of Chlorine Gas on the Lungs of Rats Without Spontaneous Lung Disease." *J. Path. Bact.* 89:307–17.

16. Arloing, F., E. Berthet, and J. Viallier. 1940. "Action of Chronic Intoxication of Chlorine Fume on Experimental Guinea Pigs." *Presse Med.* 48:361.

17. Rupp, H., and D. Henschler. 1967. "Effects of Low Chlorine and Bromine Concentrations in Man." *Int. Arch. Gew.* 23:79–90.

18. Beck, H. 1959. "Experimental Determination of the Olfactory Thresholds of Some Important Irritant Gases (Chlorine, Sulfur Dioxide, Ozone, Nitrous Gases) and Symptoms Induced in Humans by Low Concentrations." Inaugural dissertation. Wurzburg, West Germany.

19. Matt, L. 1889. "Experimental Contributions to the Theory of the Effects of Poisonous Gases on Human Beings." Inaugural dissertation, Julius-Maxmilliams-Universitat, Wurzburg, West Germany.

20. Anglen, D. 1981. Doctoral dissertation. University of Michigan, Ann Arbor, Michigan.

21. Rotman, H. H., et al. 1983. *J. Appl. Physiol.* 54:1120–24.

22. Amoore, J. E., and E. Hautala. 1983. "Odor as an Aid to Chemical Safety: Odor Thresholds Compared with Threshold Limit Values and Volatilities for 214 Industrial Chemicals in Air and Water Dilution." *J. Appl. Tox.* 3:272–90.

23. Kane, L. E., C. S. Barrow, and Y. Aleric. 1979. "A Short-Term Test to Predict Acceptable Levels of Exposure to Airborne Sensory Irritants." *Am. Ind. Hyg. Assoc. J.* 40:207–29.

24. Buckley, L. A., X. Z. Jiang, R. A. James, K. T. Morgan, and C. S. Barrow. 1984. "Respiratory Tract Lesions Induced by Sensory Irritants at the RD_{50} Concentration. *Tox. Appl. Pharm.* 74:417–29.

CHLOROBENZENE

Synonyms: monochlorobenzene, MCB, phenylchloride, benzene chloride

CAS Registry Number: 108–90–7

Molecular Weight: 112.56

Molecular Formula: ⬡–Cl

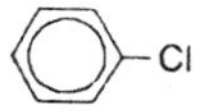

AALG: systemic toxicity—0.006 ppm (0.026 mg/m^3) 24-hour TWA

Occupational Limits: • ACGIH TLV: 75 ppm (350 mg/m^3) TWA
 • NIOSH REL: none
 • OSHA PEL: 75 ppm (350 mg/m^3) TWA

Basis for Occupational Limits: The TLV was set to prevent narcotic effects or "chronic poisoning." Although it was reported to be currently under review by the TLV committee in the 1986 documentation, it was not on the list of intended changes for either 1987–88 or 1988–89.[1–3]

Drinking Water Limits: In the ambient water quality criteria document,[4] two limits are calculated for chlorobenzene:

1. 488 μg/L, based on health effects and assuming consumption of 2 L of water and 6.5 g of fish/shellfish per day
2. 2.20 μg/L, based on undesirable taste and odor characteristics

Longer-term HAs of 4.3 mg/L and 15.0 mg/L were calculated for the 10-kg child and the 70-kg adult, respectively. Since data suitable for the calculation of 1-day and 10-day HAs were not available, it was recommended that the longer-term HA of 4.3 mg/L for the 10-kg child be used as both the 1-day and 10-day HA. Using 60 mg/kg/day as a mouse (and rat) NOAEL from the NTP bioassay, a lifetime HA of 0.3 mg/L was calculated for chlorobenzene.[5] In the "Drinking Water Criteria Document for Chlorobenzene," several possible ADIs are derived for chlorobenzene.[6] In addition, although CAG did not calculate a carcinogenic potency value for chlorobenzene, the NAS did so using the multistage model. The value derived for the upper 95% confidence estimate of lifetime cancer risk for an exposure level of 1 μg/L (unit risk) was 2.13 × 10^{-7}.[7]

Toxicity Profile

Carcinogenicity: In an NTP bioassay,[8] male and female F344/N rats and female B6C3F$_1$ mice received 0, 60, or 120 mg/kg chlorobenzene by gavage in corn oil 5 days a week for 103 weeks; male B6C3F$_1$ mice received 0, 30, or 60 mg/kg chlorobenzene in corn oil. In addition, untreated controls were also maintained throughout the study. Body weight gain was not adversely affected in either treated mice or rats, nor was survival significantly decreased in treated mice or female rats. However, survival of high-dose male rats was significantly reduced relative to vehicle controls, but no treatment-related lesions were identified as causative for this reduction in survival. No statistically significant increases in tumors were observed in mice or female rats. However, high-dose male rats had a statistically significant increase in the incidence of neoplastic nodules, but not hepatocellular carcinomas, of the liver. The incidences were as follows: untreated control, 4 of 50; vehicle control, 2 of 50; low-dose, 4 of 49; and high-dose, 8 of 49. It was concluded that the results of this study provided ''some but not clear evidence of carcinogenicity of chlorobenzene in male rats.''

Chlorobenzene has been classified in Group C, possibly carcinogenic to humans, based on the EPA weight-of-evidence classification.[6]

Mutagenicity: Chlorobenzene is reported to be nonmutagenic to *S. typhimurium*, *E. coli*, and mouse lymphoma L5178Y cells with and without metabolic activation. However, it has been reported to cause reverse mutations in *S. antibioticus* and *A. nidulans* and mitotic disturbances in *Allium cepa*.[5] Chlorobenzene was negative in the *Drosophila* sex-linked recessive lethal assay, but has been found to induce chromosomal aberrations in Chinese hamster ovary cells.[9]

Based on the IARC classification scheme, the evidence for genetic activity in short-term tests for chlorobenzene is best classified as limited.

Developmental Toxicity: John et al.[10] exposed F344 rats (n = 32 or 33/group) and New Zealand white rabbits (n = 30/group) to filtered air or 75, 210, or 590 ppm chlorobenzene for 6 hours per day on days 6–15 and days 6–18 of gestation, respectively. No statistically significant embryotoxic, fetotoxic, or teratogenic effects were observed in the rats, although slight maternal toxicity in the form of decreased body weight gain, decreased feed consumption, and increased liver weight occurred in the highest exposure group. In rabbits, an increase (not dose-related) in visceral malformations in exposed animals, but not concurrent controls, resulted in the conduct of a second study of the same design using exposure concentrations of 0, 10, 30, 75, and 590 ppm. Slight embryotoxicity, delayed skeletal development, and maternal toxicity similar to that seen in rats were observed only at the highest dose. No other significant embryotoxicity or teratogenic effects were observed.

Reproductive Toxicity: No data were found implicating chlorobenzene as a reproductive toxin.

Systemic Toxicity: The subchronic inhalation toxicity of chlorobenzene has been evaluated in several species, including rats, dogs, rabbits, and guinea pigs.[4,6,9,11] In general, the liver and kidney appear to be the target organs for chlorobenzene toxicity. Dilley[12] (cited in U.S. EPA[9]) exposed rats to 0, 75, or 250 ppm chlorobenzene for 7 hours per day, 5 days a week for 120 days. Findings at the low dose included focal lesions in the adrenal cortex and kidney tubules, congestion of the liver and kidney, and decreased SGOT. This LOAEL of 75 ppm in rats was used as the basis for calculation of an AIS (acceptable intake subchronic) criterion for chlorobenzene in the health effects assessment document.[9] In addition to hepatic and renal toxicity, toxic effects on the hematopoietic system (leukopenia) have been reported in dogs[13] (cited in NTP[8]) exposed *per os* to 272.5 mg/kg/day for 92 days and mice[14] (cited in NTP[8]) exposed via inhalation to 544 ppm for 3 weeks or 22 ppm for 3 months (7 hours per day in both cases).

In one occupational case report lacking exposure data, workers exposed to chlorobenzene for 1–2 years reported headaches, somnolence, dyspepsia, and various neurological symptoms, including spastic contractions of the finger muscles and hyperesthesia of the hands. An initial hematological examination of workers revealed leukopenia and thrombocytopenia in a number of workers; a follow-up examination several months later revealed that blood counts had returned to normal[15] (cited in U.S. EPA[5,6]).

Irritation: Chlorobenzene is reported to have an almond-like odor[1] and an odor threshold of 0.68 ppm (geometric mean of reported literature values).[16] The odor threshold in water is listed as 50 μg/L.[5]

Basis for the AALG: Chlorobenzene is used as an intermediate in chemical syntheses, as a solvent, and as a heat transfer medium. It is considered insoluble in water and has a vapor pressure of 8 mm Hg at 20°C.[1] Since inhalation is thought to be the major route of environmental exposure,[3] 80% of the contribution to total exposure is allowed from air.

AALGs were calculated for chlorobenzene based on both carcinogenicity and systemic toxicity. Because EPA has placed MCB in group C, possibly carcinogenic to humans, and CAG did not apply quantitative risk assessment to the NTP bioassay data, the AALG derivation is based on the rat NOAEL of 60 mg/kg/day MCB from the NTP bioassay data using an uncertainty factor approach. The NOAEL was adjusted for continuous exposure (60 mg/kg × 5/7) and converted to a human equivalent exposure level according to the procedures described in Appendix C, and a total uncertainty factor of 1000 (10 for interspecies variation × 10 for interindividual variation × 10 for the database factor) was applied. Application of the database factor was based on the use of an uncertainty factor approach in deriving an AALG for carcinogenicity and the use of a non-inhalation route of exposure.

Given the high degree of uncertainty associated with the carcinogenic endpoint derivation, an AALG for systemic toxicity was also derived using the rat LOAEL of 75 ppm from the data of Dilley et al.[12] The LOAEL was adjusted for continuous exposure (345 mg/m^3 × 5/7 × 7/24) and converted to a human equivalent exposure level according to the procedures described in Appendix C, and a total uncertainty factor of 5000 (10 for interindividual variation × 10 for interspecies variation × 5 for a LOAEL × 10 for less-than-lifetime exposure) was applied.

The recommended AALG is the one based on systemic toxicity because it is more conservative. However, the AALG should be considered provisional based on its group C classification (possible human carcinogen).

AALG: • carcinogenicity—0.026 ppm (0.12 mg/m^3) annual TWA
  • systemic toxicity—0.006 ppm (0.026 mg/m^3) 24-hour TWA

References

1. ACGIH. 1986. "Documentation of the Threshold Limit Values and Biological Exposure Indices." 5:123.
2. ACGIH. 1987. "Threshold Limit Values and Biological Exposure Indices for 1987–88" (Cincinnati, OH: ACGIH).
3. ACGIH. 1988. "Threshold Limit Values and Biological Exposure Indices for 1988–89" (Cincinnati, OH: ACGIH).
4. U.S. EPA. 1980. "Ambient Water Quality Criteria for Chlorinated Benzenes." EPA 440/5–80–028.
5. U.S. EPA. 1987. "Health Advisory: Chlorobenzene." Office of Drinking Water.
6. U.S. EPA. 1985. "Drinking Water Criteria Document for Chlorobenzene" (draft) (Springfield, VA: NTIS), PB86–117769.
7. Safe Drinking Water Committee. 1983. *Drinking Water and Health,* Vol. 5 (Washington, DC: National Academy Press), 22–26.
8. NTP. 1985. "Toxicology and Carcinogenesis Studies of Chlorobenzene (CAS No. 108–90–7) in F344/N Rats and B6C3F$_1$ Mice (Gavage Studies)" (Research Triangle Park, NC: National Toxicology Program), TR-261.
9. U.S. EPA. 1984. "Health Effects Assessment for Chlorobenzene." EPA/540/1–86/040.
10. John, J. A., W. C. Hayes, T. R. Hanley, K. A. Johnson, T. S. Gushow, and K. S. Rao. 1984. "Inhalation Teratology Study on Monochlorobenzene in Rats and Rabbits." *Tox. Appl. Pharm.* 76:365–73.
11. U.S. EPA. 1985. "Health Assessment Document for Chlorinated Benzenes." EPA 600/8–84–015F.
12. Dilley, J. V. 1977. "Toxic Evaluation of Inhaled Chlorobenzene." DHEW (NIOSH). Contract 210–76–0126.
13. Knapp, W., W. Busey, and W. Kundzins. 1971. "Subacute Oral Toxicity of Monochlorobenzene in Dogs and Rats." *Tox. Appl. Pharm.* 19:393(A).
14. Zub, M. 1978. "Reactivity of the White Blood Cell System to Toxic Actions of Benzene and Its Derivatives." *Acta Biol. Cracoviensia* 21:163–74.

15. Rozenbaum, N. D., R. S. Block, S. N. Bremneva, S. L. Ginzburg, and I. V. Pozhatiskii. 1947. ''Use of Chlorobenzene as a Solvent from the Standpoint of Industrial Hygiene'' (Russian). *Gig. Sanit.* 12:21–24.
16. Amoore, J. E. and E. Hautala. 1983. ''Odor as an Aid to Chemical Safety: Odor Thresholds Compared with Threshold Limit Values and Volatilities for 214 Industrial Chemicals in Air and Water Dilution.'' *J. Appl. Tox.* 3:272–90.

CHLOROFORM

Synonyms: trichloromethane

CAS Registry Number: 67–66–3

Molecular Weight: 119.38

Molecular Formula: CHCl$_3$

AALG: 10^{-6} 95% UCL—0.0045 ppb (0.022 μg/m^3) annual TWA

Occupational Limits:
- ACGIH TLV: 10 ppm (50 mg/m^3) TWA; Appendix A2 suspect human carcinogen
- NIOSH REL: 2 ppm (10 mg/m^3) ceiling (45-L sample in 60 minutes)
- OSHA PEL: 2 ppm (10 mg/m^3) TWA

Basis for Occupational Limits: The TLV was set in consideration of carcinogenic and embryotoxic effects.[1] The REL, originally 10 ppm in consideration of central nervous system toxicity, fetal abnormalities, hepatotoxicity, and irritant effects, was reduced to 2 ppm in 1976 on the basis of positive results in an NCI cancer bioassay.[2-4] OSHA recently adopted the NIOSH limit as a TWA (rather than a ceiling limit) as part of its final rule limits.[5]

Drinking Water Limits: Chloroform is the major constituent of the trihalomethanes, which also include dibromochloromethane, bromodichloromethane, and bromoform; the interim aggregate standard for trihalomethanes is 0.1 mg/L.[6] The ambient water quality criterion for chloroform is 0.19 μg/L at the 10^{-6} risk level based on consumption of 2 L of water and 6.5 g of fish/shellfish per day.[7] One-day and seven-day SNARLs of 22 mg/L and 3.2 mg/L, respectively, have been calculated for chloroform based on noncarcinogenic effects. A chronic SNARL was not calculated since chloroform is a carcinogen.[8]

Toxicity Profile

Carcinogenicity: A number of ecological and case-control studies (reviewed in detail in the HAD[4]) are suggestive of an association of increased risk of bladder, colon, and rectal cancer with exposure to chloroform in drinking water. Data on cancer risk due to inhalation exposure alone are lacking in humans. Overall, the available human data have been designated as limited or inadequate by both EPA and IARC.[3,9]

The carcinogenicity of chloroform has been evaluated in a number of studies that are most thoroughly reviewed in the HAD; those studies that serve as the basis for the quantitative risk estimates presented in the HAD are reviewed here.

In an NCI lifetime cancer bioassay[10] (cited in U.S. EPA[4]), the carcinogenicity of chloroform (98% $CHCl_3$ stabilized with 2% ethanol) was evaluated in male and female B6C3F$_1$ mice and Osborne-Mendel rats given the compound by gavage in corn oil 5 days per week. Male and female mice were exposed to 100 or 200 mg/kg/day and 200 or 400 mg/kg/day, respectively, for the first 18 weeks and the dosage was thereafter increased to 150 or 300 mg/kg/day and 250 or 500 mg/kg/day, respectively, for the remainder of the study. Male and female rats were exposed to 90 or 180 mg/kg/day and 125 or 250 mg/kg/day, respectively; after 22 weeks the dosages were reduced to 90 or 180 mg/kg/day for the female rats. Groups of matched vehicle controls and vehicle colony controls were maintained for both sexes of rats and mice throughout the study. There was a statistically significant increase in hepatocellular carcinomas in both male mice (colony control, 5 of 77; matched control, 1 of 18; low-dose, 18 of 50; and high-dose, 44 of 45) and female mice (colony control, 1 of 78; matched control, 0 of 20; low-dose, 36 of 45; and high-dose, 39 of 41). Non-neoplastic lesions attributable to treatment in mice included nodular hyperplasia and hepatocellular necrosis. There was a statistically significant treatment-related increase in renal epithelial tumors (adenomas and carcinomas) in male rats (colony control, 0 of 99; matched control, 0 of 18; low-dose, 4 of 50; and high-dose, 12 of 50). There was also an increase in thyroid (C-cell and follicular) tumors in female rats, but this was considered to be of controversial toxicologic significance based on histologic and statistical considerations. Non-neoplastic lesions attributable to treatment in rats included liver necrosis, epithelial hyperplasia of the urinary bladder, hematopoiesis in the spleen, and inflammatory pulmonary lesions characteristic of pneumonia.

Roe et al.[11] (cited in U.S. EPA[4]) conducted three bioassays of chloroform in four strains of mice (ICI, C57BL, CBA, CF1). In the first two experiments, chloroform was administered in a toothpaste vehicle; in the third experiment, the vehicle was arachis oil. The only statistically significant increase in tumors observed was for malignant kidney tumors (hypernephromas) in male ICI mice (vehicle control, 0 of 50; and 60 mg/kg/day, 9 of 48). These animals were gavaged 6 days per week for 80 weeks starting at 10 weeks of age. Treatment was followed by a 13- to 24-week observation period.

Jorgenson et al.[12] (cited in U.S. EPA[4]) exposed male Osborne-Mendel rats and female B6C3F$_1$ mice to redistilled chloroform in drinking water at 0, 200, 400, 900, and 1800 mg/L for 104 weeks. There was no increase in hepatocellular adenomas or carcinomas in mice, but there was a statistically significant increase in renal tubular cell carcinomas and adenomas in rats (control, 4 of 301; 200 mg/L, 4 of 313; 400 mg/L, 4 of 148; 900 mg/L, 3 of 48; and 1800 mg/L, 7 of 50). There were also statistically significant increases (generally comparing the control to the high-dose group) in neurofibromas, lymphomas and leukemias, adrenal cortical adenomas, adrenal pheochromocytomas, and thyroid C-cell adenomas and carcinomas.

Chloroform has been placed in group B2, probable human carcinogen, under the EPA weight-of-evidence classification. It should be noted that the EPA-CAG quantitative risk assessment for inhalation exposure is based on data from the NCI cancer bioassay reviewed above.[6]

Mutagenicity: IARC[13] recently reviewed and summarized the evidence for genotoxicity of chloroform as follows: "No data were available on the genetic and related effects of chloroform in humans. Chloroform did not induce micronuclei in bone-marrow cells of mice or DNA damage in liver or kidney cells of rats treated in vivo. It did not induce chromosomal aberrations, sister chromatid exchanges, or unscheduled DNA synthesis in human lymphocytes in vitro. Chloroform enhanced virus-induced cell transformation of Syrian hamster embryo cells. It did not induce sister chromatid exchanges or mutation in Chinese hamster cells or DNA damage in rat hepatocytes in vitro. Chloroform did not induce sex-linked recessive lethal mutations in *Drosophila* or aneuploidy, mutation, or somatic segregation in *Aspergillus*. Chloroform induced DNA damage but not mutation, aneuploidy, mitotic recombination, or gene conversion in *Saccharomyces cerevisiae*, whereas mutation, mitotic recombination, and gene conversion were induced in *S. cerevisiae* under conditions in which endogenous levels of cytochrome P450 were enhanced. Chloroform did not induce mutation or DNA damage in bacteria." In the HAD for chloroform,[4] it was stated that: "On the basis of presently available data, no definitive conclusion can be reached concerning the mutagenicity of chloroform. However, evidence from studies measuring binding to macromolecules, DNA damage, and mitotic arrest suggest that chloroform may be mutagenic."

Developmental Toxicity: Schwetz et al.[14] (cited in U.S. EPA[4]) exposed Sprague-Dawley rats (n = 20/group) to 0, 30, 100, or 300 ppm (0, 147, 489, and 1466 mg/m^3) chloroform for 7 hours per day on gestation days 6 through 15. The following outcomes at the indicated exposure levels showed statistically significant differences from controls:

- 30 ppm—decreased crown-rump length
- 100 ppm—acaudia, short tail, imperforate anus, subcutaneous edema, missing ribs, and delayed ossification of sternebrae
- 300 ppm—increase in resorptions, decrease in apparent conception rate, decreased crown-rump length

At the highest dose, subcutaneous edema and abnormalities of the skull and sternum were observed in the fetuses; however, these were not statistically significant, possibly due to the small numbers of surviving offspring. Significant maternal toxicity in the form of decreased body weight gain was observed at all exposure levels and food consumption was significantly reduced in the 100-ppm and 300-ppm groups. Based on a lack of embryotoxic and fetotoxic effects in a concurrent control group whose food intake was limited to match that of the 300-ppm group, it was concluded that chloroform itself was embryotoxic and teratogenic.

In another study, Murray et al.[15] (cited in U.S. EPA[4]) examined the effects of exposure to 100 ppm chloroform for 7 hours per day in CF1 mice exposed on days 1–7, 8–15, or 7–16 of gestation. In general, the results were similar to those of Schwetz et al.[14]: chloroform was found to interfere with pregnancy, be teratogenic, and cause embryotoxicity and maternal toxicity. Limitations in the study design did not allow for a conclusive determination of the time period during gestation when the developing fetuses were most susceptible to the effects of chloroform.

Reproductive Toxicity: Refer to the studies discussed in the previous section.

Systemic Toxicity: In humans, 10,000 to 20,000 ppm chloroform will produce clinical anesthesia. Immediate death due to chloroform anesthesia occurs as a result of ventricular fibrillation, whereas delayed anesthesia death is usually associated with liver necrosis.[1,2]

Challen[16] (cited in NIOSH[2]) reported the following effects in workers intermittently exposed to chloroform for 4 hours per day over a period of 10 to 24 months:

- 23–35 ppm—lassitude, dry mouth, and irritability
- 57–71 ppm—flatulence, lassitude, loss of appetite, and nausea

Bomski[17] (cited in NIOSH[2]) reported a study in which workers inhaling 2–205 ppm chloroform for 1 to 4 years were compared with several control groups. Significant clinical findings included increased incidence of viral hepatitis, toxic hepatitis with elevated serum gamma globulins, splenomegaly, and hepatomegaly.

Lehmann and Hasegawa[18] and Lehmann and Schmidt-Kehl[19] (cited in NIOSH[2]) reported the following effects in controlled exposure experiments with humans:

- 390 ppm for 30 minutes—light transient odor
- 680–1000 ppm for 30 minutes—moderately strong odor and taste
- 1100 ppm for 5 minutes—still stronger, permanent odor; dizziness, vertigo after 2 minutes
- 4300–5000 ppm for 20 minutes—dizziness and light intoxication

Torkelson et al.[20] (cited in U.S. EPA[4]) studied the effects of subchronic (5 days per week for 6 months) inhalation exposure to 25, 50, or 85 ppm chloroform in rats, guinea pigs, and rabbits. The NOAEL and LOAEL for liver and kidney effects in rats were found to be 25 ppm for 4 hours per day and 25 ppm for 7 hours per day, respectively. The histopathologic lesions were described as centrilobular granular degeneration of the liver and cloudy swelling of the kidneys. Results in rabbits and guinea pigs were inconsistent: similar lesions were observed in the lowest but not the two highest exposure groups.

Irritation: Refer to the human studies discussed in the previous section. Chloroform is reported to have a "characteristic pleasant, sweet odor" and an odor threshold of 0.3 mg/m^3.[1]

Basis for the AALG: Chloroform is of low water solubility (0.822 mL/100 mL water at 20°C), and is highly volatile with a vapor pressure of 158.4 mm Hg at 20°C. It is used as an intermediate in chemical manufacture and as an extractant and industrial solvent.[1] Chloroform is formed by the reaction of chlorine with organic matter in waters treated with chlorine for disinfection purposes and is also commonly present in ambient air.[4] Based on these considerations, 50% of the contribution to total exposure will be allowed from air.

Two AALGs were calculated for chloroform, one based on developmental toxicity and the other based on carcinogenicity. The basis for the developmental AALG is the rat LOAEL of 30 ppm from the study of Schwetz et al.[14] This LOAEL was adjusted for continuous exposure (146.5 × 7/24) and converted to a human equivalent exposure level according to the procedures described in Appendix C, and a total uncertainty factor of 500 (5 for a LOAEL × 10 for interindividual variation × 10 for interspecies variation) was applied. Application of the developmental uncertainty factor was considered unwarranted since fetotoxicity occurred at levels producing maternal toxicity.

The EPA-CAG has performed a quantitative risk assessment on data from the 1976 NCI bioassay. The recommended slope factor of 8.1 × 10^{-2} (mg/kg/day)$^{-1}$ presented in the IRIS database was calculated on the basis of the geometric mean of the q_1*s from the data for hepatocellular carcinomas in male and female mice (linearized multistage model with extra risk). Assuming 100% absorption, an inhalation unit risk of 2.3 × 10^{-5} was calculated, which corresponds to an inhalation q_1* of 2.3 × 10^{-5} (μg/m^3)$^{-1}$. Applying the relationship E = q_1* × d, where E is the specified level of extra risk and d is the dose, gives the estimates shown below. It is noted in the IRIS summary[6] that "the unit risk should not be used if the air concentration exceeds 400 μg/m^3, since above this concentration the slope factor may differ from that stated." This AALG should be considered provisional since it was derived from a study using a non-inhalation route of exposure.

AALG: • developmental—16 ppb (80 μg/m^3) 24-hour TWA
  • 10^{-6} 95% UCL—0.0045 ppb (0.022 μg/m^3) annual TWA
  • 10^{-5} 95% UCL—0.045 ppb (0.22 μg/m^3) annual TWA

References

1. ACGIH. 1986. "Documentation of the Threshold Limit Values and Biological Exposure Indices." 5:130–31.

2. NIOSH. 1974. "Criteria for a Recommended Standard . . . Occupational Exposure to Chloroform." DHEW (NIOSH) 75–114.

3. U.S. EPA. 1984. "Health Effects Assessment for Chloroform." EPA/540/1–86/010.

4. U.S. EPA. 1985. "Health Assessment Document for Chloroform, Final Report." EPA-600/8–84–004F.

5. OSHA. 1989. "Air Contaminants; Final Rule." *Fed. Reg.* 54:2332–2959.

6. U.S. EPA. 1988. "Chloroform; CASRN 67–66–3." IRIS (9/26/88).

7. U.S. EPA. 1980. "Ambient Water Quality Criteria for Chloroform." EPA-440/5–80–033.

8. Safe Drinking Water Committee. 1982. *Drinking Water and Health* (Washington, DC: National Academy Press), pp. 206–9.

9. IARC. 1987. "Chloroform." *IARC Monog.* suppl. 7:152–54.

10. NCI. 1976. "Report on Carcinogenesis Bioassay of Chloroform." (Springfield, VA: NTIS), PB76–264018.

11. Roe, F. J. C., A. K. Palmer, A. N. Worden, and N. J. VanAbbe. 1979. "Safety Evaluation of Toothpaste Containing Chloroform. I. Long-Term Studies in Mice." *Environ. Path. Toxicol.* 2:799–819.

12. Jorgenson, T. A., E. F. Meierhenry, C. J. Rushbrook, R. J. Bull, and M. Robinson. 1985. "Carcinogenicity of Chloroform in Drinking Water to Male Osborne-Mendel Rats and Female B6C3F$_1$ Mice." *Fund. Appl. Tox.* 5:760–69.

13. IARC. 1986. "Chloroform." *IARC Monog.* suppl. 6:155–58.

14. Schwetz, B. A., B. J. K. Leong, and P. J. Gehring. 1974. "Embryo- and Fetotoxicity of Inhaled Chloroform in Rats." *Tox. Appl. Pharm.* 28:442–51.

15. Murray, F. A., B. A. Schwetz, J. G. McBride, and R. E. Staples. 1979. "Toxicity of Inhaled Chloroform in Pregnant Mice and Their Offspring." *Tox. Appl. Pharm.* 50:151.

16. Challen, P. J. R., D. E. Hickish, and J. Bedford. 1958. "Chronic Chloroform Intoxication." *Br. J. Ind. Med.* 15:243–49.

17. Bomski, H., A. Sobolewska, and A. Strakowski. 1967. "Toxic Damage of the Liver by Chloroform in Chemical Industry Workers" (German). *Arch. Gewerbepathol. Gewerbehyg.* 24:127–34.

18. Lehmann, K. B. and T. S. Hasegawa. 1910. "Studies of the Absorption of Chlorinated Hydrocarbons in Animals and Humans" (German). *Arch. Hyg.* 72:327–42.

19. Lehmann, K. B. and L. Schmidt-Kehl. 1936. "The Thirteen Most Important Chlorinated Aliphatic Hydrocarbons from the Standpoint of Industrial Hygiene" (German). *Arch. Hyg.* 116:131–200.

20. Torkelson, T. R., F. Oyen, and V. K. Rowe. 1976. "The Toxicity of Chloroform as Determined by Single and Repeated Exposure of Laboratory Animals." *Am. Ind. Hyg. Assoc. J.* 37:697–705.

CHLOROMETHYL METHYL ETHER

Synonyms: dimethylchloroether, CMME

CAS Registry Number: 107–30–2

Molecular Weight: 80.52

Molecular Formula: $ClCH_2OCH_3$

AALG: 10^{-6} 95% UCL—6.4×10^{-8} ppb (3.0×10^{-7} $\mu g/m^3$) as an annual TWA

Occupational Limits:
- ACGIH TLV: Appendix A2 suspect human carcinogen (no TLV)
- NIOSH REL: none
- OSHA PEL: regulated as a carcinogen (no PEL)

Basis for Occupational Limits: CMME (technical grade) is considered to be a carcinogen by both the ACGIH and OSHA.[1,2] OSHA regulations concerning work practices related to CMME are given in 29 CFR 1910.1006.[3]

Drinking Water Limits: No drinking water limits were found for this substance. No criterion was derived for CMME in the ambient water quality criteria document for chloroalkyl ethers.[4]

Toxicity Profile

Carcinogenicity: Because in industrial settings BCME generally occurs as an impurity of CMME, it is difficult to epidemiologically distinguish the carcinogenic effects of one compound from the other. Figueroa et al.[5] (cited in IARC[6]) identified 4 cases of lung cancer during a 5-year period in a study of 111 CMME workers; further retrospective study resulted in the identification of a total of 14 cases in workers engaged in the production of CMME. The incidence of lung cancer was 8 times higher than expected relative to a control group with similar smoking history. Exposure ranged from 3 to 14 years in 13 of 14 cases; the workers were aged 35–55 years, which was somewhat lower than expected. Twelve of the 14 lung cancers were histologically confirmed as oat cell carcinomas. The evidence for human carcinogenicity of technical grade CMME is considered sufficient by EPA.[7]

Gargus et al.[8] studied the effect of CMME and BCME on the induction of lung adenomas in newborn Swiss ICR mice subcutaneously injected with a single dose of 125 $\mu L/kg$ CMME or 12.5 $\mu L/kg$ BCME in peanut oil. The incidence and average number of lung adenomas per mouse were as follows: controls, 14% and 0.14; BCME, 45% and 0.64; and CMME, 17% and 0.21. Although the percentage of animals with lung adenomas and the mean number of lung tumors per animal were higher in CMME-treated animals compared to controls, the difference was not statistically significant. When Van Duuren et al.[9] (cited in IARC[6]) administered CMME to female Sprague-Dawley rats and female Swiss ICR/Ha mice by subcutaneous injection weekly for 300 and 30 days, respectively, they found an increase in sarcomas at the injection site in the treated animals, but not in the controls.

Leong et al.[10] exposed male Strain A Heston mice (n = 50 per group) to urethane (130 exposures, positive controls), room air (130 exposures, negative controls), 1 ppm BCME (82 exposures), or 2 ppm CMME (101 exposures) for 6 hours per day, 5 days per week. The incidence of lung adenomas was controls 41%, urethane-treated 97%, BCME-treated 55%, and CMME-treated 50%. The increase in lung adenomas in the CMME-treated animals was not statistically significant.

In a lifetime inhalation study, Laskin et al.[11] exposed male Sprague-Dawley rats (n = 74/group) and male Syrian golden hamsters (n = 88/group) to 0 or 1 ppm CMME for 6 hours per day, 5 days per week. There were no differences between control and treated animals of either species with respect to survival or body weight gain. However, the incidence of bronchial hyperplasia was 35% in control rats compared to 59% in treated rats, and two tumors, a squamous cell carcinoma of the lung and a esthesioneuroepithelioma originating in the olfactory epithelium, were found. In treated hamsters, two tumors, an adenocarcinoma of the lung and a squamous cell papilloma of the trachea, were found.

In all of the above studies industrial grade CMME, which would have been contaminated with BCME, was used. Thus, it is not possible to determine whether the minimal effects observed were due to CMME or BCME. It is concluded in the ambient water quality criteria document that "in practical terms, commercial grade CMME must be considered as a respiratory carcinogen, although of a lower order of activity than BCME."[4]

CMME is classified in class A, human carcinogen, under the EPA weight-of-evidence classification. No oral or inhalation quantitative risk estimates were available for CMME in the IRIS database; however, it is stated in the IRIS summary[7] that the risk is "likely to be no more than that of BCME, a contaminant of CMME."

Mutagenicity: CMME is mutagenic to *E. coli* and *S. typhimurium* without metabolic activation, and chromosomal aberrations have been found in the peripheral lymphocytes of workers exposed to CMME and BCME.[7] CMME also induces oncogenic transformation of Syrian hamster embryo cells by adenovirus SA7.[12]

Developmental Toxicity: No data were found implicating CMME as a developmental toxin.

Reproductive Toxicity: No data were found implicating CMME as a reproductive toxin.

Systemic Toxicity: The LC_{50} values for 7-hour exposures are 55 ppm and 65 ppm in rats and hamsters, respectively.[11]

Irritation: Exposure to 100-ppm CMME vapor is severely irritating to the eyes and nose in humans and may be fatal within 4 hours. CMME is reported to have an "irritating" odor,[6] but no specific odor threshold was found.

Basis for the AALG: Since CMME is highly volatile and decomposes in water to formaldehyde and hydrochloric acid,[6] 80% of the contribution to total exposure is allowed from air.

The AALG for CMME is based on carcinogenicity. However, the available data do not permit a standard quantitative risk assessment to be performed because in all of the available studies the CMME was contaminated with BCME. Therefore, since CMME has been judged by the EPA to be a class A carcinogen with a potency not likely to exceed that of BCME, the AALG for CMME is based by analogy on that of BCME, i.e., since CMME is considered to be possibly as potent a carcinogen as BCME, but not more potent a carcinogen than BCME, the recommended AALG for BCME is used as the AALG for CMME. This AALG should be considered provisional since it is based by analogy on that of BCME.

AALG:
- 10^{-6} 95% UCL—6.4×10^{-8} ppb (3.0×10^{-7} $\mu g/m^3$)
- 10^{-5} 95% UCL—6.4×10^{-7} ppb (3.0×10^{-6} $\mu g/m^3$) as annual TWA

References

1. ACGIH. 1988. "Threshold Limit Values and Biological Exposure Indices for 1988–89" (Cincinnati, OH: ACGIH).
2. OSHA. 1989. "Air Contaminants; Final Rule." *Fed. Reg.* 54:2332–2959.
3. OSHA. 1988. "29 U.S. Code of Federal Regulations 1910.1006."
4. U.S. EPA. 1980. "Ambient Water Quality Criteria for Chloroalkyl Ethers." EPA 440/5–80–030.
5. Figueroa, W. G., R. Raskowski, and W. Weiss. 1973. "Lung Cancer in Chloromethyl Methyl Ether Workers." *New Engl. J. Med.* 288:1096.
6. IARC. 1974. "Chloromethyl Methyl Ether." *IARC Monog.* 4:239–45.
7. U.S. EPA. 1988. "Chloromethyl Methyl Ether; CASRN 107–30–2." IRIS (3/1/88).
8. Gargus, J. L., W. H. Reese, and H. A. Rutter. 1969. "Induction of Lung Adenomas in Newborn Mice by Bis(chloromethyl) Ether." *Tox. Appl. Pharm.* 15:92–96.
9. Van Duuren, B. L., C. Katz, B. M. Goldschmidt, K. Frenkel, and A. Sivak. 1972. "Carcinogenicity of Halo-ethers. II. Structure-Activity Relationship of Analogs of Bis(chloromethyl) Ether." *J. Natl. Cancer Inst.* 48:1431.

10. Leong, B. K. J., H. N. Macfarland, and W. H. Reese. 1971. "Induction of Lung Adenomas by Chronic Inhalation of Bis(chloromethyl) Ether." *Arch. Environ. Heal.* 22: 663–66.
11. Laskin, S., R. T. Drew, V. Cappiello, M. Kuschner, and N. Nelson. 1975. "Inhalation Carcinogenicity of α-Halo-ethers. II. Chronic Inhalation Studies with Chloromethyl Methyl Ether." *Arch. Environ. Heal.* 30:70–72.
12. IARC. 1986. "Chloromethyl Methyl Ether." *IARC Monog.* suppl. 6:159–60.

CHROMIUM AND COMPOUNDS

CAS Registry Numbers: 16065–83–3 (III), 7440–47–3 (VI)

Synonyms: metal and inorganic compounds as Cr, including both Cr(III) and Cr(VI) compounds

Molecular Weight: 51.996

Molecular Formula: Cr

AALG:
- Di- and trivalent Cr compounds, Cr metal, systemic toxicity—2.3 μg (Cr)/m^3 8-hour TWA
- Noncarcinogenic Cr(VI) compounds, systemic toxicity—0.12 μg (Cr)/m^3 8-hour TWA
- Carcinogenic Cr(VI) compounds, 95% UCL 10^{-6}—1.67×10^{-5} μg/m^3 annual TWA

Occupational Limits:

ACGIH TLV: 0.5 mg (Cr)/m^3 TWA (Cr metal, Cr(II) and Cr(III) compounds)

0.05 mg (Cr)/m^3 TWA (water-soluble Cr(VI) compounds)

0.05 mg (Cr)/m^3 TWA (certain water-insoluble Cr(VI) compounds; Appendix A1 recognized human carcinogen)

0.05 mg (Cr)/m^3 TWA (chromite ore processing; Appendix A1 recognized human carcinogen)

NIOSH REL: 0.001 mg (Cr)/m^3 TWA (carcinogenic Cr(VI) compounds)

0.025 mg (Cr)/m^3 TWA (other Cr(VI) compounds)

0.050 mg/m^3 TWA; 0.2 mg/m^3 ceiling (chromic acid (CrO$_3$))

OSHA PEL: 1 mg/10 m^3 ceiling (chromic acid/chromates)

1 mg (Cr)/m^3 TWA (chromium metal and insoluble salts)

0.5 mg (Cr)/m^3 TWA(Cr(II) and (III) salts)

Basis for Occupational Limits: The TLVs for chromium metal and di- and trivalent Cr compounds are based on the prevention of pulmonary disease and other toxic effects such as dermatitis. The TLV for water-soluble Cr compounds (e.g., chromic acid and its anhydride and the monochromates and dichromates of sodium, potassium, ammonium, lithium, cesium, and rubidium) is set to protect against irritation of the respiratory tract and possible renal and hepatic damage. The TLV for certain water-insoluble Cr compounds, e.g., sintered chromium trioxide and zinc, calcium, barium, lead, and strontium chromates, is based on respiratory cancers in workers.[1] The NIOSH and OSHA limits apply only to hexavalent chromium and chromic acid and are based on the same toxicologic effects.[2,3]

Drinking Water Limits: Under the National Interim Primary Drinking Water Regulations, the current MCL for total chromium is 0.05 mg/L. An MCLG of 0.12 mg/L has been proposed for total chromium. This latter limit is based on a provisional AADI of 0.17 mg/L with data on human exposure factored in (0.10 mg/day in the diet and 0 mg/day via air).[4]

All health advisories for chromium are based on total chromium (III and VI). Because appropriate data were not available for the calculation of a 1-day HA, it was recommended that the 10-day child HA of 1.4 mg/L be used as the 1-day HA. The 10-day HAs for Cr are 1.4 mg/L and 5.0 mg/L for the 10-kg child and the 70-kg adult, respectively. These limits were derived based on a rat NOAEL of 14.4 mg/kg/day from a 60-day exposure study. Longer-term HAs of 0.24 mg/L and 0.84 mg/L for Cr were calculated for the 10-kg child and 70-kg adult, respectively. The lifetime HA for chromium is 0.12 mg/L, which was derived from an RfD of 0.0048 mg/kg/day with an assumed relative source contribution of 71% from drinking water.[5,6]

In the "Ambient Water Quality Criteria for Chromium" document, separate criteria were derived for trivalent and hexavalent chromium. Assuming a 70-kg human drinking 2 L of water and eating 6.5 g fish/shellfish per day, the criteria for trivalent and hexavalent chromium are 170 mg/L and 0.05 mg/L, respectively.[7]

Toxicity Profile

Carcinogenicity: Epidemiologic studies of workers in the chromate, chrome-plating, and chrome pigment industries in Japan, the United States, Great Britain, and West Germany have demonstrated an association between exposure to chromium and lung cancer. However, it was not possible on the basis of human data to differentiate clearly between exposure to hexavalent and trivalent compounds.[2,8,9]

The two studies that provide the basis for the EPA quantitative risk estimates are reviewed here. Numerous other studies are reviewed in the HAD[10] and the NIOSH criteria document.[2] In a cohort study of all workers employed for greater than one year in an Ohio chromate plant from 1931 to 1949, Mancuso and Heuper[11] (cited in NIOSH[2] and U.S. EPA[10]) found that 18.2% (6 of 33) of the worker deaths were due to respiratory cancer, compared with 1.2% deaths due to respiratory cancer in the male population of the county in which the plant was located. In a follow-up study of the cohort of workers employed from 1931 to 1937, Mancuso[12] (cited in NIOSH[2] and U.S. EPA[10]) found that the percentages of cancer deaths due to lung cancer were 63.6%, 62.5%, and 58.3% for workers employed from 1931 to 1932, 1933 to 1934, and 1935 to 1937, respectively. It was found that lung cancer deaths were dose-related to trivalent, hexavalent, and total chromium exposure. However, it was not possible to distinguish the relative contributions of any given species of chromium because as exposure to one increased so did exposure to all others. It should be noted that analysis of lung tissue from chromium-exposed workers indicated that chromium is retained in the lungs for long periods of time.

Carcinogenicity studies of tri- and hexavalent chromium are reviewed extensively in the HAD for chromium. Studies with trivalent chromium compounds administered via any route and inhalation studies with hexavalent compounds have produced negative results. Several hexavalent compounds—lead chromate and its oxide, calcium chromate, sintered chromium trioxide, sodium dichromate, strontium chromate, and zinc chromate—have been shown to produce local sarcomas or lung tumors in rats at the intramuscular, subcutaneous, or intraperitoneal injection sites, or when given intratracheally or intrabronchially. Of these compounds, only calcium chromate has consistently produced tumors in rats by several routes of administration, including intratracheal instillation, intrabronchial implantation, intrapleural implant, and intramuscular injection.

Hexavalent chromium has been classified in group A, human carcinogen, and trivalent chromium in group D, unclassified substance, based on the EPA weight-of-evidence classification.[5,9]

Mutagenicity: In vitro assays using bacterial species and cultured mammalian cells have generally been negative for trivalent chromium compounds, but positive for hexavalent compounds in the absence of metabolic activation systems and negative in their presence. Based on these findings, it has been suggested that mammalian enzymes or cofactors may reduce Cr(VI) to Cr(III). Both Cr(III) and Cr(VI) interact with DNA in bacterial assays, decrease the fidelity of DNA replication, and are clastogenic to mammalian cells; in the latter effect, Cr(VI) is more active than Cr(III). The genotoxicity of chromium is reviewed in detail in the HAD.[10]

Developmental Toxicity: No data were found implicating chromium as a developmental toxin via inhalation or oral administration.[8,9] However, when administered parenterally certain chromium compounds—CrO_3, $CrCl_3$, and $Na_2Cr_2O_7$—have been reported to be teratogenic and/or embryotoxic in mice or hamsters. These studies, reviewed in the HAD,[10] usually reported fetal effects at doses that were also maternally toxic.

Reproductive Toxicity: No data were found implicating chromium as a reproductive toxin via inhalation exposure. However, testicular degeneration has been reported in rabbits receiving 2 mg/kg/day Cr(VI) or Cr(III) by intraperitoneal injection for 6 weeks;[10] in another study, rats exposed to zinc chromate and potassium chromate in the diet for an unspecified period of time became sterile[13] (cited in U.S. EPA[10]).

Systemic Toxicity: Chromium is considered to be an essential element in both animals and humans; a deficiency of chromium may result in abnormal glucose tolerance, correctable by dietary supplementation.[10]

In general, divalent and trivalent chromium compounds are less toxic than hexavalent compounds in animal studies.[10] Both trivalent and hexavalent compounds may cause allergic contact dermatitis, although Cr(VI) appears to be responsible in more reported cases. In one study[14] (cited in ACGIH[1]), it was reported that workers exposed to chromite dust showed "exaggerated pulmonary markings" on chest X-rays. In another study[15] (cited in ACGIH[1]), pulmonary disease was reported in ferrochrome alloy workers where chromium levels were 0.27 mg/m^3. However, other dusts and fumes were also present in this plant.

Occupational and case studies of workers in the chrome industry have indicated that exposure to some chromate compounds can result in pneumoconiosis, bronchitis, chronic lung congestion, liver and kidney damage, perforation of the nasal septum, and irritation of the respiratory tract. The latter two effects are discussed in the next section. With respect to liver and kidney damage resulting from chromium exposure, the association must be considered weak since it is based on a few case studies and limited acute target organ toxicity studies in animals. It was concluded in the HAD that data from the occupational studies were insufficient for development of quantitative exposure limits for the noncarcinogenic effects of chromate compounds.[10] In the criteria document for Cr(VI) compounds,[2] NIOSH indicated that the available data are insufficient to conclusively define any Cr(VI) compounds as not being capable of inducing the above effects.

Chromium dermatitis and allergic contact dermatitis are associated with exposure to both Cr(III) and Cr(VI) compounds in several areas of the chrome industry, including plating, leather tanning, and pigment manufacture.[2,10]

Irritation: Perforation of the nasal septum, nasal ulcers, and irritation of the respiratory tract are most strongly associated with the chrome-plating industry where the exposure is restricted to chromic acid mist. However, these effects have also been reported in other branches of the chromium industry using Cr(VI) compounds. Perforated nasal septa and severe irritation of the respiratory tract have been associated with chromic acid concentrations of > 0.1 mg/m^3 and 0.12 mg/m^3 for other Cr(VI) compounds.[10] It should be noted that mice exposed to 13 mg/m^3 calcium chromate dust 5 hours per day, 5 days a week for their lifetime exhibited respiratory effects, including hyperplasia and necrosis, consistent with severe local irritation of the cells lining the airways[16] (cited in U.S. EPA[10]). It is also noteworthy that poor personal hygiene, which may result in high local concentrations of chromium on the skin and mucous membranes, enhances the irritant effects of chromium on the nasal passages.

Basis for the AALG: Since Cr(III) and Cr(VI) compounds are widely distributed in the environment and the bulk of exposure is thought to come from sources other than inhalation,[3,6] 20% of the contribution to total exposure will be allowed from air.

Three AALGs are calculated for chromium: one for di- and trivalent compounds, another for evident and inferred noncarcinogenic Cr(VI) compounds, and one for evident and inferred carcinogenic Cr(VI) compounds. The latter is based on the epidemiologic studies reviewed in "Carcinogenicity" above and derived by the U.S. EPA. The noncarcinogenic AALGs are based on the TLV/42 in the case of di- and trivalent compounds and chromium metal, and on the REL/42 in the case of noncarcinogenic Cr(VI) compounds, since there is evidence for systemic as well as irritant effects in both cases.

The distinction between evident and inferred noncarcinogenic and carcinogenic Cr(VI) compounds is adopted from the NIOSH criteria document for Cr(VI) compounds. Evident noncarcinogens are those Cr(VI) compounds (sodium bichromate and chromate, and Cr(VI) oxide) for which there is evidence of noncarcinogenicity; inferred noncarcinogens are Cr(VI) compounds (lithium, potassium, rubidium, cesium, and ammonium chromates and bichromates) that have chemical and physical properties similar to the evident noncarcinogens. Likewise, the evident carcinogens are those Cr(VI) compounds (calcium chromate, sintered calcium chromate, alkaline lime roasting process residue, zinc potassium chromate, and lead chromate) for which there is strong evidence of carcinogenicity; inferred carcinogens are those Cr(VI) compounds (alkaline earth chromates and bichromates, chromyl chloride, *t*-butyl chromate, and other Cr(VI) compounds not listed under other categories) with chemical and physical properties similar to evident carcinogens, or for which there is no evidence suggestive of their noncarcinogenicity. The specific evidence and criteria for placing various Cr(VI) compounds in these categories is reviewed in the criteria document.[2]

In the IRIS database,[5] the recommended inhalation slope factor of 4.1×10^1 $(mg/kg/day)^{-1}$ is based on the linearized multistage model with extra risk applied to the data of Mancuso[12] (cited in U.S. EPA[5,10]). This corresponds to an inhalation unit risk of 1.2×10^{-2}, equivalent to an inhalation q_1^* of $1.2 \times 10^{-2} (\mu g/m^3)^{-1}$. Applying the relationship $E = q_1^* \times d$, where E is the specified level of extra risk and d is the dose, and allowing 20% of the contribution to total exposure from air, yields the limits shown below. It should be noted that the cancer mortality observed in the Mancuso study was assumed to be due to Cr(VI), which was assumed to make up no more than 1/7 of the total exposure to chromium. It is noted in the IRIS summary[5] that the "unit risk should not be used if the air concentration exceeds 0.8 $\mu g/m^3$, since above this concentration the slope factor may differ from that stated."

AALG:
- Di- and trivalent Cr compounds, Cr metal systemic toxicity—2.3 μg (Cr)/m^3 8-hour TWA
- Noncarcinogenic Cr(VI) compounds systemic toxicity—0.12 μg (Cr)/m^3 8-hour TWA
- Carcinogenic Cr(VI) compounds
  95% UCL 10^{-6}—$1.67 \times 10^{-5} \mu g/m^3$ annual TWA
  95% UCL 10^{-5}—$1.67 \times 10^{-4} \mu g/m^3$ annual TWA

References

1. ACGIH. 1986. "Documentation of the Threshold Limit Values and Biological Exposure Indices." 5:139–40.
2. NIOSH. 1975. "Criteria for a Recommended Standard . . . Occupational Exposure to Chromium (VI)." DHEW (NIOSH) 76–129.
3. ATSDR. 1987. "Toxicological Profile for Cadmium" (draft). Agency for Toxic Substances and Disease Registry (10/87).
4. U.S. EPA. 1985. "National Primary Drinking Water Regulations; Synthetic Organic Chemicals, Inorganic Chemicals, and Microorganisms; Proposed Rule." *Fed. Reg.* 50:46936–7022.
5. U.S. EPA. 1988. "Chromium (VI); CASRN 7440–47–3." IRIS (3/1/88).
6. U.S. EPA. 1987. "Health Advisory—Chromium." Office of Drinking Water (3/31).
7. U.S. EPA. 1980. "Ambient Water Quality Criteria for Chromium." EPA/440/5–80–035.
8. U.S. EPA. 1984. "Health Effects Assessment for Hexavalent Chromium." EPA/540/1–86/019.
9. U.S. EPA. 1984. "Health Effects Assessment for Trivalent Chromium." EPA/540/1–86/035.
10. U.S. EPA. 1984. "Health Assessment Document for Chromium; Final Report." EPA-600/8–83/014F.
11. Mancuso, T. F. and W. C. Heuper. 1951. "Occupational Cancer and Other Health Hazards in a Chromate Plant: A Medical Appraisal. I. Lung Cancer in Chromate Workers." *Ind. Med. Surg.* 20:358–63.
12. Mancuso, T. F. 1975. International Conference on Heavy Metals in the Environment, Toronto, Canada, Oct. 27–31.
13. Gross, W. G. and V. G. Heller. 1946. "Chromates in Animal Nutrition." *J. Ind. Hyg. Tox.* 28:52–56.
14. Parkhurst, H. J. 1925. *Arch. Dermatol. Syphilol.* 12:253–56.
15. Princi, F., L. H. Miller, A. Davis, and J. Cholak. 1962. *J. Occ. Med.* 4:301.
16. Nettesheim, P., M. G. Hanna, D. G. Doherty, R. F. Newell, and A. Hellman. 1971. "Effect of Calcium Chromate Dust, Influenza Virus, and 100 Roentgen Whole-Body X-radiation on Lung Tumor Incidence in Mice." *J. Nat. Cancer Inst.* 47:1129–44.

COBALT AND COMPOUNDS

Coverage: This assessment primarily deals with the effects of inhalation exposure to cobalt metal, dust, and fume. Information on the effects of exposure to cobalt salts, such as cobaltous chloride ($CoCl_2$, CAS 7646–79–9), cobaltous nitrate ($Co(NO_3)_2$, CAS 10141–05–6), cobaltous sulfate ($CoSO_4$, CAS 10124–43–3), cobalt carbonyl ($Co_2(CO)_4$, CAS 10210–68–1), and hydrocarbonyl ($HCo(CO)_4$, CAS 16842–03–8), is included where applicable.

CAS Registry Number: 7440–48–4

Molecular Weight: 58.93

Molecular Formula: Co

AALG: systemic toxicity (TLV/42)—0.24 $\mu g/m^3$ 8-hour TWA

Occupational Limits:
- ACGIH TLV: 0.05 mg/m^3 TWA; 0.1 mg/m^3 STEL for cobalt metal, dust, and fume as Co; 0.1 mg/m^3 TWA for cobalt carbonyl as Co; 0.1 mg/m^3 TWA for cobalt hydrocarbonyl as Co
- NIOSH REL: none recommended in most recent documentation
- OSHA PEL: 0.05 mg/m^3 TWA for metal, dust, and fume as Co; 0.1 mg/m^3 TWA for cobalt carbonyl as Co; 0.1 mg/m^3 TWA for cobalt hydrocarbonyl as Co

Basis for Occupational Limits: The ACGIH limits for cobalt metal, dust, and fume are based on prevention of pulmonary sensitization, and those for cobalt carbonyl and hydrocarbonyl are based on analogy with nickel and iron carbonyls.[1] The decision to recommend a change in the TLV-TWA for cobalt metal, dust, and fume from 0.1 mg/m^3 to 0.05 mg/m^3 was strongly influenced by the Kerfoot et al. study,[2] which is reviewed in detail in the systemic toxicity section. It is noted in the TLV documentation that these limits do not generally apply to cobalt compounds without assigned TLVs and that many of these compounds have the effect of causing polycythemia.[1] In its 1977 criteria document on tungsten and cemented tungsten carbide,[3] NIOSH noted that most occupational exposure to tungsten and its compounds occurs together with exposure to cobalt and recommended that the federal standard for cobalt (then 0.1 mg/m^3) apply to workers exposed to cemented tungsten carbide containing more than 2% cobalt. In a more recent "occupational hazard assessment" specifically dealing with cobalt exposure,[4] NIOSH put forth recommendations for engineering controls, work practices, work clothing and protective equipment, and medical surveillance, but made no recommendation for an occupational exposure limit. OSHA recently adopted the ACGIH TLV-TWAs for the cobalt compounds listed above.[5]

Drinking Water Limits: In their evaluation of cobalt in 1977, the Safe Drinking Water Committee of the NAS[6] concluded that "because the maximum no-adverse-health-effect concentration is more than an order of magnitude greater than that found in any natural-water or drinking-water supply, there appears to be no reason at present to regulate the concentration of cobalt in drinking water."

Toxicity Profile

Carcinogenicity: Cobalt and its compounds have not been adequately tested for carcinogenicity in laboratory animals. In RTECS several reports were cited in which parenteral administration of cobalt resulted in tumors at the injection site.[7] However, the significance of these findings for the general population exposed to cobalt via inhalation is unclear. In a review of the toxicity of cobalt, Carson et al.[8] concluded that "although current evidence is not conclusive, it appears that cobalt metal, cobalt salts, and cobalt carbonyl are not significant causative agents of human cancer." It is relevant to note that Wehner et al.[9] (cited in NIOSH[4]) reported that male Syrian golden hamsters exposed to 10 mg/m^3 cobalt oxide for 7 hours per day, 5 days a week for their entire lifespan had a very low tumor incidence that did not differ significantly from controls. Based on the limited available evidence, cobalt and cobalt compounds are most appropriately placed in group D ("unclassified substance") under the EPA weight-of-evidence classification.

Mutagenicity: Cobalt as cobaltous nitrate (40 μg Co/L) has been reported to markedly decrease the mitotic index in cultured human leukocytes and in diploid fibroblast cultures, but at noncytotoxic concentrations no chromosomal aberrations occurred. Cobalt as cobalt acetate (0.06 to 0.6 μg/L) has been reported to increase the frequency of diploidy formation in cultured human leukocytes.[8] Certain cobalt compounds, including $CoCl_2$, $Co(OH)_3$, $CoSO_4$, and $2CoCO_3 \cdot 3Co(OH)_2$, have been reported to be weakly positive in the *B. subtilis* rec+/rec- assay, which is a test for primary DNA damage.[4] Cobalt as cobalt chloride did not cause DNA damage in tests with two different strains of *B. subtilis*.[8] Cobalt chloride and cobalt hydroxide were also reported to be negative in spot mutation tests with two strains of *E. coli* and five strains of *Salmonella*.[4] Cobalt treatment (form not specified) has been reported to cause "sticky chromosomes" in *Allium cepa*.[8] Certain cobalt salts have also been reported to enhance the oncogenic transformation of Syrian hamster embryo cells by adenovirus SA7.[4]

Developmental Toxicity: No data were found implicating cobalt and its compounds as developmental toxins in mammals.

Reproductive Toxicity: Based on studies in which CD-1 mice were exposed to concentrations of 0–400 ppm cobaltous chloride in drinking water for 13 weeks and observed for periods of up to 20 weeks, Pedigo et al.[10] concluded that cobalt directly or indirectly interfered with spermatogenesis and localized regulatory mechanisms for testosterone synthesis. There was a significant dose-related decrease in testicular weight and sperm count in all exposure groups compared to controls accompanied by a significant increase in serum testosterone. Sperm motility and fertility were also significantly decreased in the highest exposure group relative to controls. However, based on interspecies differences and the difference in route, the implications of these findings for inhalation exposure of humans is unclear.

Systemic Toxicity: Cobalt is considered an essential micronutrient based on its necessity as a functional part of the vitamin B_{12} molecule, but it should be noted that this is its only known nutritional benefit to humans. While ruminants are able to synthesize vitamin B_{12} if provided with sufficient dietary cobalt, nonruminants require preformed vitamin B_{12} with the requirement for Co as vitamin B_{12} estimated as approximately 0.13 µg/day.[6] Birth defects in swine have been associated with dietary cobalt deficiency and decreased fertility and anemia have also been associated with cobalt deficiency.[8]

Use of cobalt therapeutically and as an additive in beer have provided data on the ingestion toxicity of cobalt in humans. Around 1965, several clusters of cases were reported in which consumption of beer to which cobalt had been added in concentrations of approximately 1.2–1.5 mg/L as a foaming agent was associated with death due to congestive heart failure (beer drinker's cardiomyopathy). It is significant that therapeutic use of cobalt in the treatment of anemias in amounts far in excess of the amount that might be expected to be consumed even by a heavy beer drinker was not associated with a similar syndrome. Subsequent animal studies demonstrated that synergistic effects occurred when cobalt was ingested in combination with ethanol.[6]

It has been reported that treatment with 200 mg/day cobalt for 44 weeks results in increased erythrocyte count in humans, and studies in animals indicate that this is a toxic rather than a beneficial effect.[8] Long-term therapeutic administration of cobalt (commercial preparation containing cobalt and iron) for the treatment of anemia has also been reported to cause goiter and decreased thyroid function in children.[6]

It is now generally recognized that the characteristic diffuse interstitial fibrosis of the lungs seen in workers in the cemented tungsten carbide industry is primarily related to cobalt exposure.[1,3,4] NIOSH noted that such lung changes have been observed in workers exposed to 0.1–0.2 mg/m^3 cobalt and that airway obstruction has been reported to occur at concentrations as low as 0.06 mg/m^3.[4] In addition, pulmonary hypersensitivity also may develop in workers exposed to cobalt and the present TLV is based primarily on consideration of this endpoint;[1,4] NIOSH noted that once sensitized, workers would probably be unable to tolerate inhalation of even "small amounts" of cobalt. It should be noted that cobalt is also a skin sensitizer and may cause allergic contact dermatitis.[4]

Studies by Kerfoot et al.[2] using miniature swine demonstrated that exposure to 0.1 mg/m^3 (then the federal standard) for a relatively short duration produced evidence of early fibrotic lung changes. Miniature swine (n = 5/group) were exposed to their assigned dose level of 0, 0.1, or 1.0 mg/m^3 pure cobalt metal (particle size 0.4 to 3.6 μm; 50% alpha and 50% beta particles) for one week (sensitization week) followed by 10 days without exposure. After this period, a regular exposure schedule of 6 hours per day, 5 days a week for 3 months was followed. During the fourth week animals in both exposed groups were lethargic and developed wheezing during exposure. There was a dose-related decrease in compliance, but at one and two months postexposure, values had returned to control levels. While there were no radiographic changes or lung lesions discernable by gross examination or light microscopy, electron micrographs revealed an increase in septal collagen in the lungs of exposed animals. The authors also reported EKG changes such as loss of QRS voltage and T-wave changes; these were interpreted as evidence of a significant decrease in the strength of ventricular contraction and repolarization abnormalities, respectively. It was also reported that there were increased leukocyte and erythrocyte counts in the high exposure group after the third week of exposure, but the blood counts had returned to normal within 3 weeks. An increase in alpha-globulins and total protein and a decrease in the albumin/globulin ratio was also reported. It should be noted that the reported effects on the hematologic parameters and cardiovascular system observed in this study are consistent with reported effects in humans exposed to cobalt and its salts by the oral route.

Irritation: Although upper respiratory tract irritation has been associated with exposure to cobalt,[8] it would generally occur only at concentrations producing other forms of systemic toxicity.

Basis for the AALG: Cobalt is used in steel alloys, jet engines, and cement carbide abrasives and tools.[1] The major sources of cobalt in the atmosphere are coal and residual fuel oil burning and attrition in the use of superalloys, hard-facing alloys, and cemented tungsten carbides; the chief form of cobalt entering the atmosphere is probably cobalt oxide.[8] The diet is thought to be the major source of cobalt relative to drinking water,[6] and the contribution of inhalation exposure has been estimated to be extremely small.[8] Based on these considerations, 20% of the contribution to total exposure will be allowed from air.

Based on available human occupational data and the Kerfoot et al. study,[2] the critical effects associated with inhalation exposure to cobalt in the form of metal, dust, and fume are pulmonary fibrosis and pulmonary sensitization. Since no NOAEL was established in the Kerfoot et al. study[2] and the aggregate occupational data support the use of the TLV as a surrogate human NOAEL, the TLV/42 was used as the basis for the AALG. It should be noted that it is unclear whether the critical effect of inhalation exposure to cobalt salts would be the same as that for cobalt metal, dust, and fume. In the absence of data to the contrary, NIOSH considered it prudent to assume inhalation exposure to cobalt salts was capable of causing pulmonary fibrosis.[4] The present AALG should only tentatively be applied to other cobalt compounds for which sufficient data are unavailable.

AALG: • systemic toxicity (TLV/42)—0.24 $\mu g/m^3$ 8-hour TWA

References

1. ACGIH. 1986. "Documentation of the Threshold Limit Values and Biological Exposure Indices." 5:144–45.
2. Kerfoot, E. J., W. G. Fredrick, and E. Domeier. 1975. "Cobalt Metal Inhalation Studies in Miniature Swine." *Am. Ind. Hyg. Assoc. J.* 36:17–25.
3. NIOSH. 1977. "Criteria for a Recommended Standard . . . Occupational Exposure to Tungsten and Cemented Tungsten Carbide." DHEW (NIOSH) 77–127.
4. NIOSH. 1982. "Occupational Hazard Assessment: Criteria for Controlling Occupational Exposure to Cobalt." DHHS (NIOSH) 82–107.
5. OSHA. 1989. "Air Contaminants; Final Rule." *Fed. Reg.* 54(12):2332–2959.
6. Safe Drinking Water Committee. 1977. *Drinking Water and Health,* Vol. 1 (Washington, DC: National Academy Press).
7. NIOSH. 1988. GF8750000. "Cobalt." RTECS, on line.
8. Carson, B. L., H. V. Ellis, and J. L. McCann. 1986. *Toxicology and Biological Monitoring of Metals in Humans* (Chelsea, MI: Lewis Publishers, Inc.), pp. 75–92.
9. Wehner, A. P., R. H. Busch, R. J. Olson, and D. K. Craig. 1977. "Chronic Inhalation of Cobalt Oxide and Cigarette Smoke by Hamsters." *Am. Ind. Hyg. Assoc. J.* 38: 338–46.
10. Pedigo, N. G., W. J. George, and M. B. Anderson. 1988. "Effects of Acute and Chronic Exposure to Cobalt on Male Reproduction in Mice." *Repro. Tox.* 2:45–53.

COPPER AND COMPOUNDS

Synonyms: includes copper fume, dusts and mists of copper salts

CAS Registry Number: 7440–50–8

Molecular Weight: 63.54

Molecular Formula: Cu

AALG: irritation (TLV-based), 8-hour TWAs
   dust and mists 0.02 mg (Cu)/m^3
   fume   0.02 mg (Cu)/m^3

Occupational Limits: • ACGIH TLV: 0.2 mg (Cu)/m^3 TWA fume; 1.0 mg (Cu)/m^3 TWA dust and mists as Cu
        • NIOSH REL: none
        • OSHA PEL: 0.1 mg (Cu)/m^3 TWA fume; 1.0 mg (Cu)/m^3 TWA dust and mists as Cu

Basis for Occupational Limits: The TLVs for copper are based on the prevention of irritant effects.[1] OSHA recently adopted the limit for copper dust and mists set by the ACGIH as part of its final rule limits.[2]

Drinking Water Limits: In 1980 the EPA concluded that there was insufficient data to derive a health-based ambient water quality criterion for copper, and a criterion based on organoleptic quality, 1 mg/L, was instead recommended. It should be noted that organoleptic data have no demonstrated relationship to adverse health effects.[3] In the "Drinking Water Criteria Document for Copper," a 1-day HA of 1.3 mg/L was recommended for both adults and children. No 10-day HAs were derived on account of insufficient data. The recommended lifetime HA was also 1.3 mg/L because acute effects from copper exposure were deemed to be the most significant concern.[4] The proposed MCLG and MCL for copper are both 1.3 mg/L.[5]

Toxicity Profile

Carcinogenicity: Data pertinent to the evaluation of the inhalation carcinogenicity of copper and its compounds are lacking. In studies designed to assess the carcinogenicity of ingested or injected copper compounds[6,7] (cited in U.S. EPA[8]), only copper 8-hydroxyquinoline gave a positive response, i.e., an increased incidence of reticulum cell sarcomas in male (but not female) B6C3F₁ mice following a single subcutaneous injection. Copper has been placed in group D, unclassified substance, under the EPA weight-of-evidence classification.[8,9]

Mutagenicity: Results of assays designed to assess the genotoxicity of copper and its compounds have been mixed; the available evidence is most thoroughly reviewed in the document, ''Summary Review of Health Effects Associated with Copper.''[8] The U.S. EPA has concluded that the available data are insufficient to draw conclusions about the mutagenicity of copper.[8]

Developmental Toxicity: Data relevant to the evaluation of the developmental toxicity of inhaled copper and its compounds are lacking. However, it has been reported that ingested copper salts in mice and injected copper salts in hamsters are able to produce teratogenic effects. It should be noted, however, that copper deficiency is also known to cause teratogenic responses in a number of species, including sheep, goats, rats, guinea pigs, dogs, and chickens.[4,8]

Reproductive Toxicity: Data relevant to the evaluation of the reproductive toxicity of inhaled copper are lacking. The spermicidal properties of copper are well established and have been applied to the prevention of pregnancy through the use of copper loops or wires placed in the uterus.[8]

Systemic Toxicity: *Essentiality*: Copper is an essential element in humans and other animals with about 2–3 mg/day required for proper nutrition. Copper is a component of several important enzymes (e.g., cytochrome oxidase and erythrocyte superoxide dismutase) and is also necessary for hemoglobin formation, carbohydrate metabolism, catecholamine biosynthesis, and cross-linking of collagen, elastin, and hair keratin. The concentration of copper in various body tissues is controlled by well-developed homeostatic mechanisms; this is thought to account for the relatively uncommon occurrence of toxicity attributable to excess copper.[4,8]

In humans, ingestion of high concentrations of copper salts produces gastric irritation resulting in salivation, nausea, vomiting, gastric pain and hemorrhage, and diarrhea. In animals exposed to copper salts in the diet, effects on the liver, kidneys, blood, G.I. tract, and brain have been reported, but the concentrations producing these effects are relatively high and of questionable relevance to the human exposure situation (only approximately 20% of inhaled copper is thought to be absorbed in humans). These studies are most thoroughly reviewed in U.S. EPA.[4,8]

Data on the inhalation toxicity of copper and its salts are generally lacking. A 50% increase in volume density of alveolar type II cells and no changes in lung lysozyme levels were reported in rabbits exposed to 0.6 mg/m³ $CuCl_2$ 6 hours per day, 5 days a week for 4 to 6 weeks[10,11] (cited in U.S. EPA[8]). Guinea pigs exposed to an atmosphere saturated with Bordeaux mixture (an aqueous solution of lime and 1.5% $CuSO_4$) three times a day (duration not reported) for 6.5 months developed micronodular lesions and small histiocytic granulomas[12] (cited in U.S. EPA[8]). Similar lesions have been found in vineyard workers exposed to Bordeaux mixture, but the interpretation of these findings is complicated by the lack of exposure data and concomitant exposure to other agents (e.g., arsenic).[8]

Occupational exposure to metallic copper dust at a concentration of approximately 0.1 mg/m^3 has been reported to produce nausea and metal fume fever, a reversible condition characterized by influenza-like symptoms[13] (cited in ACGIH[1] and U.S. EPA[8]). It is stated in the TLV documentation[1] that: "Extensive industrial experience with copper-welding operations and copper-metal refining in Great Britain support the view that no ill effects result from exposure to fumes at concentrations of up to 0.4 mg (Cu)/m^3."

Irritation: In addition to the effects cited above, copper fumes are also reported to cause upper respiratory tract irritation and a metallic or sweet taste. This latter effect was observed in welding operations at concentrations of 1–3 mg/m^3, but not at levels of 0.02–0.4 mg/m^3. Dust and mists of copper salts are also capable of producing irritation of the upper respiratory tract through inhalation and itchy eczema on skin contact, as well as eye irritation and ulceration and perforation of the nasal septum if sufficiently high concentrations are achieved.[1]

Basis for the AALG: Because copper is ubiquitous in the environment, the largest relative source contribution would be expected from food and drinking water.[4] However, since the AALG is based on irritation, no relative source contribution from air was factored into the calculation.

Given the general lack of data correlating inhalation exposure to copper with toxic effects in both animals and humans, the AALG is based on the long-standing TLVs for copper fume and dusts and mists, which are based on industrial experience. Since it was concluded in the "Drinking Water Criteria Document" that "chronic effects are not likely because the acute toxicity limits intake and because copper is an essential element controlled by a homeostatic mechanism and does not tend to bioaccumulate" and, in addition, because of the irritant effects associated with copper fume and salts, the TLV for copper fume was divided by an uncertainty factor of 10 (interindividual variation). Since there is at least one report of metal fume fever associated with exposure to copper dusts at a concentration of 0.1 mg (Cu)/m^3, the TLV for copper mists and dusts was divided by a total uncertainty factor of 50 (10 for interindividual variation × 5 for a LOAEL). Based on the limited available data on the inhalation effects of copper in humans and animal models, the AALGs should be considered provisional.

AALG: • irritation (TLV-based), 8-hour TWAs
        dust and mists   0.02 mg (Cu)/m^3
        fume   0.02 mg (Cu)/m^3

References

1. ACGIH. 1986. "Documentation of the Threshold Limit Values and Biological Exposure Indices." 5:146.

2. OSHA. 1989. "Air Contaminants; Final Rule." *Fed. Reg.* 54:2332–2959.

3. U.S. EPA. 1980. "Ambient Water Quality Criteria for Copper." EPA 440/5–80–036.

4. U.S. EPA. 1985. "Drinking Water Criteria Document for Copper" (final draft). EPA 600/X-C4–190–1.

5. U.S. EPA. 1988. "Drinking Water Regulations; Maximum Contaminant Level Goals and National Primary Drinking Water Regulations for Lead and Copper; Proposed Rule." *Fed. Reg.* 53:31516–578.

6. Bionetics Research Labs. 1968. *Evaluation of Carcinogenic, Teratogenic, and Mutagenic Activities of Selected Pesticides and Industrial Chemicals,* Vol. 1. Carcinogenic study prepared for NCI. NCI-DCCP-CG-1973–1-1.

7. Gilman, J. P. W. 1962. "Metal Carcinogenesis. II. A Study on the Carcinogenic Activity of Cobalt, Copper, Iron, and Nickel Compounds." *Cancer Res.* 22:158–66.

8. U.S. EPA. 1987. "Summary Review of the Health Effects Associated with Copper." EPA/600/8–87/001.

9. U.S. EPA. 1984. "Health Effects Assessment for Copper." EPA/540/1–86/025.

10. Johansson, A., T. Curstedt, B. Robertson, and P. Camner. 1984. "Lung Morphology and Phospholipids After Experimental Inhalation of Soluble Cadmium, Copper, and Cobalt." *Environ. Res.* 34:295–309.

11. Lundborg, M. and P. Camner. 1984. "Lysozyme Levels in Rabbit Lung After Inhalation of Nickel, Cadmium, Cobalt, and Copper Chlorides." *Environ. Res.* 34:335–42.

12. Pimental, J. C. and F. Marques. 1969. "Vineyard Sprayer's Lung: A New Occupational Disease." *Thorax* 24:670–88.

13. Gleason, R. P. 1968. "Exposure to Copper Dust." *Am. Ind. Hyg. Assoc. J.* 29:461.

CUMENE

Synonyms: isopropyl benzene

CAS Registry Number: 8–82–8

Molecular Weight: 120.19

Molecular Formula:

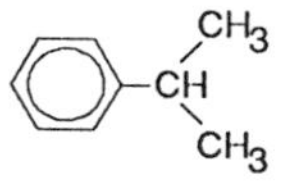

$$\text{C}_6\text{H}_5-\text{CH}(\text{CH}_3)_2$$

AALG: systemic toxicity (TLV/42)—0.6 ppm (2.9 mg/m³) 8-hour TWA

Occupational Limits:
- ACGIH TLV: 50 ppm (245 mg/m³) TWA; skin notation
- NIOSH REL: none
- OSHA PEL: 50 ppm (245 mg/m³) TWA; skin notation

Basis for Occupational Limits: The occupational limits for cumene are based on the prevention of narcosis.[1]

Drinking Water Limits: No drinking water limits were found for this substance.

Toxicity Profile

Carcinogenicity: No data were found implicating cumene as a carcinogen.

Mutagenicity: Negative results are listed for cumene in the Gene-Tox database as summarized in RTECS for both the Ames assay and *S. cerevisiae* homozygosis assay.[2]

Developmental Toxicity: No data were found implicating cumene as a developmental toxin.

Reproductive Toxicity: No data were found implicating cumene as a reproductive toxin.

Systemic Toxicity: Cumene has a relatively low order of acute toxicity based on reported lethality studies. Seven-hour and two-hour LC₅₀s of 2000 and 5020 ppm, respectively, have been reported in mice.[1,2] Smyth et al.[3] reported that exposure to

261

8000 ppm cumene for 4 hours resulted in the death of 4 of 6 rats within 14 days. These same investigators reported an oral LD_{50} of 2900 mg/kg in rats.[3]

The predominant form of systemic toxicity associated with inhalation of cumene is narcosis. Relative to benzene and toluene, this narcosis is characterized by slow induction and long duration.

In a chronic inhalation study cited in the TLV documentation and unavailable in translation, Fabre et al.[4] (cited in ACGIH[1]) reported that rats exposed to 500 ppm cumene daily for five months showed evidence of hyperemia and congestion of the lungs, liver, and kidneys with no evidence of hematologic abnormalities. In another study, repeated intubation of rats with 154 mg/kg cumene was reported to produce no adverse effects, but at 462 mg/kg an increase in kidney weight was reported[5-7] (cited in ACGIH[1]).

Irritation: Cumene is reported to be a primary skin and eye irritant.[1] No odor characteristics or odor thresholds were found for cumene.

Basis for the AALG: Based on information in the TLV documentation,[1] cumene is considered insoluble in water and its "hazards of use are minimized by virtue of its relatively high boiling point and hence low vapor pressure." Cumene is used in the production of acetone, phenol, and alpha-methyl styrene and as a solvent.[1] Based on lack of information on its environmental fate and distribution, 50% of the contribution to total exposure will be allowed from air.

Based on the limited available animal data and industrial experience that indicates no reported problems with exposures at the level of the TLV, it was considered appropriate to use the long-standing TLV as a surrogate human NOAEL. Thus, the AALG is based on the TLV/42 with a 50% relative source contribution for air. However, based on the limited available database for cumene, the AALG should be considered provisional.

AALG: • systemic toxicity (TLV/42)—0.6 ppm (2.9 mg/m³) 8-hour TWA

References

1. ACGIH. 1986. "Documentation of the Threshold Limit Values and Biological Exposure Indices." 5:151.
2. NIOSH. 1988. GR8575000. "Cumene." RTECS, on line.
3. Smyth, H. F., C. P. Carpenter, and C. S. Weil. 1951. "Range-Finding Toxicity Data: List IV." *AMA Arch. Ind. Hyg. Occ. Med.* 4:119–22.
4. Fabre, R., R. Truhaut, J. Bernuchon, and F. Loisillier. 1955. *Arch. Mal Prof. Med. Travail et Securite Sociale* 16:288.
5. Wolf, M. A., et al. 1956. *Arch. Ind. Health* 14:387.
6. Gerarde, H. W. 1956. *Arch. Ind. Health* 13:331.
7. Gerarde, H. W. 1959. *Arch. Ind. Health* 19:403.

CYCLOHEXANONE

Synonyms: pimelic ketone

CAS Registry Number: 108–94–1

Molecular Weight: 98.14

Molecular Formula: $O{=}\langle\rangle$

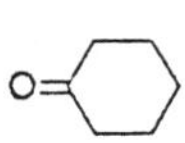

AALG: carcinogenicity—0.04 ppm (0.16 mg/m^3) annual TWA

Occupational Limits: • ACGIH TLV: 25 ppm (100 mg/m^3) TWA; skin notation
 • NIOSH REL: 25 ppm (100 mg/m^3) TWA
 • OSHA PEL: 25 ppm (100 mg/m^3) TWA; skin notation

Basis for Occupational Limits: The occupational limits for cyclohexanone are based primarily on its irritant properties with consideration of liver and kidney effects.[1,2] OSHA recently adopted the ACGIH TLV for cyclohexanone as part of its final rule limits.[3]

Drinking Water Limits: No drinking water limits were found for this substance.

Toxicity Profile

Carcinogenicity: In a two-year carcinogenicity bioassay,[4] male and female F344 rats and B6C3F$_1$ mice received cyclohexanone in their drinking water at the following levels: male and female rats—3300 and 6500 ppm, male mice—6500 and 13,000 ppm, and female mice—6500, 13,000, and 25,000 ppm. In both sexes of rats and mice, survival and body weight gain were comparable in the control and low-dose groups, but weight gain was depressed in the high exposure groups. There was a statistically significant increased incidence of the following tumor types in low-dose, but not high-dose, animals compared with controls: adenomas of the adrenal cortex in male rats, liver adenomas and carcinomas in male mice, and malignant lymphomas in female mice. In addition, there were statistically significant increases in the incidence of mammary gland fibromas and uterine endometrial stromal polyps in high-dose female rats compared with controls. No other significant neoplastic or non-neoplastic lesions attributable to cyclohexanone exposure were reported. Because the background incidence of several tumor types was high and their occurrence was limited to a single sex along with the concomitant lack of a dose-response relationship, the authors concluded that the evidence for carcinogenic activity of cyclohexanone was "marginal and the effect, if any, weak."

Based on the criteria of the EPA weight-of-evidence classification, cyclohexanone is most appropriately placed in group C, possible human carcinogen.

Mutagenicity: Cyclohexanone has been reported to be mutagenic to *S. typhimurium* in the presence of a metabolic activation system in some studies, but not in others;[4,5] to be mutagenic to *B. subtilis;* and to induce chromosomal aberrations in cultured human lymphocytes.[5] In addition, it is also reported to cause mutations and sister chromatid exchanges in Chinese hamster ovary cells.[6] However, it did not act as a transforming agent in an assay in which Syrian hamster embryo cells were used.[4]

Developmental Toxicity: In a study cited in the IRIS database[7] and apparently not published in the open literature, Schroeder exposed rats to 0, 320, 680, or 1430 ppm cyclohexanone on gestation days 9 through 16. No teratogenic effects were reported at any exposure level, and at the highest exposure level both maternal and fetal body weights were significantly decreased. The intermediate exposure level of 680 ppm was identified as a NOAEL for both maternal and fetotoxicity in the IRIS database.

Reproductive Toxicity: In a report cited in RTECS[5] and unavailable in translation (original in Russian), adverse effects on fertility were reported in rats exposed to 26 ppm cyclohexanone via inhalation for 4 hours per day on gestation days 1 through 20.

Systemic Toxicity: Treon et al.[8] exposed rabbits (n = 4/group; age, sex, and weight unspecified) to 0, 190, 309, 773, or 1414 ppm cyclohexanone for 6 hours per day, 5 days a week for 10 weeks, and 3082 ppm cyclohexanone for 3 weeks. Animals exposed at 3082 ppm exhibited weight loss, decreased body temperature, narcosis, and incoordination; two died. At 1414 ppm, signs of lethargy, as well as distended ear veins, excess salivation, and conjunctival irritation, were observed (these signs were also observed in the highest exposure group). At 309 and 773 ppm, conjunctival irritation was present also, but was less severe. Animals exposed to 190 ppm cyclohexanone exhibited no behavioral changes, but histological examination revealed slight degenerative changes in the liver and kidneys.

It is interesting to note that Rengstorff et al.[9] (cited in NIOSH[2]) reported the unexpected observation of cataract development in guinea pigs exposed cutaneously and subcutaneously to both acetone and cyclohexanone during a skin sensitization study.

Irritation: Cyclohexanone is reported to have an "acetone and peppermint odor," but no specific odor threshold was found.[1]

Cyclohexanone is considered to be a primary irritant to the eyes and skin.[1] Human subjects (n = 10) exposed to cyclohexanone for 3 to 5 minutes reported that 75 ppm was severely irritating to the eyes, nose, and throat and 50 ppm was also objectionable based on throat irritation; 25 ppm was judged by the subjects in this study to be a satisfactory limit for an 8-hour exposure[10] (cited in ACGIH[1] and NIOSH[2]).

Basis for the AALG: Cyclohexanone is considered slightly soluble in water and has a vapor pressure of 2 mm Hg at 20°C.[1] Although its widespread use as an industrial solvent and chemical intermediate would make its presence in the environment likely, specific data on its environmental fate, distribution, and occurrence and the expected contribution of various media to human exposure are lacking. Based on these considerations, 50% of the contribution to total exposure is allowed from air.

The decision on the most suitable basis for calculation of an AALG for cyclohexanone is rendered difficult for a number of reasons:

1. Evidence of carcinogenic activity is weak and limited, and the only available study did not use the inhalation route.
2. Of the two inhalation studies identified, sufficient detail was not available to allow for AALG calculation for one (i.e., the Schroeder study), and the other utilized very small group sizes and suffered from limitations in reporting of the data.
3. In view of the potential carcinogenic effect of cyclohexanone, use of the TLV as a basis for criteria derivation is difficult to justify.

Since the available inhalation studies were limited and some suggestive mutagenicity data exist, it was considered most appropriate to base the AALG on carcinogenicity, using an uncertainty factor approach. The lowest exposure level of 3300 ppm cyclohexanone in drinking water was considered to be a LOAEL for carcinogenic effects in rats and has been estimated by the U.S. EPA to correspond to a dose of 462 mg/kg/day. This dose level was converted into a human equivalent exposure level for the inhalation route according to the procedures described in Appendix C, and a total uncertainty factor of 5000 (10 for interindividual variation × 10 for interspecies variation × 5 for a LOAEL × 10 for the database factor) was applied. Application of a database factor of 10 was justified based on the use of a non-inhalation study for criteria derivation and the severity of the effect. This AALG should be considered provisional since a non-inhalation study was used as the basis for criteria derivation.

AALG: • carcinogenicity—0.04 ppm (0.16 mg/m^3) annual TWA

References

1. ACGIH. 1986. "Documentation of the Threshold Limit Values and Biological Exposure Indices." 5:159.
2. NIOSH. 1978. "Criteria for a Recommended Standard . . . Occupational Exposure to Ketones." DHEW (NIOSH) 78–173.
3. OSHA. 1989. "Air Contaminants; Final Rule." *Fed. Reg.* 54:2332–2959.

4. Lijinsky, W., and R. M. Kovatch. 1986. "Chronic Toxicity Study of Cyclohexanone in Rats and Mice." *J. Natl. Cancer Inst.* 77:941–49.
5. NIOSH. 1988. GW1050000. "Cyclohexanone." RTECS, on line.
6. Aaron, C. S., J. G. Brewen, D. G. Stetka, W. T. Bleicher, and M. C. Spahn. 1985. "Comparative Mutagenesis in Mammalian Cells (CHO) in Culture: Multiple Genetic Endpoint Analysis of Cyclohexanone *in vitro*." *Environ. Mutagen.* 7(suppl. 3):60–61.
7. U.S. EPA. 1988. "Cyclohexanone; CASRN 108–94–1." IRIS (3/1/88).
8. Treon, J. F., W. E. Crutchfield, and K. V. Kitzmiller. 1943. "The Physiological Response of Animals to Cyclohexane, Methylcyclohexane and Certain Derivatives of These Compounds. II. Inhalation." *J. Ind. Hyg. Tox.* 25:323–47.
9. Rengstorff, R. H., J. P. Petrali, and V. M. Sim. 1972. "Cataracts Induced in Guinea Pigs by Acetone, Cyclohexanone and Dimethyl Sulfoxide." *Am. J. Optom. Arch. Am. Acad. Optom.* 49:308–19.
10. Nelson, K. W., J. F. Ege, M. Ross, L. E. Woodman, and L. Silverman. 1943. "Sensory Response to Certain Industrial Solvent Vapors." *J. Ind. Hyg. Tox.* 25:282–85.

DIBUTYL PHTHALATE

Synonyms: *n*-butyl phthalate, DBP, di-*n*-butyl phthalate

CAS Registry Number: 84–74–2

Molecular Weight: 278.34

Molecular Formula:

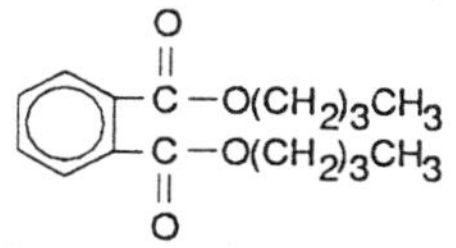

$$\text{O}$$
$$\text{C}-\text{O(CH}_2)_3\text{CH}_3$$
$$\text{C}-\text{O(CH}_2)_3\text{CH}_3$$
$$\text{O}$$

AALG: systemic toxicity—43.8 $\mu g/m^3$ (3.8 ppb) 24-hour TWA

Occupational Limits:
- ACGIH TLV: 5 mg/m^3 TWA
- NIOSH REL: none
- OSHA PEL: 5 mg/m^3 TWA

Basis for Occupational Limits: The TLV for dibutyl phthalate (DBP) is based on ''its low order of toxicity'' and on industrial experience that indicates no reports of irritation or systemic effects in humans at this level.[1]

Drinking Water Limits: The ambient water quality criterion for DBP is 34 mg/L assuming a 70-kg man consuming 2 L of water and 6.5 g fish/shellfish per day.[2] This criterion was derived using data from the Smith study,[3] reviewed in ''Systemic Toxicity'' below.

Toxicity Profile

Carcinogenicity: DBP has been classified in group D under the EPA weight-of-evidence classification because no data were found implicating it as a carcinogen.[4]

Mutagenicity: In general, DBP has been negative in standard plate reverse mutation Ames assays in strains TA98, TA100, TA1535, and TA1537 with and without metabolic activation. However, in a liquid suspension assay with 4-hour incubation, DBP was positive in strain TA100 without metabolic activation and negative with metabolic activation. It also was a direct acting mutagen in a forward mutation assay using *S. typhimurium*, but was not mutagenic to *E. coli* or *S. cerevisiae* in other studies.[4,5] With respect to mammalian cells, DBP has been reported to be mutagenic in the mouse lymphoma forward mutation assay only in the presence of a metabolic activation system.[4] There is limited evidence to indicate that DBP will induce chromosomal aberrations in Chinese hamster fibroblasts, but it did not induce aberrations in cultured human leukocytes.[4,5] There is some evidence that DBP induces peroxisome proliferation.[4]

Developmental Toxicity: In a study in which the teratogenicity of several phthalate esters was investigated, Singh et al.[6] exposed Sprague-Dawley rats (n = 5/group) to 0.305, 0.610, or 1.017 mL/kg DBP (purity unspecified) via intraperitoneal injection on gestation days 5, 10, and 15. These levels were 1/10, 1/5, and 1/3 of the i.p. LD_{50}, respectively. Control groups consisted of untreated animals and those injected with distilled water, normal saline, or cottonseed oil on the same schedule as treated animals. Compared with untreated controls, there was a statistically significant decrease in fetal weight in all of the exposed groups and in the rats injected with normal saline. In addition, there was a definite increase in resorptions in the highest treated group relative to all of the control groups. There was a dose-related increase in skeletal abnormalities (based on percentage figures) in the DBP-exposed groups; in the cottonseed oil and normal saline treated groups, there was also an increase in skeletal abnormalities which was not as high as that seen in the DBP-treated animals. Malformations most frequently seen included absence of tail, twisted hind legs, elongated and fused ribs, and abnormal skull bones. No mention was made of maternal toxicity in this study.

Shiota and Nishimura[7] exposed ICR mice (n = 7–21/group) to 0, 0.05, 0.1, 0.2, 0.4, or 1.0% DBP (purity unspecified) in the diet throughout their entire period of gestation. Maternal toxicity in the form of decreased body weight gain was observed only at the highest exposure level (2100 mg/kg/day). There was also a statistically significant increase in resorptions in the highest exposure group, as well as a significant decrease in male, but not female, fetal weight at the next highest dose (660 mg/kg/day). In addition, there was a significant increase in exencephaly in the highest dose group. A NOAEL of 370 mg/kg/day for developmental effects in mice is suggested by this study.

Nikonorow et al.[8] evaluated the effects of exposure to DBP before or during gestation in female Wistar rats (n = 10/group). Administration of 0, 120, or 600 mg/kg/day DBP in olive oil by gavage for 3 months prior to gestation had no effect on reproductive performance or fetal development. However, in animals exposed throughout gestation to 600 mg/kg/day DBP, there was a significant increase in fetal resorptions and decrease in fetal weight. Placental weights were significantly decreased in both exposure groups. No statistically significant increase in malformations was reported in this study. A LOAEL for developmental effects in rats of 120 mg/kg/day DBP is suggested by this study.

Available data indicate that DBP is capable of inducing developmental effects in laboratory animals at levels that do not produce overt maternal toxicity. However, in a safety assessment of the use of alkyl phthalates in cosmetic preparations, it was considered that the available data were inadequate to conclude that DBP is a proven teratogen.[5]

Reproductive Toxicity: Testicular toxicity in the form of seminiferous tubular atrophy has been reported in several studies in which rats were exposed to DBP at doses of 2000 mg/kg or higher both by gavage and in the diet for several days.[9,10,11] It has been found that these changes are accompanied by high testosterone and low zinc concentrations in testicular tissue.[9] In addition, there appears to be a marked difference in species sensitivity to DBP-induced testicular toxicity. Gray et al.[11] reported that exposure to 2000 mg/kg DBP in corn oil by gavage for 7 to 9 days produced severe seminiferous tubular atrophy in rats and guinea pigs, but caused only focal atrophy in mice and no detectable effect in hamsters.

The toxicity of DBP has been evaluated in several multigeneration reproduction studies summarized as follows in Elder[5]: "The dietary administration of DBP, in doses of 10 and 100 mg/kg per day, to two mouse strains for three generations increased the formation of renal cysts in the F_1 and F_2.[12] In another three-generation reproduction study, female rats were dosed daily for 6 weeks with 50% DBP solution in oil, at a dose of 1 mL/kg, and then were paired with untreated males. The offspring were bred to produce two additional generations; it is not known whether the second and third generations were dosed with DBP. No impairment of reproductive performance was noted. Development, growth, and fertility were normal for all three generations."[13]

Systemic Toxicity: Data on the inhalation toxicity of DBP are limited. In acute studies, a 2-hour LC_{50} of 2200 ppm has been reported in mice.[14] It has been reported that exposure to 1.5 mg/m^3 DBP for 6 hours per day, 6 days a week for one month produced no adverse effects on male rats compared to unexposed controls[15] (cited in Elder[5]).

In the only available chronic inhalation study, Kawano[16] exposed rats to 0, 0.5, or 50 mg/m^3 DBP via inhalation for 6 hours per day, 7 days a week for 6 months. Rats in the highest exposure group had significantly decreased body weight gain and significantly increased lung and brain weights. Significantly increased percentages of PMNs (neutrophils) occurred in both exposure groups. In addition, rats in the high exposure group had abnormal clinical chemistry determinations including high blood urea nitrogen (BUN), low cholesterol, and in some cases, high transaminase activity. Similar findings were evident in the low exposure group, but the changes were generally of decreased magnitude. A LOAEL of 0.5 mg/m^3 DBP for systemic toxicity in rats is suggested by this study.

The acute oral toxicity of DBP in rats is quite low, with reported oral LD_{50}s in the range of 8 to 25 g/kg.[5] Subacute studies in rats have indicated that DBP is capable of inducing hepatic cytochrome P450 activity.[5] In a 3-month study in which rats were gavaged five times a week with 120 mg/kg or 1200 mg/kg DBP in olive oil, there was a statistically significant increase in relative liver weights in both exposure groups with no histologic evidence of pathologic changes.[8]

The strongest available chronic (not lifetime) study of DBP toxicity was conducted by Smith.[3] In this study, male Sprague-Dawley rats (n = 10/group) were fed either a basal diet or the basal diet supplemented with 0.01, 0.05, 0.25, or 1.25% DBP (purity unspecified) for one year. One-half of the rats at the highest exposure level died within the first week of the study, but at concentrations of 0.25% and lower there were no differences between treated and control animals in survival or growth. Hematologic studies and gross and histological examination at terminal sacrifice revealed no differences in treated and control animals. A NOAEL of 0.25% DBP in the diet of rats is suggested by this study. The U.S. EPA has used this study as the basis of the oral RfD; it is estimated that 0.25% DBP in the diet corresponds to a dose of 125 mg/kg/day.[4]

Irritation: Based on its widespread use in cosmetic preparations, DBP has been evaluated for potential irritancy in both animals and humans. These studies have generally revealed minimal potential for dermal and ocular irritation.[5]

Basis for the AALG: DBP is considered insoluble in water and has a low vapor pressure (less than 0.1 mm Hg at 20°C). On account of its low vapor pressure, the TLV committee concluded that DBP would pose minimal inhalation hazard and that inhalation of significant amounts would occur only by spray or mist exposure.[1] In a safety evaluation, it was concluded that phthalates are ubiquitous in the environment and human exposure from a variety of sources is likely.[5] Based on these considerations, 20% of the contribution to total exposure is allowed from air for DBP.

AALGs for DBP were derived based on both systemic and developmental toxicity. AALGs for systemic toxicity were derived based on both the Smith ingestion study in rats[3] and the Kawano inhalation study in rats.[16]

The LOAEL of 0.5 mg/m^3 for systemic toxicity from the inhalation study in rats was adjusted for continuous exposure (0.5 mg/m^3 × 6/24) and converted to a human equivalent LOAEL for the inhalation route, and a total uncertainty factor of 5000 applied (10 for interindividual variation × 10 for interspecies variation × 10 for a less-than-lifetime study × 5 for a LOAEL).

The NOAEL of 125 mg/kg/day for systemic toxicity from the rat ingestion study was converted to a human equivalent NOAEL for the inhalation route based on the procedures described in Appendix C, and a total uncertainty factor of 2000 was applied (10 for interindividual variation × 10 for interspecies variation × 10 for a less-than-lifetime study × 2 for the database factor). Application of the database factor was based on the use of the ingestion route of exposure.

Two studies were used to derive AALGs for developmental toxicity. The NOAEL of 370 mg/kg/day from the Shiota and Nishimura study[7] was converted to a human equivalent NOAEL for the inhalation route based on the procedures described in Appendix C, and a total uncertainty factor of 1000 was applied (10 for interindividual variation × 10 for interspecies variation × 5 for the developmental uncertainty factor × 2 for the database factor). The LOAEL of 120 mg/kg/day from the study of Nikonorow et al.[8] was converted to a human equivalent LOAEL for the inhalation route based on the procedures described in Appendix C, and a total uncertainty factor of 5000 was applied (10 for interindividual variation × 10 for interspecies variation × 5 for the developmental uncertainty factor × 5 for a LOAEL × 2 for the database factor). AALGs were derived based on both these studies because, although the Shiota and Nishimura study[7] was stronger and established a NOAEL, the LOAEL from the Nikonorow et al. study[8] was for a different species and was more conservative. Application of the database factor in each of the latter two cases was based on the use of the ingestion route of exposure.

The choice of a recommended AALG for DBP is complicated by several factors. Ideally the recommended AALG derived from an animal study should meet the following criteria:

1. inhalation route
2. strong study design
3. if more than one AALG has been derived, the most conservative figure is usually recommended.

In the case of DBP, the AALG derived from the inhalation study meets the first and third criteria. However, with respect to the second criterion, several questions arise, principally because the original article is in Japanese with only the abstract and the tables and figures in English. It was therefore difficult to determine certain key points, e.g., methods of pathologic examination and changes observed, numbers of animals of each sex per group, treatment of the control group(s), and whether the changes observed at the 0.5 mg/m^3 exposure level were statistically significant compared to the control group (based on examination of graphic data, they appeared to be). Given these uncertainties, complete translation and reevaluation of this study is warranted before recommending this as the final AALG. Unfortunately translation of this article was beyond the scope of this project. Thus, the most conservative AALG from the best designed study, the AALG for systemic toxicity based on the Smith study, is recommended as the AALG. Although this AALG was not as conservative as that based on the Nikonorow et al. study[8], it is considered that, given the limitations of the latter study, the recommended AALG will offer adequate protection against potential developmental effects.

AALG:
- systemic toxicity—43.8 μg/m^3 (3.8 ppb) 24-hour TWA from Smith study
- systemic toxicity—0.011 μg/m^3 (0.001 ppb) 24-hour TWA from Kawano study
- developmental toxicity—220 μg/m^3 (20 ppb) 24-hour TWA from Shiota and Nishimura study
  —14.4 μg/m^3 (1.3 ppb) 24-hour TWA from Nikonorow et al. study

References

1. ACGIH. 1986. ''Documentation of the Threshold Limit Values and Biological Exposure Indices.'' 5:176–77.

2. U.S. EPA. 1980. "Ambient Water Quality Criteria for Phthalate Esters." EPA 440/5–80–067.

3. Smith, C. C. 1953. "Toxicity of Butyl Sterate, Dibutyl Sebacate, Dibutyl Phthalate, and Methoxyethyl Oleate." *AMA Arch. Ind. Hyg. Occ. Med.* 7:310–18.

4. U.S. EPA. 1988. "Dibutyl Phthalate." IRIS summary (12/88).

5. Elder, R. L. (Ed.). 1985. "Final Report on the Safety Assessment of Dibutyl Phthalate, Dimethyl Phthalate, and Diethyl Phthalate." *J. Am. Coll. Tox.* 4:267–303.

6. Singh, A. R., W. H. Lawrence, and J. Autian. 1972. "Teratogenicity of Phthalate Esters in Rats." *J. Pharm. Sci.* 61:51–5.

7. Shiota, K. and H. Nishimura. 1982. "Teratogenicity of Di(2-ethylhexyl) Phthalate (DEHP) and Di-*n*-butyl Phthalate (DBP) in Mice." *Env. Health Perspect.* 45:65–70.

8. Nikonorow, M., H. Mazur, and H. Piekacz. 1973. "Effect of Orally Administered Plasticizers and Polyvinyl Chloride Stabilizers in the Rat." *Tox. Appl. Pharm.* 26:253–59.

9. Oishi, S. and K. Hiraga. 1980. "Testicular Atrophy Induced by Phthalic Acid Esters: Effect on Testosterone and Zinc Concentrations." *Tox. Appl. Pharm.* 53:35–41.

10. Foster, P. M. D., L. V. Thomas, M. W. Cook, and S. D. Gangolli. 1980. "Study of the Testicular Effects and Changes in Zinc Excretion Produced by Some *n*-Alkyl Phthalates in the Rat." *Tox. Appl. Pharm.* 54:392–98.

11. Gray, T. J. B., I. R. Rowland, P. M. D. Foster, and S. D. Gangolli. 1982. "Species Differences in the Testicular Toxicity of Phthalate Esters." *Tox. Letters* 11:141–47.

12. Onda, S., H. Kodama, N. Yamada, and H. Ota. 1974. "Studies on the Toxic Effect of Phthalate. III. Teratogenicity in Mice." *Jap. J. Hyg.* 29:177.

13. Bornmann, G., A. Loeser, K. Mikulicz, and K. Ritter. 1956. "Behavior of the Organism as Influenced by Various Plasticizers." *Z. Lebensm. Unters. Forsch.* 103:413–24.

14. NIOSH. 1988. TI087500. "Phthalic Acid, Dibutyl Ester." RTECS, on line.

15. Nishiyama, K., Y. Suzuki, M. Kawano, and F. Okamoto. 1977. "Effect of Phthalic Acid Esters on Biological Systems. Effects on Rats." *Shikoku Eisei Gakkai Zasshi* 22:19–21.

16. Kawano, M. 1980. "Toxicological Studies on Phthalate Esters. I. Inhalation Effects of Dibutyl Phthalate on Rats." *Jap. J. Hyg.* 35:684–92.

1,2-DICHLOROPROPANE

Synonyms: propylene dichloride, DCP

CAS Registry Number: 78–87–5

Molecular Weight: 112.99

Molecular Formula: $CH_2Cl-CHCl-CH_3$

AALG: 10^{-6} 95% UCL—0.01 ppb (0.046 $\mu g/m^3$) as an annual TWA

Occupational Limits:
- ACGIH TLV: 75 ppm (347 mg/m^3) TWA; 110 ppm (508 mg/m^3) STEL
- NIOSH REL: none
- OSHA PEL: 75 ppm (347 mg/m^3) TWA; 110 ppm (508 mg/m^3) STEL

Basis for Occupational Limits: The TLVs are based on the prevention of hepatotoxicity, observed in inhalation studies in mice (see "Systemic Toxicity" below). It is noted in the TLV documentation that: "TLVs for hepatotoxic halogenated compounds, however, have generally been substantially reduced from the values in effect when the 75-ppm limit for propylene dichloride was adopted. Reconsideration of the TLV-TWA of 75 ppm and the TLV-STEL of 110 ppm would seem to be in order."[1] OSHA recently adopted the TLV-STEL as part of its final rule limits.[2]

Drinking Water Limits: In 1980, the U.S. EPA concluded that data were insufficient for the derivation of an ambient water quality criterion.[3] The 10-day HA for a 10-kg child is 0.090 mg/L; sufficient data from which to calculate 1-day and longer-term HAs were not available.[4] An MCLG of 6 $\mu g/L$ corresponding to the 10^{-5} upper bound cancer risk level was proposed for DCP. A cancer risk level was used as opposed to setting the MCLG at zero because chronic toxicity data for this compound are very limited and a DWEL had not been established for DCP.[5]

Toxicity Profile

Carcinogenicity: In an NTP bioassay,[6] male and female $B6C3F_1$ mice and female F344/N rats received 0, 125, or 250 mg/kg DCP by gavage in corn oil 5 days a week for 103 weeks; male F344/N rats under the same exposure conditions received 0, 62.5, or 125 mg/kg DCP. There was a dose-related increase in liver adenomas in both male and female mice along with an increased frequency of liver carcinomas.

Non-neoplastic lesions included hepatocytomegaly and hepatic necrosis in male, but not female, mice. In female rats, there was a dose-related increase in mammary gland carcinomas compared with historical controls for the laboratory at which the bioassay was performed and for all other laboratories combined; most of these tumors were found at the end of the study. In the NTP bioassay report, it was concluded that there was no evidence of a carcinogenic response in male rats, *equivocal evidence* of carcinogenicity in female rats (marginal increase in borderline malignant lesions concurrent with decreased survival and reduced body weight gain) and *some evidence* of carcinogenicity in male and female B6C3F$_1$ mice. In the NTP bioassay program, "*equivocal evidence of carcinogenicity* is demonstrated by studies that are interpreted as showing a chemically-related marginal increase in neoplasms. *Some evidence of carcinogenicity* is demonstrated by studies that are interpreted as showing a chemically-related increased incidence of benign neoplasms, studies that exhibit marginal increases in neoplasms of several organs/tissues, or studies that exhibit a slight increase in uncommon malignant or benign neoplasms."[6]

CAG has classified 1,2-dichloropropane in group B2, probable human carcinogen, based on the EPA weight-of-evidence classification.[4]

Mutagenicity: Positive results have been reported in reverse mutation assays with *S. typhimurium* and in forward mutation assays with *A. nidulans*.[7] DCP has also been reported to induce chromosomal aberrations and sister chromatid exchanges in cultured Chinese hamster ovary cells,[6] but it was negative in a *Drosophila* sex-linked recessive lethal assay.[8] IARC concluded that the evidence for genetic activity for 1,2-dichloropropane is limited.[8]

Developmental Toxicity: No data were found implicating DCP as a developmental toxin.

Reproductive Toxicity: No data were found implicating DCP as a reproductive toxin.

Systemic Toxicity: The liver is the primary target organ for DCP toxicity in animals exposed via ingestion or inhalation. Acute poisoning via ingestion in humans has been reported to result in centrilobular and mediolobular necrosis.[6,7]

Heppel et al.[9,10] (cited in U.S. EPA[7]) studied the effects of subchronic inhalation exposure to DCP (purity unknown) in several species. Dogs and guinea pigs were reported to be the least sensitive and mice the most sensitive. Of mice exposed at 400 ppm DCP for 4 to 7 hours per day, 5 days a week for 37 exposures, most died during the course of the study (hepatomas were reported in the few survivors), and in another group that received 14 to 28 exposures the findings included liver congestion and fatty degeneration, liver centrilobular necrosis, and fatty degeneration of the kidneys. LC$_{50}$ values of approximately 3000 ppm for rats following an 8-hour exposure and 720 ppm for mice following a 10-hour exposure have been reported.[8] Additional studies on the toxicity of DCP via ingestion are most thoroughly reviewed in the "Drinking Water Criteria Document."[7] Subchronic and chronic inhalation studies sufficient for the derivation of an AALG are lacking.

Irritation: DCP is reported to be mildly irritating to the skin and, when undiluted, moderately irritating to the eyes with no serious or permanent injury.[1] DCP is reported to have a "chloroform-like" odor[1] and an odor threshold of 0.25 ppm (geometric mean of several literature values) has been reported.[11]

Basis for the AALG: DCP is slightly soluble in water (2.7 g/L at 20°C) and has a vapor pressure of 40 mm Hg at 20°C.[8] DCP is used as a solvent in a number of applications, a scavenging agent for gasoline, a soil fumigant, an intermediate in chemical manufacture, and as an insecticide.[1] It has been found in both drinking and surface waters and in ambient air.[8] Since there is insufficient data to determine the major source of DCP exposure in humans,[4] 50% of the contribution to total exposure will be allowed from air.

Data correlating inhalation exposure with toxic effects are lacking for humans and limited for experimental animals. Based on available data, hepatic effects including carcinogenesis are the most relevant endpoints for the AALG. Since it is the only chronic study available and the principal target organ, the liver, is the same for both ingestion and inhalation exposure,[6,7] the data for hepatocellular adenomas and carcinomas in male and female mice is used as the basis for the AALG. Given its B2 designation under the EPA weight-of-evidence classification,[4] the quantitative risk assessment approach was used to derive the AALG. Using combined liver adenoma/carcinoma incidence for mice, human q_1^* values were calculated as 6.33 $\times$ 10^{-2} (mg/kg/day)$^{-1}$ and 2.25 $\times$ 10^{-2} (mg/kg/day)$^{-1}$ from the data for male and female mice, respectively, by the U.S. EPA Carcinogen Assessment Group.[7] Using the geometric mean of these two slope values, 3.77 $\times$ 10^{-2} (mg/kg/day)$^{-1}$, and converting it to an inhalation unit risk assuming a 70-kg human breathing 20 m^3 and 100% absorption, yields a figure of 1.08 $\times$ 10^{-5}, which corresponds to an inhalation q_1^* of 1.08 $\times$ 10^{-5} (μg/m^3)$^{-1}$. Applying the relation E = q_1^* $\times$ d, where E is the specified level of extra risk and d is the dose, and allowing a relative source contribution of 50%, gives the figures shown below.

This AALG should be considered provisional because it was derived from non-inhalation data and a higher degree of uncertainty than usual is associated with route-to-route extrapolation of this type.

AALG: • 10^{-6} 95% UCL—0.01 ppb (0.046 μg/m^3) annual TWA
• 10^{-5} 95% UCL—0.1 ppb (0.46 μg/m^3) annual TWA

References

1. ACGIH. 1986. "Documentation of the Threshold Limit Values and Biological Exposure Indices." 5:501.
2. OSHA. 1989. "Air Contaminants; Final Rule." *Fed. Reg.* 54:2332–2959.

3. U.S. EPA. 1980. "Ambient Water Quality Criteria for Dichloropropane and Dichloropropene." EPA 440/5–80–043.

4. U.S. EPA. 1987. "Health Advisory for 1,2-Dichloropropane." Office of Drinking Water (3/31).

5. U.S. EPA. 1985. "National Primary Drinking Water Regulations; Synthetic Organic Chemicals, Inorganic Chemicals, and Microorganisms; Proposed Rule." *Fed. Reg.* 50:46935–7025.

6. NTP. 1986. "Toxicology and Carcinogenesis Studies of 1,2-Dichloropropane (Propylene Dichloride) (CAS No. 78–87–5) in F344/N Rats and B6C3F$_1$ Mice (Gavage Studies)." (Research Triangle Park, NC: National Toxicology Program), TR-263.

7. U.S. EPA. 1985. "Drinking Water Criteria Document for 1,2-Dichloropropane" (final draft). (Springfield, VA: NTIS), PB86–117850.

8. IARC. 1986. "1,2-Dichloropropane." *IARC Monog.* 41:131–47.

9. Heppel, L. A., P. A. Neal, B. Highman, and V. T. Porterfield. 1946. "Toxicology of 1,2-Dichloropropane (Propylene Dichloride). I. Studies on Effects of Daily Inhalations." *J. Ind. Hyg. Tox.* 28:1–8.

10. Heppel, L. A., B. Highman, and E. G. Peake. 1948. "Toxicology of 1,2-Dichloropropane (Propylene Dichloride). IV. Effects of Repeated Exposures to a Low Concentration of the Vapor." *J. Ind. Hyg. Tox.* 30:189–91.

11. Amoore, J. E. and E. Hautala. 1983. "Odor as an Aid to Chemical Safety: Odor Thresholds Compared with Threshold Limit Values and Volatilities for 214 Industrial Chemicals in Air and Water Dilution." *J. Appl. Tox.* 3:272–90.

DIETHANOLAMINE

Synonyms: 2,2'-iminodiethanol, DEA

CAS Registry Number: 111–42–2

Molecular Weight: 105.14

Molecular Formula: $HO(CH_2)_2NH(CH_2)_2OH$

AALG: systemic toxicity—7 $\mu g/m^3$ (2 ppb) 24-hour TWA

Occupational Limits:
- ACGIH TLV: 3 ppm (15 mg/m^3) TWA
- NIOSH REL: none
- OSHA PEL: 3 ppm (15 mg/m^3) TWA

Basis for Occupational Limits: The TLV for diethanolamine was derived from a 90-day feeding study in rats in which a NOAEL of 20 mg/kg/day was reported and by analogy with ethanolamine.[1] OSHA recently adopted the TLV-TWA as part of its final rule limits.[2]

Drinking Water Limits: No drinking water limits were found for this substance.

Toxicity Profile

Carcinogenicity: Diethanolamine is currently under evaluation in the National Toxicology Program with prechronic studies in progress (skin painting and drinking water, rats and mice) and two-year studies with quality assessment in progress (gavage, rats and mice).[3]

Mutagenicity: In NTP studies, DEA was nonmutagenic to *S. typhimurium* strains TA98, TA100, TA1535, and TA1537 with and without metabolic activation.[3]

Developmental Toxicity: No data were found implicating diethanolamine as a developmental toxin.

Reproductive Toxicity: No data were found implicating diethanolamine as a reproductive toxin.

Systemic Toxicity: Data on the systemic toxicity of diethanolamine are limited to acute and subchronic studies, mainly using the ingestion route. DEA is of relatively low acute toxicity based on reported LD_{50} values, 710 mg/kg for rats, 3300 mg/kg for mice, and 2200 mg/kg for rabbits.[5]

Results of the only inhalation studies found on DEA were available only as an abstract in which it was reported that short-term exposure to 200-ppm DEA vapor or 1400-ppm DEA aerosol caused respiratory distress and some deaths in male rats. Continuous exposure to DEA vapor at a concentration of 25 ppm for 216 hours increased liver and kidney weights, elevated serum glutamic-oxaloacetic transaminase (SGOT), and elevated blood urea nitrogen (BUN). In addition, exposure to 6 ppm DEA for 13 weeks on a "workday schedule" caused decreased growth rates, increased lung and kidney weights, and some deaths among exposed male rats.[6]

Smyth et al.[7] reported the results of a single subchronic feeding study in which rats (n = 10/group, sex unspecified) received DEA in the diet at levels ranging from 5 to 680 mg/kg/day. The lowest daily dose causing altered liver or kidney weights was 90 mg/kg and the lowest dose causing microscopic lesions (in liver, kidney, spleen, or testes) was 170 mg/kg. A NOAEL of 20 mg/kg/day was determined from this study.

It is interesting to note that in a safety assessment of ethanolamines it was concluded that although they were safe for use in cosmetics under specified conditions, both triethanolamine and DEA should not be used in products containing N-nitrosating agents because nitrosation of these ethanolamines may result in the formation of N-nitrosodiethanolamine, which is carcinogenic in laboratory animals.[4]

Irritation: Only mild skin irritation has been reported in clinical tests of cosmetic products containing concentrations of greater than 5% DEA; no data were found implicating DEA as a respiratory tract irritant.[4] DEA has been reported to have a "mild ammoniacal" odor and an odor threshold of 0.27 ppm (geometric mean of reported literature values).[8,9]

Basis for the AALG: Because DEA is very soluble in water, has a low vapor pressure (less than 0.01 mm Hg at 20°C), and has a number of uses (e.g., as a detergent in shampoos, paints, cutting oils, and other cleaners, and as an intermediate in the manufacture of plasticizers and resins) that would favor its release to surface and ground waters,[1,4] 20% of the contribution to total exposure will be allowed from air.

The AALG for DEA is based on the rat NOAEL of 20 mg/kg/day from the study of Smyth et al.[7] and converted to a human equivalent NOAEL according to the procedures described in Appendix C with a total uncertainty factor of 2000 (10 for interindividual variation × 10 for interspecies variation × 10 for a less-than-lifetime study × 2 for the database factor). Application of the database factor was considered relevant based on the use of the non-inhalation route of exposure. This AALG should be considered provisional pending availability of results from the NTP bioassay and because its derivation was based on a study in which the route of exposure was not inhalation.

The results of the subchronic inhalation study reported by Hartung et al.[6] were considered inappropriate for criterion derivation for two reasons:

1. The reporting of results was very limited with no description of control groups.
2. The 6 ppm exposure level is a free-standing frank effect level (FEL) considered unsuitable for criterion derivation by the U.S. EPA (for details on this consult the methodology section in Part I).

AALG: • systemic toxicity—7 $\mu g/m^3$ (2 ppb) 24-hour TWA

References

1. ACGIH. 1986. "Documentation of the Threshold Limit Values and Biological Exposure Indices." 5:197.
2. OSHA. 1989. "Air Contaminants; Final Rule." *Fed. Reg.* 54:2332–2959.
3. NTP. 1988. "Management Status Report." National Toxicology Program, Research Triangle Park, NC (2/8).
4. Beyer, K. H., W. F. Bergfeld, W. O. Berndt, R. K. Boutwell, W. W. Carlton, D. K. Hoffmann, and A. L. Schroeter. 1983. "Final Report on the Safety Assessment of Triethanolamine, Diethanolamine, and Monoethanolamine." *J. Am. Coll. Tox.* 2:183–235.
5. NIOSH. 1988. KL2975000. "Ethanol, 2,2′-iminodi." RTECS, on line.
6. Hartung, R., L. K. Rigas, and H. H. Cornish. 1970. "Acute and Chronic Toxicity of Diethanolamine." *Tox. Appl. Pharm.* 17:308.
7. Smyth, H. F., C. P. Carpenter, and C. S. Weil. 1951. "Range-Finding Toxicity Data: List IV." *AMA Arch. Ind. Hyg. Occ. Med.* 4:119–22.
8. Windolz, M. (Ed). 1976. *The Merck Index* (Rahway, NJ: Merck and Co., Inc.), p. 410 (No. 3078).
9. Amoore, J. E. and E. Hautala. 1983. "Odor as an Aid to Chemical Safety: Odor Thresholds Compared with Threshold Limit Values and Volatilities for 214 Industrial Chemicals in Air and Water Dilution." *J. Appl. Tox.* 3:272–90.

DIISOCYANATE DIPHENYL METHANE

Synonyms: methylene bisphenyl isocyanate, MDI

CAS Registry Number: 101–68–8

Molecular Weight: 250.25

Molecular Formula: $O=C=N-\langle\bigcirc\rangle-CH_2-\langle\bigcirc\rangle-N=C=O$

AALG: systemic toxicity (TLV-based)—0.06 ppb (0.6 $\mu g/m^3$) 8-hour TWA

Occupational Limits:
- ACGIH TLV: 0.005 ppm (0.055 mg/m^3) TWA
- NIOSH REL: 0.005 ppm (0.055 mg/m^3) TWA; 0.02 ppm (0.2 mg/m^3) ceiling
- OSHA PEL: 0.02 ppm (0.2 mg/m^3) ceiling

Basis for Occupational Limits: The occupational limits for MDI are based on irritation, sensitization, and other respiratory effects.[1-3]

Drinking Water Limits: No drinking water limits were found for this substance.

Toxicity Profile

Carcinogenicity: No data were found implicating MDI as a carcinogen.

Mutagenicity: MDI has been reported to be mutagenic to *S. typhimurium* in the presence of a metabolic activation system.[4]

Developmental Toxicity: No data were found implicating MDI as a developmental toxin.

Reproductive Toxicity: No data were found implicating MDI as a reproductive toxin.

Systemic Toxicity: Based on occupational exposure data and controlled studies in humans, the principal hazards associated with MDI exposure are direct irritant effects and sensitization or allergic responses; the former are discussed in the next section. Respiratory sensitization to MDI has been reported in humans occupationally exposed[1,2] and early symptoms may resemble those of a cold or mild hay fever with asthmatic-like respiratory symptoms developing later.[5] Konzen et al.[6] reported that workers exposed to 1.3 ppm MDI/minute (total time 10 minutes) developed an antigenic response to MDI, but those exposed to 0.9 ppm MDI/minute did not. O'Brien et al.[7] (cited in Woolrich[5]) reported that exposure to toluene diisocyanate (TDI) may also result in sensitization to MDI on account of their structural similarity.

Irritation: MDI is a direct irritant to the respiratory tract. Eye, nose, or throat irritation was reported in 34 of 35 workers in a plant where MDI concentrations were in the range of 0.01 to 0.15 mg/m^3 based on area samples and 0.12 to 0.27 mg/m^3 in the breathing zone of several sprayers. In addition, almost half of these workers experienced shortness of breath, wheezing, and chest tightness.[2]

Although at least one source has reported MDI to be odorless,[1] in another source MDI was reported to have an odor threshold of 0.4 ppm.[5]

Basis for the AALG: MDI is considered to have a "low, but significant, vapor pressure" (0.00014 mm Hg at 25°C, 0.001 mm Hg at 40°C, 1 mm Hg at 170°C, and 10 mm Hg at 205°C) and is relatively insoluble in water.[1] It is stated in the TLV documentation that: "The physical characteristic of low, but significant, vapor pressures presents both a vapor and particulate exposure in one of the current applications of MDI, namely, in foam or film coating of surfaces by spray-gun techniques." MDI has commercial applications in the area of polyurethane and polyisocyanurate manufacture.[5] Based on lack of information on its environmental fate and distribution and the expected contribution of various media to human exposure, 50% of the contribution to total exposure will be allowed from air.

The AALG is based on the TLV/42 because this limit is based on the known sensitizing effects of MDI, an endpoint of considerable relevance to the highly heterogeneous general population. Based on available human data (e.g., Konzen et al.[6]), the TLV is a reasonable surrogate human NOAEL. However, because allergic sensitization is an endpoint for which it is difficult to establish a threshold, this AALG should be considered provisional.

AALG: • systemic toxicity (TLV-based)—0.06 ppb (0.6 μg/m^3) 8-hour TWA

References

1. ACGIH. 1986. "Documentation of the Threshold Limit Values and Biological Exposure Indices." 5:389 (update).
2. NIOSH. 1978. "Criteria for a Recommended Standard . . . Occupational Exposure to Diisocyanates." DHEW (NIOSH) 78–215.
3. OSHA. 1989. "Air Contaminants; Final Rule." Fed. Reg. 54:2332–2959.
4. NIOSH. 1988. NQ9350000. "Isocyanic Acid, Methylene Di-*p*-phenylene Ester." RTECS, on line.

5. Woolrich, P. F. 1982. "Toxicology, Industrial Hygiene, and Medical Control of TDI, MDI, and PMPPI." *Am. Ind. Hyg. Assoc. J.* 43:89–97.

6. Konzen, R. B., B. F. Craft, L. D. Scheel, and C. H. Gorski. 1966. "Human Response to Low Concentrations of p,p-Diphenylmethane Diisocyanate (MDI)." *Am. Ind. Hyg. Assoc. J.* 27:121–27.

7. O'Brien, I. M., M. G. Harries, P. S. Burge, and J. Pepys. 1979. "Toluene Diisocyanate-Induced Asthma. I. Reactions to TDI, MDI, HDI, and Histamine." *Clin. Allergy* 9:1–6.

DIMETHYLAMINE

Synonyms: *N*-methylmethanamine, DMA

CAS Registry Number: 124–40–3

Molecular Weight: 45.08

Molecular Formula: $(CH_3)_2NH$

AALG: systemic toxicity—2.0 $\mu g/m^3$ (1.1 ppb) 24-hour TWA

Occupational Limits:
- ACGIH TLV: 10 ppm (18 mg/m^3) TWA
- NIOSH REL: none
- OSHA PEL: 10 ppm (18 mg/m^3) TWA

Basis for Occupational Limits: The TLV is based on prevention of potential irritant effects and systemic toxicity.[1,2]

Drinking Water Limits: No drinking water limits were found for this substance.

Toxicity Profile

Carcinogenicity: No data were found implicating DMA as a carcinogen.

Mutagenicity: In a report cited in RTECS, DMA was reported to have caused chromosomal aberrations (cell type not specified) in rats exposed to 50 $\mu g/m^3$ via inhalation (duration not specified).[3]

Developmental Toxicity: No data were found implicating DMA as a developmental toxin.

Reproductive Toxicity: No data were found implicating DMA as a reproductive toxin.

Systemic Toxicity: DMA has been reported to have chemical and toxicologic properties similar to ammonia, but it is a stronger base and more potent local irritant.[1]

Coon et al.[4] exposed rats (n = 15, sex and strain not specified), Princeton-derived guinea pigs (n = 15, sex not specified), male New Zealand white rabbits (n = 3), male beagle dogs (n = 2), and male squirrel monkeys (n = 3) to 9 mg/m^3

(5 ppm) DMA (99% purity) via inhalation continuously for 90 days. No mortality or signs of toxicity were observed and hematologic parameters were normal. However, histological examination revealed interstitial inflammatory changes in the lungs of all species and dilation of the bronchi in rabbits (3 of 3) and monkeys (2 of 3).

In a study cited in the TLV documentation (unpublished data, Dow Chemical Co., 1964), rats, mice, guinea pigs, and rabbits, as well as a single monkey, were exposed to 97 or 183 ppm dimethylamine via inhalation for 7 hours per day, 5 days a week for 18 to 20 weeks. Corneal injury (nature not specified) occurred in guinea pigs and rabbits after 9 days of exposure. In addition, centrilobular fatty degeneration and hepatic necrosis were reported in all species except the monkey, as well as degeneration of the renal tubular epithelium in the male rabbit at the higher concentration and the male monkey at the lower concentration.[1]

Irritation: DMA has been reported to have a "strong ammoniacal" or "fish-like" odor and an odor threshold of 0.6 ppm. Exposure to "high" concentrations of the vapor has been reported to result in signs of mucous membrane and eye irritation in rats.[1]

Steinhagen et al.[5] evaluated the sensory irritation potential of DMA in male F344 rats and male Swiss-Webster mice and reported $RD_{50}s$ (concentration that causes a 50% decrease in respiratory rate) of 573 and 511 ppm, respectively. Based on suggested procedures for establishing human sensory irritation thresholds from animal RD_{50} data (0.01 RD_{50}: minimal or no sensory irritation expected; 0.1 RD_{50}: irritating, but tolerable), the authors suggested a threshold within the range of 5 to 51 ppm.

Basis for the AALG: DMA may be either a liquid or gas at "ordinary" temperatures (vapor pressure 1080 mm Hg at 20°C) and is freely soluble in water. DMA is used as a solvent and a rubber accelerator and is used in pharmaceutical preparations and textile chemicals and in the synthesis of dimethylacetamide and dimethylformamide.[1] Since there is a lack of information on its environmental fate and distribution and the expected contribution of various media to human exposure, 50% of the contribution to total exposure will be allowed from air.

The AALG is based on the Coon et al. study[4] because it uses continuous exposure which is more relevant for the general population and because, although a single exposure level was used, it is the lowest reported LOAEL. The calculations were made assuming a 350-g rat because no weights were given in the paper and because use of the rat results in the most conservative estimated exposure. The rat LOAEL was converted to a human equivalent LOAEL according to the procedures described in Appendix C of the methodology document and a total uncertainty factor of 5000 was applied (10 for interindividual variation × 10 for interspecies variation × 5 for a LOAEL × 10 for less-than-lifetime exposure). Note that the TLV was considered unsuitable as a basis for AALG derivation because it was twice as high as the animal LOAEL and no human data were cited in support of it.

AALG: • systemic toxicity—2.0 μg/m³ (1.1 ppb) 24-hour TWA

References

1. ACGIH. 1986. "Documentation of the Threshold Limit Values and Biological Exposure Indices." 5:206.
2. OSHA. 1989. "Air Contaminants; Final Rule." *Fed. Reg.* 54:2332–2959.
3. NIOSH. 1988. IP8750000. "Dimethylamine." RTECS, on line.
4. Coon, R. A., R. A. Jones, L. J. Jenkins, and J. Siegel. 1970. "Animal Inhalation Studies on Ammonia, Ethylene Glycol, Formaldehyde, Dimethylamine, and Ethanol." *Tox. Appl. Pharm.* 16:646–55.
5. Steinhagen, W. H., J. A. Swenberg, and C. S. Barrow. 1982. "Acute Inhalation Toxicity and Sensory Irritation of Dimethylamine." *Am Ind. Hyg. Assoc. J.* 43:411–17.

DIMETHYL PHTHALATE

Synonyms: DMP, dimethyl 1,2-benzenedicarboxylate

Molecular Formula:

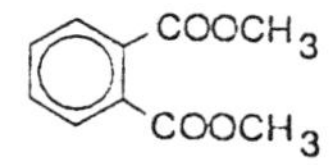

CAS Registry Number: 131–11–3

Molecular Weight: 194.19

AALG: systemic toxicity—0.088 ppm (0.7 mg/m^3) 24-hour TWA

Occupational Limits:
- ACGIH TLV: 5 mg/m^3 TWA
- NIOSH REL: none
- OSHA PEL: 5 mg/m^3 TWA

Basis for Occupational Limits: The TLV for DMP is not based on prevention of toxic effects, but rather "to control the excess mist." It should be noted that unless heat is applied, inhalation exposure to DMP is via spray or mist.[1]

Drinking Water Limits: The ambient water quality criterion for dimethyl phthalate is 313 mg/L assuming a 70-kg man consuming 2 L of water and 6.5 g fish/shellfish per day.[2] This criterion was derived from the NOAEL for effects on growth and systemic toxicity of 2% DMP in the diet of rats for 2 years. This study is reviewed under "Systemic Toxicity."

Toxicity Profile

Carcinogenicity: No data were found implicating DMP as a carcinogen. However, it is on test in the NTP bioassay program with 2-year skin painting studies in mice and rats in progress.[3]

Mutagenicity: DMP has been reported to be mutagenic to *S. typhimurium* depending on test conditions. It was positive without metabolic activation and negative with metabolic activation with strain TA100 in a liquid suspension assay with 4-hour incubation. In a modified Ames assay with histidine and biotin incorporated into the bottom agar, DMP was negative in strain TA98 with and without metabolic activation and in strain TA100 with metabolic activation, but it was positive in strain TA100 without metabolic activation and a dose-response relationship was observed.[4] DMP has been reported to produce chromosomal aberrations in rat skin cells,[5] but did not produce them in cultured human leukocytes at a dose of 0.25 mg/mL.[4]

Developmental Toxicity: Singh et al.[6] studied the developmental toxicity of phthalate esters in Sprague-Dawley rats which were administered the compounds by intraperitoneal injection. Groups of five rats were untreated, injected with distilled water, normal saline, or cottonseed oil (controls), or 0.338, 0.675, or 1.125 mL/kg DMP on gestation days 5, 10, and 15. The doses of DMP used were 1/10, 1/5, and 1/3 of the previously determined i.p. LD_{50} for this compound. No effects on fertility (as evidenced by corpora lutea:implantation site ratio) were noted, however fetal weights were significantly lower in DMP- and saline-treated rats compared to untreated controls. There was also a dose-dependent increase in skeletal abnormalities (e.g., absence of tail and leg bones, abnormal or incomplete skull bones, incomplete leg bones, and elongated and fused ribs) as well as an increase in gross abnormalities such as hematomas, anophthalmia, absence of tail, and twisted limbs.

While the above study is of interest and points to a need for more extensive evaluation of the developmental toxicity of DMP, it is unsuitable for criteria derivation because of the route of exposure. In addition, various limitations of the study design, (e.g., small group sizes, lack of historical or concurrent positive controls, and no evaluation of maternal toxicity) limit the conclusions which may be drawn from this study.

Reproductive Toxicity: No data were found implicating DMP as a reproductive toxin.

Systemic Toxicity: Data on the inhalation toxicity of DMP is almost totally lacking. In a single exposure, the lowest concentration reported to produce signs of toxicity in cats was 10,000 ppm.[1] Other reported acute exposure data have also indicated that DMP is of low acute toxicity with reported oral LD_{50}s of 6800, 6800, 2400, and 4400 mg/kg in mice, rats, guinea pigs, and rabbits, respectively.[5]

Lehman[7] exposed female rats (n = 10/group) to 0, 2.0, 4.0, or 8.0% DMP in the diet for 2 years. Although no differences in mortality were observed between control and treated groups, slight but significant decreases in growth occurred in the two highest exposure groups. Histopathologic evidence of chronic nephritis was found in rats in the highest exposure group. A NOAEL of 2.0% DMP in the diet of rats is suggested by this study.

In a report from the Russian literature cited in the TLV documentation, worker exposure to mixed phthalate esters (concentration range 1.7 to 66 mg/m³) as well as tri-*ortho*-cresyl phosphate resulted in pain, numbness, and spasms of the upper and lower extremities after 6 or 7 years of employment; clinical examination of workers revealed polyneuritis and altered vestibular function[8] (cited in ACGIH[1]).

Experience with DMP use as a mosquito repellent during World War II resulted in few reports of toxic effects. However, DMP may be absorbed through intact human skin and the metabolites appear in the urine.[2,4]

Irritation: Based on its widespread use in cosmetic preparations (e.g., as a plasticizer in fingernail products, a solvent, and as a perfume fixative), DMP has been extensively evaluated for potential irritancy in both animals and humans. These studies have generally revealed minimal potential for dermal and ocular irritation.[4] Although it may be absorbed through the skin, DMP was not reported to cause irritation or sensitization when used as an insect repellent during World War II.[1]

DMP is reported to have none to slight odor and no odor threshold was found.[4]

Basis for the AALG: The water solubility (0.4% at 32°C) and volatility (less than 0.1 mm Hg at 20°C) of DMP are both low.[1] DMP is used in cosmetics, as a solvent and plasticizer for nitrocellulose and cellulose acetate, in the manufacture of varnishes and plastics, in insecticides and insect repellents, and in adhesives used as components of packaging material in contact with food. Phthalates are ubiquitous in the environment and exposure of humans by various routes including oral, dermal, and inhalation is thought to be likely.[4] Based on these considerations, 20% of the contribution to total exposure will be allowed from air.

The AALG for DMP is based on systemic toxicity, using the NOAEL of 2% DMP in the diet of rats suggested by the study of Lehman.[7] This NOAEL was converted to a figure in mg/kg/day assuming a 350-g rat and using the empirically derived food factor, i.e., a factor of 0.05 or 5% which is the proportion or percentage of the body weight consumed per day in food for the rat, suggested by the U.S. EPA as described in Appendix C. This figure was then converted to a human equivalent NOAEL for the inhalation route (as described in Appendix C) and a total uncertainty factor of 1000 applied (10 for interindividual variation $\times$ 10 for interspecies variation $\times$ 10 for the database factor). In this case, a database factor of 10 is warranted based on the limited reporting of data in this study and the use of only female animals, as well as the serious nature of other toxic effects (developmental and mutagenic) reported and the use of the non-inhalation route of exposure.

Since suggestive mutagenicity and developmental toxicity data exist for DMP and the AALG is based on non-inhalation data, the AALG should be considered provisional. Note that the TLV was considered unsuitable for criteria derivation because it is apparently not based on prevention of toxic effects.

AALG: • systemic toxicity—0.088 ppm (0.7 mg/m^3) 24-hour TWA

References

1. ACGIH. 1986. "Documentation of the Threshold Limit Values and Biological Exposure Indices." 5:211.
2. U.S. EPA. 1980. "Ambient Water Quality Criteria for Phthalate Esters." EPA 440/5-80-067.
3. NTP. 1988. "Management Status Report." National Toxicology Program, Research Triangle Park, NC (10/7).

4. Elder, R. L. (Ed.). 1985. "Final Report on the Safety Assessment of Dibutyl Phthalate, Dimethyl Phthalate, and Diethyl Phthalate." *J. Am. Coll. Tox.* 4:267–303.
5. NIOSH. 1988. TI1575000. "Phthalic Acid, Dimethyl Ester." RTECS, on line.
6. Singh, A. R., W. H. Lawrence, and J. Autian. 1972. "Teratogenicity of Phthalate Esters in Rats." *J. Pharm. Sci.* 61:51–55.
7. Lehman, A. J. 1955. "Insect Repellants." *Assoc. Food Drug Offic. U.S. Quart. Bull.* 19:87–99.
8. Milkov, L. E., M. V. Aldyreva, T. B. Popova, K. A. Lopukhova, Y. L. Makarenko, L. M. Malyar, and T. K. Shakhova. 1973. "Health Status of Workers Exposed to Phthalate Plasticizers in the Manufacture of Artificial Leather and Films Based on PVC Resins." *Env. Health Perspect.* 3:175–78. Translated from: *Gigiena Truda.* 13:14 (1969).

1,4-DIOXANE

Synonyms: dioxane, *p*-dioxane, 1,4-diethylene dioxide

CAS Registry Number: 123–91–1

Molecular Weight: 88.10

Molecular Formula: $O(CH_2CH_2)_2O$

AALG: 10^{-6} 95% UCL—0.044 ppb (0.16 $\mu g/m^3$) annual TWA

Occupational Limits:
- ACGIH TLV: 25 ppm (90 mg/m^3) TWA; skin notation (technical grade)
- NIOSH REL: 1 ppm (3.6 mg/m^3) ceiling (30 minutes)
- OSHA PEL: 25 ppm (90 mg/m^3) TWA; skin notation

Basis for Occupational Limits: The TLV is based on hepatotoxic and nephrotoxic effects observed in workers and confirmed in animal studies.[1] The NIOSH REL is "based on the belief that dioxane can cause tumors in exposed workers and on the belief that information allowing the derivation of a safe exposure limit is not now available."[2] The limit of 1 ppm is given as the lowest concentration reliably measured over a short period of time. The TLV documentation gives the following explanation of the TLV committee's decision not to base its limit on carcinogenicity: "In view of the findings of tumors in both liver and lung at or near the 10,000 ppm dietary level, and the lack of such findings at inhalation exposure levels slightly above 100 ppm for two years, dioxane has been classed by the TLV Committee as an animal tumorigen of such low potential as to be of no practical significance as an occupational carcinogen at or around the former TLV of 50 ppm."[1] OSHA recently adopted the ACGIH TLV as part of its final rule limits.[3]

Drinking Water Limits: A 1-day HA for a 10-kg child consuming 1 L of water per day was calculated as 4.12 mg/L based on a rabbit LOAEL for renal and hepatotoxicity. In the absence of an acceptable study on which to base a 10-day HA, the 1-day HA was divided by 10 to give 0.412 mg/L. No longer-term or lifetime HAs were recommended on account of the carcinogenic potential of dioxane. Using the linearized multistage model, the EPA estimated a criterion of 7 $\mu g/L$ at the 10^{-6} risk level assuming a 70-kg man consuming 2 L of water per day.[4]

Toxicity Profile

Carcinogenicity: In an NCI bioassay,[5] male and female Osborne-Mendel rats (n = 35/sex/group) and B6C3F$_1$ mice (n = 50/sex/group) were exposed to 0, 0.5, or 1.0% (v/v) dioxane in drinking water for 110 and 90 weeks, respectively. There were statistically significant increases in squamous cell carcinomas of the nasal turbinate in male and female rats, as well as significant dose-related increases in hepatocellular adenomas and carcinomas in female rats. There were also statistically significant dose-related increases in hepatocellular carcinomas in both male and female mice. Although survival was decreased in treated rats and female mice relative to control animals, it was concluded that sufficient numbers of animals survived to allow evaluation of carcinogenicity. These results are consistent with the findings of at least six earlier studies in which dioxane was reported to induce tumors, primarily in the liver and nasal cavity, in rats and mice given the chemical in drinking water (lowest level, 5000 ppm). These studies are reviewed in the NIOSH criteria document.[2] In addition, dioxane has been shown to induce pulmonary tumors in mice following i.p. injection.[2]

Torkelson et al.[6] exposed male and female Wistar strain rats to filtered air (n = 192/sex) or 111 ppm dioxane (n = 288/sex) for 7 hours per day, 5 days a week for 2 years. There were no statistically significant differences between control and treated animals with respect to tumor induction, growth, mortality, or clinical chemistry and hematologic parameters. However, this study is not sufficient to establish the noncarcinogenicity of dioxane by the inhalation route because only a single dose level was used and it was below the maximally tolerated dose (MTD).

Dioxane has been classified in group B2, probable human carcinogen, under the EPA weight-of-evidence classification.[4]

Mutagenicity: The genotoxic activity of 1,4-dioxane in short-term tests was recently reviewed by IARC.[7] Dioxane was nonmutagenic to *S. typhimurium* strains TA98, TA100, TA1535, TA1537, and TA1538 with and without the addition of a metabolic activation system. It did not induce aneuploidy in *S. cerevisiae* and was negative in the *Drosophila* sex-linked recessive lethal assay. However, it did induce chromosomal aberrations in plants and DNA damage (strand breaks, cross links) in rat hepatocytes in vitro.

Developmental Toxicity: Giavini et al.[8] exposed Sprague-Dawley rats to 0, 0.25, 0.5, or 1.0 mL/kg/day dioxane by gavage (diluent was water) on days 6 through 15 of gestation. No evidence of gross, skeletal, or visceral malformations were found in any of the fetuses examined. Embryotoxicity, attributed to maternal toxicity by the authors, was observed only at the highest dose.

Reproductive Toxicity: No data were found implicating dioxane as a reproductive toxin.

Systemic Toxicity: Liver and kidney damage are the predominant adverse effects in workers poisoned by dioxane.[2] In one case report, severe liver and kidney damage and death were associated with exposure to 208 to 650 ppm dioxane for one week[9] (cited in NIOSH[2]). In experimental animals the major target organ toxicity also involves the liver and kidney, and observed lesions include centrilobular hepatocellular and renal tubular epithelial degeneration and necrosis.[2]

In a lifetime ingestion study, Kociba et al.[10] exposed male and female Sherman strain rats (n = 60/sex/group) to 0, 0.01, 0.1, and 1.0% dioxane in drinking water. Hepatocellular and nasal tumors were observed at the highest dose. Dose-dependent hepatocellular and renal tubular degenerative changes were observed at the middle and high concentrations, but not at the low concentration.

Irritation: Exposure to 5500 ppm dioxane for 1 minute is reported to cause vertigo and irritation of the eyes, nose, and throat in humans. Exposure to 300 ppm for 15 minutes causes similar irritation.[2] Dioxane is reported to have an "ethereal" odor and an odor threshold of 170 ppm.[1]

Basis for the AALG: Dioxane is considered to be miscible in water at all concentrations and has a vapor pressure of 30 mm Hg at 20°C. It is used as a solvent for cellulose acetate, resins, oils, and waxes. Dioxane has been reported to occur in both surface and ground waters, and data is lacking on its occurrence in food and ambient air.[4] Since there is no information on the expected contribution of various media to human exposure, 50% of the contribution to total exposure will be allowed from air.

Based on the well-documented carcinogenic effects of dioxane in experimental animals, the AALG is based on this endpoint. The U.S. EPA Carcinogen Assessment Group has evaluated the carcinogenic risk associated with ingestion of dioxane using the linearized multistage model with extra risk applied to the NCI bioassay data; this evaluation is presented in the IRIS database.[11] The recommended oral slope factor of 1.1×10^{-2} (mg/kg/day)$^{-1}$ is based on the incidence of squamous cell carcinomas of the nasal turbinates in male Osborne-Mendel rats (raw data: controls 0 of 33, 240 mg/kg/day 12 of 24, and 530 mg/kg/day 16 of 33; the corresponding human equivalent doses in mg/kg/day are 0, 48, and 106). If the oral slope factor is used to derive a unit risk for inhalation assuming a 70-kg human breathing 20 m^3 air/day and 100% absorption, the resultant unit risk is 3.15×10^{-6}, which corresponds to an inhalation q_1^* of 3.15×10^{-6} (μg/m^3)$^{-1}$. It should be noted that 100% absorption is assumed in the absence of actual experimental data and is considered reasonable in this particular case since dioxane has been reported to be readily absorbed through the lungs in mammals.[4] Applying the relationship E = $q_1^* \times$ d, where E is the specified level of extra risk and d is the dose, yields the figures shown below. This AALG should be considered provisional since it is derived from a non-inhalation study and there is a high degree of uncertainty associated with this type of extrapolation.

AALG: • 10^{-6} 95% UCL—0.044 ppb (0.16 μg/m³) annual TWA
• 10^{-5} 95% UCL—0.44 ppb (1.6 μg/m³) annual TWA

Note: In a 1989 report on the effects of 1,4-dioxane prepared by Hartung for Gelman Sciences,[12] it was pointed out that target organ toxicity in the liver and kidney has occurred at lower doses than those producing tumors in animal studies. In addition, the levels of 1,4-dioxane exposure required to produce tumors in animal models are associated with saturation enzyme kinetics. Hartung has pointed out that the available evidence suggests that 1,4-dioxane does not act as a somatic mutagen, i.e., it does not produce mutations in short-term test systems or alkylate DNA, and it may act as a promotor or an epigenetic carcinogen. The significance of this hypothesis is that it implies the existence of a threshold for the observed carcinogenic effects of 1,4-dioxane which, based upon current practice, would favor the use of an uncertainty factor approach for criteria derivation. It is our recommendation that future assessments of 1,4-dioxane reconsider this issue as new data become available and the issue of thresholds for epigenetic carcinogens is further refined.

References

1. ACGIH. 1986. "Documentation of the Threshold Limit Values and Biological Exposure Indices." 5:217–18.
2. NIOSH. 1977. "Criteria for a Recommended Standard . . . Occupational Exposure to Dioxane." DHEW (NIOSH) 77–226.
3. OSHA. 1989. "Air Contaminants; Final Rule." *Fed. Reg.* 54:2332–2959.
4. U.S. EPA. 1987. "Health Advisory for *p*-Dioxane." Office of Drinking Water (3/31).
5. NCI. 1978. "Bioassay of 1,4-Dioxane for Possible Carcinogenicity." TR-80, DHEW (NIH) 78–1330.
6. Torkelson, T. R., B. J. K. Leong, R. J. Kociba, W. A. Richter, and P. J. Gehring. "1,4-Dioxane. II. Results of a 2-Year Inhalation Study in Rats." *Tox. Appl. Pharm.* 30:287–98.
7. IARC. 1986. "1,4-Dioxane." *IARC Monog.* suppl. 6:272–74.
8. Giavini, E., C. Vismara, and M. L. Broccia. 1985. "Teratogenesis Study of Dioxane in Rats." *Tox. Letters* 26:85–88.
9. Johnstone, R. T. 1959. "Death Due to Dioxane?" *Arch. Ind. Health* 20:445–47.
10. Kociba, R. J., S. B. McCollister, C. Park, T. R. Torkelson, and P. J. Gehring. 1974. "1,4-Dioxane. I. Results of a 2-Year Ingestion Study in Rats." *Tox. Appl. Pharm.* 30:275–86.
11. U.S. EPA. 1988. "1,4-Dioxane; CASRN 123–91–1." IRIS (12/1/88).
12. Hartung, R. 1989. "Health and Environmental Effects Assessment for 1,4-Dioxane." Prepared for Gelman Sciences, Ann Arbor, MI (4/15).

EPICHLOROHYDRIN

Synonyms: 1-chloro-2,3-epoxypropane

CAS Registry Number: 106–89–8

Molecular Weight: 92.53

Molecular Formula: $\overset{O}{\triangleright}\!\!-CH_2Cl$

AALG: 10^{-6} UCL—0.66 µg/m^3 (0.17 ppb) annual TWA

Occupational Limits:
- ACGIH TLV: 2 ppm (8 mg/m^3) TWA; skin notation
- NIOSH REL: minimize occupational exposure (no numerical limit derived)
- OSHA PEL: 2 ppm (8 mg/m) TWA; skin notation

Basis for Occupational Limits: The TLV is based on protection against respiratory and eye irritation and liver and kidney effects.[1] The NIOSH recommendation is based on the carcinogenic and mutagenic properties of epichlorohydrin as well as evidence for effects on fertility in male laboratory animals.[2] OSHA recently adopted the TLV-TWA for epichlorohydrin as part of its final rule limits.[3]

Drinking Water Limits: Since sufficient data were not available to calculate a 1-day HA, it was recommended that the 10-day HA of 0.14 mg/L for a 10-kg child be used as a conservative estimate. Data sufficient for the calculation of longer-term HAs were not available also and it was recommended that the DWEL of 0.07 mg/L be used as a conservative approximation. A lifetime HA was not recommended due to epichlorohydrin's group B2 (probable human carcinogen) designation under the EPA weight-of-evidence classification.[4] Quantitative risk estimates based on a 70-kg man consuming 2 L of water per day are given in the "Drinking Water Criteria Document."[5] An MCLG of zero has been proposed for epichlorohydrin.[6]

Toxicity Profile

Carcinogenicity: Epichlorohydrin is a weak contact carcinogen that appears to cause no metastases, i.e., tumors appear only at the site of initial contact. This has been demonstrated by the occurrence of nasal tumors alone in rat inhalation studies[7] (cited in U.S. EPA[8]), forestomach tumors only in rat drinking water and gavage studies[9–11] (cited in U.S. EPA[8]), and injection (subcutaneous) site sarcomas in mice[12] (cited in U.S. EPA[8]). Skin painting studies in mice have indicated that epichlorohydrin is a tumor initiator, requiring complementation by a promotor for activity in the test system used[12] (cited in U.S. EPA[8]). These studies and others relevant to the carcinogenicity of epichlorohydrin are reviewed in the HAD.[8] The rat inhalation study is reviewed in detail since it provides the basis for the quantitative risk assessment described in the HAD.

In separate experiments, Laskin et al.[7] (cited in U.S. EPA[8]) exposed male non-inbred Sprague-Dawley rats (aged 8 weeks) to 100 ppm epichlorohydrin for 30 daily 6-hour periods followed by lifetime observation; initially 40 rats were used, followed by an additional group of 100 animals. High early mortality attributable to respiratory disease occurred in both untreated and air-exposed controls. Squamous cell carcinomas of the nasal cavity occurred in 15 of 140 exposed animals, with none seen in the controls (n = 100). Non-neoplastic lesions observed included severe inflammatory changes in the respiratory tract and renal damage. In a second study, Laskin et al. exposed male non-inbred Sprague-Dawley rats (n = 100/group) to 10 or 30 ppm epichlorohydrin 6 hours per day, 5 days a week for their entire lifetime and compared them to concurrent untreated (n = 50) and air-exposed (n = 100) controls. A single squamous cell carcinoma of the nasal cavity was found in a 30-ppm rat and none in any of the other groups. The incidence of squamous cell metaplasia (transformation of nasal mucosal epithelium into stratified squamous epithelium) was 0% in controls, and 2% and 4% in the 10- and 30-ppm animals, respectively. Early mortality was high in all groups and kidney damage (degenerative changes in renal tubules) occurred with a dose-related trend.

Negative epidemiologic studies in humans occupationally exposed to epichlorohydrin do not provide evidence of the noncarcinogenicity of epichlorohydrin due to design limitations.[8] Epichlorohydrin has been placed in group B2, probable human carcinogen, under the EPA weight-of-evidence classification.[4]

Mutagenicity: Epichlorohydrin has been shown to induce gene mutations in bacteria, yeast, *Drosophila,* and cultured mammalian cells, as well as chromosomal aberrations and sister chromatid exchanges in cultured human cells and in mammalian cell assays, both in vitro and in vivo. It is particularly noteworthy that occupational exposure to epichlorohydrin is associated with an increased frequency of chromosomal aberrations in the peripheral lymphocytes of workers.[8]

Developmental Toxicity: Studies in which rats and rabbits were exposed to epichlorohydrin via inhalation[13] (cited in U.S. EPA[8]) and mice and rats exposed by gavage[14] (cited in U.S. EPA[8]) have indicated that epichlorohydrin has little or no potential for causing fetotoxicity or teratogenic effects.

Reproductive Toxicity: Exposure to epichlorohydrin by gavage[15] (cited in U.S. EPA[8]) or via inhalation[16] (cited in U.S. EPA[8]) causes sterility in male rats which is generally reversible, but may become irreversible if exposure levels are sufficiently high and/or long in duration. In a more recent study, John et al.[17] reported transient infertility in male Sprague-Dawley rats exposed to 50 ppm epichlorohydrin

6 hours per day, 5 days a week for 10 weeks, but not in rats exposed to 5 ppm. No effects on estrous cycle, pregnancy rate, parturition, or numbers and viability of offspring were reported in female rats under similar exposure conditions. It should be noted that the effects on male fertility observed with epichlorohydrin may be due to its metabolite, α-chlorohydrin, which is known for its antifertility effects.[8]

Systemic Toxicity: Epichlorohydrin is moderately toxic to laboratory animals exposed via ingestion, dermal absorption, and inhalation. Acute high-level exposures (above 325 ppm) may result in death due to central nervous system depression and respiratory paralysis. Under conditions of chronic or subchronic inhalation exposure, the most sensitive organs and tissues (in descending order) are the nasal mucosa, kidneys, liver, and cardiovascular system. Refer to the description of the study by Laskin et al.[7] (cited in U.S. EPA[8]) in "Carcinogenicity" above for further detail on the types of non-neoplastic lesions induced by long-term low-level exposure.

Transient exposure to concentrations of epichlorohydrin greater than 100 ppm have been reported to cause lung edema and kidney lesions in occupationally exposed individuals.[18] In addition, epichlorohydrin is reported to induce skin sensitization (allergy) in both humans and laboratory animals.[8]

Irritation: Epichlorohydrin is irritating to the eyes, skin, and mucous membranes.[1,8,18] Exposure to 20 ppm epichlorohydrin has been reported to cause transient burning of the eyes and nasal passages in humans; exposure to 40 ppm resulted in eye and nasal irritation lasting 48 hours.[18] Epichlorohydrin is reported to have a "chloroform-like" odor and an odor threshold of 0.93 ppm (geometric mean of reported literature values).[19]

Basis for the AALG: Epichlorohydrin is considered slightly soluble in water and has a vapor pressure of 13 mm Hg at 20°C. It is a major raw material for the manufacture of epoxy and phenoxy resins and is also used as a solvent, in the synthesis of glycerol, and in the rubber and paper industries.[1] Since the major route of exposure to epichlorohydrin in humans is thought to be the respiratory tract and it is not expected to be persistent in the environment,[8] 80% of the contribution to total exposure is allowed from air.

Given its demonstrated carcinogenicity in laboratory animals exposed by multiple routes, the AALG for epichlorohydrin is based on carcinogenicity. In the HAD for epichlorohydrin,[7] several data sets were used to calculate slope coefficients for epichlorohydrin; however, the AALG is based on the quantitative risk assessment on the Laskin et al.[7] (cited in U.S. EPA[8]) rat inhalation data because of the relevance of the route. Application of the linearized multistage model to this data with conversion to human equivalent doses resulted in a q_1^* of 1.2×10^{-6} ($\mu g/m^3)^{-1}$. Although other methods were also used to estimate q_1^*, CAG considered the above to be the "best" estimate and it is the slope factor used here. Applying the relationship $E = q_1^* \times d$, where E is the specified level of extra risk and d is the dose, yields the estimates shown below. It is stated in the IRIS summary[20] that:

"The unit risk value should not be used if the air concentration exceeds 8×10^3 $\mu g/m^3$, since above this concentration the slope factor may differ from that stated above."

AALG: • 10^{-6} UCL—0.66 $\mu g/m^3$ (0.17 ppb) annual TWA
        • 10^{-5} UCL—6.6 $\mu g/m^3$ (1.7 ppb) annual TWA

References

1. ACGIH. 1986. "Documentation of the Threshold Limit Values and Biological Exposure Indices." 5:233.
2. NIOSH. 1978. "Epichlorohydrin. Current Intelligence Bulletin 30." DHEW (NIOSH) 79–105.
3. OSHA. 1989. "Air Contaminants; Final Rule." *Fed. Reg.* 54:2332–2959.
4. U.S. EPA. 1978. "Health Advisory for Epichlorohydrin." Office of Drinking Water (3/31).
5. U.S. EPA. 1985. "Drinking Water Criteria Document for Epichlorohydrin" (final draft). EPA-600/X-84–200–1.
6. U.S. EPA. 1985. "National Primary Drinking Water Regulations; Synthetic Organic Chemicals, Inorganic Chemicals, and Microorganisms; Proposed Rule." *Fed. Reg.* 50:46935–7025.
7. Laskin, S., A. R. Sellakumar, M. Kuschner, N. Nelson, S. La Mendola, G. M. Rusch, G. V. Katz, N. C. Dulak, and R. E. Albert. 1980. "Inhalation Carcinogenicity of Epichlorohydrin in Non-Inbred Sprague-Dawley Rats." *JNCI* 65:751–55.
8. U.S. EPA. 1984. "Health Assessment Document for Epichlorohydrin." EPA-600/8–83–032F.
9. Konishi, T., A. Kawabata, A. Denda, T. Ikeda, H. Katada, H. Maruyama, and R. Higashiguchi. 1980. "Forestomach Tumors Induced by Orally Administered Epichlorohydrin in Male Wistar Rats." *Gann* 71:922–23.
10. Kawabata, A. 1981. "Studies on the Carcinogenic Activity of Epichlorohydrin by Oral Administration in Male Wistar Rats." *Nara Igaku Zasshi* 32:270–80.
11. Van Esch, G. J. 1982. "Induction of Preneoplastic Lesions in the Forestomach of Rats After Oral Administration of 1-Chloro-2,3-epoxypropane." Preliminary report no. 627805–005.
12. Van Duuren, B. L., B. M. Goldschmidt, C. Katz, I. Seidman, and J. S. Paul. 1974. "Carcinogenic Activity of Alkylating Agents." *JNCI* 53:695–700.
13. Pilny, M. K., T. S. Lederer, J. S. Murray, M. S. Deacon, T. R. Hanley, J. F. Quast, and J. A. John. 1979. "Epichlorohydrin—Subchronic Studies. IV. The Effects of Maternally Inhaled Epichlorohydrin on Rat and Rabbit Embryonal and Fetal Development." Dow Chemical USA, Toxicology Research Laboratory, Health and Environmental Sciences, Midland, MI (unpublished).
14. Marks, T. A., F. S. Gerling, and R. E. Staples. 1982. "Teratogenic Evaluation of Epichlorohydrin in the Mouse and Rat and Glycidol in the Mouse." *J. Tox. Env. Health* 9:87–96.
15. Copper, E. R., A. R. Jones, and H. Jackson. 1974. "Effects of α-Chlorohydrin and Related Compounds on the Reproductive Organs and Fertility of the Male Rat." *J. Reprod. Fert.* 39:379–86.

16. John, J. A., J. F. Quast, F. J. Murray, J. W. Henck, L. G. Calhoun, J. S. Murray, M. K. Pilny, M. S. Deacon, A. A. Crawford, and B. A. Schwetz. 1979. "The Effects of Inhaled Epichlorohydrin on the Semen of Rabbits and on the Fertility of Male and Female Rats." Dow Chemical USA, Toxicology Research Laboratory, Health and Environmental Sciences, Midland, MI (unpublished).

17. John, J. A., J. F. Quast, F. J. Murray, L. G. Calhoun, and R. E. Staples. 1983. "Inhalation Toxicity of Epichlorohydrin: Effects on Fertility in Rats and Rabbits." *Tox. Appl. Pharm.* 68:415–23.

18. NIOSH. 1976. "Criteria for a Recommended Standard . . . Occupational Exposure to Epichlorohydrin." DHEW (NIOSH) 76–202.

19. Amoore, J. E. and E. Hautala. 1983. "Odor as an Aid to Chemical Safety: Odor Thresholds Compared with Threshold Limit Values and Volatilities for 214 Industrial Chemicals in Air and Water Dilution." *J. Appl. Tox.* 3:272–90.

20. U.S. EPA. 1989. "Epichlorohydrin; CASRN 106–89–8." IRIS (1/1/89).

ETHYLENEDIAMINE

Synonyms: 1,2-ethanediamine, 1,2-diaminoethane

CAS Registry Number: 107–15–3

Molecular Weight: 60.10

Molecular Formula: $NH_2CH_2CH_2NH_2$

AALG: systemic toxicity (TLV-based)—0.12 ppm (0.29 mg/m^3) 8-hour TWA

Occupational Limits:
- ACGIH TLV: 10 ppm (25 mg/m^3) TWA
- NIOSH REL: none
- OSHA PEL: 10 ppm (25 mg/m^3) TWA

Basis for Occupational Limits: The occupational limits are based on the prevention of irritation and possible systemic effects and on the reduction of the incidence of hypersensitivity responses.[1,2]

Drinking Water Limits: No drinking water limits were found for this substance.

Toxicity Profile

Carcinogenicity: No data were found implicating ethylenediamine as a carcinogen.

Mutagenicity: In a report cited in RTECS, ethylenediamine was reported to induce mutations in *S. typhimurium* with and without metabolic activation.[3]

Developmental Toxicity: No data were found implicating ethylenediamine as a developmental toxin.

Reproductive Toxicity: No data were found implicating ethylenediamine as a reproductive toxin.

Systemic Toxicity: Studies in occupationally exposed individuals have provided some evidence that, in addition to causing irritation, ethylene amines are allergenic and it was suggested that respiratory effects would be expected in hypersensitive individuals.[1]

Pozzani and Carpenter[4] exposed male and female Sherman strain rats (n = 15/sex/group) to 59, 132, 225, or 484 ppm ethylenediamine (measured concentra-

303

tions) for 30 daily exposures (duration not specified) with concurrent controls for each exposure group. All animals in the highest exposure group died within 20 days of the initial exposure and histological and gross examination of tissues revealed damage to the lungs, kidneys, and liver. These changes included cloudy swelling of the liver, congestion of the lungs, and cloudy swelling with degenerative changes in the kidney convoluted tubules. Survival was also significantly decreased in the 225-ppm rats and surviving animals showed significantly lower weight gain and higher liver and kidney weights with pathological changes similar to the highest exposure level animals. No mortality or pathologic changes related to ethylenediamine exposure were reported in the 132-ppm rats, although a slight depilatory effect (which had marked occurrence in the two highest exposure groups) still persisted. The authors considered 132 ppm to be the NOAEL in this study.

Irritation: Ethylenediamine has been reported to have an "ammoniacal" odor[1] and an odor threshold of 1.0 ppm has been reported.[5]

Dernehl[6] (cited in ACGIH[1]) reported that short-term exposure to 200 or 400 ppm ethylenediamine resulted in irritation of the nasal mucosa and upper respiratory tract with asthmatic symptoms in humans, but 100 ppm was without effect.

Basis for the AALG: Ethylenediamine is freely soluble in water and has a vapor pressure of 10 mm Hg at 20°C. It is used as a solvent, a stabilizer for rubber latex, and an inhibitor in antifreeze solutions.[1] Based on lack of information on its environmental fate and distribution and the expected contribution of various media to human exposure, 50% of the contribution to total exposure will be allowed from air.

The AALG is based on the long-standing TLV divided by 42 because the endpoints on which the TLV is based are relevant for the general population and available evidence indicates that the TLV might reasonably correspond to a human NOAEL. However, since there is evidence that ethylenediamine may induce allergic sensitization (the TLV is partially based on minimization of this effect) and this is an endpoint for which it is difficult to establish a threshold, the AALG should be considered provisional.

AALG: • systemic toxicity (TLV-based)—0.12 ppm (0.29 mg/m^3) 8-hour TWA

References

1. ACGIH. 1986. "Documentation of the Threshold Limit Values and Biological Exposure Indices." 5:249.
2. OSHA. 1989. "Air Contaminants; Final Rule." *Fed. Reg.* 54:2332–2959.
3. NIOSH. 1988. KH8575000. "1,2-Ethanediamine." RTECS, on line.
4. Pozzani, U. C. and C. P. Carpenter. 1954. "Response of Rats to Repeated Inhalation of Ethylenediamine Vapors." *AMA Arch. Ind. Hyg. Occ. Med.* 9:223–26.
5. Amoore, J. E. and E. Hautala. 1983. "Odor as an Aid to Chemical Safety: Odor Thresholds Compared with Threshold Limit Values and Volatilities for 214 Industrial Chemicals in Air and Water Dilution." *J. Appl. Tox.* 3:272–90.
6. Dernehl, C. U. 1951. "Clinical Experiences with Exposures to Ethylene Amines." *Ind. Med.* 29:541–46.

ETHYL ACRYLATE

Synonyms: ethoxycarbonylethylene, ethyl-2-propanoate

CAS Registry Number: 140–88–5

Molecular Weight: 100.11

Molecular Formula: $CH_2 = CHC(O)OCH_2CH_3$

AALG: irritation (cumulative effects)—16 ppb (0.065 mg/m^3) 24-hour TWA

Occupational Limits:
- ACGIH TLV: 5 ppm (20 mg/m^3) TWA; 15 ppm (61 mg/m^3) STEL; Appendix A2, suspect human carcinogen
- NIOSH REL: none
- OSHA PEL: 5 ppm (20 mg/m^3) TWA; 25 ppm (103 mg/m^3) STEL; skin notation

Basis for Occupational Limits: The TLV-TWA was set to minimize irritant effects associated with chronic exposure to ethyl acrylate; in recent revisions a STEL based on prevention of irritation was also added, and the skin notation deleted.[1,2] In these same revisions, ethyl acrylate was added to Appendix A2 (suspected human carcinogen) based on the occurrence of forestomach tumors in rats and mice in an NTP gavage bioassay.[2,3] However, it is stated in the TLV documentation[2] that "the Committee believes that the NTP gavage study used an experimental exposure procedure to administer ethyl acrylate to test animals that is inconsistent with worker exposures to this chemical. Therefore, the Committee also believes that in this case, cancer is not the limiting factor in the recommendation of the TLV for ethyl acrylate." OSHA recently adopted the above limits as part of its final rule limits.[4]

Drinking Water Limits: No drinking water limits were found for this substance.

Toxicity Profile

Carcinogenicity: Ethyl acrylate has been tested for carcinogenicity in mice and rats administered the compound by gavage and via inhalation and in mice only by skin application. In the skin application experiments, 25 μL undiluted ethyl acrylate was applied to the back skin of male C3H/HeJ mice 3 times a week for life (total dose, approximately 770 mg/kg body weight). No treatment-related tumors were found in either the ethyl acrylate–treated animals or in acetone-treated controls upon histological examination of skin tissue, nor was there any difference in survival. However, it should be noted that the issue of possible loss due to volatilization or polymerization was not addressed[5] (cited in IARC[6]).

In an NTP bioassay,[3] male and female F344 rats and B6C3F$_1$ mice (n = 50/sex/group) received 0, 100, or 200 mg/kg ethyl acrylate (purity 99–99.5%, stabilized with 15 mg/kg hydroquinone monomethyl ether) in 5 mL/kg (rats) or 10 mL/kg (mice) corn oil by gavage 5 days per week for 103 weeks. Statistically significant dose-related increases in squamous cell carcinomas or papillomas of the forestomach were found in both sexes of each species. In addition, a dose-related increased incidence of hyperkeratosis, hyperplasia, and inflammation of the forestomach was observed in both species. It is relevant to note that the cancers occurred at the site of deposition of ethyl acrylate in the forestomach.

Miller et al.[7] exposed male and female B6C3F$_1$ mice (n = 105/sex/group) and F344 rats (n = 115/sex/group) to 0, 100, 310, or 920 mg/m^3 (0, 25, 75, or 225 ppm) ethyl acrylate via inhalation for 6 hours per day, 5 days per week for 27 months in the case of the control and low and medium exposure groups; exposure at the highest concentration was stopped after 6 months on account of decreased body weight gain, but the animals were observed for 27 months. No treatment-related increases in neoplasms were found in either species, except for thyroid follicular adenomas in high-dose male mice relative to concurrent, but not historical, controls. With respect to other signs of toxicity, mean body weight gains in each sex of both species were significantly reduced in the moderate- and high-dose groups throughout the study; also, dose-dependent increases in glandular hyperplasia and respiratory metaplasia of the olfactory mucosa were observed in mice and rats of both sexes at all dose levels. Respiratory metaplasia was defined in this study as loss of olfactory epithelium with replacement by ciliated respiratory epithelium similar to that lining the remainder of the upper respiratory tract and tracheobronchial tree. It is relevant that no changes were observed at any time at any concentration in the ciliated respiratory epithelium lining the respiratory tract either anterior or posterior to the affected areas of the olfactory epithelium. The only major criticism that might be leveled at this study is that the highest exposure level was above the maximally tolerated dose and that a lower dose tolerable for lifetime exposure without severe (greater than 10%) weight loss should have been used.

It was concluded by IARC[6] that there was sufficient evidence of ethyl acrylate carcinogenicity in animals, but no human data relevant to ethyl acrylate carcinogenicity were available at the time of the evaluation.

In a study cited in the TLV documentation[8] (cited in ACGIH[2]), it was reported that an excess of colon and rectum cancers (p = 0.0001) was found in a group of 3934 workers exposed to ethyl acrylate prior to 1946 compared with unexposed workers. However, this study is confounded by the fact that there was concurrent exposure to a number of other chemicals (unquantified); ethyl acrylate comprised approximately 10% of all the chemicals manufactured at the plant. The U.S. EPA has placed ethyl acrylate in group B2, probable human carcinogen, based on the EPA weight-of-evidence classification.

Mutagenicity: Ethyl acrylate was nonmutagenic in several strains of *S. typhimurium* with and without metabolic activation under a variety of assay conditions. It was also negative in the *Drosophila* sex-linked recessive lethal assay, but it did induce a dose-related increase in chromosomal aberrations in cultured Chinese hamster ovary cells and increase the number of micronucleated polychromic erythrocytes in bone marrow of mice injected intraperitoneally. IARC considered the evidence of genetic activity for ethyl acrylate in short-term tests to be limited.[6]

Developmental Toxicity: The developmental toxicity of ethyl acrylate has been evaluated in two studies[9,10] (cited in IARC[6]) summarized as follows by IARC: "In one experiment in rats, oral administration of ethyl acrylate produced signs of embryotoxicity and fetotoxicity at mildly maternally toxic doses but did not increase fetal malformation. It was not embryotoxic, fetotoxic, or teratogenic to rats at an airborne concentration that produced slight maternal toxicity (150 ppm)."

Reproductive Toxicity: No data were found implicating ethyl acrylate as a reproductive toxin.

Systemic Toxicity: Prolonged exposure to 50–75 ppm ethyl acrylate has been reported to cause drowsiness, headache, and nausea in workers[11] (cited in ACGIH[1,2] and IARC[6]).

In a 30-day study in which F344 rats and B6C3F$_1$ mice were exposed to 25, 75, or 225 ppm ethyl acrylate vapor for 6 hours per day, decreased body weight gain and degenerative inflammatory and metaplastic changes in the nasal turbinate were found at the two highest doses, but not at the lowest dose[12] (cited in ACGIH[1]). However, in another study (unpublished), degeneration of the olfactory nasal mucosa was reported in rats and mice exposed to 25 ppm ethyl acrylate 6 hours per day, 5 days per week over a period of 6 months.[1]

In a follow-up study designed to establish a NOAEL for changes in the olfactory epithelium, Miller et al.[7] exposed rats and mice to 0 (n = 80/sex/group) or 5 ppm (n = 90/sex/group) ethyl acrylate for 6 hours per day, 5 days a week for 24 months. No neoplastic or metaplastic changes were found in the olfactory epithelium or other tissues of either species. For manifestations of ethyl acrylate toxicity associated with lifetime exposure, see the studies cited in "Carcinogenicity," above.

Several studies have been conducted by NTP to examine the mechanisms of ethyl acrylate-induced gastric toxicity.[13–15] The following quotations summarize the results of these studies:

The absence of systemic toxicity plus the dependency of gastric lesions on the gavage route of administration suggest that the EtAc-induced gastric lesions may be a consequence of localized hemodynamic changes, specifically those characteristic of a classical immediate inflammatory response to an injurious agent at the site of administration.[13]

Gastric toxicity may be attributed to the intact ester molecule or to metabolite(s) other than products of carboxylesterase-mediated hydrolysis (acrylic acid and alcohol) and . . . is dependent upon both acrylate ester concentration in dose vehicle and the lipophilicity of the dose vehicle (corn oil vs water).[14]

Foreign body reaction, which was present in all animals 4 weeks postexposure, appeared to have resulted from entrapment of hair and/or feed particles in forestomach lesions in the course of healing. It is speculated that the increased cell proliferation and the induced foreign body reactions may contribute to the previously demonstrated carcinogenic effect of EtAc on the rat forestomach.[15]

Irritation: Ethyl acrylate is generally regarded as a strong irritant to the skin, eyes, mucous membrane, and respiratory tract.[1,5] It is reported to have an acrid odor; the odor threshold for a 100% response is given as 5 ppb, and for a 50% response as 1 ppb.[1]

Basis for the AALG: Ethyl acrylate is only slightly soluble and tends to polymerize in water; it has a vapor pressure of 29 mm Hg at 20°C.[1,5] No mention was found of ethyl acrylate in drinking water in any of the references reviewed. Ethyl acrylate was listed on the FDA "generally recognized as safe" (GRAS) list for use as a synthetic flavoring substance and food adjuvant[16] (cited in IARC[6]), and it was its use in this capacity as well as the production volume that originally prompted its testing in the NTP bioassay program.[3] It should be noted that its use as a food additive has been steadily decreasing since the early 1980s.[3] Given these considerations, 80% of the contribution to total exposure will be allowed from air.

It was deemed inappropriate to set the AALG (set in consideration of inhalation exposure) on the basis of carcinogenicity for the following reasons:

1. negative results in lifetime inhalation bioassays
2. evidence indicating that the carcinogenic effect observed in the forestomach of rats and mice is more a consequence of the irritant properties of ethyl acrylate and subsequent events occurring during the healing process
3. nonrelevance of the gavage route of exposure to the most likely route of human exposure

Given that the AALG should not be set on the basis of carcinogenicity, two possible bases remain: the occupational limit (TLV) or the rat and mouse NOAEL of 5 ppm from the lifetime study of Miller et al.[7] Although ethyl acrylate is an irritant, some of the effects observed in humans and animals (see "Systemic Toxicity," above) are indicative of cumulative systemic toxicity. Thus, it is treated as such, rather than as an irritant, in the assignment of uncertainty factors, i.e., TLV/42 rather than TLV/10 is used. The animal-based AALGs are calculated for both rats and mice, assuming a weight of 350 and 30 grams, respectively, and using a continuous exposure adjustment (5 ppm × 6/24 × 5/7) with a total uncertainty factor of 100 (10 for interspecies variation × 10 for interindividual variation).

AALG: • TLV-based—95 ppb (0.390 mg/m^3) 24-hour TWA
• mouse NOAEL—32 ppb (0.133 mg/m^3) 24-hour TWA
• rat NOAEL—16 ppb (0.065 mg/m^3) 24-hour TWA

References

1. ACGIH. 1986. "Documentation of the Threshold Limit Values and Biological Exposure Indices." 5:240–1.
2. ACGIH. 1988. "Documentation of the Threshold Limit Values and Biological Exposure Indices." 240(88)-242(88) (revision of the preceding).
3. NTP. 1983. "Carcinogenesis Studies of Ethyl Acrylate (CAS No. 140–88–5) in F344/N Rats and B6C3F$_1$ Mice (Gavage Studies)." (Research Triangle Park, NC: National Toxicology Program), TR-259.
4. OSHA. 1989. "Air Contaminants; Final Rule." *Fed. Reg.* 54:2332–2959.
5. DePass, L.R., E.H. Fowler, D.R. Meckley, and C.S. Weil. 1984. "Dermal Oncogenicity Bioassays of Acrylic Acid, Ethyl Acrylate, and Butyl Acrylate." *J. Tox. Environ. Health* 14:115–20.
6. IARC. 1986. "Ethyl acrylate." *IARC Monog.* 39:81–98.
7. Miller, R.R., J.T. Young, R.J. Kociba, D.G. Keyes, K.M. Bodner, L.L. Calhoun, and J.A. Ayres. 1985. "Chronic Toxicity and Oncogenicity Bioassay of Inhaled Ethyl Acrylate in Fischer 344 Rats and B6C3F$_1$ Mice." *Drug Chem. Tox.* 8:1–42.
8. U.S. EPA. 1986. "The Health and Environmental Effects of Acrylate and Methacrylate Chemicals and the Acrylate/Methacrylate Category." Health and Environmental Review Division Position Paper 45. Office of Toxic Substances, U.S. EPA, Washington, DC (March 1986).
9. Pietrowicz, D., A. Owecka, and B. Baranski. 1980. "Disturbances in Rats' Embryonal Development due to Ethyl Acrylate." *Zwierzeta Lab.* 17:67–72.
10. Murray, J.S., R.R. Miller, M.M. Deacon, T.R. Hanley, W.C. Hayes, K.S. Rao, and J.A. John. 1981. "Teratological Evaluation of Inhaled Ethyl Acrylate in Rats." *Tox. Appl. Pharm.* 60:106–11.
11. Nemec, J. and W. Bauer. 1978. "Acrylic Acid and Derivates." In *Kirk-Othmer Encyclopedia of Chemical Technology,* 3rd ed., H.F. Mark, D.F. Othmer, C.G. Overberger, and G.T. Seaborg, Eds., Vol. 1 (New York, NY: John Wiley & Sons), pp. 330–54.
12. Miller, R. et al. 1980. "Ethyl Acrylate Vapor Inhalation Study: 3-Month Sacrifice of Rats and 6-Month Sacrifice of Rats and Mice." Presented at the 19th annual SOT meeting, Washington, DC (abstract).
13. Ghanayem, B.I., R.R. Maronpot, and H.B. Matthews. 1985. "Ethyl Acrylate–Induced Gastric Toxicity. I. Effect of Single and Repetitive Dosing." *Tox. Appl. Pharm.* 80: 323–35.
14. Ghanayem, B.I., R.R. Maronpot, and H.B. Matthews. 1985. "Ethyl Acrylate–Induced Gastric Toxicity. II. Structure-Toxicity Relationships and Mechanism." *Tox. Appl. Pharm.* 80:336–44.
15. Ghanayem, B.I., R.R. Maronpot, and H.B. Matthews. 1986. "Ethyl Acrylate–Induced Gastric Toxicity. III. Development and Recovery of Lesions." *Tox. Appl. Pharm.* 83:576–83.
16. U.S. Food and Drug Administration. 1984. Food and Drugs. U.S. Code of Fed. Regul., Title 40, Parts 175.515, 175.105, 175.300, 176.170, 176.180, 177.1010, 178.3790, pp. 49, 133, 145, 188, 198.

ETHYL BENZENE

Synonyms: phenylethylene

CAS Registry Number: 100–41–4

Molecular Weight: 106.16

Molecular Formula: $CH_3CH_2-\langle\bigcirc\rangle$

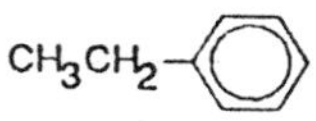

AALG: systemic toxicity—0.03 ppm (0.13 mg/m^3) 24-hour TWA

Occupational Limits:
- ACGIH TLV: 100 ppm (435 mg/m^3) TWA; 125 ppm (545 mg/m^3) STEL
- NIOSH REL: none
- OSHA PEL: 100 ppm (435 mg/m^3) TWA; 125 ppm (545 mg/m^3) STEL

Basis for Occupational Limits: The TLVs are based on skin and eye irritation since "no systemic effects can be expected at levels producing distinctly disagreeable skin and eye irritation."[1] OSHA recently adopted the TLV-STEL for ethyl benzene as part of its final rule limits.[2]

Drinking Water Limits: The ambient water quality criterion of 1.4 mg/L for a 70-kg man consuming 2 L of water and 6.5 g fish/shellfish per day was derived from the TLV.[3] The 1-day and 10-day HAs for a 10-kg child are 32 mg/L and 3.2 mg/L, respectively. Since data sufficient for the derivation of a longer-term HA were unavailable, it was recommended that the DWEL be adjusted for 10 kg as a conservative estimate. The lifetime HA is 0.68 mg/L.[4] An MCLG of 0.68 mg/L has been proposed by the EPA.[5]

Toxicity Profile

Carcinogenicity: No data were found implicating ethyl benzene as a carcinogen, and it has been placed in group D, unclassified substance, based on the EPA weight-of-evidence classification.[6] Ethyl benzene is in the NTP bioassay program and has been assigned to a laboratory for toxicity studies (inhalation, rats and mice).[7]

Mutagenicity: Both ethyl benzene and its common metabolites have produced negative responses in the Ames assay using several strains of *S. typhimurium* with and without metabolic activation.[4,6] Negative findings were also reported in assays measuring gene conversion in yeasts, frequency of recessive lethal mutations in *Drosophila,* and chromosome damage in rat liver RL_4 epithelial cells.[4] Negative results have also been reported for an oncogenic transformation assay (Syrian hamster embryo cells by adenovirus SA7) in the Gene-Tox database[8] as summarized in RTECS. However, a marginally positive response at the highest dose level was reported for the induction of sister chromatid exchanges in human whole blood lymphocytes.[6]

Developmental Toxicity: Hardin et al.[9] (cited in U.S. EPA[4-6]) exposed rats and rabbits to 0, 100, or 1000 ppm ethyl benzene via inhalation for 6 to 7 hours per day on gestation days 1 through 19 and 1 through 24, respectively. Although there was a statistically significant decrease in the number of live rabbit kits per litter at both exposure levels, there was no increase in dead and resorbed fetuses compared with matched controls; there was no evidence of maternal toxicity or teratogenicity in rabbits. In rats there was a statistically significant increase in extra ribs at both dose levels; however, this finding was considered only suggestive (vs indicative) of teratogenic activity by the authors. It should be noted that extra ribs are considered by many developmental toxicologists to be a variation rather than a malformation.[10] Maternal toxicity in the form of increased spleen, liver, and kidney weights was manifest in the 1000-ppm rats.

Reproductive Toxicity: No data indicative of reproductive toxicity were found for ethyl benzene; however, the findings of Wolf et al.[11] (cited in ACGIH[1] and U.S. EPA[5]) with respect to testicular degeneration should be noted.

Systemic Toxicity: Data on the effects of chronic inhalation exposure to ethyl benzene in animals are lacking. It is relevant to note, however, that the target organs for acute high-level exposure (on the order of 5000 to 10,000 ppm) are the lungs and central nervous system, but longer term, lower level exposure tends to result in effects on the visceral organs[1] (see Wolf et al.[11]).

Wolf et al.[11] studied the effects of subchronic inhalation exposure to ethyl benzene in guinea pigs, monkeys, rabbits, and rats exposed 7 hours per day, 5 days a week for periods of up to 6 months at concentrations ranging from 400 to 2200 ppm. Slight increases in liver and kidney weights occurred at 400, 600, and 1250 ppm in male and female rats, at 600 and 1250 ppm in guinea pigs, and at 600 ppm in rhesus monkeys. Slight testicular degeneration was reported at 600 ppm in monkeys and rabbits. The authors concluded that the LOAEL for the most sensitive species, rats, was 400 ppm.

Data on the systemic toxicity of ethyl benzene to humans are generally lacking. It is worthwhile to note that the greatest potential for exposure to ethyl benzene is use of technical grade xylene, which may contain up to 20% ethyl benzene.[1]

Irritation: The following irritant effects with associated exposure levels have been reported for ethyl benzene:[1]

- 5000 ppm—intolerable irritation
- 2000 ppm—immediate severe eye irritation and lacrimation with moderate nasal irritation, which may subside with continued exposure
- 1000 ppm—irritation and lacrimation with rapidly developing tolerance
- 200 ppm—transient eye irritation

Bardodej and Bardodejova[12] identified a NOAEL of 100 ppm for ethyl benzene–induced eye irritation in humans (subjects, 18 healthy male volunteers) in a single 8-hour exposure. It was also reported that the absorption efficiency of ethyl benzene was 64%.

Ethyl benzene is reported to have an "aromatic" odor[1] and an odor threshold of approximately 0.46 ppm.[5]

Basis for the AALG: Ethyl benzene is used primarily as a solvent and chemical intermediate in styrene production. It is considered to be volatile (vapor pressure = 7.1 mm Hg at 20°C) and only slightly soluble in water (14 mg/100 mL at 15°C) and is apparently an infrequent and very low level contaminant in drinking water, based on national surveys.[1,5] Based on consideration of its physical properties and uses, 80% of the contribution to total exposure will be allowed from air.

Given that systemic effects have been reported in animals exposed to ethyl benzene via inhalation in subchronic studies, the recommended AALG is based on this endpoint. The AALG is derived from the animal LOAEL of 400 ppm for liver and kidney effects in rats (the most sensitive species tested in the Wolf et al. study[11]). The rat LOAEL of 400 ppm was adjusted for continuous exposure (400 ppm × 7/24 × 5/7) and converted to a human equivalent LOAEL according to the procedures described in Appendix C, and a total uncertainty factor of 5000 (5 for a LOAEL × 10 for human interindividual variation × 10 for interspecies variation × 10 for a less-than-chronic study) was applied.

From studies by Engstrom et al.[13] (cited in U.S. EPA[14]), it is reported that the major metabolites of ethyl benzene differ for humans and rodents, i.e., the major metabolites in humans (mandelic acid and phenylglyoxylic acid) are minor metabolites in rats and rabbits. On account of this, and because a NOAEL for irritation has been established in humans, an AALG based on irritation was calculated for comparative purposes. A total uncertainty factor of 10 (for interindividual variation) was applied to the NOAEL of 100 ppm from the Bardodej and Bardodejova[12] study; no relative source contribution was allowed from air since the endpoint used was irritation. The resultant figure is 10 ppm.

The recommended AALG should be considered provisional pending availability of results of the animal inhalation bioassay data from the NTP program.

AALG: • systemic toxicity—0.03 ppm (0.13 mg/m^3) 24-hour TWA

References

1. ACGIH. 1986. "Documentation of the Threshold Limit Values and Biological Exposure Indices." 5:244.
2. OSHA. 1989. "Air Contaminants; Final Rule." *Fed. Reg.* 54:2332–2959.
3. U.S. EPA. 1980. "Ambient Water Quality Criteria for Ethyl Benzene." EPA/440/5–80–048.
4. U.S. EPA. 1987. "Health Advisory for Ethyl Benzene." Office of Drinking Water (3/31).
5. U.S. EPA. 1985. "Drinking Water Criteria Document for Ethyl Benzene" (final draft). EPA-600/X-84–163.
6. U.S. EPA. 1984. "Health Effects Assessment for Ethyl Benzene." EPA/540/1–86/008.
7. NTP. 1988. "Management Status Report." National Toxicology Program, Research Triangle Park, NC (1/21/88).
8. NIOSH. 1988. DA0700000. "Benzene, ethyl-." RTECS, on line.
9. Hardin, B.D., G.P. Bond, M.R. Sikov, F.D. Andrew, R.P. Beliles, and R.W. Niemeier. 1981. "Testing of Selected Workplace Chemicals for Teratogenic Potential." *Scand. J. Work. Environ. Health* 7(suppl. 4):66–75.
10. U.S. EPA. 1986. "Guidelines for the Health Assessment of Suspect Developmental Toxicants." *Fed. Reg.* 51:34027–40.
11. Wolf, M.A., V.K. Rowe, D.D. McCollister, R.L. Hollingsworth, and F. Oyen. 1956. "Toxicological Studies of Certain Alkylated Benzenes and Benzene." *Arch. Ind. Health* 14:387–98.
12. Bardodej, Z., and E. Bardodejova. 1970. "Biotransformation of Ethylbenzene, Styrene, and α-Methylstyrene in Man." *Am. Ind. Hyg. Assoc. J.* 31:206–9.
13. Engstrom, K., V. Riihimaki, and A. Laine. 1984. "Urinary Disposition of Ethyl Benzene and *m*-Xylene in Man Following Separate and Combined Exposure." *Int. Arch. Occup. Environ. Health* 54:355–63.
14. U.S. EPA. 1989. "Ethylbenzene; CASRN 100–41–4." IRIS (1/1/89).

ETHYLENE DICHLORIDE

Synonyms: 1,2-dichloroethane

CAS Registry Number: 107–06–2

Molecular Weight: 98.96

Molecular Formula: CH_2ClCH_2Cl

AALG: 10^{-6} 95% UCL—0.008 ppb (0.03 $\mu g/m^3$) annual TWA

Occupational Limits:
- ACGIH TLV: 10 ppm (40 mg/m^3) TWA
- NIOSH REL: 1 ppm (4 mg/m^3) TWA; 2 ppm (8 mg/m^3) 15-minute ceiling
- OSHA PEL: 1 ppm (4 mg/m^3) TWA; 2 ppm (8 mg/m^3) 15-minute ceiling

Basis for Occupational Limits: The TLV is based on adverse nervous system and liver effects, whereas the NIOSH limit takes into account carcinogenicity as well as effects on other organ systems.[1-3] OSHA recently adopted the NIOSH limits as part of its final rule limits.[4]

Drinking Water Limits: Since appropriate data for the derivation of 1-day and 10-day health advisories were lacking, it was recommended that the child longer-term HA of 0.74 mg/L be used as a conservative estimate of these limits. The adult longer-term HA is 2.6 mg/L; no lifetime HA was calculated because of a lack of appropriate data.[5] The ambient water quality criterion for EDC was based on the NCI bioassay data and is 0.94 $\mu g/L$ at the 10^{-6} risk level, assuming consumption of 2 L of water and 6.5 g of fish/shellfish per day.[6] The final MCLG for EDC is 0 mg/L and the MCL is 0.005 mg/L.[7]

Toxicity Profile

Carcinogenicity: No epidemiologic data relevant to EDC was found by the EPA in its literature review for the HAD.[8]

The carcinogenicity of EDC has been evaluated in animal models using multiple routes of exposure, including skin painting, ingestion, inhalation, and parenteral administration. In one study, intraperitoneal administration of EDC to Strain A mice resulted in a nonstatistically significant increase in the incidence of lung adenomas[9] (cited in U.S. EPA[8]), and in a two-stage skin painting bioassay a significant increase in benign lung tumors was reported, although no increase in skin tumors was observed[10] (cited in U.S. EPA[8]).

In an NCI bioassay,[11] male and female Osborne-Mendel rats and B6C3F$_1$ mice (n = 50/sex/group) were gavaged with EDC in corn oil for 78 weeks. The mice were observed for an additional 12–13 weeks, and the rats for 32 weeks; however, all high-dose male and female rats died after 23 and 15 weeks of observation, respectively. The TWA doses in this bioassay were as follows: rats, 47 and 95 mg/kg/day; male mice, 97 and 195 mg/kg/day; and female mice, 149 and 299 mg/kg/day. There was a statistically significant increase in squamous cell carcinomas of the forestomach and circulatory system hemangiosarcomas in male rats and adenocarcinomas of the mammary gland in female rats. In mice there was a statistically significant increase in mammary gland adenocarcinomas and endometrial stromal polyps or sarcomas in female mice and alveolar/bronchiolar adenomas in both sexes.

Inhalation bioassays conducted with EDC have been negative. An early study by Spencer et al.[12] (cited in U.S. EPA[8]) found no increase in tumors in treated animals, compared with controls, when Wistar rats (n = 15/sex/group) received 151 seven-hour exposures to 200 ppm EDC over a period of 212 days.

Maltoni et al.[13] (cited in U.S. EPA[8]) exposed male and female Sprague-Dawley rats and Swiss mice (n = 90/sex/group for both species) to 0, 5, 10, 50, or 150–250 ppm EDC via inhalation for 7 hours per day, 5 days a week for 78 weeks and observed them up to 104 weeks of age. In both species, exposures were reduced to 150 ppm in the high-exposure groups after a few weeks because of severe toxic effects. No statistically significant increases in tumors were observed in any of the groups. There was, however, an increase in nonmalignant mammary fibromas and fibroadenomas in rats in all exposure groups that was statistically significant compared with controls in, but not outside, the exposure chamber. The authors state that this effect was age-correlated, but in the HAD the effect was attributed to differences in survival rates within the groups.

EDC has been placed in group B2, probable human carcinogen, based on the EPA weight-of-evidence classification.[14]

Mutagenicity: EDC is mutagenic to bacteria, plants, *Drosophila,* and cultured Chinese hamster ovary cells; the response is generally enhanced in the presence of a metabolic activation system.[8] Metabolites of EDC have been reported to form DNA adducts following in vivo or in vitro exposures.[14] At the time the HAD was published (1985), the EPA considered that data on the ability of EDC to cause chromosomal effects and germinal mutations to be limited and insufficient to draw conclusions pertaining to these endpoints.[8]

Developmental Toxicity: The results of teratogenicity and reproductive studies in animal models that are reviewed in the HAD indicate that EDC has minimal potential for producing adverse reproductive and fetal outcomes except at maternally toxic levels. However, it was thought that human data and further animals studies were needed to establish this conclusively.[8]

Reproductive Toxicity: Refer to the preceding section.

Systemic Toxicity: The systemic toxicity of EDC in animals and humans is summarized as follows in the HAD:[8]

> The effects of acute inhalation exposure to EDC are similar in humans and animals. Immediate symptoms of toxicity are CNS depression and irritation of the respiratory tract and eyes. Death was usually ascribed to respiratory and circulatory failure, and pathologic examinations typically revealed congestion, degeneration, necrosis, and hemorrhagic lesions of most internal organs (e.g., liver, kidneys, spleen, lungs, respiratory tract, gastrointestinal tract).
>
> Several papers describing human health surveys appear in the literature; adverse effects are largely associated with the gastrointestinal and nervous systems. The exposure information in these studies is not well documented, but taken together indicate that adverse effects have likely occurred in humans at EDC levels below 100 ppm, although probably not as low as 10 ppm. Subtle neurological effects (e.g., fatigue, irritability, sleeplessness) may be more prevalent than overt symptoms of CNS toxicity at lower concentrations. Studies with multiple species of animals have shown that subchronic or chronic exposure to EDC at vapor concentrations of <100 ppm did not produce treatment-related adverse effects on survival, growth, hematology, clinical chemistry, organ weight, or histology. Toxic effects were apparent at higher concentrations and were exposure-related; exposure to 400–500 ppm produced high mortality and histopathological alterations in rodents within a few exposures.

Irritation: Refer to the first paragraph in the extract in the preceding section. The odor of EDC is described as "typical of chlorinated hydrocarbons," and an odor threshold of 6 ppm has been reported.[1,15]

Basis for the AALG: EDC is soluble in water to the extent of 8820 mg/L at 20°C and is volatile with a vapor pressure of 64 mm Hg at 20°C.[5] In the TLV documentation,[1] the following uses are listed for EDC: intermediate in vinyl chloride manufacture; scavenger in leaded gasoline; degreasing agent; fumigant; solvent; and used in paint removers, wetting and penetrating agents, ore flotation, soaps, and scouring components. In the health advisory,[5] it was concluded that (1) EDC is released mainly to the air, and releases associated with metal working operations occur in all industrialized areas; (2) due to its limited releases, EDC is a relatively uncommon environmental contaminant, and unlike other VOCs, the highest concentrations have been reported in surface waters rather than ground waters; and (3) exposure for the majority of the general population occurs mainly through air. Based on these considerations, 80% of the contribution to total exposure will be allowed from air.

The AALG for EDC is based on carcinogenicity, using the data for hemangiosarcomas in male Osborne-Mendel rats from the NCI gavage bioassay in the linearized multistage model. In the IRIS database,[14] the inhalation (and oral) slope factor and inhalation unit risk are given as 9.1×10^{-2} (mg/kg/day)$^{-1}$ and 2.6×10^{-5}, respectively. This corresponds to an inhalation q_1^* of 2.6×10^{-5} (μg/

$m^3)^{-1}$. Applying the relation $E = q_1^* \times d$, where E is the specified level of extra risk and d is the dose, and allowing an 80% relative source contribution from air, gives the estimates shown below. It is stated in IRIS[14] that "the unit risk should not be used if the air concentration exceeds 400 $\mu g/m^3$, since above this concentration the slope factor may differ from that stated."

In the HAD, the EPA discusses in considerable detail pharmacokinetic comparisons between the oral and inhalation routes of exposure. Unfortunately, no convincing evidence exists to resolve or account for the cancer outcome differences between routes or to allow the EPA to describe in quantitative terms the range of uncertainty for cancer risk as bounded by the estimates based on the positive NCI study and the negative Maltoni et al. study.[13] Of concern is that the highest dose used in the Maltoni et al. rat study was approximately biologically equivalent (metabolically) to the low dose in the NCI bioassay. The low exposure group in the NCI bioassay displayed a 19% incidence of hemangiosarcomas, but the rats exposed via the inhalation route did not develop this tumor. If dose were the principal risk factor, then one would have expected some incidence of this, the most sensitive tumor type in the inhalation study.[16] Because the pharmacokinetic discrepancies between the two studies are presently unable to be resolved, and because the limit given below was derived from a non-inhalation study, the AALG should be considered provisional.

AALG: • 10^{-6} 95% UCL—0.008 ppb (0.03 $\mu g/m^3$) annual TWA
 • 10^{-5} 95% UCL—0.076 ppb (0.31 $\mu g/m^3$) annual TWA

References

1. ACGIH. 1986. "Documentation of the Threshold Limit Values and Biological Exposure Indices." 5:252–3.
2. NIOSH. 1985. "NIOSH Recommendations for Occupational Safety and Health Standards." *Morb. Mort. Wkly. Rpt.* suppl. 34:1s–34s.
3. NIOSH. 1978. "Ethylene dichloride. Current Intelligence Bulletin No. 25." DHEW (NIOSH) 78–149.
4. OSHA. 1989. "Air Contaminants; Final Rule." *Fed. Reg.* 54:2332–2959.
5. U.S. EPA. 1987. "Health Advisory—1,2-Dichloroethane." Office of Drinking Water (3/31).
6. U.S. EPA. 1980. "Ambient Water Quality Criteria Document for Chlorinated Ethanes." EPA/440/5–80–029.
7. U.S. EPA. 1987. "Drinking Water Regulations; Public Notification; Final Rule." *Fed. Reg.* 52:20959–21393.
8. U.S. EPA. 1985. "Health Assessment Document for 1,2-Dichloroethane (Ethylene Dichloride)." EPA/600/8–84/006F.
9. Theiss, J., C. Stoner, M. Schimkin, and E.L. Weisburger. 1977. "Test for Carcinogenicity of Organic Contaminants of United States Drinking Water by Pulmonary Tumor Response in Strain A Mice." *Cancer Res.* 37:2717–20.
10. Van Duuren, B., B. Goldschmidt, G. Loewengart, A. Smith, S. Mechionne, I. Seld-

man, and D. Roth. 1979. "Carcinogenicity of Halogenated Olefinic and Aliphatic Hydrocarbons in Mice." *J. Natl. Cancer Inst.* 63:1433–9.

11. NCI. 1978. "Bioassay of 1,2-Dichloroethane for Possible Carcinogenicity." NCI-TR-55. DHEW (NIH) 78–1361.

12. Spencer, H.C., V.K. Rowe, E.M. Adams, D.D. McCollister, and D.D. Irish. 1951. "Vapor Toxicity of Ethylene Dichloride Determined by Experiments on Laboratory Animals." *Ind. Hyg. Occup. Med.* 4:482.

13. Maltoni, C., L. Valgimigli, and C. Scarnato. 1980. "Long-Term Carcinogenicity Bioassays on Ethylene Dichloride Administered by Inhalation to Rats and Mice." In *Ethylene Dichloride: A Potential Health Risk?* B.N. Ames, P. Infante, and R. Reitz, Eds., Banbury Report No. 5 (Cold Spring Harbor, NY: Cold Spring Harbor Laboratory), pp. 3–33.

14. U.S. EPA. 1989. "1,2-Dichloroethane; CASRN 107–06–02." IRIS (5/1/89).

15. NIOSH. 1976. "Criteria for a Recommended Standard . . . Occupational Exposure to Ethylene Dichloride." DHEW (NIOSH) 76–139.

16. Spreafico, F., E. Zuccato, F. Marcuri, M. Sironi, S. Paglialunga, M. Madonna, and E. Mussini. 1980. "Pharmacokinetics of Ethylene Dichloride in Rats Treated by Different Routes and Its Long-Term Inhalation Toxicity." In *Ethylene Dichloride: A Potential Health Risk?* B.N. Ames, P. Infante, and R. Reitz, Eds., Banbury Report No. 5 (Cold Spring Harbor, NY: Cold Spring Harbor Laboratory), pp. 107–33.

ETHYLENE GLYCOL

Synonyms: 1,2-ethanediol

CAS Registry Number: 107–21–1

Molecular Weight: 62.07

Molecular Formula: $HOCH_2$-CH_2OH

AALG: irritation—0.12 ppm (0.3 mg/m^3) as a 24-hour TWA

Occupational Limits:
- ACGIH TLV: 50 ppm (125 mg/m^3) ceiling (vapor and mist)
- NIOSH REL: none
- OSHA PEL: 50 ppm (125 mg/m^3) ceiling (vapor and mist)

Basis for Occupational Limits: The TLV is set to minimize irritation of the respiratory tract.[1] OSHA recently adopted the TLV for ethylene glycol as part of its final rule limits.[2]

Drinking Water Limits: The HAs for ethylene glycol are as follows:[3]

- 1-day HA, 10-kg child—18.86 mg/L
- 10-day and longer-term HA, 10-kg child—5.5 mg/L
- longer-term HA, 70-kg adult—19.25 mg/L
- lifetime HA—7 mg/L

Toxicity Profile

Carcinogenicity: DePass et al.[4] (cited in U.S. EPA[3]) exposed male and female CD-1 mice (n = 80/sex/group) and F344 rats (n = 130/sex/group) to 0, 0.04, 0.2, and 1.0 g/kg/day ethylene glycol in the diet for 24 months. No carcinogenic response was observed in either species, and in addition, there was no clinical or histological evidence of toxicity attributable to ethylene glycol in mice. Nephrotoxicity occurred in rats, with oxalate nephrosis being the primary cause of death in animals at the high-dose level. Male rats were the most sensitive to the effects of ethylene glycol; at the high-dose level, increased water intake, increased BUN and creatinine, reduced erythrocyte count, reduced hematocrit and hemoglobin, increased neutrophil count, increased urinary volume, and reduced urine specific gravity and pH were observed. Calcium oxalate crystals were found with increased frequency at the medium dose; however, 200 mg/kg/day was considered to be a NOAEL for renal effects in rats and was used as the basis for calculation of the oral reference dose by the U.S. EPA.[5]

Ethylene glycol is under study in the NTP bioassay program (mice, in feed) with histopathology for the two-year studies in progress.[6] Ethylene glycol has been placed in group D, unclassified substance, under the EPA weight-of-evidence classification.[3]

Mutagenicity: Negative findings were reported in a dominant lethal assay in which rats were exposed to ethylene glycol in the diet[7] (cited in U.S. EPA[3]). In the Gene-Tox database as summarized in RTECS, the following findings are listed: (1) negative for mutagenesis in *S. typhimurium*, aneuploidy in *N. crassa*, and oncogenic transformation in Syrian hamster embryo cells by adenovirus SA7; and (2) inconclusive for whole chromosome loss and nondisjunction in *D. melanogaster*.[8]

Developmental Toxicity: Statistically significant increases in skeletal abnormalities (primarily rib and craniofacial) have been reported in a continuous breeding study using mice given ethylene glycol in drinking water[9] (cited in U.S. EPA[3]) and in another study in which mice and rats were given ethylene glycol by gavage on days 6 through 15 of gestation[10] (cited in U.S. EPA[3]). These effects were reported at or below levels producing maternal toxicity, and fetotoxicity in the form of reduced fetal weight gain was also reported.

Reproductive Toxicity: In a continuous breeding study in which mice were exposed to 0.25, 0.5, and 1% ethylene glycol in drinking water, decreased fertility (statistically significant) was reported at the highest dose, but not at the lower levels[9] (cited in U.S. EPA[3]). No effects on fertility or offspring development were reported in a three-generation reproduction study in which F344 rats were fed ethylene glycol at 0, 0.04, 0.2, or 1.0 g/kg/day in their diet[7] (cited in U.S. EPA[3]).

Systemic Toxicity: In laboratory animals, ingestion of ethylene glycol may produce C.N.S. depression and a characteristic type of nephrotoxicity described in the study of DePass et al.[4] (cited in U.S. EPA[3]). Histological lesions may include tubular cell hyperplasia, tubular dilation, and peritubular nephritis with oxalate crystal formation. Ethylene glycol poisoning in humans has been reported to result in a variety of C.N.S. and behavioral disturbances, including numbness, visual problems, dizziness, headache, and lethargy at approximately 1000 mg/kg and ataxia, somnolence, and disorientation at approximately 3000 mg/kg.[3]

Irritation: Data on inhalation exposure of laboratory animals to ethylene glycol appear to be quite limited, and the effects confined to eye irritation.[1] For example, Coon et al.[11] (cited in ACGIH[1]) reported that exposure to 12 mg/m^3 (4.7 ppm) ethylene glycol 24 hours a day for 90 days produced moderate to severe eye irritation in rats and rabbits.

In a controlled exposure study using human volunteers, Willis et al.[12] exposed subjects to an average concentration of 30 mg/m^3 (12 ppm) ethylene glycol 20 to 22 hours per day for a month. No signs of systemic toxicity were observed, nor were any abnormalities revealed by clinical chemistry determinations and blood and urine analysis; however, there were some complaints of throat irritation, mild headache, and low backache. Additional studies on the effects of short-term, higher-level exposures revealed the following: (1) reports of upper respiratory tract (URT) irritation increased when the concentration was raised to 60 ppm, and (2) 80 ppm was intolerable because of severe URT irritation (burning along the trachea and burning cough). The authors concluded that "the irritative effects appear to rule out the possibility that a harmful amount of ethylene glycol could be absorbed from the respiratory tract of a healthy individual who was unaware that he had been exposed to this substance."

Basis for the AALG: Since the AALG for ethylene glycol is based on irritation, no relative source contribution was allocated for inhalation. It should be noted that the low vapor pressure of ethylene glycol renders it very unlikely that excessive exposure could occur via inhalation at room temperature. In addition, the irritant response to ethylene glycol in humans is sufficiently severe and occurs at a low enough concentration so that it is highly unlikely that an individual would remain in contact with it long enough for systemic toxicity to result (assuming exposure is via inhalation only).[1]

The AALG for ethylene glycol is based on irritation. The level of 12 ppm (30 mg/m^3) from the study of Willis et al.[12] is used as the LOAEL with a total uncertainty factor of 50 (5 for a LOAEL $\times$ 10 for interindividual variation). Irritation is only appropriate as an endpoint for inhalation exposure; the endpoint of greatest concern for ingestion exposure is nephrotoxicity or C.N.S. depression, depending on the species.

AALG: • irritation—0.24 ppm (0.15 mg/m^3) as a 24-hour TWA

References

1. ACGIH. 1986. "Documentation of the Threshold Limit Values and Biological Exposure Indices." 5:253.
2. OSHA. 1989. "Air Contaminants; Final Rule." *Fed. Reg.* 54:2332–2959.
3. U.S. EPA. 1987. "Health Advisory for Ethylene Glycol." Office of Drinking Water (3/31).
4. DePass, L.R., R.H. Garman, M.D. Woodside, W.E. Giddens, R.R. Maronpot, and C.S. Weil. 1986. "Chronic Toxicity and Oncogenicity Studies of Ethylene Glycol in Rats and Mice." *Fund. Appl. Tox.* 7:547–65.
5. U.S. EPA. 1988. "Ethylene Glycol; CASRN 107–21–1." IRIS (3/1/88).
6. NTP. 1988. "Management Status Report." National Toxicology Program, Research Triangle Park, NC (1/21/88).

7. DePass, L.R., M.D. Woodside, R.R. Maronpot, and C.S. Weil. 1986. "Three-Generation Reproduction and Dominant Lethal Mutagenesis Studies of Ethylene Glycol in the Rat." *Fund. Appl. Tox.* 7:566–72.

8. NIOSH. 1988. KW2975000. "Ethylene Glycol." RTECS, on line.

9. Lamb, J.C., R.R. Maronpot, D.K. Gulati, V.S. Russell, L. Hommel-Barnes, and P.S. Sabharwal. 1985. "Reproductive and Developmental Toxicity of Ethylene Glycol in the Mouse." *Tox. Appl. Pharm.* 81:173–201.

10. Price, C.J., C.A. Kimmel, R.W. Tyl, and M.C. Marr. 1985. "The Developmental Toxicity of Ethylene Glycol in Rats and Mice." *Tox. Appl. Pharm.* 81:113–27.

11. Coon, R.A., R.A. Jones, L.J. Jenkins, and J. Seigl. 1970. *Tox. Appl. Pharm.* 16:646.

12. Willis, J.H., F. Coulston, E.S. Harris, E.W. McChesney, J.C. Russell, and D.M. Serrone. 1974. "Inhalation of Aerosolized Ethylene Glycol by Man." *Clin. Tox.* 7: 463–76.

ETHYLENE OXIDE

Synonyms: oxirane

CAS Registry Number: 75–21–8

Molecular Weight: 44.65

Molecular Formula: $H_2C - CH_2$ over O

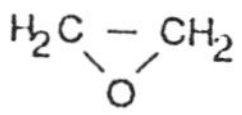

AALG: 10^{-6} 95% UCL—4.4 × 10^{-3} ppb (8.0 × 10^{-3} μg/m^3) annual TWA

Occupational Limits:
- ACGIH TLV: 1 ppm (2 mg/m^3) TWA; Appendix A2, suspect human carcinogen
- NIOSH REL: <0.1 ppm (<0.2 mg/m^3) TWA; 5 ppm (10 mg/m^3) ceiling (10 minutes per day)
- OSHA PEL: none listed in the Z table (1/19/89)

Basis for Occupational Limits: All of the occupational limits are based on consideration of suggestive epidemiologic evidence and positive results from animal bioassays indicative of carcinogenic activity for ethylene oxide.[1-3] In addition, evidence of mutagenicity and adverse reproductive effects were also considered in setting the NIOSH limits.[2] Ethylene oxide is regulated as a carcinogen by OSHA, and regulations governing worker protection and work practices are given in 29 CFR 1910.1047.[4]

Drinking Water Limits: No drinking water limits were found for this substance.

Toxicity Profile

Carcinogenicity: An increased risk of cancer has been demonstrated in three different epidemiologic studies of workers exposed to ethylene oxide. Two of these studies reported an excess risk of leukemia among ethylene oxide production workers (reported to have an excess risk of stomach cancer also) and ethylene oxide sterilization workers[5-7] (cited in U.S. EPA[8]). In another study, ethylene oxide exposure was significantly correlated with increased mortality due to pancreatic cancer and Hodgkin's disease[9] (cited in U.S. EPA[8]). Concurrent exposure to other chemicals, some of which are recognized carcinogens, weakens the conclusions which may be drawn from these studies; the epidemiologic evidence is regarded as limited.[8]

Ethylene oxide has been shown to be carcinogenic in animal models via inhalation, ingestion (gavage), and subcutaneous injection. The inhalation data of Snellings et al.[10] provide the basis for the quantitative risk estimates and will be reviewed here; additional studies relevant to carcinogenicity are reviewed in the health assessment document.[8]

Snellings et al.[10] exposed male and female F344 rats (n = 120/sex/group) to untreated air or 10, 33, or 100 ppm ethylene oxide via inhalation for 6 hours per day, 5 days a week for 2 years. Statistically significant increases in several tumor types occurred, including a dose-related increase in mononuclear cell leukemia in female rats, peritoneal mesotheliomas in male rats in the high and medium exposure groups, and subcutaneous fibromas also in male rats. In addition, increases in brain neoplasms were found in both sexes, and formation of pituitary adenomas appeared to be accelerated in female rats. It is relevant to note that rats in this study became infected with sialodacryoadenitis virus after 15 months of exposure and, apparently on account of this, exhibited increased mortality in the high-dose groups; however, there is no evidence to indicate that the increased incidence of tumors seen was related to the viral infection.[8]

Data from the study of Lynch et al.[11] (cited in IARC[12]) support the results of Snellings et al.[10] since an increased incidence of mononuclear cell leukemia, peritoneal mesotheliomas, and gliomas of the brain were observed. In this study male F344 rats were exposed to filtered air or 50 or 100 ppm ethylene oxide via inhalation for 7 hours per day, 5 days a week over a period of 2 years. Non-neoplastic lesions included inflammatory lesions of the respiratory tract, bronchiectasis, bronchial epithelial hyperplasia, and multifocal and nodular cortical hyperplasia of the adrenal gland.

In an NTP bioassay,[13] male and female B6C3F$_1$ mice (n = 50/sex/group) were exposed to 0 (chamber control), 50, or 100 ppm ethylene oxide (>99% purity) 6 hours per day, 5 days a week for 102 weeks. Survival and body weight gain were not adversely affected by ethylene oxide exposure, and no treatment-related nonneoplastic lesions were observed. There was clear evidence of carcinogenic activity; the following tumor types were considered related to ethylene oxide exposure in both male and female mice: alveolar/bronchiolar adenoma, alveolar/bronchiolar carcinoma, harderian gland papillary cystadenoma, and harderian gland papillary cystadenocarcinoma. In addition, the following tumor types were also considered to be treatment-related in female mice only: malignant lymphoma, uterine adenocarcinoma, and mammary gland adenocarcinoma or adenosquamous carcinoma.

Ethylene oxide has been placed in group B1 (bordering on group B2) under the EPA weight-of-evidence classification.[8]

Mutagenicity: Ethylene oxide is an alkylating agent and has been demonstrated to be a direct-acting mutagen in a number of bacterial and mammalian test systems. It is also capable of inducing chromosomal aberrations in higher plants, insects, and rodents. In addition, ethylene oxide is distributed to gonadal tissue, where it reacts with DNA and can cause heritable mutations in intact rodents.[8] IARC has also reviewed the evidence for genetic activity of ethylene oxide in short-term tests and has judged it to be sufficient.[12]

Developmental Toxicity: In an epidemiologic study of reproductive outcomes in nurses exposed to ethylene oxide, an increased risk of spontaneous abortion was reported[14] (cited in U.S. EPA[8]). Although this study suffered from experimental design limitations, it was concluded in the HAD that the data were suggestive of an association between spontaneous abortion and ethylene oxide exposure and that more study was warranted.[8]

The teratogenicity and reproductive toxicity of ethylene oxide have been evaluated in one primate (cynomolgus monkeys) and three rodent (mice, rats, and rabbits) species via inhalation and intravenous administration. These studies are reviewed in detail in the HAD,[8] and the following conclusion was reached:

> Ethylene oxide produces adverse reproductive and teratogenic effects in both females (maternal toxicity, depression of fetal weight gain, fetal death, and fetal malformation) and males (reduced sperm numbers and sperm motility) if the concentration of the chemical reaching the target organ is sufficiently high or if exposure at lower levels is sufficiently long. Thus, the experiments in which ethylene oxide was injected intravenously have produced more detrimental effects than the short-term inhalation experiments. Even short-term inhalation experiments, however, have resulted in suggestive evidence of detrimental effects. The levels needed to produce the developmental effects approach or equal the levels needed to produce toxicity in the dams.

Reproductive Toxicity: The reproductive toxicity of ethylene oxide is discussed under "Developmental Toxicity," above.

Systemic Toxicity: Human case studies suggest that ethylene oxide may cause neurological effects (e.g., headache, vomiting, sensorimotor neuropathy, seizures), respiratory tract irritation, ocular effects (e.g., cataracts), and hematological effects (e.g., reduced hemoglobin and leukocytosis). However, occupational cohort studies have not generally revealed noncarcinogenic effects from low-level exposure.[1,8]

The principal effect observed in animal models exposed acutely and subchronically to ethylene oxide by a variety of routes of exposure is neurotoxicity, often manifest as hindquarter neuropathy. However, congestive and degenerative changes have also been reported in the lungs, liver, and kidneys.[1,8]

Irritation: High concentrations of ethylene oxide vapor are irritating to the skin and respiratory tract, and aqueous solutions of ethylene oxide are severely irritating to the skin. Ethylene oxide is reported to have an "etherlike" odor that is only detectable at high concentrations (on the order of 700 ppm).[1]

Basis for the AALG: Ethylene oxide is miscible with water and is a gas at room temperature. The major use of ethylene oxide is as a chemical intermediate in the manufacture of materials such as antifreeze, polyester resins, non-ionic surfactants, and specialty solvents.[1] Approximately 0.02% of the ethylene oxide produced in the U.S. is used as a gaseous sterilant in hospitals.[13] Based on these considerations, 80% of the contribution to total exposure will be allowed from air.

The AALG for ethylene oxide is based on carcinogenicity. In the HAD, quantitative risk assessment using the linearized multistage model was performed on data from the studies of Snellings et al.[10] and Lynch et al.[11] (cited in OSHA[4]). The estimates presented here are based on the more conservative slope estimates derived from the combined brain glioma and peritoneal mesothelioma data in female rats in the Snellings et al. study. The upper 95% confidence estimate on the slope, q_1*, was 3.5×10^{-1} $(mg/kg/day)^{-1}$ or 1.0×10^{-4} $(\mu g/m^3)^{-1}$ (assuming a 70-kg man breathing 20 m^3 air per day). Applying the relation $E = q_1* \times d$, where E is the specified level of extra risk and d is the dose, and allowing 80% of the contribution to total exposure from air, gives the estimates shown below.

AALG: • 10^{-6} 95% UCL—4.4×10^{-3} ppb (8.0×10^{-3} $\mu g/m^3$) annual TWA
  • 10^{-5} 95% UCL—4.4×10^{-2} ppb (8.0×10^{-2} $\mu g/m^3$) annual TWA

References

1. ACGIH. 1986. "Documentation of the Threshold Limit Values and Biological Exposure Indices." 5:256–7.
2. NIOSH. 1985. "NIOSH Recommendations for Occupational Safety and Health Standards." *Morb. Mort. Wkly. Rpt.* 34:5s–31s.
3. NIOSH. 1981. "Current Intelligence Bulletin 35: Ethylene Oxide." DHHS (NIOSH) 81–130.
4. OSHA. 1988. 29 U.S. Code of Federal Regulations 1910.1047.
5. Hogstedt, C., O. Rohlen, B.S. Berndtson, O. Axelson, and L. Ehrenberg. 1979. "A Cohort Study of Mortality and Cancer Incidence in Ethylene Oxide Production Workers." *Br. J. Ind. Med.* 36:276–80.
6. Hogstedt, C., N. Malmgrist, and B. Wadman. 1979. "Leukemia in Workers Exposed to Ethylene Oxide." *JAMA* 241:1132–3.
7. Hogstedt, C., L. Aringer, and A. Gustavasson. 1984. "Ethylene Oxide and Cancer: A Literature Review and Follow-up of Two Epidemiologic Studies" (Swedish). *Arbete Halsa* 49:1–32.
8. U.S. EPA. 1985. "Health Assessment Document for Ethylene Oxide." EPA/600/8–84/009F.
9. Morgan, R.W., K.W. Claxton, B.J. Divine, S.D. Kaplan, and V.B. Harris. 1981. "Mortality Among Ethylene Oxide Workers." *J. Occ. Med.* 23:767–70.
10. Snellings, W.M., C.S. Weil, and R.R. Marenpot. 1984. "A Two-Year Inhalation Study of the Carcinogenic Potential of Ethylene Oxide in Fischer 344 Rats." *Tox. Appl. Pharm.* 75:105–17.
11. Lynch, D.W., T.R. Lewis, W.J. Moorman, J.R. Burg, D.H. Groth, A. Khan, L.J. Ackerman, and B.Y. Cockrell. 1984. "Carcinogenic and Toxicological Effects of Inhaled Ethylene Oxide and Propylene Oxide." *Tox. App. Pharm.* 76:69–84.
12. IARC. 1985. "Ethylene Oxide." *IARC Monog.* 36:189–226.

13. NTP. 1987. "Toxicology and Carcinogenesis Studies of Ethylene Oxide (CAS No. 75–21–8) in $B6C3F_1$ Mice (Inhalation Studies)." National Toxicology Program, TR-326.

14. Hemminki, K., P. Utanaen, I. Saloniemi, M.L. Hiemi, and H. Vainio. 1982. "Spontaneous Abortions in Hospital Staff Engaged in Sterilising Instruments with Chemical Agents." *Br. Med. J.* 285:1461–63.

ETHYLENE THIOUREA

Synonyms: 2-imidazolidenethione, ETU

CAS Registry Number: 96–45–7

Molecular Weight: 102.17

Molecular Formula:

AALG: thyroid effects—0.042 ppb (0.175 $\mu g/m^3$) 24-hour TWA

Occupational Limits:
- ACGIH TLV: none
- NIOSH REL: minimize worker exposure
- OSHA PEL: none

Basis for Occupational Limits: The NIOSH recommendation is based on consideration of carcinogenic and teratogenic effects in animal models.[1]

Drinking Water Limits: The following HAs (draft) have been recommended for ETU:[2]

- 1-day, child—data insufficient, use of 10-day HA recommended
- 10-day, child—0.25 mg/L
- longer-term, child—0.125 mg/L
- longer-term, adult—0.44 mg/L
- lifetime—none recommended because of the B2 classification (probable human carcinogen) of ETU under the EPA weight-of-evidence classification

Toxicity Profile

Carcinogenicity: Ethylene thiourea has been shown to be carcinogenic in both sexes and multiple strains of rats and mice exposed via the diet; data on the inhalation carcinogenicity of ETU are lacking.[2]

Dietary exposure to ETU has been shown to produce thyroid follicular cell neoplasia as well as hyperplasia and hypertrophy of the thyroid gland in rats in at least three studies[3,4] (cited in Paynter et al.[5] and Weisburger et al.[6]). Graham et al.[4] (cited in Paynter et al.[5]) exposed male and female CD rats to 0, 5, 25, 125, 250, or

500 ppm ETU in the diet for two years. Statistically significant increases in thyroid adenomas and carcinoma/adenocarcinoma occurred in both sexes at the 250- and 500-ppm exposure levels. Uptake of iodine (I-131) was increased in animals exposed to 5 ppm ETU and decreased in animals exposed to 500 ppm ETU. In addition, thyroid-to-body weight ratios were increased in 250- and 500-ppm exposure groups, and there was histological evidence of thyroid hyperplasia in all other groups. A lifetime exposure LOAEL of 5 ppm ETU in the diet for cellular changes in the thyroid gland was suggested by this study. This corresponds to an estimated (by the U.S. EPA) daily intake of 0.25 mg/kg/day.

In another study, Ulland et al.[7] (cited in IARC[8]) also reported the occurrence of thyroid follicular cell neoplasia in male and female CD rats exposed to 0, 175, or 350 ppm ETU in the diet for 18 months and observed for an additional 6 months. In addition, hyperplastic liver nodules and pulmonary metastases also occurred in some animals.

Recently, evidence has been presented in both an EPA external review draft document[9] and a journal article[5] indicating that thyroid follicular cell neoplasia of the type produced by ETU and other compounds may occur via a threshold mechanism. Briefly, it appears that thyroid follicular cell tumors result from prolonged disturbances of thyroid-pituitary hormonal feedback in cases where levels of circulating thyroid hormone are reduced and levels of thyroid stimulating hormone (TSH) are elevated. It has also been shown that these effects are reversible, that prolonged exposure is required before the induction of neoplasia, and that neoplasia induced by ETU and certain other goiterogenic agents is preventable if animals are concurrently treated with thyroid hormones.[5]

Innes et al.[10] (cited in U.S. EPA[2] and IARC[8]) exposed male and female (C57BL/6 × C3H/Anf)F_1 and (C57BL/6 × AKR)F_1 mice to 0 or 215 mg/kg/day ETU via ingestion for 18 months (single daily dose of 215 mg/kg in gelatin capsules by stomach tube from 7 to 28 days of age, followed by dietary administration of 646 ppm for the remainder of the study). A statistically significant increase in hepatomas occurred in both sexes and strains of treated mice with the following incidences: (C57BL/6 × C3H/Anf)F_1 mice—male controls 8 of 79 and treated 14 of 16, female controls 0 of 87 and treated 18 of 18; (C57BL/6 × AKR)F_1 mice—male controls 5 of 90 and treated 18 of 18, female controls 1 of 82 and treated 9 of 16.

Ethylene thiourea is currently under evaluation in two-year studies in the NTP bioassay program (feed, rats and mice) with the pathology working group in progress.[11] ETU has been placed in group B2, probable human carcinogen, under the EPA weight-of-evidence classification.[2]

Mutagenicity: The genotoxicity of ETU has been evaluated in a large number of test systems encompassing a broad range of endpoints and multiple levels of phylogenetic complexity. There is limited evidence for gene mutation: negative to weakly positive results in bacteria, negative in *Drosophila,* and mixed and conflicting results in cultured mammalian cells. ETU does not induce chromosomal effects, e.g., aberrations and SCEs, in cultured mammalian cells or in vivo; results of tests designed to detect DNA damage in bacteria, yeast, and cultured human cells have been mixed. However, ETU has been reported to induce oncogenic transformation in baby hamster kidney (BHK21) cells in two tests, but not in Syrian hamster embryo cells infected with adenovirus SA7.[9]

Developmental Toxicity: ETU has been demonstrated to be teratogenic in rats and other species at levels where no maternal toxicity was manifest. In addition, in utero exposure to ETU may result in delayed growth and fetal resorption. Based on available data, the rat is the most sensitive species, and the characteristic spectrum of malformations involving the central nervous system (e.g., hydrocephaly, exencephaly, meningocele, cleft palate, etc.) is a consequence of the particular sensitivity of neuroblast cells (neuronal cells in vitro) to the action of ETU. The teratogenicity of ETU and related compounds (including mechanistic and distribution studies) has recently been well reviewed by Khera;[12] only those studies relevant to AALG derivation are reviewed here.

The inhalation developmental toxicity of ETU has been evaluated in a single study available only as an abstract. Dilley et al.[13] exposed rats to 0 or 120.4 ± 8.0 mg/m^3 ETU for 7 hours per day on gestation days 7 through 14. No maternal toxicity or malformations were reported; however, fetal body weights were significantly reduced and there was an increase in fetal resorptions.

Khera[14] exposed Wistar rats and New Zealand white rabbits to 0, 5, 10, 20, 40, or 80 mg/kg ETU in distilled water by gavage (1) from 21–42 days before conception to gestation day 15 in rats, (2) on gestation days 6 to 15 or 7 to 20 in rats, and (3) on gestation days 7 to 20 in rabbits. In all rat experiments, ETU induced a dose-dependent increase in malformations, including meningoencephalocele, meningorrhagia, meningorrhea, hydrocephalus, obliterated neural canal, abnormal pelvic limb posture with equinovarus, and short or kinky tail, in groups exposed to 10 mg/kg or higher doses of ETU. Maternal toxicity (lethality) occurred only in the highest dose group, and no adverse effects on fetal survival were reported; however, fetal growth was significantly retarded in the 40 and 80 mg/kg/day exposure groups. The only effects reported in rabbits were an increased incidence of resorption sites and decreased brain weights (with no evidence of malformations or maternal toxicity) in the highest exposure group. A NOAEL of 5 mg/kg/day for malformations in rats was suggested by this study.

Reproductive Toxicity: No data were found implicating ETU as a reproductive toxin.

Systemic Toxicity: ETU is a potent goiterogenic agent, and the thyroid is the primary target organ for ETU toxicity in subchronic and chronic ingestion studies in rats, the most extensively studied species; inhalation studies in laboratory animals are lacking. The lowest LOAEL for effects on the thyroid gland in a lifetime study is 5 ppm (corresponding to a dose of 0.25 mg/kg/day) from the study of Graham et al.[4] (cited in Paynter et al.[5]). The physiological basis for the goiterogenic effects of ETU is extensively reviewed in U.S. EPA[9].

No data on the health effects of ETU in humans were found in the available literature.

Irritation: No data were found implicating ETU as an irritant.

Basis for the AALG: The most likely route of exposure for the general population to ETU is via ingestion of fruits and vegetables sprayed with ethylene bisdithiocarbamate fungicides because ETU is both a metabolite and a degradation product of these compounds. Air has the potential to form a larger relative source contribution in cases of workers manufacturing, spraying, and handling these fungicides or in rubber-manufacturing workers exposed due to ETU use in curing and vulcanizing rubber. This may also apply to populations living in the immediate vicinity of fungicide application or rubber manufacturing areas. The biological half-life for ETU in natural waters is estimated to be 1–3 hours, and it is reported to degrade to low or undetectable levels in 2–3 weeks when sprayed on crops.[12] Based on these considerations, 20% of the contribution to total exposure was allowed from air in the calculations shown below. However, the relative source contribution from air might be increased in the groups noted above.

AALGs were calculated on the basis of three different endpoints—carcinogenicity (hepatomas) in mice, thyroid follicular cell hyperplasia and other alterations in rats, and teratogenic effects in rats—because of the diverse nature of the endpoints under consideration. All of these AALGs should be considered provisional because they were based on ingestion doses converted to inhalation doses using the procedures described in Appendix C.

The AALG for carcinogenicity is based on data for hepatomas in male mice from the study of Innes et al.[10] (cited in U.S. EPA[2] and IARC[8]). The U.S. EPA Carcinogen Assessment Group reported that the most conservative sex-species $q_1{}^*$ (converted to a human equivalent $q_1{}^*$) was 0.14 $(mg/kg/day)^{-1}$.[15] Assuming 100% absorption (in the absence of inhalation absorption data) and a 70-kg human breathing 20 m^3 of air per day, this $q_1{}^*$ yields an inhalation unit risk of 4.0×10^{-5}, which corresponds to an inhalation $q_1{}^*$ of 4.0×10^{-5} $(\mu g/m^3)^{-1}$. Applying the relation $E = q_1{}^* \times d$, where E is the specified level of extra risk and d is the dose, gives the estimates shown below.

The AALG for effects on the thyroid gland was based on the LOAEL of 5 ppm (0.25 mg/kg/day) from the lifetime ingestion study in rats by Graham et al.[4] (cited in Paynter et al.[5]). The LOAEL was converted to a human equivalent LOAEL, and a total uncertainty factor of 1000 was applied (10 for interspecies variation × 10 for interindividual variation × 5 for a LOAEL × 2 for the database factor). It was considered inappropriate to apply the linearized multistage model to the data for thyroid follicular cell carcinogenesis from the study of Graham et al. because a preponderance of the evidence indicates that this type of carcinogenic effect occurs via a threshold mechanism,[5,9] and one of the assumptions implicit in the use of the multistage model is that carcinogenesis is a nonthreshold effect.

The AALG for developmental toxicity is based on the NOAEL of 5 mg/kg/day for teratogenic effects (primarily neuroterata at doses where maternal toxicity was not evident) in rats from the study by Khera.[14] The NOAEL was converted to a human equivalent NOAEL, and a total uncertainty factor of 1000 was applied (10 for interspecies variation $\times$ 10 for interindividual variation $\times$ 5 for developmental uncertainty factor based on fetal effects occurring below levels producing maternal toxicity $\times$ 2 for the database factor).

In general, when AALGs are derived for multiple endpoints, the most conservative figure is usually recommended as the final AALG. However, in the case of ETU, the AALG derived for thyroid effects in the rat is recommended as the final AALG. The rationale for this choice is that the study from which the AALG for thyroid effects in rats was derived was an exceptionally well-designed investigation compared to the Innes et al. study,[10] and the more conservative figures derived from the Innes et al. study reflect to some extent the methodology used to derive them, i.e., quantitative risk assessment using the linearized multistage model tends to result in very conservative figures. Strengths of the Graham et al. study[11] relative to the Innes et al. study include the following:

1. The Graham et al. study was a full lifetime study (vs 18 months for the Innes et al. study).
2. Examination of tissues and use of multiple endpoints of thyroid function were extensive in the Graham et al. study.
3. The Graham et al. study used an extensive number and broad range of dose levels, whereas the Innes et al. study used only two, including the control group. This is significant, given the use of quantitative risk assessment, because the use of only two dose levels precludes the application of a true goodness-of-fit test to the data, i.e., it is not possible to appropriately evaluate the fit of the model to the data.

AALG:
- 10^{-6} 95% UCL—0.0012 ppb (0.005 $\mu g/m^3$) annual TWA
- 10^{-5} 95% UCL—0.012 ppb (0.05 $\mu g/m^3$) annual TWA
- thyroid effects—0.042 ppb (0.175 $\mu g/m^3$) 24-hour TWA
- developmental—0.71 ppb (3.0 $\mu g/m^3$) 24-hour TWA

References

1. NIOSH. 1978. "Ethylene Thiourea. Current Intelligence Bulletin No. 22." DHEW (NIOSH) 78–144.
2. U.S. EPA. 1987. "Health Advisory for Ethylene Thiourea." Office of Drinking Water (draft 8/87).
3. Gak, J.C., C. Graillot, and R. Truhaut. 1976. "Difference de sensibilite du hamster et du rat vis-a-vis des effect de l'administration de long-term de l'ethylene thiourea." *Eur. J. Tox.* 9:303–12.
4. Graham, S.L., K.J. Davis, W.H. Hansen, and C.H. Graham. 1975. "Effects of Prolonged Ethylene Thiourea Ingestion on the Thyroid of the Rat." *Food Cosmet. Tox.* 13:493–99.
5. Paynter, O.E., G.J. Burin, R.B. Jaeger, and C.A. Gregorio. 1988. "Goitrogens and

Thyroid Follicular Cell Neoplasia: Evidence for a Threshold Process." *Reg. Tox. Pharm.* 8:102–19.

6. Weisburger, E.K., B.M. Ulland, J. Nam, J.J. Gart, and J.H. Weisburger. 1981. "Carcinogenicity Tests of Certain Environmental and Industrial Chemicals." *J. Natl. Cancer Inst.* 67:75–88.

7. Ulland, B.M., J.H. Weisburger, E.K. Weisburger, J.M. Rice, and R. Cypher. 1972. "Thyroid Cancer in Rats from Ethylene Thiourea Intake." *J. Natl. Cancer Inst.* 49: 583–84.

8. IARC. 1974. "Ethylenethiourea." *IARC Monog.* 7:45–52.

9. U.S. EPA. 1988. "Thyroid Follicular Cell Carcinogenesis: Mechanistic and Science Policy Considerations" (external review draft). EPA/625/3–88/014A.

10. Innes, J.R.M., B.M. Ulland, M.G. Valero, L. Petrucelli, L. Fishbein, E.R. Hart, A.J. Pallotta, R.R. Bates, H.L. Falk, J.J. Gart, M. Klein, I. Mitchell, and J. Peters. 1969. "Bioassay of Pesticides and Industrial Chemicals for Tumorigenicity in Mice: A Preliminary Note." *J. Natl. Cancer Inst.* 42:1101–14.

11. NTP. 1988. "Management Status Report." National Toxicology Program, Research Triangle Park, NC (5/11/88).

12. Khera, K.S. "Ethylene Thiourea: A Review of Teratogenicity and Distribution Studies and an Assessment of Reproduction Risk." *CRC Crit. Rev. Tox.* 18:129–52.

13. Dilley, J.V., N. Chernoff, D. Kay, N. Winslow, and G.W. Newell. 1977. "Inhalation Teratology Studies of Five Chemicals in Rats" (abstract). *Tox. Appl. Pharm.* 41:196.

14. Khera, K.S. 1973. "Ethylene Thiourea: Teratology Study in Rats and Rabbits." *Teratology* 7:243–52.

15. U.S. EPA. 1987. "Dietary and Worker Exposure and Risk Analysis for Mancozeb and Ethylene Thiourea (ETU)" (4/1/87).

2-ETHYLHEXYL ACRYLATE

Synonyms: octyl acrylate, 2-ethylhexyl 2-propenoate

CAS Registry Number: 103–11–7

Molecular Weight: 184.31

Molecular Formula: $CH_2C(O)OCH_2CH(C_2H_5)C_4H_9$

AALG: LD_{50}-derived—0.065 ppb (0.49 $\mu g/m^3$) 24-hour TWA

Occupational Limits: • ACGIH TLV: none
• NIOSH REL: none
• OSHA PEL: none

Drinking Water Limits: No drinking water limits were found for this substance.

Toxicity Profile

Carcinogenicity: In a skin-painting study in mice in which the oncogenicity of several acrylates was evaluated, DePass et al.[1] treated the clipped backskin of C3H/HeJ male mice (n = 40/group) with one "brushful" of either acetone or 75% 2-ethylhexyl acrylate (99% purity) three times a week, starting at 7 to 10 weeks of age and continuing for up to two years. The authors attempted to quantify the dose received by weighing the sample bottle after each day's applications and found that the amounts used varied significantly; it was estimated that the approximate dose per mouse per application was 20 mg. The first tumor was observed after 11 months, at which time 30 treated animals were still alive. After 1.5 years, 15 mice were still alive and a total of six skin tumors—two squamous cell carcinomas and four squamous cell papillomas—had occurred; no tumors were found in acetone-treated controls and no treated animals remained alive after two years. There was an increase in chronic nephritis in treated (68%) compared with control (15%) mice, which the authors interpreted as a chemically-induced exacerbation of a disease process common in aging mice.

Although the above study does indicate that 2-ethylhexyl acrylate is carcinogenic when applied to the skin of mice, no sound conclusions can be reached about the dose that caused this effect because of the imprecise methods used to quantify the dose and because no provision was made to prevent or quantify loss due to volatilization of the chemical from the skin.

Mutagenicity: No data were found implicating 2-ethylhexyl acrylate as a mutagen.

Reproductive Toxicity: No data were found implicating 2-ethylhexyl acrylate as a reproductive toxin.

Developmental Toxicity: No data were found implicating 2-ethylhexyl acrylate as a developmental toxin.

Systemic Toxicity: The acute oral LD_{50} for rats has been reported as 5600 mg/kg for 2-ethylhexyl acrylate.[2] Note the renal effects reported in the skin-painting study described in "Carcinogenicity," above.

Irritation: Acrylate monomers are, in general, irritating to the skin, eyes, and mucous membranes;[3] however, with the exception of RTECS references to skin and eye irritancy tests in rabbits, no specific data on the irritant properties of 2-ethylhexyl acrylate were found.[4]

In the material safety data sheet for 2-ethyl hexyl acrylate, it is listed as being both a respiratory tract irritant and a skin irritant, but no specific concentrations were associated with these effects. However, Union Carbide has recommended a limit of 5 ppm as a TWA, but the basis for this recommendation was not provided. Ethyl hexyl acrylate is also reported to have a "pleasant" odor, but no specific odor threshold was found.[5]

Basis for the AALG: 2-Ethyl hexyl acrylate is of relatively low water solubility (0.01% by weight) and has a vapor pressure of 0.15 mm Hg at 20°C.[5] Based on the lack of data on the environmental fate and distribution of 2-ethylhexyl acrylate and the expected contribution to total exposure from various media, 50% of the contribution to total exposure will be allowed from air.

No suitable basis for AALG derivation exists for this compound based on generally accepted criteria. The skin-painting study of DePass et al.[1] is unsuitable as a basis for criteria derivation, not only because a single dose level was used and the route is inappropriate, but also because of the problems with dose quantification discussed above. One possible alternative methodology to derive an *interim provisional* criterion is to derive a surrogate animal NOAEL based on the LD_{50}, convert this to a human equivalent NOAEL for inhalation, and apply appropriate uncertainty factors. Based on the statistical analysis of the relationship of oral LD_{50}s to the corresponding chronic NOAEL for a large number of compounds, Layton et al.[6] recommended that the LD_{50} be multiplied by a factor between 1.0×10^{-5} and 5.0×10^{-6} (equivalent to dividing by a factor between 100,000 and 200,000) to convert an LD_{50} to a corresponding surrogate animal NOAEL. In the case of 2-ethylhexyl acrylate the LD_{50}, 5600 mg/kg, was divided by the larger factor (200,000) since there is data indicating that it has carcinogenic potential. This surrogate animal NOAEL was then converted to a human equivalent NOAEL for the inhalation route according to the procedures described in Appendix C and divided by an uncertainty factor of 100 (10 for interindividual variation $\times$ 10 for interspecies variation).

AALG: • LD_{50}-derived—0.065 ppb (0.49 µg/m³) 24-hour TWA

Note: This AALG is provisional, and further appropriate testing would most likely result in a less conservative number. It is strongly recommended that a full literature search be conducted on this compound to identify a more suitable basis for AALG derivation and/or aid in planning appropriate animal studies.

References

1. DePass, L.R., R.R. Maronpot, and C.S. Weil. 1985. "Dermal Oncogenicity Bioassays of Monofunctional and Multifunctional Acrylates and Acrylate-Based Oligomers." *J. Tox. Env. Heal.* 16:55–60.
2. Autian, J. 1975. "Structure-Toxicity Relationships of Acrylic Monomers." *Env. Heal. Perspect.* 11:141–52.
3. Tanii, H., and K. Hashimoto. 1982. "Structure-Toxicity Relationship of Acrylates and Methacrylates." *Tox. Letters* 11:125–29.
4. NIOSH. 1988. AT0855000. "Acrylic Acid, 2-Ethylhexyl Ester." RTECS, on line.
5. Union Carbide. 1985. "Material Safety Data Sheet for 2-Ethyl Hexyl Acrylate." (Danbury, CT: Union Carbide Corporation).
6. Layton, D.W., B.J. Mallon, D.H. Rosenblatt, and M.J. Small. 1987. "Deriving Allowable Daily Intakes for Systemic Toxicants Lacking Chronic Toxicity Data." *Reg. Tox. Pharm.* 7:96–112.

FORMALDEHYDE

Synonyms: HCHO

CAS Registry Number: 50–00–0

Molecular Weight: 30.03

Molecular Formula: $H_2C = O$

AALG: 10^{-6} 95% UCL—0.05 ppb (0.062 $\mu g/m^3$) annual TWA

Occupational Limits:
- ACGIH TLV: 1 ppm (1.5 mg/m^3) TWA; 2 ppm (3.0 mg/m^3) STEL; Appendix A2, suspect human carcinogen
- NIOSH REL: limit to lowest feasible level
- OSHA PEL: 3 ppm (4.5 mg/m^3) TWA; 5 ppm (7.5 mg/m^3) acceptable ceiling; 10 ppm (15 mg/m^3) acceptable maximum peak above the acceptance ceiling concentration for an 8-hour shift

Basis for Occupational Limits: The A2 designation of formaldehyde is based on the occurrence of nasal tumors in rats, and carcinogenicity is also the basis for the NIOSH recommendation.[1-3] It is stated in the TLV documentation that the "TLV of 1.0 ppm is adequate and no serious or persistent adverse effects should develop"; however, it is acknowledged that hypersensitive persons may experience irritant effects.[1] The above OSHA limits were current as of July 1988, and regulations governing work practices related to formaldehyde may be found in 29 CFR 1910.1048.[4]

Drinking Water Limits: No drinking water limits were found for this substance.

Toxicity Profile

Carcinogenicity: Results of multiple epidemiologic studies (over 25) are reviewed in the document "Assessment of Health Risks to Garment Workers and Certain Home Residents from Exposure to Formaldehyde."[5] Results from nine of the studies suggest that lung, nasopharyngeal, sinonasal, and oro-hypo-pharyngeal cancers are associated with formaldehyde exposure. Although several of these studies attempted to control for confounding due to smoking, alcohol consumption, and other factors, the overall evidence of carcinogenicity has been designated as "limited" because various alternative explanations for the findings could not be adequately addressed.[5]

Sufficient evidence exists that formaldehyde is carcinogenic via the inhalation route in experimental animals based on the occurrence of nasal squamous cell carcinomas (a rare tumor type) in both sexes of F344 rats, multiple rat strains, and mice. Generally, however, tumors are not observed beyond the initial site of nasal contact.

The study of Kerns et al.,[6] which provides the basis for the quantitative risk assessment, will be discussed here. Other relevant studies are reviewed by NIOSH[3] and IARC[7]. Kerns et al. exposed male and female F344 rats and (C57BL/6 × C3H)F1 mice (n = 120/sex/group) to 0, 2.0, 5.6, and 14.3 ppm HCHO gas 6 hours per day, 5 days a week for 24 months, followed by up to 6-months observation. Squamous cell carcinomas of the nasal cavity were found in both sexes of rats in the two highest dose groups (5.6 ppm—males 1 of 119, females 1 of 116; 14.3 ppm—males 51 of 117, females 52 of 115), but not in the controls or low dose animals. In mice two nasal squamous cell carcinomas were observed only at the highest dose. It is speculated that the differential susceptibility of mice and rats is in part accounted for by the greater decrease (on the order of 50%) in respiratory minute volume (RMV) of mice in response to irritants. If target tissue doses are calculated based on this information, then the high dose in mice approximates the moderate dose in rats, and the incidences of squamous cell carcinoma are the same at the high dose in mice and the moderate dose in rats.[5] The only other tumor type observed was benign polypoid adenomas in rats at all dose levels. There was also a dose-dependent increase in mice and rats in the incidence and severity of non-neoplastic nasal lesions including squamous metaplasia, epithelial dysplasia, and purulent rhinitis.[6]

Formaldehyde is most appropriately placed in group B1, probable human carcinogen, under the EPA weight-of-evidence classification.

Mutagenicity: Formaldehyde has been reported to be genotoxic in a variety of test systems encompassing a broad range of endpoints and levels of phylogenetic complexity. It has been shown to be mutagenic in multiple bacterial test systems (usually, but not always, without the addition of a metabolic activation system), fungi, and *Drosophila*. Formaldehyde also induces SCEs and chromosomal aberrations and causes adduct formation with DNA and proteins in vivo and in vitro. Formaldehyde inhibits DNA repair in cultured human cells, and positive results have been reported in cell transformation assays.[5,7]

Developmental Toxicity: Based on a survey of the literature up to 1984, it was concluded in the EPA document[5] that

1. there was no evidence of teratogenicity in rodents in ingestion or dermal application studies at maternally nontoxic levels;
2. inhalation studies in rodents have reported fetotoxic but not teratogenic effects[8] (cited in U.S. EPA[5]), and further study is warranted;
3. data relevant to developmental effects in humans are lacking.

Reproductive Toxicity: Menstrual irregularities and ovarian cysts as well as an increased incidence of low birth weight babies and miscarriages have been reported in some studies. Although these reports suffer from limitations in study design that preclude drawing causal inferences, they do indicate a need for further study.[5]

Systemic Toxicity: In humans, formaldehyde is a primary skin sensitizing agent inducing allergic contact dermatitis by a delayed-type (Type IV) hypersensitivity mechanism.

Some studies have indicated that HCHO exposure may be linked to neurobehavioral effects such as thirst, dizziness, apathy, inability to concentrate, sleep disturbances, etc. However, sufficient evidence to draw firm conclusions is lacking.

It was concluded in the EPA document[5] that: "there is no convincing evidence in experimental animals that inhalation exposure causes significant primary toxicologic effects in organs other than the upper respiratory tract."

Asthmatics appear to display an enhanced susceptibility to formaldehyde, and concentrations at or below 5 ppm are capable of inducing wheezing and chest tightness.[5]

Irritation: Formaldehyde is irritating to the eyes, nose, and throat, and although no clear thresholds have been established, most people experience these effects in the range of 0.1 to 3 ppm. Slightly higher concentrations—around 5 ppm—will result in coughing and possibly a feeling of chest tightness. Concentrations in excess of 50 ppm can cause severe injury to the airways and alveoli, producing pneumonia and pulmonary edema.

Basis for the AALG: Formaldehyde is used in the manufacture of a variety of resins and other chemicals, as a preservative, hardening and reducing agent, corrosion inhibitor, sterilizing agent, and in embalming fluid.[1] Since formaldehyde occurs as a gas at room temperature and undergoes rapid biodegradation in water and soil,[9] 80% of the contribution to total exposure will be allowed from air.

The AALG for formaldehyde is based on carcinogenicity, using the Kerns et al.[6] data for squamous cell carcinomas of the nasal turbinates in rats (males and females combined) in the linearized multistage model. The q_1^* based on this data and assuming a 70-kg human breathing 20 m^3 of air per day is 1.3×10^{-5} $(\mu g/m^3)^{-1}$. Applying the relationship $E = q_1^* \times d$, where E is the specified level of extra risk and d is the dose, gives the estimates below. There is, however, considerable nonlinearity of response in the data; the possible physiological basis for this in light of the irritant properties of formaldehyde (i.e., effects on animal breathing rate, cell proliferation, and the mucociliary layer) is discussed in detail in the EPA document.[5]

AALG:
- 10^{-6} 95% UCL—0.05 ppb (0.062 $\mu g/m^3$) annual TWA
- 10^{-5} 95% UCL—0.51 ppb (0.62 $\mu g/m^3$) annual TWA

References

1. ACGIH. 1986. "Documentation of the Threshold Limit Values and Biological Exposure Indices." 5:276–77.
2. NIOSH. 1976. "Criteria for a Recommended Standard . . . Occupational Exposure to Formaldehyde." DHEW (NIOSH) 77–126.
3. NIOSH. 1981. "Formaldehyde: Evidence of Carcinogenicity. Current Intelligence Bulletin No. 34." DHHS (NIOSH) 86–122.
4. OSHA. 1988. 29 Code of Federal Regulations 1910.1048.
5. U.S. EPA. 1987. "Assessment of Health Risks to Garment Workers and Certain Home Residents from Exposure to Formaldehyde." U.S. EPA Office of Pesticides and Toxic Substances (4/87).
6. Kerns, W.D., K.L. Pavkov, D.J. Donofrio, E.J. Gralla, and J.A. Swenberg. 1983. "Carcinogenicity of Formaldehyde in Rats and Mice after Long-Term Inhalation Exposure." *Cancer Res.* 43:4382–92.
7. IARC. 1986. "Formaldehyde." *IARC Monog.* suppl. 6:321–24.
8. Ulsamer, A.G., J.R. Beall, H.K. Kang, and J.A. Frazier. 1984. "Overview of Health Effects of Formaldehyde." In *Hazard Assessment of Chemicals—Current Developments,* J. Saxsena, Ed. (Academic Press: New York, NY), p. 337–400.
9. IARC. 1982. "Formaldehyde." *IARC Monog.* 29:345–89.

n-HEXANE

Synonyms: hexane

CAS Registry Number: 110–54–3

Molecular Weight: 86.17

Molecular Formula: $CH_3(CH_2)_4CH_3$

AALG: systemic toxicity—0.14 ppm (0.38 mg/m^3) 24-hour TWA

Occupational Limits:
- ACGIH TLV: 50 ppm (180 mg/m^3) TWA
- NIOSH REL: 100 ppm (360 mg/m^3) TWA
- OSHA PEL: 50 ppm (180 mg/m^3) TWA

Basis for Occupational Limits: All of the occupational limits are based on neurotoxicity. However, both the ACGIH and OSHA limits specify *n*-hexane, whereas the NIOSH limit does not distinguish between *n*-hexane and other hexane isomers.[1-3] OSHA recently adopted the TLV-TWA for *n*-hexane as part of its final rule limits.[4]

Drinking Water Limits: The 1-day HA for a 10-kg child is 12.9 mg/L; it was recommended that the longer-term HA for a child, 4 mg/L, also be used as the 10-day (child) HA because data sufficient to derive a 10-day HA was not available. The adult longer-term HA is 14.3 mg/L. No lifetime HA was recommended on account of insufficient data.[5]

Toxicity Profile

Carcinogenicity: No data were found implicating *n*-hexane as a carcinogen.

Mutagenicity: *n*-Hexane has been reported to induce chromosomal aberrations in cultured hamster fibroblasts.[6]

Developmental Toxicity: Bus et al.[7] (cited in U.S. EPA[5]) reported no significant embryofetal toxicity or teratogenic effects in F344 rats exposed to 1000 ppm *n*-hexane via inhalation 6 hours per day on gestation days 8 to 12, 12 to 16, or 8 to 16 compared to controls. However, postnatal growth of pups exposed to 1000 ppm *n*-hexane on gestation days 8 to 16 was significantly depressed relative to controls up to 3 weeks after birth, but was not significantly different at 7 weeks after birth.

No signs of neurotoxicity were observed in the offspring of treated dams. In a gavage study in mice, *n*-hexane was reported to be nonteratogenic at maternally toxic doses[8] (cited in U.S. EPA[5]).

Reproductive Toxicity: Krasavage et al.[9] reported that male rats gavaged 5 days per week for 90 days with 6.6 mmol/kg *n*-hexane developed atrophy of the testicular germinal epithelium. Such effects were not observed, however, in a subchronic inhalation study using male and female F344 rats exposed to concentrations up to 10,000 ppm *n*-hexane.[10]

Systemic Toxicity: Chronic exposure to *n*-hexane in an occupational setting has been associated with the development of polyneuropathy and sensorimotor poly-neuropathy at exposures in the range of 500 to 2000 ppm and possibly lower.[1,2] Clinical signs observed included paresthesia in the distal extremities, muscular weakness, blurred vision, headache, and fatigue. Short-term exposure to hexane (10 minutes) at 2000 ppm has been reported to cause no effects, but exposure at 5000 ppm (10 minutes) produced marked vertigo. It should be noted that many occupational case reports involving hexane exposure do not give information regarding the purity of the hexane and that commercial grade hexane may contain 20–80% *n*-hexane isomer.[1]

Animal studies with *n*-hexane, using inhalation as well as other routes of exposure, have confirmed the effects observed in humans and also seem to indicate that the neurotoxic properties of *n*-hexane are not shared by other hexane isomers.[1,2,5] In addition, studies in rats indicate that the neurotoxicity of *n*-hexane is dependent on the amount metabolized to a known neurotoxic metabolite, 2,5-hexanedione.[9]

Miyagaki[11] (cited in NIOSH[2]) exposed Swiss-strain male mice (n = 10/group) to 0, 100, 250, 500, 1000, or 2000 ppm commercial grade hexane (65–70% *n*-hexane) 24 hours per day, 6 days a week for 1 year. The following findings were reported:

- 1000 and 2000 ppm—marked abnormal posture and muscular atrophy, severe abnormal responses in multiple electromyographic tests
- 500 ppm—abnormal posture and muscular atrophy
- 250 ppm—slightly abnormal posture and lesser degree of muscular atrophy, slightly abnormal response in a single electromyographic test
- 100 ppm—no apparent effects.

A NOAEL of 100 ppm is suggested by this study.

Cavender et al.[10] performed a subchronic inhalation study in which male and female F344 rats were exposed to 0, 3000, 6500, or 10,000 ppm *n*-hexane (99.5% purity) 6 hours per day, 5 days a week for 13 weeks. No signs of neurotoxicity were observed even in the highest exposure group, although the body weight gain of the male rats was significantly less than that of the controls. In teased tibial nerve fiber preparations, axonopathy was observed in 4 of 5 male rats in the 10,000-ppm group and 1 of 5 male rats examined in the 6500-ppm group. The authors attributed the minimal changes observed even at high concentrations in their intermittent exposure study (relative to severe effects seen in rats continuously exposed to 400–600 ppm *n*-hexane for up 162 days[12]) to the relatively short half-life in blood and peripheral nerve tissue of *n*-hexane compared with 2,5-hexanedione. It was felt that "in continuous exposures 2,5-hexanedione may build to an effective neurotoxic concentration and since exposure is continuous there is no recovery during each day or week."[10]

Irritation: *n*-Hexane has been reported to cause eye and throat irritation at high concentrations (1400–1500 ppm),[1] and an odor threshold of 130 ppm (geometric mean of reported literature values) has been reported.[13]

Basis for the AALG: Commercial grades of hexane are used as solvents for glues, varnishes, cements, and inks.[5] Various hexane isomers occur as constituents of petroleum products such as gasoline. *n*-Hexane is considered insoluble in water (23 mg/L) and is volatile with a vapor pressure of 124 mm Hg at 20°C.[1] Based on lack of data on its environmental fate and distribution and the expected contribution of various media to human exposure, 50% of the contribution to total exposure will be allowed from air.

The AALG is based on neurotoxicity, using the NOAEL of 100 ppm suggested by the study of Miyagaki[11] (cited in NIOSH[2]). This NOAEL was selected even though the hexane used in the Miyagaki study was only approximately 70% *n*-hexane (the remainder was various other hexane isomers) for several reasons:

1. Sufficient exposure levels were used to define a range of severity of neurotoxic effects and both a LOAEL and a NOAEL for those effects.
2. Exposure was almost continuous.
3. The study was of longer duration (1 year vs 90 days) than other available studies.

Since the hexane used in this study was 70% *n*-hexane the NOAEL of 100 ppm was adjusted to 70 ppm (100 × 0.7) and was further adjusted for continuous exposure (70 ppm × 6/7) to give a NOAEL of 60 ppm. This was converted to a human equivalent exposure level according to the procedures described in Appendix C and divided by a total uncertainty factor of 1000 (10 for interindividual variation × 10 for interspecies variation × 10 for less-than-chronic [lifetime] exposure).

AALG: • systemic toxicity—0.14 ppm (0.38 mg/m^3) 24-hour TWA

References

1. ACGIH. 1986. "Documentation of the Threshold Limit Values and Biological Exposure Indices." 5:305–6.

2. NIOSH. 1977. "Criteria for a Recommended Standard . . . Occupational Exposure to Alkanes (C5-C8). DHEW (NIOSH) 77–151.
3. NIOSH. 1985. "NIOSH Recommendations for Occupational Safety and Health Standards." *Morb. Mort. Wkly. Rpt.* 34:5s-31s.
4. OSHA. 1989. "Air Contaminants; Final Rule." *Fed. Reg.* 54:2332–2959.
5. U.S. EPA. 1987. "Health Advisory for *n*-Hexane." Office of Drinking Water (3/87).
6. NIOSH. 1988. MN9275000. "Hexane." RTECS, on line.
7. Bus, J.S., E.L. White, R.W. Tyl, and C.S. Barron. 1979. "Prenatal Toxicity and Metabolism of *n*-Hexane in F344 Rats after Inhalation Exposure." *Tox. Appl. Pharm.* 61:414–22.
8. Marks, T.A., P.W. Fisher, and E. Staples. 1980. "Influence of *n*-Hexane on Embryo and Fetal Development in Mice." *Drug Chem. Tox.* 3:393–406.
9. Krasavage, W.J., J.L. O'Donoghue, G.D. DiVincenzo, and C.J. Terhaar. 1980. "The Relative Neurotoxicity of Methyl-*n*-butyl Ketone, *n*-Hexane, and Their Metabolites." *Tox. Appl. Pharm.* 52:433–41.
10. Cavender, F.L., H.W. Casey, H. Salem, D.G. Graham, J.A. Swenberg, and E.G. Gralla. 1984. "A 13-Week Vapor Inhalation Study of *n*-Hexane in Rats with Emphasis on Neurotoxic Effects." *Fund. Appl. Tox.* 4:191–201.
11. Miyagaki, H. 1967. "Electrophysiological Studies on the Peripheral Neurotoxicity of *n*-Hexane" (Japanese). *Jpn. J. Ind. Health* 9:660–71.
12. Schaumburg, H.H., and P.S. Spencer. 1976. "Degeneration in Central and Peripheral Nervous Systems Produced by Pure *n*-Hexane: An Experimental Study." *Brain* 99:183–92.
13. Amoore, J.E., and E. Hautala. 1983. "Odor as an Aid to Chemical Safety: Odor Thresholds Compared with Threshold Limit Values and Volatilities for 214 Industrial Chemicals in Air and Water Dilution." *J. Appl. Tox.* 3:272–90.

HYDRAZINE

Synonyms: diamine

CAS Registry Number: 302–01–2

Molecular Weight: 32.05

Molecular Formula: $H_2N\text{-}NH_2$

AALG: 10^{-6} 95% UCL—7.7×10^{-5} ppb (1.0×10^{-4} $\mu g/m^3$) as an annual TWA

Occupational Limits:
- ACGIH TLV: 0.1 ppm (0.1 mg/m^3) TWA; Appendix A2, suspect human carcinogen; skin notation
- NIOSH REL: 0.03 ppm (0.04 mg/m^3) ceiling (2-hour)
- OSHA PEL: 0.1 ppm (0.1 mg/m^3) TWA; skin notation

Basis for Occupational Limits: The occupational limits for hydrazine are based on carcinogenicity;[1,2] the lower REL recommended by NIOSH is based on the detection limit of the best available analytical method.[2] OSHA recently lowered the PEL to 0.1 ppm as a TWA as part of its final rule limits.[3]

Drinking Water Limits: No drinking water limits were found for this substance.

Toxicity Profile

Carcinogenicity: Hydrazine and its sulfate salts have been shown to be carcinogenic to mice via ingestion (liver, mammary, and lung tumors) and intraperitoneal injection (lung tumors, leukemias, and sarcomas) and to rats via ingestion (lung and liver tumors).[4] These studies have been well reviewed by NIOSH[2] and IARC[5] and will not be discussed further here.

Vernot et al.[6] evaluated the carcinogenicity of inhaled hydrazine (97% purity) in male and female F344 rats (controls, n = 150/sex; treated, n = 100/sex/group), female C57BL/6 mice (controls, n = 800; treated, n = 400/group), male Syrian golden hamsters (n = 200/group), and male and female Beagle dogs (n = 4/sex/group) exposed for 6 hours per day, 5 days a week for 1 year and held for 18, 15, 12, and 38 months, respectively. The exposure levels used were 0.05, 0.25, 1.0, and 5.0 ppm for the rats; 0.05, 0.25, and 1.0 ppm for the mice; 0.25, 1.0, and 5.0 ppm for the hamsters; and 0.25 and 1.0 ppm for the dogs. In rats, the most common tumor type was nasal adenomatous polyps (a benign tumor type), which was significantly elevated at the highest dose level in female rats and at the two highest dose levels in male rats. Although not statistically significant, there was also an increased incidence of nasal squamous cell carcinomas and papillomas at the highest exposure level; none was observed in controls or other exposure groups. The incidence of thyroid carcinomas was also significantly elevated in the highest exposure group of male rats. The major non-neoplastic lesions in rats were inflammatory changes and metaplasia of the respiratory epithelium in the nose, larynx, and trachea, occurring mainly at the highest dose. In hamsters, there was also a statistically significant increase in nasal adenomatous polyps at the highest dose level as well as smaller (not statistically significant) increases in tumors of the thyroid, colon, and stomach. The major non-neoplastic lesion in hamsters was a dose-related generalized amyloidosis occurring in the liver, kidneys, thyroid, and adrenal glands, which the authors interpreted as an exposure-related acceleration of common aging changes in hamsters. In mice, an increase in lung adenomas, considered to be of borderline statistical significance, was reported at the highest dose level, and no other neoplastic or non-neoplastic lesions were observed. No neoplastic or non-neoplastic lesions considered to be consistently related to hydrazine exposure were observed in the dogs in this study.

In its recent reevaluation, IARC concluded that there was inadequate evidence for hydrazine carcinogenicity in humans and sufficient evidence in laboratory animals.[4] Based on the EPA weight-of-evidence classification, hydrazine is most appropriately placed in group B2, probable human carcinogen.

Mutagenicity: Positive results have been reported for hydrazine in the following assays or for the following endpoints: mutagenic in *S. typhimurium* and *E. coli* with metabolic activation, mutagenic in *S. cerevisiae*, alterations in DNA synthesis and repair in cultured mammalian cells, nondisjunction in *Drosophila*, and oncogenic transformation in various cultured mammalian cell lines.[7]

Developmental Toxicity: In a study cited in RTECS and unavailable in translation (original in Russian), it was reported that rats exposed to 4 mg/m^3 hydrazine for 2 hours per day on gestation days 7 through 20 or to 1 mg/m^3 hydrazine continuously on gestation days 1 through 11 experienced decreased fertility and embryotoxicity.[7]

Reproductive Toxicity: No data were found implicating hydrazine as a reproductive toxin.

Systemic Toxicity: Based on limited reports, it appears that hydrazine is capable of causing allergic sensitization in occupationally exposed individuals.[2]

One-hour LC_{50}s of 570 ppm and 252 ppm have been reported for hydrazine in rats and mice, respectively.[2] In long-term studies, Haun and Kinkead[8] (cited in NIOSH[2]) exposed male Beagles (n = 8/group), female Rhesus monkeys (n = 4/group), male Sprague-Dawley rats (n = 50/group), and female ICR mice (n = 40/group) to 1.0 or 5.0 ppm hydrazine for 6 hours per day, 5 days a week for 6 months and to 0.2 or 1.0 ppm continuously also for 6 months; unexposed controls were also included for each species. Mortality was increased in a dose-related manner in all groups of exposed mice and in the high-dose dogs exposed continuously. Histologic examination of tissues from these animals revealed moderate to severe fatty degeneration of the liver, considered to be the cause of death. There was also evidence of slight fatty degeneration in the livers of both control and treated monkeys, but it was somewhat more severe in the treated animals. The only effect reported in rats was a dose-related decrease in growth. In dogs, dose-related hematotoxicity in the form of decreased hematocrit, hemoglobin, and erythrocyte count with evidence of increased erythropoietic activity was reported.

Irritation: Hydrazine has been reported to have an ''ammonia-like'' odor[1] and an odor threshold of 3 to 4 ppm.[2] Hydrazine is irritating to the eyes and mucous membranes of the respiratory tract in humans (no exposure levels given), and liquid hydrazine may produce chemical burns upon skin contact.[2]

Basis for the AALG: Hydrazine is miscible with water and has a vapor pressure of 10.4 mm Hg at 20°C. In the IARC review,[5] the following was concluded:

1. Hydrazine discharged into aqueous media would be present only briefly since it would rapidly react with oxygen.
2. Exhaust gases of hydrazine and its derivatives used as rocket fuel produce only trace quantities of unreacted hydrazine.
3. Use of hydrazine as a chemical intermediate would be unlikely to result in its occurrence in the environment.
4. Hydrazine is produced naturally by certain nitrogen-fixing bacteria.

Based on these considerations, 50% of the contribution to total exposure will be allowed from air.

The AALG for hydrazine is based on carcinogenicity using data from the studies of MacEwen et al.[9] recently published in the open literature as Vernot et al.[6] In the IRIS database,[10] the recommended risk estimates are based on the application of the linearized multistage model with extra risk to the data for nasal cavity adenomas or adenocarcinomas in male F344 rats. The specific dose-response data used was the following: control, 0 of 149; 1.0 ppm, 11 of 98; and 5.0 ppm, 72 of 99. It should be noted that the data for the two lowest exposure levels, 0.05 and 0.25 ppm, were excluded from the model. The inhalation unit risk based on this data was 4.9×10^{-3}, which corresponds to an inhalation q_1* of 4.9×10^{-3} $(\mu g/m^3)^{-1}$. Applying the relation $E = q_1$* $\times$ d, where E is the specified level of extra risk and d is the dose, and allowing a 50% relative source contribution from air, yields the estimates shown below. It is stated in the IRIS assessment that: ''the unit risk should not be used if the air concentration exceeds 2 $\mu g/m^3$, since above this concentration the slope factor may differ from that stated.''

AALG: • 10^{-6} 95% UCL—7.7×10^{-5} ppb (1.0×10^{-4} $\mu g/m^3$) as an annual TWA

- 10^{-5} 95% UCL—7.7 $\times$ 10^{-4} ppb (1.0 $\times$ 10^{-3} $\mu g/m^3$) as an annual TWA

References

1. ACGIH. 1986. "Documentation of the Threshold Limit Values and Biological Exposure Indices." 5:310.
2. NIOSH. 1978. "Criteria for a Recommended Standard . . . Occupational Exposure to Hydrazines." DHEW (NIOSH) 78–172.
3. OSHA. 1989. "Air Contaminants; Final Rule." *Fed. Reg.* 54:2332–2959.
4. IARC. 1987. "Hydrazine (Group 2B)." *IARC Monog.* suppl. 7:223–24.
5. IARC. 1974. "Hydrazine." *IARC Monog.* 4:127–36.
6. Vernot, E.H., J.D. MacEwen, R.H. Bruner, C.C. Haun, E.R. Kinkead, D.E. Prentice, A. Hall, R.E. Schmidt, R.L. Eason, G.B. Hubbard, and J.T. Young. 1985. "Long-Term Inhalation Toxicity of Hydrazine." *Fund. Appl. Tox.* 5:1050–64.
7. NIOSH. 1988. MU7175000. "Hydrazine." RTECS, on line.
8. Haun, C.C., and E.R. Kinkead. 1973. "Chronic Inhalation Toxicity of Hydrazine." In: *Proceedings of the 4th Annual Conference on Environmental Toxicology,* Wright-Patterson Air Force Base, Ohio, AMRL-TR-73–125, NTIS AD781–031 (Springfield, VA: NTIS), pp. 351–65.
9. MacEwen, J.D., E.H. Vernot, C.C. Haun, E.R. Kinkead, and A. Hall. 1981. "Chronic Inhalation Toxicity of Hydrazine: Oncogenic Effects." Air Force Aerospace Medical Research Laboratory, Wright-Patterson Air Force Base, Dayton, OH.
10. U.S. EPA. 1989. "Hydrazine/Hydrazine Sulfate; CASRN 302–01–2." IRIS (6/1/89).

HYDROCHLORIC ACID

Synonyms: hydrogen chloride. (Hydrochloric acid, produced when the gas hydrogen chloride is dissolved in water, is usually available as 20% or 38% HCl).

CAS Registry Number: 7647–01–0

Molecular Weight: 36.47

Molecular Formula: HCl

AALG: irritation—0.1 ppm (0.15 mg/m^3) 24-hour TWA

Occupational Limits:
- ACGIH TLV: 5 ppm (7 mg/m^3) ceiling
- NIOSH REL: none
- OSHA PEL: 5 ppm (7 mg/m^3) ceiling

Basis for Occupational Limits: In the TLV documentation, it is stated that the ceiling limit is "sufficiently low to prevent toxic injury from exposure to hydrogen chloride, but on the borderline of severe irritation."[1]

Drinking Water Limits: Although no drinking water limits were found for hydrochloric acid, chlorine (usually in the form of hypochlorite or hydrochloric acid in the United States) is used for water disinfection and treatment of sewage effluent.[2]

Toxicity Profile

Carcinogenicity: No data were found implicating hydrochloric acid as a carcinogen.

Mutagenicity: Negative results are listed for hydrochloric acid in the cell transformation assay using Syrian hamster embryo cells and adenovirus SA7 in the Gene-Tox database as summarized in RTECS. Hydrochloric acid is reported to alter DNA repair in *E. coli* and to cause chromosome loss and nondisjunction in *D. melanogaster* exposed to an oral dose of 100 ppm hydrochloric acid or 100 ppm hydrogen chloride via inhalation for 24 hours.[3]

Developmental Toxicity: In a report cited in RTECS and unavailable in translation, exposure of rats to 302 ppm hydrogen chloride on day 1 of gestation resulted in fetotoxicity and specific developmental abnormalities.[3]

Reproductive Toxicity: No data were found implicating hydrochloric acid as a reproductive toxin.

Systemic Toxicity: Data on the systemic effects of repeated exposure to hydrochloric acid are very limited in both man and experimental animals, and it is believed that "exposure to hydrogen chloride does not result in effects on organs some distance from the portal of entry."[4]

One-hour LC_{50}s of 3124 ppm and 1108 ppm have been reported in rats and mice, respectively.[3] In most animal lethality studies, death has been attributed to respiratory injury, e.g., pulmonary edema, emphysema, and atelectasis, with secondary changes such as passive congestion of the liver, intestine, and kidneys.[4] Machle et al.[5] (cited in International Program on Chemical Safety[4]) reported no adverse effects evident at necropsy in rabbits or guinea pigs exposed to 34 ppm hydrogen chloride for 6 hours per day, 5 days a week for 4 weeks and killed several months later.

Elfimova[6] (cited in International Program on Chemical Safety[4]) reported the results of tests designed to measure reflex neurological changes in humans, including optical chronaxie, blood vessel tone, dark adaptation, and respiratory changes. It was found that concentrations of 0.4 to 1.01 ppm shifted the value for optical chronaxie while 0.13 to 0.27 ppm did not, and a threshold of 0.4 ppm was statistically determined for this test. Changes in blood vessel tone were reported at levels above 0.34 ppm and thresholds for changes in dark adaptation and respiration were given as 0.13 ppm and 0.07 to 0.13 ppm, respectively.

Irritation: It is generally accepted that the major effects of hydrogen chloride in both humans and laboratory animals are those of local irritation. In laboratory animals, hydrogen chloride has been shown to cause irritation of the conjunctiva and corneal injury as well as irritation of the skin and upper and lower respiratory tract.[2] Barrow et al.[7] (cited in International Program on Chemical Safety[4]), in studying the reflex decrease in respiratory rate in mice exposed to irritants, determined the RD_{50} (concentration causing a 50% decrease in respiratory rate) in mice to be 309 ppm and on the basis of this endpoint reported hydrogen chloride to be 33 times less irritating than chlorine (RD_{50} of 9.3 ppm).

It has been reported that exposure to 50–100 ppm HCl for one hour was barely tolerable to workers and 35 ppm caused irritation of the throat in a short time period[8] (cited in ACGIH[1]). It has also been suggested that adaptation may occur to lower levels of exposure that might be initially irritating, e.g., 10 ppm. However, long-term exposure to levels in this range may also result in erosion of the incisolabial surfaces of the teeth.[4] Two other sources have indicated that concentrations of 5 ppm HCl or higher are immediately irritating[9,10] (cited in ACGIH[1]).

Hydrogen chloride is reported to have a "characteristic, suffocating, pungent" odor[1] and an odor threshold (geometric mean of several reported literature values) of 0.77 ppm.[11] In the course of correlating concurrent environmental HCl concentrations in the field with the subjective reactions of trained industrial hygienists, it was reported that there was (1) no reaction at 0.06 to 1.8 ppm, (2) minimum reaction at 0.07 to 2.17 ppm, (3) obvious perception at 1.9 to 8.6 ppm, and (4) strong reaction at 5.6 to 22.1 ppm[12] (cited in International Program on Chemical Safety[4]).

Basis for the AALG: Since the AALG for hydrogen chloride is based on irritation, no relative source contribution from inhalation was factored into the calculations. The reason is that allocation of a proportion of exposure to a given source is not relevant when the effects of exposure are not cumulative, as is the case with irritant effects.

The AALG for hydrogen chloride is based on a composite human LOAEL of 5 ppm, since several sources[9,10,12] (cited in ACGIH[1] and International Program on Chemical Safety[4]) as well as industrial experience[1] indicate that this is the approximate beginning of the concentration range in which irritant effects are experienced by workers. This LOAEL was divided by a total uncertainty factor of 50 (10 for interindividual variation × 5 for a LOAEL).

For comparative purposes only, an AALG was also calculated on the basis of the RD_{50} in mice multiplied by a factor of 10^{-3} as suggested by Kane et al.[13] to derive an exposure level at which no effects would be expected. Using the RD_{50} of 309 ppm reported by Barrow et al.[7] (cited in International Program on Chemical Safety[4]) yields a figure of 0.309 ppm.

AALG: • irritation—0.1 ppm (0.15 mg/m^3) 24-hour TWA

References

1. ACGIH. 1986. "Documentation of the Threshold Limit Values and Biological Exposure Indices." 5:313.
2. Safe Drinking Water Committee. 1986. *Drinking Water and Health,* Vol 7 (Washington, DC: National Academy Press).
3. NIOSH. 1988. MW9610000. "Hydrogen Chloride." RTECS, on line.
4. International Program on Chemical Safety. 1982. "Chlorine and Hydrogen Chloride." Environmental Health Criteria 21 (Geneva: World Health Organization).
5. Machle, W., K. V. Kitzmiller, E. W. Scott, and J. F. Treon. 1942. "The Effect of the Inhalation of Hydrogen Chloride." *J. Ind. Hyg. Tox.* 24:222–25.
6. Elfimova, E. W. 1964. "Data for the Hygienic Evaluation of Hydrochloric Acid Aerosol (Hydrochloride Gas) as an Atmospheric Pollutant." In: V. A. Rjazanov, *Limits of Allowable Concentrations of Atmospheric Pollutants* (translated from the Russian in: B. S. Levine [ed.], *A Survey of Russian Literature on Air Pollution and Related Occupational Diseases,* Vol. 4).
7. Barrow, C. S., Y. Alarie, J. C. Warrick, and M. F. Stock. 1977. "Comparison of the Sensory Irritation Response in Mice to Chlorine and Hydrogen Chloride." *Arch. Env. Health* 31:68–76.
8. Henderson, Y., and H. W. Haggard. 1943. *Noxious Gases* (New York: Reinhold Publishing Corporation), p. 126.

9. Elkins, H. B. 1959. *The Chemistry of Industrial Toxicology* (New York: John Wiley & Sons), p. 79.

10. Patty, F. A. 1963. *Industrial Hygiene and Toxicology,* Vol. II (New York: Wiley-Interscience), p. 851.

11. Amoore, J. E., and E. Hautala. 1983. "Odor as an Aid to Chemical Safety: Odor Thresholds Compared with Threshold Limit Values and Volatilities for 214 Industrial Chemicals in Air and Water Dilution." *J. Appl. Tox.* 3:272–90.

12. NAS/NRC. 1976. *Medical and Biological Effects of Environmental Pollutants, Chlorine and Hydrogen Chloride* (Washington, DC: National Academy of Sciences–National Research Council), pp. 62, 92–144.

13. Kane, L. E., C. S. Barrow, and Y. Aleric. 1979. "A Short-Term Test to Predict Acceptable Levels of Exposure to Airborne Sensory Irritants." *Am. Ind. Hyg. Assoc. J.* 40:207–29.

HYDROGEN CYANIDE

Synonyms: hydrocyanic acid, HCN

CAS Registry Number: 74–90–8

Molecular Weight: 27.03

Molecular Formula: HCN

AALG: systemic toxicity—0.09 ppm (0.095 mg/m^3) ceiling

Occupational Limits:
- ACGIH TLV: 10 ppm (11 mg/m^3) ceiling; skin notation
- NIOSH REL: 4.7 ppm (5 mg/m^3) ceiling
- OSHA PEL: 4.7 ppm (5 mg/m^3) STEL; skin notation

Basis for Occupational Limits: The TLV is based on the prevention of acute cyanide poisoning.[1] In addition to acute toxic effects, the NIOSH limit is also based on consideration of the irritant properties of cyanide salts and the subjective symptoms reported by workers in one study[2] (cited in NIOSH[3]), who were exposed to concentrations between 4 and 12 ppm for several years.[3] OSHA recently adopted the NIOSH recommended limit as a STEL (15 minutes) as part of its final rule limits.[4]

Drinking Water Limits: The ambient water quality criterion for cyanide is 3.77 mg/L, assuming a 70-kg human consuming 2 L of water and 6.5 g of fish/shellfish per day.[5] The 1-day and 10-day health advisories for the 10-kg child are the same—0.22 mg/L. The lifetime HA for cyanide, 154 µg/L, is based on a rat NOAEL obtained in a lifetime study in which rats were fed a diet fumigated with HCN to give the specified concentrations in the diet.[6]

Toxicity Profile

Carcinogenicity: No data were found implicating HCN as a carcinogen.

Mutagenicity: HCN gas was reported to be "marginally" mutagenic to *S. typhimurium* strain TA100, but addition of S9 mix decreased the activity.[6]

Developmental Toxicity: No data were found implicating HCN as a developmental toxin.

Reproductive Toxicity: No data were found implicating HCN as a reproductive toxin.

Systemic Toxicity: At the cellular level, cyanide inhibits cytochrome c oxidase, the terminal enzyme in the electron transport chain, decreasing cellular ability to utilize oxygen and producing histotoxic hypoxia.[7]

In general, clinically recognized cyanide poisoning is a result of acute exposure, and as the following human data indicate, both the concentration and duration of exposure are consequential in determining the effects observed.[2] The effects associated with a given concentration of HCN in air are as follows:[1]

- 270 ppm—immediately fatal
- 181 ppm—fatal after 10 minutes
- 110–135 ppm—fatal after 30–60 minutes or later, or dangerous to life
- 45–54 ppm—tolerated for 30 to 60 minutes without immediate or late effects
- 18–36 ppm-slight symptoms after several hours

The physiologic consequences of cyanide poisoning include metabolic acidosis (resulting from inhibition of oxidative metabolism and subsequent utilization of the glycolytic pathway) and high concentrations of oxyhemoglobin in the venous blood (which accounts for the characteristic brick red color of the skin). Other symptoms of poisoning include hyperventilation, vomiting, rapid and irregular heart rate, convulsions, and ultimately death without effective treatment.[4]

There are far fewer data on chronic low-level exposure to cyanide than on acute exposure. El Ghawabi et al.[2] (cited in ACGIH[1] and NIOSH[3]) studied Egyptian electroplating workers exposed to concentrations of cyanide between 4 and 12 ppm for periods of up to 15 years (most exposures on the order of 7 years). The workers reported an increase in symptoms such as headache, weakness, changes in taste and smell, irritation of the throat, vomiting, effort dyspnea, lacrimation, colic, and mental instability. Although other clinical findings were unremarkable, enlarged thyroid glands were reported in 20 of 36 workers. In addition, uptake of radioactive iodine by the thyroid gland was significantly elevated in exposed workers relative to controls after 4 and 24, but not 72, hours.

The acute and chronic effects of cyanide exposure in animals are most thoroughly reviewed in the NIOSH criteria document[3] and the ambient water quality criteria document.[8]

Irritation: Some investigators have stated that gaseous HCN is itself not irritating to the skin and mucous membranes of the respiratory tract[9] (cited in NIOSH[3]). However, there are other reports in which low-level exposure to HCN vapor caused reddening of the skin or blotchy eruptions[10,11] (cited in NIOSH[3]), and cyanide salt aerosols are known to cause skin and upper respiratory tract irritation.

HCN is reported to have an odor resembling bitter almonds to individuals genetically able to perceive it,[3] and an odor threshold of 0.58 ppm (geometric mean of reported literature values) has been reported.[12]

Basis for the AALG: Hydrogen cyanide is used in the manufacture of various synthetic fibers, plastics, cyanide salts, and nitriles and as a fumigant. It may also be produced during the refining of petroleum, electroplating, metallurgy, and photographic development.[1] The largest sources of cyanide in air and water are automobile emissions and industrial discharges, respectively.[13] Hydrogen cyanide is considered to be miscible with water and is a colorless gas at room temperature.[1] Since the AALG applies only to hydrogen cyanide and not cyanide salts, 80% of the contribution to total exposure will be allowed from air.

Given the uncertainty associated with data concerning the correlation of exposure and effect in the case of long-term, low-level exposures and based on the lack of adequate inhalation studies in animals and humans, the more conservative NIOSH limit (REL/42) was selected as the basis for the AALG. This AALG should be considered provisional since it is based on an occupational limit.

AALG: • systemic toxicity—0.09 ppm (0.095 mg/m^3) ceiling

References

1. ACGIH. 1986. "Documentation of the Threshold Limit Values and Biological Exposure Indices." 5:314.
2. El Ghawabi, S.H., M.A. Gaafar, A.A. El-Saharti, S.H. Amed, K.K. Malash, and R. Fanes. 1975. "Chronic Cyanide Exposure: A Clinical, Radioisotope, and Laboratory Study." *Br. J. Ind. Med.* 32:215–19.
3. NIOSH. 1976. "Criteria for a Recommended Standard . . . Occupational Exposure to Hydrogen Cyanide and Cyanide Salts." DHEW (NIOSH) 77–108.
4. OSHA. 1989. "Air Contaminants; Final Rule." *Fed. Reg.* 54:2332–2959.
5. U.S. EPA. 1982. "Ambient Water Quality Criteria for Cyanides, with Errata for Ambient Water Quality Criteria Documents." Environmental Criteria and Assessment Office, Cincinnati, OH (6/9/81, updated 2/23/82).
6. U.S. EPA. 1987. "Health Advisory for Cyanide." Office of Drinking Water (3/31).
7. Klassen, C.D., M.O. Amdur, and J. Doull, Eds. 1986. *Toxicology—The Basic Science of Poisons,* 3rd ed. (New York, NY: Macmillan Publishing Company), p. 241.
8. U.S. EPA. 1980. "Ambient Water Quality Criteria for Cyanides." EPA-440/5–80–037.
9. Wolfsie, J.H., and C.B. Shaffer. 1959. "Hydrogen Cyanide—Hazards, Toxicology, Prevention, and Management of Poisoning." *J. Occup. Med.* 1:281–88.
10. Hamilton, A., and H.L. Hardy. 1949. *Industrial Toxicology,* 2nd ed. (New York, NY: Paul B. Hoeber, Inc.), pp 248–62.
11. Williams, C.L. 1931. "Fumigants." *Public Health Rep.* 46:1013–31.
12. Amoore, J.E., and E. Hautala. 1983. "Odor as an Aid to Chemical Safety: Odor Thresholds Compared with Threshold Limit Values and Volatilities for 214 Industrial Chemicals in Air and Water Dilution." *J. Appl. Tox.* 3:272–90.
13. ATSDR. 1988. "Toxicological Profile for Cyanide" (draft). Agency for Toxic Substances and Disease Registry.

HYDROQUINONE

Synonyms: 1,4-benzenediol, *p*-dihydroxybenzene, *p*-hydrophenol

CAS Registry Number: 123–31–9

Molecular Weight: 110.11

Molecular Formula: HO–⟨◯⟩–OH

AALG: 10^{-6} 95% UCL—0.0046 ppb (0.021 $\mu g/m^3$) annual TWA

Occupational Limits:
- ACGIH TLV: 2 mg/m^3 TWA
- NIOSH REL: 2 mg/m^3 ceiling (15 minutes)
- OSHA PEL: 2 mg/m^3 TWA

Basis for Occupational Limits: The occupational limits for hydroquinone are based on prevention of characteristic skin and eye changes that occur following chronic exposure (usually greater than 5 years).[1,2]

Drinking Water Limits: No drinking water limits were found for this substance.

Toxicity Profile

Carcinogenicity: The carcinogenicity of hydroquinone has been evaluated in animal models exposed dermally, parenterally, and by the oral route. It did not act as an initiator in skin-painting studies in mice[3] (cited in IARC[4]), but it did increase the incidence of bladder carcinomas in mice when implanted in the bladder in cholesterol pellets[5] (cited in IARC[4]).

In an NTP bioassay,[6] male and female B6C3F$_1$ mice and F344/N rats (n = 65/sex/species/group) were gavaged with 0, 50, or 100 mg/kg and 0, 25, or 50 mg/kg hydroquinone (>99% purity) in water, respectively, 5 days a week for 103 weeks. An interim sacrifice of 10 mice and 10 rats from all groups was performed at 15 months. There were no treatment-related effects on survival for rats or mice. Body weight gain was decreased in both male and female high-dose mice compared with vehicle controls; body weight gain was decreased in both groups of exposed male rats compared with vehicle controls but was similar in female rats. There was a statistically significant increase in renal tubular cell adenomas in male rats (control, 0 of 55; low, 4 of 55; and high, 8 of 55) and mononuclear cell leukemia in female rats (control, 9 of 55; low, 15 of 55; and high, 22 of 55). No significant treatment-related neoplasms were observed in male mice, but there was a statistically significant increase in hepatocellular adenomas or carcinomas (combined) in female mice (control, 3 of 55; low, 16 of 55; and high, 13 of 55). Significant treatment-related non-neoplastic effects in mice included thyroid gland follicular cell hyperplasia in both males and females and hepatic proliferative lesions in males only. It was concluded in the NTP report that there was *some evidence* of carcinogenic activity in male and female rats and female mice. The data was interpreted as "showing a chemically-related increased incidence of neoplasms (malignant, benign, or combined) in which the strength of the response is less than that required for clear evidence."

It is the judgement here that hydroquinone is most appropriately placed in group B2, probable human carcinogen, based on the EPA weight-of-evidence classification.

Mutagenicity: Based on citations in RTECS, hydroquinone has been reported to be mutagenic to *S. typhimurium* and in mouse lymphocytes and rabbit bone marrow cells. In addition, positive results have been reported in two mouse micronucleus tests, and it was also reported to induce mutations and sister chromatid exchanges in cultured human lymphocytes.[7] Hydroquinone has also been reported to cause abnormal cell divisions in the chromosomes of plants, mice, rats, hamsters, and chick fibroblasts.[2]

In conjunction with the NTP bioassay,[6] genetic toxicity tests were performed and the following results reported:

Hydroquinone was not mutagenic in *S. typhimurium* strains TA98, TA100, TA1535, or TA1537 with or without exogenous metabolic activation. It induced trifluorothymidine (Tft) resistance in mouse L5178Y/TK lymphoma cells in the presence and absence of metabolic activation. An equivocal response was obtained in tests for induction of sex-linked recessive lethal mutations in *Drosophila* administered hydroquinone by feeding. Hydroquinone induced sister chromatid exchanges (SCEs) in CHO cells both with and without exogenous metabolic activation and caused chromosomal aberrations in the presence of activation.

Developmental Toxicity: No data were found implicating hydroquinone as a teratogen. Embryotoxicity (increased fetal resorption) was reported in a study in which rats were fed relatively large amounts of hydroquinone in their diet, but not in another study where lower levels were used[8,9] (cited in NIOSH[2]).

Reproductive Toxicity: Disruption of the estrous cycle has been reported in rats given oral doses of 100 or 200 mg/kg/day hydroquinone, and adverse effects on the fertility of male rats have been noted following subcutaneous injection[10,11] (cited in NIOSH[2]). The relevance of these findings to human inhalation exposure, however, is uncertain in view of the relatively high doses used and the difference in route.

Systemic Toxicity: The primary effects reported in humans exposed to hydroquinone involve ocular and dermal toxicity,[1,2] except in cases of poisoning via ingestion where effects on the C.N.S. were reported.[2]

Anderson and Oglesby[12] reported in detail the ocular manifestations of worker exposure to quinone vapor and hydroquinone dust. Initially it is characterized by staining and pigmentation of the superficial layers of the cornea (which will regress if exposure ceases), followed by the gradual development of changes in lens curvature leading to astigmatism and impaired vision over a period of years. It was also reported by Oglesby et al.[13] (cited in NIOSH[2]) that control of hydroquinone dust to levels of 1–4 mg/m^3 resulted in mild, reversible eye injury and no systemic toxicity, and it is stated in the TLV documentation that later studies have confirmed this finding, i.e., no systemic effects arise at a level of 2 mg hydroquinone dust per m^3.[1]

Dermal contact with 5% hydroquinone cream has resulted in skin irritation, allergic sensitization, dermatitis, and depigmentation in some individuals, but 2% hydroquinone cream is associated with few or none of these effects.[2]

Data on the effects of oral exposure to hydroquinone in humans are limited. Based on case reports, fatalities have been associated with ingestion of 5–12 g, and ingestion of 1 g has been reported to produce tinnitus, nausea, dizziness, vomiting, muscular twitching, headache, dyspnea, cyanosis, delirium, and collapse.[4] In a controlled human study, however, ingestion of 0.1–0.15 g of hydroquinone 3 times a day (approximately 4.3 to 6.4 mg/kg/day assuming a 70-kg man) for 3 to 5 months resulted in "no significant pathologic changes in blood or urine" in 17 male volunteers[14] (cited in IARC[4]).

No data pertinent to the inhalation toxicity of hydroquinone in animal models were found. However, in subchronic gavage studies conducted by NTP prior to the two-year bioassay, dose levels of 200 and 400 mg/kg hydroquinone in corn oil were associated with C.N.S. effects including tremors and convulsions.[6]

Irritation: Refer to the discussion in the preceding section.

Basis for the AALG: Hydroquinone is soluble in water (9.4 g/100 mL at 28.5°C) and has a low vapor pressure (less than 0.001 mm Hg at 20°C). Hydroquinone is used as an antioxidant, as an inhibitor or stabilizer in paints, fuels, oils, and polymers, and as a dye intermediate. Based on lack of information on its environmental fate and distribution and the contribution of various media to human exposure, 50% of the contribution to total exposure will be allowed from air.

The AALG for hydroquinone is based on carcinogenicity. The linearized multistage model was applied to three data sets from the NTP bioassay: renal tubular cell adenomas in male rats, mononuclear cell leukemia in female rats, and liver adenomas and carcinomas (combined) in female mice. The q_1* of 8.44×10^{-2} (mg/kg/day)$^{-1}$ from the data set for mononuclear cell leukemia was selected for use in AALG derivation because it was the most conservative. The inhalation equivalent q_1* is 2.41×10^{-5} (μg/m^3)$^{-1}$ assuming a 70-kg human breathing 20 m^3 of air per day and complete absorption. Applying the relation $E = q_1$* $\times$ d, where E is the specified level of extra risk and d is the dose, and allowing a 50% relative source contribution from air, gives the estimates shown below. This AALG should be considered provisional because it was derived based on data from a non-inhalation study.

Since no animal inhalation studies or human epidemiologic studies were available for hydroquinone, it was considered relevant to calculate an AALG based on the TLV—supported by extensive industrial experience—for comparative purposes only. If the TLV/42 is used with a 50% relative source contribution, the resultant figure is 0.024 mg/m^3.

AALG: • 10^{-6} 95% UCL—0.0046 ppb (0.021 μg/m^3) annual TWA 10^{-5}
  • 95% UCL—0.046 ppb (0.21 μg/m^3) annual TWA

References

1. ACGIH. 1986. "Documentation of the Threshold Limit Values and Biological Exposure Indices." 5:319.
2. NIOSH. 1978. "Criteria for a Recommended Standard . . . Occupational Exposure to Hydroquinone." DHEW (NIOSH) 78–155.
3. Roe, F.J.C., and M.H. Salaman. 1955. "Further Studies on Incomplete Carcinogenesis: Triethylene Melamine (T.E.M.), 1,2-Benzanthracene and B-propiolactone as Initiators of Skin Tumor Formation in the Mouse." *Brit. J. Cancer* 9:177–203.
4. IARC. 1977. "Dihydroxybenzenes." *IARC Monog.* 15:155–75.
5. Boyland, E., E.R. Busby, C.E. Dukes, P.L. Grover, and D. Manson. 1964. "Further Experiments on Implantation of Materials into the Urinary Bladder of Mice." *Brit. J. Cancer* 18:575–81.
6. NTP. 1989. "Toxicology and Carcinogenesis Studies of Hydroquinone (CAS No. 123–31–9) in F344/N Rats and B6C3F$_1$ Mice (Gavage Studies)" (galley draft). NTP-TR 366.
7. NIOSH. 1988. MX3500000. "Hydroquinone." RTECS, on line.
8. Telford, I.R., C.S. Woodruff, and R.H. Linford. 1962. "Fetal Resorption in the Rat as Influenced by Certain Antioxidants." *Am. J. Anat.* 110:29–36.
9. Ames, S.R., M.L. Ludwig, W.J. Swanson, and P.L. Harris. 1956. "Effect of DPPD, Methylene Blue, BHT, and Hydroquinone on Reproductive Process in the Rat." *Proc. Soc. Exp. Biol. Med.* 93:39–42.
10. Racz, G., J. Fuzi, G. Kemeny, and Z. Kisgyorgy. 1958. "The Effect of Hydroquinone and Phlorizin on the Sexual Cycle of White Rats" (Hungarian). *Orv. Szemle* 5:65–67.
11. Skalka, P. 1964. "The Influence of Hydroquinone on the Fertility of Male Rats" (Czech). *Sb. Vys. Sk. Zemed. Brne. Rada. B.* 12:491–94.
12. Anderson, B., and F. Oglesby. 1958. "Corneal Changes from Quinone-Hydroquinone Exposure." *AMA Arch. Ophthmol.* 59:495–501.
13. Oglesby, F.L., J.H. Sterner, and B. Anderson. 1947. "Quinone Vapors and Their Harmful Effects; II. Plant Exposures Associated with Eye Injuries." *J. Ind. Hyg. Tox.* 29:74–84.
14. Carlson, A.J., and N.R. Brewer. 1953. "Toxicity Studies on Hydroquinone." *Proc. Soc. Exp. Biol. (NY)* 84:684–88.

ISOBUTYRALDEHYDE

Synonyms: 2-methyl propanal, isobutyl aldehyde

CAS Registry Number: 78–84–2

Molecular Weight: 74.11

Molecular Formula: $(CH_3)_2CHCHO$

AALG: LD_{50}-derived—0.16 ppb (0.49 $\mu g/m^3$) 24-hour TWA

Occupational Limits:
  - ACGIH TLV: none
  - NIOSH REL: none
  - OSHA PEL: none

Drinking Water Limits: No drinking water limits were found for this substance.

Toxicity Profile

Carcinogenicity: No data were found implicating isobutyraldehyde as a carcinogen.

Mutagenicity: No data were found implicating isobutyraldehyde as a mutagen.

Developmental Toxicity: No data were found implicating isobutyraldehyde as a developmental toxin.

Reproductive Toxicity: No data were found implicating isobutyraldehyde as a reproductive toxin.

Systemic Toxicity: Only acute toxicity data and limited subacute inhalation data were available for this compound. An LD_{50} of 2810 mg/kg was reported for rats, and an LC_{50} of 39,500 mg/m^3 (2 hours) was reported for mice.[1]

Gage[2] reported the results of subacute inhalation exposure to a large number of compounds including isobutyraldehyde. Four male and four female rats were given 12 six-hour exposures to 1000 ppm isobutyraldehyde, and the only adverse effect reported was slight nasal irritation. No gross or histologic changes attributable to treatment were observed at necropsy. It should be noted that although the studies conducted by Gage are useful for exposure level selection for longer-term studies,

365

reporting of the results was too limited in the case of most compounds to justify use of the results in criteria derivation.

Irritation: In standard rabbit skin and eye irritation tests, isobutyraldehyde was reported to cause severe skin and eye irritation.[1] Isobutyraldehyde is reported to have a pungent odor that is detectable at a concentration of approximately 0.2 ppm.[3]

In the material safety data sheet for isobutyraldehyde, the vapors are reported to be irritating to the eyes and respiratory tract, but no specific concentrations correlated with these effects were given. Acute overexposure is also reported to cause chest discomfort, headache, nausea, and loss of consciousness.[3]

Basis for the AALG: Isobutyraldehyde is soluble in water to the extent of 6.5% by weight (at 20°C) and is volatile with a vapor pressure of 138 mm Hg at 20°C.[3] Based on the lack of data on its environmental fate and distribution and the expected contribution of various media to human exposure, 50% of the contribution to total exposure will be allowed from air.

No suitable basis for AALG derivation exists for this compound based on generally accepted criteria. The most viable alternative is to derive a surrogate animal NOAEL based on LC_{50} or LD_{50} data, convert this to a human equivalent NOAEL for inhalation, and apply appropriate uncertainty factors. Based on the statistical analysis of the relationship of oral LD_{50}s to their corresponding chronic NOAELs for a large number of compounds, Layton et al.[4] recommended that the LD_{50} be multiplied by a factor between 1.0×10^{-5} and 5.0×10^{-6} (equivalent to dividing by a factor between 100,000 and 200,000) to convert an LD_{50} to a corresponding surrogate animal NOAEL. In the case of isobutyraldehyde both a rat LD_{50} and a mouse LC_{50} were available. It should be noted that while an LC_{50} would be a better basis for an air criterion than an LD_{50}, an analysis of the relationship between LC_{50}s and their corresponding NOAELs, similar to that done by Layton et al., is not yet available. Therefore, the rat LD_{50} of 2810 mg/kg was divided by 100,000, converted to a human equivalent surrogate NOAEL for the inhalation route according to the procedures described in Appendix C, and divided by an uncertainty factor of 100 (10 for interspecies variation × 10 for interindividual variation). For comparative purposes only, the mouse LC_{50} of 39,500 mg/m^3 was also divided by 100,000, converted to a human equivalent surrogate NOAEL for the inhalation route according to the procedures described in Appendix C, and divided by an uncertainty factor of 100 (10 for interspecies variation × 10 for interindividual variation). Note that the AALG derived from the rat LD_{50}, 0.49 μg/m^3 (0.16 ppb), is more conservative than that derived from the mouse LC_{50}, 9.0 μg/m^3 (3.0 ppb).

AALG: • LD_{50}-derived—0.16 ppb (0.49 μg/m^3) 24-hour TWA

Note: This AALG is provisional and further appropriate testing would most likely result in a less conservative number. It is strongly recommended that a full literature search be conducted on this compound to identify a more suitable basis for AALG derivation and/or aid in planning appropriate animal studies.

References

1. NIOSH. 1988. NQ4025000. "Isobutyraldehyde." RTECS, on line.
2. Gage, JC. 1970. "The Subacute Inhalation Toxicity of 109 Industrial Chemicals." *Br. J. Ind. Med.* 27:1–18.
3. Union Carbide. 1985. "Material Safety Data Sheet for Isobutyraldehyde." Union Carbide Corporation, Danbury, CT.
4. Layton, D.W., B.J. Mallon, D.H. Rosenblatt, and M.J. Small. 1987. "Deriving Allowable Daily Intakes for Systemic Toxicants Lacking Chronic Toxicity Data." *Reg. Tox. Pharm.* 7:96–112.

ISOPHORONE

Synonyms: 3,5,5-trimethyl-2-cyclohexen-1-one

CAS Registry Number: 78–59–1

Molecular Weight: 138.2

Molecular Formula:

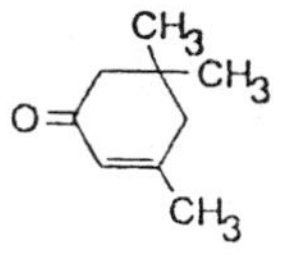

AALG: carcinogenicity—0.01 ppm (0.06 mg/m^3) annual TWA

Occupational Limits:
- ACGIH TLV: 5 ppm (28 mg/m^3) ceiling
- NIOSH REL: 4 ppm (23 mg/m^3) TWA
- OSHA PEL: 4 ppm (23 mg/m^3) TWA

Basis for Occupational Limits: The occupational limits are based on prevention of irritant and narcotic effects and subjective symptoms such as fatigue, nausea, and headaches.[1,2] OSHA recently adopted the NIOSH REL as part of its final rule limits.[3]

Drinking Water Limits: An ambient water quality criterion of 5.2 mg/L was derived for isophorone, assuming a 70-kg human consuming 2 L of water and 6.5 g of fish/shellfish per day.[4]

Toxicity Profile

Carcinogenicity: In an NTP bioassay,[5] male and female B6C3F$_1$ mice and F344/N rats (n = 50/sex/group/species) were gavaged with 0, 250, or 500 mg/kg isophorone (96–98% pure) in corn oil 5 days a week for 103 weeks. The major effects in rats involved the kidneys. In female rats, this was limited to a moderate increase in nephropathy, but male rats showed increased incidences of a variety of proliferative renal lesions (tubular cell hyperplasia, tubular cell adenomas and adenocarcinomas, and epithelial hyperplasia of the renal pelvis) as well as mineralization of the medullary collecting ducts and severe nephropathy in low-dose male rats. In addition, carcinomas of the preputial gland increased in high-dose male rats.

The major effects seen in mice involved the liver, and as was the case with rats, female animals were essentially unaffected. Non-neoplastic lesions included coagulative necrosis and hepatocytomegaly, and in high-dose male mice there was an increased incidence of hepatocellular adenomas and carcinomas.

In addition, there was an increased incidence of mesenchymal tumors of the integumentary system (fibromas, fibrosarcomas, neurofibrosarcomas, sarcomas) in high-dose male mice and an increased incidence of lymphomas or leukemias in low-dose male mice.

Based on the results of this bioassay, NTP concluded the following:

1. There was *some evidence* of carcinogenicity for isophorone in male F344/N rats, based on the occurrence of renal tubular cell adenomas and carcinomas and preputial gland carcinomas.
2. There was *equivocal evidence* of carcinogenicity for isophorone in male B6C3F$_1$ mice, based on the occurrence of hepatocellular adenomas and carcinomas, mesenchymal tumors of the integumentary system, and malignant lymphomas.
3. There was no evidence of carcinogenicity for isophorone in female B6C3F$_1$ mice or female F344/N rats.

Some evidence of carcinogenicity is defined by NTP to be "demonstrated by studies that are interpreted as showing a chemically-related increased incidence of benign neoplasms, studies that exhibit marginal increases in neoplasms of several organs/ tissues, or studies that exhibit a slight increase in uncommon malignant or benign neoplasms."[5] *Equivocal evidence* of carcinogenicity is defined by NTP to be "demonstrated by studies that are interpreted as showing a chemically-related marginal increase in neoplasms."[5]

In a paper describing the results of this same bioassay,[6] it was stated that "the results of current in vitro studies do not suggest a strong mutagenic capability and tend to suggest that isophorone may be acting through some type of promoter action to produce the effects noted in male rats and male mice."

Based on the criteria of the EPA weight-of-evidence classification for carcinogenicity, it is the judgement here that isophorone is most appropriately placed in group C, possible human carcinogen.

Mutagenicity: The results of a series of short-term tests conducted under the National Toxicology Program were summarized as follows in the technical report:[5]

Isophorone was not mutagenic in strains TA100, TA1535, TA1537, or TA98 of *Salmonella typhimurium* in the presence or absence of Aroclor 1254–induced male Sprague-Dawley rat or male Syrian hamster liver S9. Isophorone was weakly mutagenic in the mouse L5178Y/TK$^{+/-}$ assay in the absence of S9; it was not tested in the presence of S9. Isophorone induced sister chromatid exchanges in the absence of S9 in Chinese hamster ovary cells; it did not induce sister chromatid exchanges in the presence of Aroclor 1254–induced male rat liver S9, and it did not induce chromosomal aberrations in Chinese hamster ovary cells in the presence or absence of S9.

Developmental Toxicity: No data were found implicating isophorone as a developmental toxin.

Reproductive Toxicity: No data were found implicating isophorone as a reproductive toxin.

Systemic Toxicity: Data on the systemic toxicity of inhaled isophorone are limited in both humans and experimental animals. Smyth and Seaton in early work reported that at high concentrations isophorone caused death by narcosis in rats[7] (cited in NIOSH[2]). Smyth et al.[8] (cited in NIOSH[2]) exposed groups of 10 male rats and 10 male guinea pigs to isophorone via inhalation for 8 hours per day, 5 days a week for 6 weeks at concentrations of 25, 50, 100, 200, and 500 ppm. Growth was depressed and deaths due to isophorone exposure were reported in the 100-, 200-, and 500-ppm exposure groups. In the highest exposure group, chronic conjunctivitis and nasal irritation were observed. The authors reported that the effects on the two species were similar, although rats were slightly more sensitive. The lungs and kidneys of both species were affected by isophorone exposure in a dose-related manner, but the kidneys were somewhat more seriously affected. Histopathological lesions included congestion of the kidneys with dilation of Bowman's capsule and cloudy swelling of the convoluted tubules and congestion of the lungs with desquamation of the bronchial epithelium and evidence of pneumonia in some cases. A NOAEL of 25 ppm was identified from this study.

Smyth and Seaton[7] (cited in NIOSH[2]) exposed humans to the vapors of isophorone at concentrations of 40, 85, 200, and 400 ppm for a few minutes. Concentrations of 200 and 400 ppm were highly irritating, and subjective complaints of nausea, headache, dizziness, inebriation, and a feeling of suffocation were reported. Signs of irritation and narcotic effects, although still present at 40 and 85 ppm, were much less severe.

Based on work by Rowe and Wolf[9] (cited in U.S. EPA[4]), it has been concluded that the concentrations of isophorone reported in the papers by Smyth and Seaton[7] (cited in NIOSH[2]) and Smyth et al.[8] (cited in NIOSH[2]) were not accurate and that the isophorone used was probably an impure commercial product. These findings severely limit the value of these studies for criteria derivation.

It should be noted that in a communication to the TLV committee, it was reported that occupational exposure to isophorone levels in the range of 5 to 8 ppm for a month resulted in complaints of fatigue and malaise from the workers. When the levels in the work area were lowered to 1 to 4 ppm, no further symptoms were reported.[1]

Irritation: Isophorone has been reported to have a "camphor-like" odor[1] and an odor threshold of 0.2 ppm (geometric mean of reported literature values).[10]

Silverman et al.[11] (cited in NIOSH[2]) exposed human subjects (n = 12) to various concentrations of several ketones for periods of 15 minutes. Isophorone was reported to be irritating to the eyes, nose, and throat at 25 ppm, and 10 ppm was judged to be the highest satisfactory working concentration.

Basis for the AALG: Isophorone is considered very slightly soluble in water and has a vapor pressure of 0.2 mm Hg at 20°C. It is used as a solvent for vinyl resins, cellulose esters, and similar substances, and in pesticides. Since isophorone has been detected in effluents from manufacturing facilities and in municipal drinking water supplies, potential for exposure of the general public exists.[6] No data on isophorone concentrations in ambient air were found. Based on these considerations, 50% of the contribution to total exposure will be allowed from air.

Based on its class C designation and limited evidence of mutagenic activity (as defined by IARC criteria), the decision was made not to apply quantitative risk assessment to data from the NTP bioassay. Instead an uncertainty factor approach was used, with 250 mg/kg taken as the LOAEL for renal tubular cell adenomas and carcinomas in male F344/N rats. This LOAEL was selected because the evidence for carcinogenicity was greater for this species and tumor type, based on NTP criteria. This LOAEL was adjusted for continuous exposure (250 mg/kg × 5/7) and converted to a human equivalent LOAEL according to the procedures described in Appendix C; a total uncertainty factor of 5000 was applied (10 for interspecies variation × 10 for interindividual variation × 5 for a LOAEL × 10 for the database factor). Application of the database factor was considered relevant in this case because the study on which the AALG is based used a non-inhalation route of exposure and the endpoint was renal tubular cell adenomas and carcinomas. This AALG should be considered provisional since it is based on a non-inhalation study. Isophorone is currently being evaluated by CAG for quantitative risk assessment purposes,[12] and the basis for the AALG should be reevaluated when this information becomes available.

Since no long-term inhalation studies were available for isophorone and evidence of carcinogenicity and mutagenicity for this compound is limited, a limit based on the NIOSH REL was calculated for comparative purposes only. Using the REL/42 and the 50% relative source contribution given above yields a figure of 0.27 mg/m^3 (0.048 ppm).

AALG: • carcinogenicity—0.01 ppm (0.06 mg/m^3) annual TWA

References

1. ACGIH. 1986. "Documentation of the Threshold Limit Values and Biological Exposure Indices." 5:333.
2. NIOSH. 1978. "Criteria for a Recommended Standard . . . Occupational Exposure to Ketones." DHEW (NIOSH) 78–173.
3. OSHA. 1989. "Air Contaminants; Final Rule." *Fed. Reg.* 54:2332–2959.
4. U.S. EPA. 1980. "Ambient Water Quality Criteria for Isophorone." EPA 440/5–80–056.
5. NTP. 1986. "Toxicology and Carcinogenesis Studies of Isophorone (CAS No. 78–59–1) in F344/N Rats and B6C3F₁ Mice (Gavage Studies)." NTP-TR-291.
6. Bucher, J.R., J. Huff, and W.M. Kluwe. 1986. "Toxicology and Carcinogenesis Studies of Isophorone in F344 Rats and B6C3F₁ Mice." *Toxicology* 39:207–19.

7. Smyth, H.F., and J. Seaton. 1940. "Acute Response of Guinea Pigs and Rats to Inhalation of the Vapors of Isophorone." *J. Ind. Hyg. Tox.* 22:477–83.

8. Smyth, H.F., J. Seaton, and L. Fischer. 1942. "Response of Guinea Pigs and Rats to Repeated Inhalation of Vapors of Mesityl Oxide and Isophorone." *J. Ind. Hyg. Tox.* 24:46–50.

9. Rowe, V.K., and M.A. Wolf. 1963. "Ketones." In *Industrial Hygiene and Toxicology*, 2nd ed., F.A. Patty, Ed. (New York, NY: Interscience Publishers).

10. Amoore, J.E., and E. Hautala. 1983. "Odor as an Aid to Chemical Safety: Odor Thresholds Compared with Threshold Limit Values and Volatilities for 214 Industrial Chemicals in Air and Water Dilution." *J. Appl. Tox.* 3:272–90.

11. Silverman, L., H.F. Schultz, and M.W. First. 1946. "Further Studies on Sensory Responses to Certain Industrial Solvent Vapors." *J. Ind. Hyg. Tox.* 28:262–66.

12. U.S. EPA. 1989. "Isophorone; CASRN 78–59–1." IRIS (6/30/88).

ISOPROPYL ALCOHOL

Synonyms: 2-propanol, isopropanol

CAS Registry Number: 67–63–0

Molecular Weight: 60.09

Molecular Formula: $CH_3CH(OH)CH_3$

AALG: irritation (TLV-based)—8 ppm (19.7 mg/m^3) 8-hour TWA

Occupational Limits:
- ACGIH TLV: 400 ppm (980 mg/m^3) TWA; 500 ppm (1225 mg/m^3) STEL
- NIOSH REL: 400 ppm (980 mg/m^3) TWA; 800 ppm (1970 mg/m^3) ceiling (15 minutes)
- OSHA PEL: 400 ppm (980 mg/m^3) TWA

Basis for Occupational Limits: The occupational limits are based primarily on prevention of eye, nose, and throat irritation, and the TLV committee considered that exposure at these levels would not produce detectable narcotic effects.[1,2] OSHA has recently proposed to adopt the ACGIH TLV-STEL for isopropanol.[3]

Drinking Water Limits: No drinking water limits were found for this substance.

Toxicity Profile

Carcinogenicity: Epidemiologic studies have shown an excess risk of cancer of the paranasal sinuses and possibly laryngeal cancer in workers engaged in the manufacture of isopropyl alcohol by the strong acid process[4] (cited in IARC[5]), and the degree of evidence for this association was considered to be sufficient by IARC. Data on the manufacture of isopropanol by the weak acid process were not sufficient in humans to make an evaluation.[5]

Various studies have been conducted to evaluate the carcinogenic potential of isopropyl oils formed during the manufacture of isopropanol by both processes in mice and dogs using the inhalation, dermal, and parenteral routes of exposure. These studies are reviewed thoroughly in the NIOSH criteria document and by IARC.[2,6] However, all of these studies were judged inadequate to draw inferences pertaining to carcinogenicity of isopropyl oils by IARC.[5] Data on the carcinogenicity of isopropyl alcohol in animals were also inadequate to draw conclusions.[5]

Isopropyl alcohol is most appropriately placed in group D ("unclassified substance") under the EPA weight-of-evidence classification.

Mutagenicity: Chromosomal aberrations (cell type unspecified) have been reported in rats exposed to 1.03 mg/m³ isopropanol for 16 weeks (from Russian literature).[7] However, negative results are listed in the Gene-Tox database as summarized in RTECS for oncogenic transformation of Syrian hamster embryo cells by adenovirus SA7 and in the *N. crassa* aneuploidy test.[7]

Developmental Toxicity: In a report from the Russian literature summarized in RTECS and unavailable in translation, ingestion of isopropanol on days 1 through 20 of gestation in rats resulted in decreased litter size at a dose of 5040 mg/kg and increased pre-implantation mortality at a dose of 20,160 mg/kg.[7]

Reproductive Toxicity: In the same report cited above, exposure of female rats to 11,340 mg/kg isopropanol via ingestion for 45 days prior to mating resulted in changes in the estrus cycle. Exposure of male rats to 6480 mg/kg and female rats to 32.4 mg/kg isopropanol via ingestion for 26 weeks prior to mating resulted in decreased growth in the newborn and an increase in fetal death, respectively.[7]

Systemic Toxicity: Based on acute toxicity data in animals, isopropanol is of low toxicity via both the ingestion and inhalation routes. LD_{50}s of 3600, 5045, and 6410 mg/kg have been reported in mice, rats, and rabbits, respectively.[7] An 8-hour LC_{50} of 12,000 ppm has also been reported in rats.[1]

Isopropanol exerts its primary effects on the central nervous system in both humans and laboratory animals, exerting a more potent narcotic effect than ethanol. However, because it does not produce exhilaration and euphoria and tastes foul, it has generally not found wide use by the general public. Isopropanol, like ethanol, is oxidized primarily by alcohol dehydrogenase in humans.[8]

Rigorous, long-term epidemiologic studies have not been conducted on the effects of occupational exposure to ethanol alone. However, Willis et al.[9] (cited in NIOSH[2]) exposed healthy male volunteers (n = 8/group) to 0, 2.6, or 6.4 mg/kg/day isopropanol via ingestion for 6 weeks and found no evidence of ocular effects or hepatotoxicity (the latter as measured by bromosulfophthalein clearance and serum enzyme levels). It should be noted that no reports of narcotic effects due to inhalation exposure alone were found by NIOSH in its literature review. Narcotic effects have been reported in cases of combined dermal and inhalation exposure when isopropanol was sponged on patients to reduce fever. However, in these cases exposure concentrations were quite high due to use in a very confined, poorly ventilated area.[2]

Although isopropanol is of low toxicity by itself, it is known to potentiate the hepatotoxicity of carbon tetrachloride and possibly other compounds as well.[8]

Irritation: Nelson et al.[10] evaluated the sensory irritation properties of isopropanol and a number of other solvents in a controlled human exposure study (n = 10) in which subjects inhaled various concentrations of isopropanol for 3- to 5-minute periods. Mild irritation of the eyes, nose, and throat were reported at 400 ppm and these effects were more pronounced at 800 ppm. It should be noted that these concentrations were nominally, not analytically, determined. Isopropyl alcohol has an odor similar to rubbing alcohol (which is 70% isopropanol) and odor thresholds ranging from 40 to 200 ppm have been reported.[2]

Isopropyl alcohol as a 70% solution in water is widely used as a mild skin disinfectant and astringent and has been characterized as nonirritating to the skin in this form and only mildly irritating to the eyes.[1]

Basis for the AALG: Isopropanol is miscible with water and somewhat volatile (vapor pressure = 33 mm Hg at 20°C) and is in wide use in consumer products, e.g., rubbing alcohol.[1] However, since the AALG for isopropanol is based on irritation, no relative source contribution from inhalation was factored in the calculation. The reason is that allocation of a proportion of exposure to a given source is not relevant when the effects of exposure are not cumulative, as is the case with irritant effects.

The AALG is based on the irritant effects of isopropyl alcohol using the long-standing TLV as a starting point. This TLV was divided by a total uncertainty factor of 50 (10 for interindividual variation × 5 for a LOAEL) since the TLV of 400 ppm was identified as a LOAEL for irritant effects in the study of Smyth et al.[10] This AALG should be considered provisional, since it is based on an occupational limit. It should be noted that although narcotic effects may result from ingestion of isopropanol, no reports of narcosis due to inhalation exposure alone have been found.

AALG: • irritation—8 ppm (19.7 mg/m^3) 8-hour TWA

References

1. ACGIH. 1986. "Documentation of the Threshold Limit Values and Biological Exposure Indices." 5:337.
2. NIOSH. 1976. "Criteria for a Recommended Standard . . . Occupational Exposure to Isopropyl Alcohol. DHEW (NIOSH) 76–142.
3. OSHA. 1988. "Air Contaminants; Proposed Rule." *Fed. Reg.* 53:20959–21393.
4. Weil, C. S., H. F. Smyth, and T. W. Nale. 1952. "Quest for a Suspected Industrial Carcinogen." *Arch. Ind. Hyg. Occ. Med.* 5:535–47.
5. IARC. 1987. "Isopropyl Alcohol Manufacture (Strong Acid Process), Isopropyl Alcohol and Isopropyl Oils." *IARC Monog.* suppl. 7:229.
6. IARC. 1977. "Isopropyl Alcohol Manufacture (Strong Acid Process), Isopropyl Alcohol and Isopropyl Oils." *IARC Monog.* 15:223–43.
7. NIOSH. 1988. NT8050000. "Isopropyl Alcohol." RTECS, on line.
8. Wimer, W. W., J. A. Russell, and H. L. Kaplan. 1983. *Alcohols Toxicology* (Park Ridge, NJ: Noyes Data Corporation), pp. 48–55.
9. Willis, J. H., E. D. Jameson, and F. Coulston. 1969. "Effects on Man of Daily Ingestion of Small Doses of Isopropyl Alcohol." *Tox. Appl. Pharm.* 15:560–85.
10. Nelson, K. W., J. F. Ege, M. Ross, R. Woodman, and L. Silverman. 1943. "Sensory Response to Certain Industrial Solvent Vapors." *J. Ind. Hyg. Tox.* 25:282–85.

LEAD AND COMPOUNDS

Synonyms: "Lead and compounds" includes inorganic compounds, dust, and fume, excluding lead arsenate and lead chromate

CAS Registry Number: 7439–92–1

Molecular Weight: 207.19

Molecular Formula: Pb

AALG: 1.5 μg (Pb)/m^3 as a monthly TWA

Occupational Limits:
- ACGIH TLV: 0.15 mg (Pb)/m^3 TWA
- NIOSH REL: 0.10 mg (Pb)/m^3 TWA
- OSHA PEL: 0.05 mg (Pb)/m^3 TWA

Basis for Occupational Limits: All of the occupational limits were set in consideration of systemic toxicity, principally effects on the blood, nervous system, and kidney.[1-3] The lower limit proposed by NIOSH is apparently based on reports of decreased nerve conduction velocities and other effects in workers with blood leads between 0.05 and 0.07 mg/100 grams.[3] With respect to the difference between the ACGIH and NIOSH limits, the following quotation from the TLV documentation[1] is relevant:

> The Committee is not convinced that the biochemical changes found due to low level lead absorption are incompatible with good health. It has not adopted, or proposed, a biologic TLV for lead, nor has it accepted the NIOSH hypothesis that an air TLV must be set at a level at which most workers (i.e., 90–95%) do not exceed a specified biologic TLV.

OSHA regulations governing work practices concerning occupational lead exposure are given in 29 CFR 1910.1025.[4]

Drinking Water Limits: The ambient water quality criterion for lead in drinking water is 0.05 mg/L, assuming consumption of 2 L of water and 6.5 g fish/shellfish per day.[5] Concern has been expressed that this level may not be sufficiently protective for fetuses and young children in view of exposure from other sources.[6] The U.S. EPA recently set an MCLG of 0 and proposed a final MCL of 0.005 mg/L for lead in drinking water.[7]

Toxicity Profile

Carcinogenicity: The issue of lead carcinogenicity is controversial, and it is summarized as follows in the 1986 "Air Quality Criteria for Lead" document:[8]

> It is hard to draw clear conclusions concerning what role lead may play in the induction of human neoplasia. Epidemiologic studies of lead-exposed workers provide no definitive findings. However, statistically significant elevations in respiratory tract and digestive system cancer in workers exposed to lead and other agents warrant some concern. Since lead acetate can produce renal tumors in some experimental animals, it seems reasonable to conclude that at least this particular lead compound should be regarded as a carcinogen and prudent to treat it as if it were also a human carcinogen (as concluded by the International Agency for Research on Cancer). However, this statement is qualified by noting that lead has been observed to increase tumorigenesis rates in animals only at relatively high concentrations, and therefore it does not appear to be a potent carcinogen. In vitro studies further support the genotoxic and carcinogenic role of lead, but also indicate that lead is not potent in these systems either.

The evidence for lead carcinogenicity is reviewed in depth in Chapter 12 of the "Air Quality Criteria for Lead" document.[9]

More recently the EPA published a document in which the carcinogenic potential of lead via oral exposure was examined[10] (cited in U.S. EPA[11]). Based on this more recent review, it was concluded that the evidence for carcinogenicity of lead in animal models is sufficient: statistically significant increases in renal tumors were observed in ten rat bioassays and one mouse bioassay in which the animals were exposed to soluble lead salts either in the diet or by subcutaneous injection. Lead and compounds (inorganic) have consequently been placed in group B2, probable human carcinogen, under the EPA weight-of-evidence classification.

Mutagenicity: Reports of the induction of chromosomal aberrations in lymphocytes from humans exposed to lead are conflicting and difficult to evaluate on account of various factors, including absence of sufficient evidence of lead intoxication, no dose-response relationship, lack of information on lymphocyte culture times, and concurrent exposures to other substances. There is at least one apparently well-conducted study where lead was reported to induce SCEs in lymphocytes of exposed workers. It is relevant to note that lead acetate has been found to induce chromosomal aberrations in cultured human lymphocytes and in cynomolgus monkeys treated chronically.[9]

Lead has been reported to induce cell transformation and to affect DNA to DNA and DNA to RNA transcription. Together with evidence of chromosomal effects, this has been suggested to indicate initiating activity. Based on indications that lead has proliferative activity, i.e., it increases DNA, RNA, and protein synthesis in the kidney, it has been suggested that lead also may have promotional activity.[9]

Developmental Toxicity: Based on human and animal data, lead is readily transferred across the placenta. The effects of various exposure levels of lead in experimental animals may be summarized as follows: In rodent studies, fetotoxic effects have occurred in animals at chronic exposures of 600–800 ppm inorganic lead in the diet and more subtle effects (behavioral and neurological) appear to have been observed at 5–10 ppm lead in the drinking water and at levels of 10 $\mu g/m^3$ lead in air. Teratogenic effects appear to be confined to studies where the dams were given lead by injection.[9]

Based on the results of recent well-conducted epidemiologic studies, which are supported by animal studies, it was concluded in the "Air Quality Criteria for Lead" document[8] that

> fetal exposure to lead at relatively low and prevalent concentrations can have undesirable effects on infant mental development, length of gestation, and possibly other aspects of fetal development . . . perinatal blood lead levels at least as low as 10–15 $\mu g/dL$ clearly warrant concern for deleterious effects on early postnatal as well as prenatal development.

Reproductive Toxicity: Data on the effects of lead on maternal reproductive function (e.g., ovarian function) are lacking. In males, adverse effects on sperm (asthenospermia, hypospermia, and teratospermia) and decreased function of the prostate and/or seminal vesicles (as measured by seminal plasma constituents) have been reported to result from chronic lead exposure with blood lead levels of 40–50 $\mu g/dL$.[9]

Systemic Toxicity: The basis for lead toxicity is its ability to bind to ligating groups in physiologically critical biomolecules, which may then disrupt their function by competing with essential metal ions for binding sites, inhibiting enzyme activity, and altering ion transport.

Although lead induces toxicity in a number of organ systems, hematological, neurological, and renal effects are generally considered the most significant because they are induced at lower levels of exposure. Lead inhibits the heme synthesis pathway at several specific steps and also causes disturbances in globin biosynthesis. The resultant anemia is characterized as mildly hypochromic and normocytic with reticulocytosis due to shortened survival time of erythrocytes, which are more fragile and have increased osmotic resistance. In addition, lead alters vitamin D metabolism secondary to heme synthesis effects at blood lead levels below 30 $\mu g/dL$.

In adults, central and peripheral nerve dysfunction occurs at blood lead levels as low as 40–60 $\mu g/dL$, and slowed nerve conduction velocities in peripheral nerves at levels as low as 30–50 $\mu g/dL$ with no clear threshold for these effects. In children, there is evidence for decrements in IQ and behavioral (reaction time and psychomotor performance) and electrophysiological (altered EEG patterns and peripheral nerve conduction velocities) effects at blood lead levels as low as 15–30 $\mu g/dL$ and possibly lower.

In occupational settings, lead-induced nephropathy has been associated with blood lead levels between 40 and 100 μg/dL. In children, lead-induced nephropathy is less common and may occur only at blood lead levels in excess of 100 μg/dL.

Lead also has effects on the cardiovascular system, liver, gastrointestinal tract, and endocrine system, but these systems are generally not as sensitive to the effects of lead as the hematological and nervous systems. However, some recent evidence suggests significant elevations in blood pressure in animals and humans may occur following chronic low-level lead exposure.

In laboratory animals, lead increases susceptibility to endotoxins and infectious agents, and the macrophage has been suggested as the primary target cell. However, these effects remain largely uninvestigated in humans.[8,9]

Irritation: Lead is not generally regarded as an irritant.

Basis for the AALG: Determination of an acceptable air guideline for lead is complicated by two factors:

1. Lead has been so extensively studied that in recent years much controversy has arisen about which effects are most significant and sensitive.
2. Lead is ubiquitous in the environment, and there is a high level of "background" exposure, primarily via food, that must be taken into consideration in establishing a guideline.[12]

The national ambient air quality standard (NAAQS) of 1.5 μg/m^3 is the most appropriate AALG, based on these factors as well as its legal status, i.e., a standard rather than a guideline.

EPA used the following assumptions to derive the NAAQS for lead:

1. Children aged 1–5 years were identified as the most sensitive population.
2. The lowest detectable adverse health effect was identified as increased erythrocyte protoporphyrin.
3. The average population blood lead level that would provide sufficient protection to children aged 1–5 years was set as 15 μg/dL.
4. The relationship of air lead to blood lead was assumed to be a 1:2 ratio, i.e., for each 1 μg (Pb)/m^3 increase there would be a concomitant increase of 2 μg (Pb)/dL of blood.
5. It was assumed that 12 μg (Pb)/dL came from nonair sources, leaving 3 μg Pb/dL to be allowed from air.

Thus a standard of 1.5 μg/m^3 was derived.[13–15]

With respect to the recent classification of inorganic lead in group B2 of the EPA weight-of-evidence classification and the influence this may have on criteria derivation, it should be noted that the U.S. EPA Carcinogen Assessment Group has recommended that current quantitative risk assessment techniques not be applied to lead. The following was stated in the IRIS database:[11]

Quantifying lead's cancer risk involves many uncertainties, some of which may be unique to lead. Age, health, nutritional state, body burden, and exposure duration influence the absorption, release, and excretion of lead. It is also felt that current knowledge of lead pharmacokinetics indicates that an estimate of derived by standard procedures would not truly describe the potential risk.

The AALG should be considered provisional since the current national ambient air quality standard (NAAQS) is under review to determine if revision is warranted.[10]

AALG: • 1.5 μg (Pb)/m^3 as a monthly average

References

1. ACGIH. 1986. Documentation of the threshold limit values and biological exposure indices. 5:343–5.
2. NIOSH. 1985. NIOSH Recommendations for occupational safety and health standards. MMWR 34(Suppl.): 5s–31s.
3. NIOSH. 1978. Criteria for a recommended standard . . . occupational exposure to inorganic lead, revised criteria—1978. DHEW(NIOSH) 78–158.
4. OSHA. 1988. 29 Code of Federal Regulations 1910.1025.
5. U.S. EPA. 1980. Ambient water quality criteria for lead. EPA 440/5–80–057.
6. Safe Drinking Water Committee, NAS 1983. Drinking water and health, vol. 5 (National Academy Press: Washington, DC).
7. U.S. EPA. 1988. National primary drinking water regulations; lead and copper; proposed rule. Fed. Reg. 53:31516–78.
8. U.S. EPA. 1986. Air quality criteria for lead, volume I of IV. EPA-600/8–83/028AF.
9. U.S. EPA. 1986. Air quality criteria for lead, volume III of IV. EPA-600/8–83/028CF.
10. U.S. EPA. 1989. Lead and compounds (inorganic); CASRN 7439-92-1. Integrated Risk Information System (IRIS). 6/1/89.
11. U.S. EPA. 1987. Preliminary review of the carcinogenic potential of lead associated with oral exposure. Prepared by the Office of Health and Environmental Assessment, Carcinogen Assessment Group, Washington, DC, for the Office of Drinking Water, Office of Solid Waste and the Office of Emergency and Remedial Response (Superfund). OHEA-C-267. Internal Review Draft.
12. U.S. EPA. 1984. Health effects assessment for lead. EPA-540/1–86/055.
13. U.S. EPA. 1986. Air quality criteria for lead, volume II of IV. EPA-600/8–83/028BF.
14. U.S. EPA. 1986. Air quality criteria for lead, volume IV of IV. EPA-600/8–83/028DF.
15. Calabrese, E.J. 1978. Methodological approaches to deriving environmental and occupational health standards. (John Wiley & Sons: New York, NY).

MALEIC ANHYDRIDE

Synonyms: 2,5-furandione

CAS Registry Number: 108–31–6

Molecular Weight: 98.06

Molecular Formula:

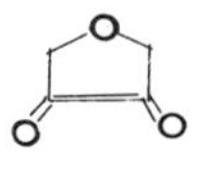

AALG: irritation (TLV-based)—5 ppb (20 μg/m^3) 8-hour TWA

Occupational Limits:
- ACGIH TLV: 0.25 ppm (1 mg/m^3) TWA
- NIOSH REL: none
- OSHA PEL: 0.25 ppm (1 mg/m^3) TWA

Basis for Occupational Limits: The TLV and PEL are based on protection against respiratory irritation in non-sensitized workers and on analogy with phthalic anhydride.[1]

Drinking Water Limits: No drinking water limits were found for this substance.

Toxicity Profile

Carcinogenicity: Dickens and Jones[2] reported that male rats injected subcutaneously with 1 mg of maleic anhydride in arachis oil twice a week for 61 weeks and observed for up to 106 weeks developed injection-site fibrosarcomas (numbers not reported). In a CIIT-sponsored chronic feeding study[3] (cited in Short et al.[4,5]), maleic anhydride was not carcinogenic to rats. Maleic anhydride is most appropriately placed in group D, unclassified substance, under the EPA weight-of-evidence classification.

Mutagenicity: Maleic anhydride has been reported to be mutagenic to hamster lung cells in culture.[6]

Developmental Toxicity: Short et al.[4] exposed CD rats (n = 19–23/group) to 0, 30, 90, or 140 mg/kg/day maleic anhydride by gavage in corn oil on days 6 through 15 of gestation. No significant maternal toxicity, fetotoxicity, or evidence of teratogenic effects were observed.

Reproductive Toxicity: Short et al.[4] also reported no treatment-related effects on reproduction at doses of up to 55 mg/kg/day in a multigeneration study in CD rats in which males and females received 0, 20, 55, or 150 mg/kg/day maleic anhydride by gavage in corn oil. However, the highest dose produced increased mortality due to renal tubular necrosis in both males and females.

Systemic Toxicity: Maleic anhydride has been reported to cause sensitization of the skin and respiratory tract. In addition, it is noted in the TLV documentation that there is a much greater frequency of chronic bronchitis among workers exposed to maleic anhydride over many years compared to workers with similar exposure to phthalic anhydride.[1] Other than the renal toxicity observed at high doses in the Short et al. study,[4] no reports of systemic toxicity in laboratory animals were found.

Irritation: Maleic anhydride is recognized to be an irritant to the skin, eyes, and respiratory tract, and the following exposure-response relationships for humans are cited in the TLV documentation:[1]

- 2.5 ppm and higher—immediate severe irritant response
- 1.5–2 ppm—nasal irritation in one minute, followed by eye irritation within 15 minutes (in unadapted workers)
- 0.25–0.38 ppm—minimal concentration of maleic anhydride causing conjunctival and upper respiratory tract irritation (Soviet data source)

The threshold concentration for irritation has been variously reported as 0.5 ppm and 0.22 ppm.[1] The odor threshold for maleic anhydride is reported to be 0.32 ppm.[7]

Short et al.[5] exposed male and female CD rats (n = 15/sex/group), Engle hamsters (n = 15/sex/group) and rhesus monkeys (n = 3/sex/group) to room air or 1.1, 3.3, or 9.8 mg/m³ maleic anhydride via inhalation 6 hours per day, 5 days a week for 6 months. Signs of nasal and ocular irritation, including discharge, sneezing, gasping, and coughing, occurred in a dose-related pattern at each exposure level in all three species. However, there was no evidence of treatment-related systemic toxicity based on hematology, clinical chemistry, urinalysis, or necropsy findings. In addition, pulmonary function tests on monkeys prior to exposure and at 3 and 6 months revealed no treatment-related changes. Histological examination of nasal tissues revealed hyperplastic and metaplastic changes in the rodent species and inflammatory changes only in the monkeys; in all cases, these lesions were judged to be reversible by the authors.

Basis for the AALG: Maleic anhydride is soluble and hydrolyzes slowly in water. It has a vapor pressure of 0.1 mm Hg at 25°C. The following uses are listed for maleic anhydride in the TLV documentation: in the manufacture of polyester and alkyl coating resins; as a starting material in the synthesis of fumaric and tartaric acids and maleic hydrazide; as a pesticide ingredient and preservative in oils and fats; and in paper sizing and permanent press fabrics.[1] Since the AALG for maleic anhydride is based on irritation, and irritant effects tend to be noncumulative and are more concentration-dependent than time-dependent, no relative source contribution from air was factored into the calculation.

The AALG for maleic anhydride is based on irritation (TLV/50) (5 for a LOAEL × 10 Interindividual variation). The TLV is treated as a human LOAEL, based on the correlation of exposure and effect data in humans cited in the TLV documentation.[1] Further support for the validity of this treatment of maleic anhydride is provided by the data from the Short et al. study[5] indicating no systemic toxicity and occurrence of only reversible irritant effects in all three species exposed both at the TLV and 10 times the TLV. The AALG should be considered provisional since it is based on an occupational limit and because there is at least one report (in the form of a private communication to the TLV committee) indicating that allergic sensitization may be associated with exposure to maleic anhydride in humans.

AALG: • irritation (TLV-based)—5 ppb (20 $\mu g/m^3$) 8-hour TWA

References

1. ACGIH. 1986. "Documentation of the Threshold Limit Values and Biological Exposure Indices." 5:353.
2. Dickens, F., and H.E.H. Jones. 1963. "Further Studies on the Carcinogenic and Growth-Inhibitory Activity of Lactones and Related Substances." *Br. J. Cancer* 17:100–108.
3. Chemical Industry Institute of Toxicology. 1983. "Chronic Dietary Administration of Maleic Anhydride—Final Report." CIIT Docket No. 114N3. Research Triangle Park, NC.
4. Short, R.D., F.R. Johannsen, G.J. Levinskas, D.E. Rodwell, and J.L. Schardein. 1986. "Teratology and Multigeneration Reproduction Studies with Maleic Anhydride in Rats." *Fund. Appl. Tox.* 7:359–66.
5. Short, R.D., F.R. Johannsen, and C.E. Ulrich. 1988. "A 6-Month Multispecies Inhalation Study with Maleic Anhydride." *Fund. Appl. Tox.* 10:517–24.
6. NIOSH. 1988. ON3675000. "Maleic Anhydride." RTECS, on line.
7. Amoore, J.E., and E. Hautala. 1983. "Odor as an Aid to Chemical Safety: Odor Thresholds Compared with Threshold Limit Values and Volatilities for 214 Industrial Chemicals in Air and Water Dilution." *J. Appl. Tox.* 3:272–90.

MANCOZEB

Synonyms: manganese zinc complex ethylenebisdithiocarbamic acid, vondozeb, dithane M-45

CAS Registry Number: 8018–01–7

Molecular Weight: 541.03

Molecular Formula: C_4-H_6-Mn-N_2-S_4.C_4-H_6-N_2-S_4-Zn

AALG: developmental toxicity—0.017 mg/m^3 24-hour TWA

Occupational Limits:
- ACGIH TLV: none
- NIOSH REL: none
- OSHA PEL: none

Drinking Water Limits: No drinking water limits were found for this substance.

Toxicity Profile

Carcinogenicity: No data were found implicating mancozeb as a carcinogen. However, it is significant to note that ETU, which is classified as a B2, probable human carcinogen, occurs as a contaminant during mancozeb manufacture, a degradation product during storage and following application, and a metabolite following ingestion of mancozeb. The U.S. EPA has estimated that "the average consumer in the U.S. population receives a dietary exposure of 3.6×10^{-5} mg/kg/day of ETU from the conversion of mancozeb residues on crops." Using these data together with data on the in vivo metabolism of mancozeb and the q_1* for ETU, the U.S. EPA estimated a total dietary oncogenic risk of 2.2×10^{-5} for exposure to mancozeb, above the generally accepted de minimis level of risk of 10^{-6}.[1]

Mutagenicity: The following short-term test results are reported for mancozeb in the Gene-Tox data base as summarized in RTECS:[2] positive in *S. cerevisiae* homozygosis and gene conversion and negative for in vitro unscheduled DNA synthesis in human fibroblasts. In addition, mancozeb is reported to cause chromosomal aberrations in cultured human lymphocytes and in rats exposed orally and parenterally and to cause mutations in *A. nidulans* and *S. cerevisiae* without metabolic activation.[2]

Developmental Toxicity: Larsson et al.[3] exposed NMRI mice and Sprague-Dawley rats to 0, 380, 730, or 1330 mg/kg mancozeb (20% Mn, 2.5% Zn, 80% active compound—technical grade) on day 9 or 13 and day 11 of gestation in mice and rats, respectively. No treatment-related effects on the dams or offspring were found in the mice. However, gross malformations (cleft palate, brachygnathia, hydrocephaly, short tail, etc.) occurred in 25% of the surviving fetuses exposed to the highest dose of mancozeb. No reference was made to maternal toxicity associated with mancozeb exposure in the rats. Based on the similarity of the malformations reported at the highest dose in this study with those attributable to ETU in other studies, it is possible that these malformations may have been due to ETU present either as a contaminant, degradation product, or metabolite.

Lu and Kennedy[4] exposed Crl:CD rats (n = 27–38/group) to mancozeb (80% active ingredients, 20% inert ingredients) via inhalation at concentrations of 0, 1, 17, 55, 110, 890, or 1890/(reduced to 500 after 1 day due to high mortality) mg/m^3 for 6 hours per day on gestation days 6 through 15; note that the three highest and lowest exposure groups were part of separate studies with separate control groups for each. No treatment-related increase in malformations occurred in this study; however, embryofetal toxicity in the form of a significantly increased incidence of totally resorbed litters, external hemorrhage, and wavy ribs occurred at concentrations of 55 mg/m^3 and higher. Maternal toxicity was observed as follows: at 500 to 1890 mg/m^3, decreased body weight gain, hind limb paralysis, general debilitation, and death were reported; at 55 and 110 mg/m^3, decreased body weight gain and hind limb weakness. Thus, in this study, embryofetal toxicity (with no malformations) occurred only at maternally toxic levels, and a NOAEL of 17 mg/m^3 is suggested for both maternal and embryofetal toxicity.

Reproductive Toxicity: No data were found implicating mancozeb as a reproductive toxin.

Systemic Toxicity: The acute toxicity of insoluble EBDC salts such as mancozeb is generally considered to be low: an oral LD_{50} of 12.8 grams/kg in rats has been reported.[2] It has been suggested that the effects observed at high concentrations are at least partially attributable to the metal component.[5]

Data on the subchronic and chronic toxicity of mancozeb are quite limited. In a review in *Drinking Water and Health,* Volume 1, several NOAELs and LOAELs for thyroid hyperplasia in rats exposed to mancozeb (listed as dithane M-45, a synonym for mancozeb given in RTECS) in the diet.[5] The NOAELs ranged from 100 to 1000 ppm and the LOAELs from 1000 to 2510 ppm. These figures were quoted from an EPA document[6] and had originally come from unpublished reports by various corporate entities.

The Lu and Kennedy study[4] suggests that short-term inhalation exposure to mancozeb causes a dose-related decrease in body weight gain and hind limb paralysis and weakness in rats.

Irritation: No data were found implicating mancozeb as an irritant.

Basis for the AALG: Because of its use as an agricultural fungicide, the major source of exposure to mancozeb for the general population is likely to be the diet.[1] Therefore, 20% of the contribution to total exposure will be allowed from air. It should be noted that a larger proportion of exposure might occur via inhalation in certain population subgroups such as pesticide applicators and mixer/loaders and individuals living in the immediate vicinity of application areas.

The AALG for mancozeb is based on the developmental toxicity study of Lu and Kennedy since it used the inhalation route and sufficient dose levels to define a range of maternal and fetal responses. The NOAEL of 17 mg/m^3 was adjusted for continuous exposure (17 mg/m^3 × 6/24), converted to a human equivalent NOAEL for the inhalation route based on the procedures described in Appendix C, and divided by a total uncertainty factor of 100 (10 for interindividual variation × 10 for interspecies variation). It was considered inappropriate to apply the developmental uncertainty factor since no teratogenic effects were observed and embryofetal toxicity occurred only at exposure levels producing overt maternal toxicity.

This AALG should be considered provisional on account of possible oncogenic hazards and thyroid effects for mancozeb that have not yet been adequately defined. The EBDC pesticides, including mancozeb, are currently the subject of an EPA Special Review with particular reference to potential oncogenic and teratogenic hazards.[1] It is strongly recommended that this compound be reevaluated when the results of this review become available.

AALG: • developmental toxicity—0.017 mg/m^3 24-hour TWA

References

1. U.S. EPA. 1987. "EBDC Pesticides: Initiation of Special Review." *Fed. Reg.* 52: 27172–177.
2. NIOSH. 1988. ZB3200000. "Vondozeb." RTECS, on line.
3. Larsson, K.S., C. Arnander, E. Cekanova, and M. Kjellberg. 1976. "Studies of Teratogenic Effects of the Dithiocarbamates Maneb, Mancozeb, and Propineb." *Teratology* 14:171–84.
4. Lu, M-H., and G.L. Kennedy. 1986. "Teratogenic Evaluation of Mancozeb in the Rat Following Inhalation Exposure." *Tox. Appl. Pharm.* 84:355–68.
5. Safe Drinking Water Committee. 1977. *Drinking Water and Health,* Vol. 1 (Washington, DC: National Academy Press), pp. 650–60.
6. U.S. EPA. 1973. "The Toxicology and Environmental Hazards of Ethylenebisdithiocarbamate Fungicides and Ethylene Thiourea." Report of the Special Pesticide Review Group (privileged document).

MANEB

Synonyms: manganese ethylene-1,2-bisdithiocarbamate

CAS Registry Number: 12427–38–2

Molecular Weight: 265.3

Molecular Formula:

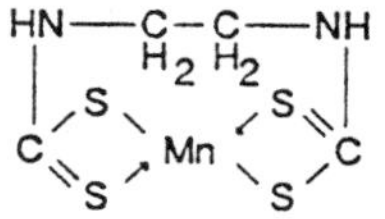

```
HN———C – C———NH
       H₂ H₂
     |  S       S  |
     C ⁄    ⟍  ⁄    ⟍ C
       ⟍S  Mn  S ⁄
```

AALG: systemic toxicity (thyroid hyperplasia)—1.75 $\mu g/m^3$ 24-hour TWA

Occupational Limits:
- ACGIH TLV: none
- NIOSH REL: none
- OSHA PEL: none

Drinking Water Limits: In 1977, the Safe Drinking Water Committee of the NAS calculated a SNARL (time period unspecified) of 35 $\mu g/L$ for maneb based on the usual assumption of a 20% relative source contribution from drinking water; an alternative SNARL of 1.75 $\mu g/L$ was also calculated based on a 1% contribution from drinking water.[1]

Toxicity Profile

Carcinogenicity: In 1976, IARC evaluated the evidence for maneb carcinogenicity.[2] It was reported in one study[3] (cited in IARC[2]) that maneb induced lung adenomas in two strains of mice given 500 mg/kg maneb in 6 weekly doses by gavage and killed at intervals of up to 9 months. In another study[4] (cited in IARC[2]), however, maneb did not cause an increased incidence of tumors in either of two strains of mice exposed by gavage (46.4 mg/kg) and later in the diet (158 mg/kg of diet) until 78 weeks of age. This is the same study in which ETU was found to produce liver tumors in mice. The exposure level used in this negative study was at the maximum tolerated dose (MTD) for infant and juvenile mice but may not have been the MTD for adult mice.

In another study[5] (cited in IARC[2]) in which two strains of mice were given a single subcutaneous injection of 100 mg/kg maneb at 4 weeks of age and observed until 78 weeks of age, no increase in tumor incidence was found. However, it should be noted that a negative result in this type of study does not constitute

393

adequate evidence to determine a lack of carcinogenic activity.[2] Although other studies were cited in the IARC review in which rats were exposed to maneb by ingestion and injection, these studies were considered inadequate to evaluate carcinogenic effects on account of poor survival.[2]

IARC concluded that these studies in aggregate were not sufficient to evaluate the carcinogenicity of maneb,[2] and no carcinogenicity studies were found for maneb that postdated the IARC review. Based on this information, maneb is most appropriately placed in group D (unclassified substance) under the EPA weight-of-evidence classification.

It is significant to note that ETU, classified as a B2 (probable human carcinogen), occurs as a contaminant during maneb manufacture, a degradation product during storage and following application, and a metabolite following ingestion of maneb.[1,2] In addition, there are reports indicating that maneb on raw agricultural commodities would be expected to break down to ETU during cooking.[2] For further information on ETU carcinogenicity, refer to the chemical-specific evaluation for ETU.

Mutagenicity: The following short-term test results are reported for maneb in the Gene-Tox database as summarized in RTECS:[6] positive *S. cerevisiae* homozygosis, weakly positive in vitro unscheduled DNA synthesis in human fibroblasts, and negative *S. cerevisiae* gene conversion. In addition, maneb is reported to cause chromosomal aberrations in cultured hamster lung cells, gene conversion and mitotic recombination in *S. cerevisiae,* and mutation in *S. cerevisiae* without metabolic activation.[2]

Developmental Toxicity: Petrova-Vergieva and Ivanova-Tchemishanska[7] demonstrated a clear dose-response relationship for the occurrence of malformations (mainly of the brain, facial bones, limbs, and tail) in the offspring of albino rats (n = approximately 10/group) exposed to single oral doses of 0, 0.5, 1.0, 2.0, or 4.0 g/kg maneb on days 11 or 13 of gestation. A NOAEL of 0.5 g/kg maneb as a single oral dose is suggested by this study. In follow-up studies, groups of approximately 10 rats were given 0, 125, 250, or 500 mg/kg maneb on gestation days 2 to 21. No evidence of malformations or embryotoxicity was found in any of these studies. It should be noted that maternal toxicity was either not evaluated or not reported in this study.

Larsson et al.[8] confirmed the findings of the above study using Sprague-Dawley rats. In addition, they demonstrated that the teratogenic effect of maneb in rats could be reduced or eliminated by concomitant administration of zinc acetate and that the zinc-containing fungicides, mancozeb and propineb, were less teratogenic than maneb. Both of these findings lend support to the hypothesis that the teratogenic effect of maneb is related to zinc deficiency.

Larsson et al.[8] also evaluated the teratogenicity of maneb in NMRI mice exposed to 400, 770, or 1420 mg/kg maneb by gavage on day 9 or 13 of gestation and found no evidence of maternal toxicity, fetotoxicity, or malformations attributable to maneb exposure. Chernoff et al.[9] exposed CD-1 mice to 0, 375, 750, or

1500 mg/kg maneb by gavage (in water) on gestation days 7 through 16. There was a significant dose-related increase in the maternal liver-body weight ratio and decrease in the number of caudal ossification centers in the fetuses, but no malformations were observed.

Reproductive Toxicity: Marcon[10] (cited in IARC[2]) demonstrated a reversible effect on fertility in both male and female juvenile rats exposed to maneb via ingestion at a dose of 50 mg/kg/day for 4 weeks.

Systemic Toxicity: The acute toxicity of insoluble EBDC salts such as maneb is generally considered to be low: LD_{50}s in the range of 4 to 7 g/kg have been reported.[2] It has been suggested that the effects observed at high concentrations are at least partly attributable to the metal component.[1]

Sobotka et al.[11] evaluated the behavioral and neuroendocrine effects of maneb exposure during the postnatal period of life in male Osborne-Mendel rats using dietary levels of 0, 0.5, 1.0, and 10 ppm maneb. Within each dietary exposure level, the following exposure groups were set up:

- Preweaning exposure group. Dams were fed the diet containing maneb from birth through weaning, and after weaning at 28 days of age the pups were switched over to a control diet.
- Lifetime exposure group. Offspring were exposed to maneb in the preweaning period as described above, and exposure continued after weaning until the end of the experiment when the animals were 6 months of age.
- Postweaning exposure group. Offspring were not exposed to maneb during the preweaning period but were started on the maneb containing diets after weaning and kept on these diets until the end of the experiment.

Preweaning exposure to maneb decreased the exploratory activity of 30-day-old weanling rats in a dose-related manner in two different tests. This decrease in exploratory activity was statistically significant at 0.5 ppm in one test and 10 ppm in the other test; insufficient animals were available for testing in the 1.0-ppm group. Learning ability was enhanced in an operant conditioning procedure in adult rats exposed to maneb from the preweaning and lifetime exposure groups, but not in the postweaning exposure group. The authors suggested the following explanation for this finding of enhanced learning ability: ''. . . the presence of maneb during early life of the infants may have effectively created a state of stress. Maneb was found capable of increasing adrenocortical activity under certain conditions. Moderate stress during infancy has repeatedly been found to enhance adult learning ability.''

Neuroendocrine effects unrelated to the behavioral changes were also found. Regional brain cholinesterase activity was decreased in rats exposed during the preweaning and postweaning periods, but not in rats exposed during their whole lifetime (up to age 6 months). In addition, adult plasma corticosterone levels were increased in the postweaning exposure group, but not in the other two exposure groups. No obvious toxicity in the form of clinical signs or altered body or organ weights occurred in any of the maneb-exposed animals. A LOAEL of 0.5 ppm maneb in the diet for certain types of behavioral and neuroendocrine effects is suggested by this study.

In addition to behavioral and neuroendocrine effects, it has also been suggested that maneb may exert effects on the immune system. It has been reported that rats exposed to maneb 5 times a week for 4.5 months at a level of 150 mg/kg experienced a decreased resistance to infection[12] (cited in IARC[2]).

Like ETU, maneb has been shown to produce subtle alterations in thyroid function and thyroid hyperplasia. Based on several subchronic studies in rats, a NOAEL of 100 ppm maneb in the diet has been suggested for these effects.[1] It is interesting to note that in a study in which rats were given either ETU (concentration: 5–500 ppm) or maneb (concentration: 250–1500 ppm) in the diet, results obtained indicated that ETU as a contaminant of maneb alone could not account for the effects of maneb on the thyroid, and ETU or other antithyroid metabolites of ETU also contribute to the observed effects.[1,13]

Data on the inhalation toxicity of maneb are very limited. Matokhnyuk[14] (cited in Lu and Kennedy[15]) reported that rats exposed to 30 mg/m^3 maneb (hours per day not specified) for one month experienced decreased body weight gain. After a single exposure to 700 mg/m^3, lethargy, humped back, rough coat, tremors, and ataxia were observed.

Basis for the AALG: Because of its use as an agricultural fungicide, the major source of exposure for the general population is likely to be the diet.[16] Therefore, 20% of the contribution to total exposure will be allowed from air. It should be noted that a larger proportion of exposure might occur via inhalation in certain population subgroups such as pesticide applicators or mixer/loaders or individuals living in the immediate vicinity of application areas.

The profile of toxicologic effects associated with maneb is sufficiently complex and, in certain instances, conflicting to justify calculation of limits based on several different endpoints. However, it should be noted that since no inhalation data for maneb exist that are suitable for criteria derivation, any limit derived here is likely to have an unusually high degree of uncertainty associated with it and should be considered provisional.

Although some studies exist that provide evidence suggestive of carcinogenic effects and some degree of genotoxicity associated with maneb, negative results were reported in the one study[4] (cited in IARC[2]) which most closely conformed to generally accepted guidelines for carcinogenicity testing, e.g., close to lifetime exposure and exposure at or near the MTD. Given the conflicting results and limitations of the studies designed to assess maneb carcinogenicity, no sound basis exists for calculation of an AALG based on carcinogenicity for maneb at this time. However, since ETU, a known contaminant, degradation product, and metabolite of maneb, is carcinogenic in laboratory animals, consideration might be given to calculating an AALG based on carcinogenicity if sufficiently detailed exposure information were available to demonstrate a carcinogenic risk for maneb based on ETU exposure.

Because exposure to maneb via ingestion has been found to induce terata in rats in at least two studies, this endpoint could serve as a suitable basis for criteria derivation. In the Larsson et al. study,[8] 400 mg/kg as a single oral dose by gavage on day 11 of gestation was defined as a NOAEL for malformations and as a LOAEL for increased incidence of resorptions. In the study by Petrova-Vergieva and Ivanova-Tchemishanska,[7] 500 mg/kg as a single oral dose on day 11 or 13 of gestation was established as the NOAEL for malformations, but there was no significant increase in resorptions at this dose. Petrova-Vergieva and Ivanova-Tchemishanska also found no increase in malformations or evidence of fetotoxicity when rats (n = approximately 10/group) were treated with 125, 250, or 500 mg/kg/day maneb on gestation days 2 through 21, indicating that maneb did not exert a cumulative effect. Since no increase in resorptions was found in this study at 500 mg/kg, the validity of this finding in the other study may be open to question. Therefore, 500 mg/kg was taken as a NOAEL for malformations in rats for purposes of AALG derivation. There was no adjustment for continuous exposure because further studies indicated that the effect was not cumulative. This NOAEL was converted to a human equivalent inhalation exposure level according to the procedures described in Appendix C, and a total uncertainty factor of 1000 was applied (10 for interindividual variation × 10 for interspecies variation × 5 for the developmental uncertainty factor × 2 for the database factor). The developmental uncertainty factor was applied because no evidence of maternal toxicity was reported in either of the studies at levels producing an obvious increase in malformations. Application of the database factor was justified based on use of the non-inhalation route of exposure.

Two different endpoints indicative of systemic toxicity are available for maneb: effects on the thyroid gland and behavioral alterations. Maneb—in common with other EBDC fungicides, e.g., zineb and mancozeb, as well as ETU—has been shown to induce thyroid hyperplasia in several subchronic feeding studies in rats. At least three such studies are cited by the Safe Drinking Water Committee;[1] however, none of the original reports were available because they were either published in the foreign literature with no translation available (one case) or were private communications to the U.S. EPA (the other two cases). However, since the same NOAEL of 100 ppm maneb in the diet of rats (estimated to correspond to a dose of 5 mg/kg/day) for thyroid hyperplasia was reported for all three studies, this NOAEL may reasonably serve as a starting point for AALG derivation. It was converted to a human equivalent inhalation exposure level according to the procedures described in Appendix C, and a total uncertainty factor of 2000 was applied (10 for interspecies variation × 10 for interindividual variation × 10 for a less-than-chronic study × 2 for the database factor). Application of the database factor was considered appropriate based on use of the non-inhalation route of exposure.

An AALG for maneb could also be calculated based on neuroendocrine and behavioral effects observed in rats in the Sobotka et al. study.[11] The LOAEL of 0.5 ppm in the diet was converted to a figure in mg/kg/day based on the methods used by the U.S. EPA, described in Appendix C. Because no weight data were given in the paper, a weight of 100 g for these rats was assumed based on their age. Since these effects were demonstrated in young animals and may be significant for human children, the animal LOAEL was converted to a human equivalent LOAEL for a 20-kg human child breathing 6.9 m^3/day, and a total uncertainty factor of 1000 was applied (10 for interindividual variation × 10 for interspecies variation × 5 for a LOAEL × 2 for the database factor). Application of the database factor was considered appropriate based on use of the non-inhalation route of exposure.

Although the neuroendocrine and behavioral effects observed in the study of Sobotka et al.[11] are the most sensitive endpoints available, interpretation and extrapolation of these findings to humans is difficult. The main reason for the difficulty is that the behavioral changes observed were not correlated with pathologic or biochemical lesions and hence it is uncertain whether the changes observed represent a form of permanent injury, temporary reversible effects, or merely an adaptive response that is not truly indicative of injury. For these reasons, the AALG based on thyroid hyperplasia is recommended for general use, but more comprehensive investigation of the potential for maneb exposure to cause neurobehavioral toxicity is recommended.

These AALGs should be considered provisional since they are derived from non-inhalation data and there is a potential oncogenic hazard from exposure to maneb.

AALG: • developmental toxicity—300 μg/m^3 24-hour TWA
• behavioral/neuroendocrine effects—0.0067 μg/m^3 24-hour TWA
• thyroid hyperplasia—1.75 μg/m^3 24-hour TWA

Note: The EBDC pesticides, including maneb, are currently the subject of an EPA Special Review with particular reference to potential oncogenic and teratogenic hazards.[16] It is strongly recommended that this compound be reevaluated when the results of this review become available.

References

1. Safe Drinking Water Committee. 1977. *Drinking Water and Health,* Vol. 1 (Washington, DC: National Academy Press), pp. 650–60.
2. IARC. 1976. "Maneb." *IARC Monog.* 12:137–49.
3. Balin, P.N. 1970. "Experimental Data on the Blastomogenic Activity of the Fungicide Maneb." *Vrach. Delo.* 4:21–24.
4. Innes, J.R.M., B.M. Ulland, M.G. Valerio, L. Petrucelli, L. Fishbein, E.R. Hart, A.J. Pallotta, R.R. Bates, H.L. Falk, J.J. Gart, M. Klein, I. Mitchell, and J. Peters. 1969. "Bioassay of Pesticides and Industrial Chemicals for Tumorigenicity in Mice: A Preliminary Note." *J. Natl. Cancer Inst.* 42:1101–14.
5. NTIS. 1968. "Evaluation of Carcinogenic, Teratogenic, and Mutagenic Activities of Selected Pesticides and Industrial Chemicals, Vol. 1, Carcinogenic Study." U.S. Department of Commerce, Washington, DC.

6. NIOSH. 1988. OP0700000. "Manganese, (ethylenebis(dithio- carbamato))." RTECS, on line.

7. Petrova-Vergieva, T., and L. Ivanova-Tchemishanska. 1973. "Assessment of the Teratogenic Activity of Dithiocarbamate Fungicides." *Fd. Cosmet. Tox.* 11:239–44.

8. Larsson, K.S., C. Arnander, E. Cekanova, and M. Kjellberg. 1976. "Studies of Teratogenic Effects of the Dithiocarbamates Maneb, Mancozeb, and Propineb." *Teratology* 14:171–84.

9. Chernoff, N., R.J. Kaviock, E.H. Rogers, B.D. Carver, and S. Murray. 1979. "Perinatal Toxicity of Maneb, Ethylene Thiourea, and Ethylenebisisothiocyanate Sulfide in Rodents." *J. Tox. Env. Heal.* 5:821–34.

10. Marcon, L.V. 1969. "The Effect of Maneb on the Embryonal Development of Generative Function of Rats." *Farmakol. i Toksikol.* 32:731–32.

11. Sobotka, T.J., R.E. Brodie, and M.P. Cook. 1972. "Behavioral and Neuroendocrine Effects in Rats of Postnatal Exposure to Low Dietary Levels of Maneb." *Devel. Psychobiol.* 5:137–48.

12. Olefir, A.I. 1973. "Effect of Chronic Carbamate Pesticide Poisoning on Immunologic Reactivity and Infection Resistance." *Vrach. Delo.* 8:137–40. *Pesticide Abstracts* 7:95, 74–0429.

13. Sobotka, T. 1971. "Comparative Effects of 60-Day Feeding of Maneb and ETU on Thyroid Electrophoretic Patterns in Rats." *Fd. Cosmet. Tox.* 9:537–40.

14. Matokhnyuk, L.A. 1971. "Toxicity of the Fungicide Maneb on Inhalation." *Hyg. Sanit.* 36:195–99.

15. Lu, M-H., and G.L. Kennedy. 1986. "Teratogenic Evaluation of Mancozeb in the Rat Following Inhalation Exposure." *Tox. App. Pharm.* 84:355–68.

16. U.S. EPA. 1987. "EBDC Pesticides; Initiation of Special Review." *Fed. Reg.* 53: 27172–177.

MANGANESE AND COMPOUNDS

Synonyms: "Manganese and compounds" includes dust and compounds of inorganic manganese including manganese dioxide (MnO_2, CAS 1313–13–9) and manganese sulfate ($MnSO_4$, CAS 7785–87–7)

CAS Registry Number: 7439–96–5

Molecular Weight: 54.94 (MnO_2, 86.94; $MnSO_4$, 151.00)

Molecular Formula: Mn

AALG:
- Manganese and compounds, including $MnSO_4$ (systemic toxicity)—0.012 μg/m³ 24-hour TWA
- Manganese dioxide (systemic toxicity)—0.154 μg/m³ 24-hour TWA

Occupational Limits:
- ACGIH TLV: dust and compounds—5 mg (Mn)/m³ TWA; fume—1 mg (Mn)/m³ TWA, 3 mg (Mn)/m³ STEL
- NIOSH REL: none
- OSHA PEL: dust and compounds—5 mg (Mn)/m³ TWA; fume—1 mg (Mn)/m³ TWA, 3 mg (Mn)/m³ STEL

Basis for Occupational Limits: The occupational limits are based on protection against chronic manganism, a disease affecting the central nervous system and characterized by languor, weakness in the legs, masklike face, slow monotonous voice, muscular twitching, and parkinsonian gait. It is noted in the TLV documentation that workers exposed to manganese fume develop signs of manganism at exposure levels below those causing the development of manganism in workers exposed to manganese dust.[1] OSHA recently adopted the ACGIH TLVs as part of its final rule limits.[2]

Drinking Water Limits: In a review of manganese, the Safe Drinking Water Committee stated that the major importance of manganese in drinking water relates to undesirable taste and discoloration rather than toxicity. Such problems are thought to arise at concentrations greater than 0.05 mg/L, which was also the limit recommended by the U.S. Public Health Service in 1962.[3]

Toxicity Profile

Carcinogenicity: Data on the carcinogenicity of manganese to humans are lacking, and studies in animals are generally limited to parenteral routes of administration. Intraperitoneal administration of manganese sulfate using the Strain A mouse lung tumor system produced results suggestive (rather than indicative) of carcinogenic activity.[4] Manganese sulfate is currently under evaluation in the NTP bioassay program (rats and mice, in feed) with histopathology in progress.[5] Manganese dichloride was reported to cause an increased incidence of lymphosarcomas in DBA/1 mice following multiple subcutaneous or intraperitoneal injections; however, the reporting of the data was limited. In other studies, intramuscular injection of MnO_2 or manganese powder did not produce a carcinogenic response in male and female F344 rats or female Swiss mice, nor was there a positive response in a study in which F344 rats were given manganese powder by the oral route.[4] Manganese and its compounds are most appropriately placed in group D, unclassified substance, under the EPA weight-of-evidence classification.[6]

Mutagenicity: In the HAD for manganese,[4] it was concluded that "data are both insufficient and inadequate at this time [1984] to reach a conclusion about the mutagenic potential of manganese." In the EPA Gene-Tox database as summarized in RTECS, positive results are listed for the *B. subtilis* rec assay and the *S. cerevisiae* reversion assay, but a negative response is listed for the *S. cerevisiae* gene conversion assay for manganese sulfate.[7]

Developmental Toxicity: Based on two studies available only as abstracts, prenatal exposure to manganese may result in behavioral abnormalities in the adult offspring of mice. Of particular interest (because of the route of exposure) is the study of Massaro et al.[8] (cited in U.S. EPA[4]) in which female mice were exposed to 0 or 48.9 ± 7.5 mg/m^3 MnO_2 dust continuously on days 0 through 18 of gestation. As adults, pups from exposed mothers were deficient in open-field, exploratory, and rotarod (balance and coordination) behavioral performance tests, and normal offspring fostered to exposed mothers also had decreased rotarod performance (indicating adverse effects due to postpartum exposure). It is relevant to note that manganese administered as manganese sulfate in a single intraperitoneal dose to QS mice has been reported to cause neural tube defects.[9]

In the HAD,[4] it is noted that manganese deficiency during gestation in experimental animals results in a variety of developmental defects, including reduced coordination, bone and growth deficiencies, reproductive difficulties, and C.N.S. changes, which are a consequence of decreased chondroitin sulfate formation and delayed otolith calcification.

Reproductive Toxicity: The reproductive toxicity of manganese is summarized as follows in the HAD:[4] "Reports of impotence in a majority of patients with chronic manganese poisoning are common; however, no other supporting human data are available. Existing animal data addressing reproductive failure in males describe long-term dietary exposure to manganese. Results show that dietary levels up to 1004 ppm as $MnSO_4 \cdot 7H_2O$ and up to 3550 ppm as Mn_3O_4 were almost without effect on reproductive performance. However, these and other observations need to be verified using well-defined reproductive testing protocols.''

Systemic Toxicity: *Essentiality*: Manganese has been identified as an essential element in laboratory animals and presumably humans (although no well-defined deficiency syndromes have been documented). The concentration of manganese in various tissues is controlled by well-developed homeostatic mechanisms, and the major effects associated with manganese deficiency include skeletal abnormalities, impaired growth, ataxia in neonates, and defects in lipid and carbohydrate metabolism.[4]

The major effects of manganese exposure in humans involve the C.N.S. and respiratory tract. The neurological syndrome resulting from chronic manganese exposure (usually in an occupational setting) has been termed *manganism* and is characterized by psychiatric disturbances during the early stages, followed by neurologic signs similar to those seen in Parkinson's disease (e.g., muscular weakness, tremors, awkward speech, and loss of skilled movement). The correlation between exposure and response for the neurological effects of manganese and the limitations of this data are described in the HAD:[4]

These reports [of single cases or clusters of cases] include no longitudinal studies and are therefore not adequate to identify a dose-response relationship, but do permit the identification of the lowest-observed effect level (LOEL). The full clinical picture of chronic manganese poisoning is reported less frequently at exposure levels below 5 mg/m^3. The reports of a few early signs of manganism in workers exposed to 0.3 to 5 mg/m^3 suggest 0.3 mg/m^3 (300 $\mu g/m^3$) as the LOEL. The data available for identifying effect levels below 0.3 mg/m_3 is equivocal or inadequate. This is further complicated by the fact that good biological indicators of manganese exposure are not presently available. Also, there are neither human nor animal data suggesting the rate of absorption of manganese through the lung; therefore, extrapolating from other routes of exposure would be difficult.

It has generally proved difficult to induce diseases similar to chronic manganism in laboratory animals exposed via ingestion. In a study in which monkeys and rats were exposed to 0.0116, 0.1125, or 1.152 mg/m^3 Mn_3O_4 continuously for 9 months, no treatment-related effects on C.N.S. or pulmonary function, EMG, clinical chemistry, hematology, or histology were reported, although tissue levels of manganese were elevated[10,11,12] (cited in U.S. EPA[4,13]). In other studies, Nishiyama et al.[14,15] (cited in U.S. EPA[13]) exposed monkeys to 0.7 or 3.0 mg/m^3 MnO_2 for 22 hours per day for 5 or 10 months. Pulmonary congestion developed in both groups of animals exposed for 5 months but was less severe and occurred later in the lower exposure group. In animals exposed for 10 months to 3 mg/m^3, there were elevated levels of SGOT, SGPT, monoamine oxidase, calcium, and manganese relative to controls; in addition, two of three monkeys developed mild tremors of the fingers and decreased pinch force as well as reduced dexterity of the upper limbs. No such neurological signs were noted in the low-dose group exposed for 10 months, and clinical chemistries and tissue levels of manganese were only slightly affected. In a subsequent report[16] (cited in U.S. EPA[13]), it was indicated that "pathologic changes" occurred in the lungs of low-dose monkeys exposed for 10 months. In the Health Effects Assessment Document,[13] this is stated to be the lowest reported LOAEL (0.7 mg/m^3) for pulmonary effects. In spite of the difficulty in inducing effects similar to chronic manganism in laboratory animals, it should be noted that animal studies do support the hypothesis that manganese neurotoxicity is attributable to altered neurotransmitter metabolism involving the central dopaminergic system.[4]

Available human and animal data seem to indicate that pulmonary toxicity is the critical effect (the adverse health effect that occurs at the lowest level of exposure) for manganese. Saric and Lucic-Palaic[17] (cited in U.S. EPA[4]) reported an increased prevalence of chronic bronchitis in workers exposed to 0.4–16 mg/m^3 manganese relative to controls, but not in workers exposed to 0.005–0.04 mg/m^3 manganese. The conclusions that may be drawn from this study are limited on account of the broad range of exposure levels and the effects of simultaneous exposure to other substances such as silica. Nogawa et al.[18] (cited in U.S. EPA[4]) evaluated ventilatory function and the occurrence of subjective symptoms in two groups of children: one group of 1258 children attending a school 100 m from a ferromanganese plant and a similar group of 648 students 7 km away from the plant. The EPA estimated the emissions from the ferromanganese plant to correspond to exposure levels of 3–11 μg/m^3 (0.003–0.011 mg/m^3). An increased prevalence of respiratory symptoms, including sputum, wheezing, and sore throat, and lower mean values on objective tests of lung function were reported in children at the school closer to the plant. This association is considered fairly strong due to the large numbers of children involved, high participation rate (over 97%), and documented monitoring of settled manganese over several years. Studies in laboratory animals indicate that the pulmonary effects of manganese may be attributable to disruption of normal mechanisms of lung clearance.[4]

Irritation: Manganese is not generally recognized as an irritant.

Basis for the AALG: Manganese is used primarily as an alloy with other metals and in the manufacture of steel to impart hardness.[1] Because manganese is ubiquitous in the environment and exposure is thought to occur mainly via the diet,[4] 20% of the contribution to total exposure is allowed from air.

Given that pulmonary toxicity is the critical effect for manganese, this is the endpoint on which the AALG is based. Pending results of the NTP carcinogenesis bioassay, it is recommended that manganese sulfate (MnSO$_4$) be included with manganese and compounds. The limit for manganese and compounds is based on the study of Nogawa et al.,[18] taking the lower estimated exposure limit of 3 μg/m^3 as the LOAEL, and the limit for manganese dioxide is based on the study of

Nishiyama et al.,[14,15] taking 0.7 mg/m³ (10-month exposure) as the LOAEL. A total uncertainty factor of 50 was used with the LOAEL of 3 µg/m³ (5 for a LOAEL × 10 for interindividual variation). The LOAEL of 0.7 mg/m³ was converted to a human exposure level according to the procedures described in Appendix C, and a total uncertainty factor of 5000 was applied (5 for a LOAEL × 10 for interspecies variation × 10 for interindividual variation × 10 for a less-than-lifetime study). It is recommended that the more conservative limit for manganese and compounds also be used for manganese dioxide until longer-term inhalation studies defining a NOAEL become available.

AALG: • Manganese and compounds, including MnSO₄ (systemic toxicity)— 0.012 µg/m³ 24-hour TWA
• Manganese dioxide (systemic toxicity)—0.154 µg/m³ 24-hour TWA

References

1. ACGIH. 1986. "Documentation of the Threshold Limit Values and Biological Exposure Indices." 5:354–55.
2. OSHA. 1989. "Air Contaminants; Final Rule." *Fed. Reg.* 54:2332–2959.
3. Safe Drinking Water Committee. 1977. *Drinking Water and Health,* Vol. 1 (Washington, DC: National Academy Press).
4. U.S. EPA. 1984. "Health Assessment Document for Manganese." EPA-600/8–83–013F.
5. NTP. 1988. "Management Status Report." National Toxicology Program, Research Triangle Park, NC (1/21).
6. U.S. EPA. 1988. "Manganese; CASRN 7439–96–5." IRIS (12/1/88).
7. NIOSH. 1988. OP1050000. "Manganese (II) Sulfate (1:1)." RTECS, on line.
8. Massaro, E.J., R.B. D'Agostino, C. Stineman, J.B. Morganti, and B.A. Lown. 1980. "Alterations in Behavior of Adult Offspring of Female Mice Exposed to MnO₂ Dust during Gestation." *Fed. Proc.* 39:623.
9. Webster, W.S. 1986. "Manganese and Neural Tube Defects." *Teratology* 33:6B.
10. Ulrich, C.E., W. Rinehart, and W. Busey. 1979. "Evaluation of the Chronic Inhalation Toxicity of a Manganese Oxide Aerosol. I. Introduction, Experimental Design, and Aerosol Generation Methods." *Am. Ind. Hyg. Assoc. J.* 40:238–44.
11. Ulrich, C.E., W. Rinehart, W. Busey, and M.A. Dorato. 1979. "Evaluation of the Chronic Inhalation Toxicity of a Manganese Aerosol. II. Clinical Observations, Hematology, Clinical Chemistry, and Histopathology." *Am. Ind. Hyg. Assoc. J.* 40:322–29.
12. Ulrich, C.E., W. Rinehart, and M. Brandt. 1979. "Evaluation of the Chronic Inhalation Toxicity of a Manganese Aerosol. III. Pulmonary Function, Electromyograms, Limb Tremor, and Tissue Manganese Data." *Am. Ind. Hyg. Assoc. J.* 40:349–53.
13. U.S. EPA. 1984. "Health Effects Assessment for Manganese (and Compounds)." EPA/540/1–86/057.
14. Nishiyama, K., Y. Suzuki, N. Fujii, H. Yano, T. Mikai, and K. Ohmishi. 1975. "Effect of Long-Term Inhalation of Manganese Dusts. II. Continuous Observation of the Respiratory Organs in Monkeys and Mice." *Jap. J. Hyg.* 30:117.
15. Nishiyama, K., Y. Suzuki, N. Fujii, H. Yano, K. Ohnishi, and T. Miyai. 1977. "Bio-

chemical Changes and Manganese Distribution in Monkeys Exposed to Manganese Dioxide Dust.'' *Tokushima J. Exp. Med.* 24:137.

16. Suzuki, Y., N. Fujii, H. Yano, T. Ohkita, A. Ichikawa, and K. Nishiyama. 1978. ''Effects of the Inhalation of Manganese Dioxide Dust on Monkey Lungs.'' *Tokushima J. Exp. Med.* 25:119–25.

17. Saric, M., and S. Lucic-Palaic. 1977. ''Possible Synergism of Exposure to Airborne Manganese and Smoking Habit in Occurrence of Respiratory Symptoms.'' In *Inhaled Particles IV*, W.H. Walton, Ed. (New York, NY: Pergamon Press), pp. 773–79.

18. Nogawa, K., E. Kobayashi, and M. Sakamoto. 1973. ''Epidemiological Studies on Disturbance of Respiratory System Caused by Manganese Air Pollution. Report 1: Effects on Respiratory System of Junior High School Students.'' *Jap. J. Pub. Health* 20:315–26.

MERCURY AND COMPOUNDS

Synonyms: "Mercury and compounds" includes mercury vapor, alkyl, aryl and inorganic mercury compounds

CAS Registry Number: 7439–97–6

Molecular Weight: 200.59 (Hg)

Molecular Formula: Hg

AALG: systemic toxicity (TLV-based)
- mercury vapor   0.95 μg (Hg)/m³ 8-hour TWA
- inorganic mercury 0.48 μg (Hg)/m³ 8-hour TWA
- alkyl mercury   0.05 μg (Hg)/m³ 8-hour TWA

Occupational Limits:
- ACGIH TLV: aryl and inorganic mercury compounds— 0.10 mg (Hg)/m³ TWA; mercury vapor—0.05 mg (Hg)/ m³ TWA; alkyl mercury—0.01 mg (Hg)/m³ TWA, 0.03 mg (Hg)/m³ STEL; skin notation (all forms)
- NIOSH REL: inorganic mercury—0.05 mg (Hg)/m³ TWA
- OSHA PEL: inorganic mercury—0.1 mg (Hg)/m³ ceiling; mercury vapor—0.05 mg (Hg)/m³ TWA; alkyl mercury—0.01 mg (Hg)/m³ TWA, 0.03 mg (Hg)/m³ STEL; skin notation (all forms)

Basis for Occupational Limits: All of the occupational limits are based on neurotoxicity, the most sensitive endpoint for mercury toxicity.[1-3] The rationale for the different TLVs for the various forms of mercury is the sequentially greater penetration of the inorganic and aryl compounds, vapor, and alkyl forms of mercury through the blood-brain barrier.[1] OSHA recently adopted the ACGIH limits for mercury vapor and alkyl mercury as part of its final rule limits.[4]

Drinking Water Limits: The ambient water quality criterion for mercury is 144 ng/L assuming consumption of 2 L of water and 6.5 g of fish/shellfish per day.[5] Since data were insufficient to calculate 1-day, 10-day, and longer-term HAs, it was recommended that the modified DWEL of 1.58 μg/L be used as a conservative approximation to these limits. Using the DWEL of 5.5 μg/L and a relative source contribution of 20%, a lifetime HA of 1.1 μg/L was calculated.[6] More recently, a lifetime HA of 2 μg/L was recommended for inorganic mercury, and an MCLG of 2 μg/L has been proposed for inorganic mercury.[7]

Toxicity Profile

Carcinogenicity: There is no evidence implicating mercury (all forms) as a carcinogen based on epidemiologic studies; however, the populations studied may have been too small to detect such effects.[8,9]

Mitsumori et al.[10] (cited in U.S. EPA[9]) exposed ICR mice to 0, 15, or 30 ppm methyl mercury in the diet for periods of up to 78 weeks. Excess mortality in the high-dose group precluded evaluation of carcinogenicity; however, in the low-dose group 13 of 16 males surviving after 53 weeks developed renal tumors (primarily adenocarcinomas) compared with 1 of 37 controls. No tumors were observed in the control or low-dose females. Mercuric chloride is currently under evaluation in the NTP bioassay program (histopathology in progress).[11] Mercury has been placed in group D, unclassified substance, under the EPA weight-of-evidence classification.[12]

Mutagenicity: Data on the mutagenicity of mercury are generally lacking, particularly for the inorganic and vapor forms. Methyl mercury has been reported to block mitosis in plant cells and human cells treated in vitro and to induce chromosome breaks in plant cells and mutations in *Drosophila*. In one study, a positive correlation was found between blood concentrations of methyl mercury and the frequency of chromosome breaks in lymphocytes of 23 Swedish subjects consuming high fish diets.[9]

Developmental Toxicity: Data on the developmental effects of inhaled mercury are generally lacking in both animals and humans. However, concerning ingestion of methyl mercury, usually from contaminated fish or seed grain, the U.S. EPA[8] concluded the following:

> Prenatal exposures at the lowest recorded levels produce signs of psychomotor retardation in infants. Recent studies indicate that male infants may be more sensitive than females at these low levels of maternal dietary intake. Substantially higher prenatal exposures, some only occurring in the last trimester of pregnancy, produce a severe form of cerebral palsy. Although detailed mechanisms of methyl mercury poisoning are not known, prenatal effects appear to be due to a derangement of the normal processes of growth and development of the central nervous system. Ongoing studies indicate (based on estimated blood levels in the mother during pregnancy) that the fetus is about three times more sensitive than the adult to methyl mercury exposures.

Reproductive Toxicity: Case reports and epidemiologic studies on occupationally-exposed women from several countries (not the U.S.) have indicated increased frequencies of menstrual disturbances and spontaneous abortions. However, insufficient details were presented in these reports to establish exposure-response patterns, and based on the occurrence of overt signs of mercury poisoning in some instances, exposure levels were probably fairly high.[8]

Systemic Toxicity: The types of systemic toxicity observed with mercury exposure are dependent on the chemical form: the three major forms of mercury discussed here are mercury vapor, inorganic mercurial compounds, and methyl mercury.

Based on occupational studies, the chronic effects of exposure to mercury vapor involve mainly the nervous system and kidney. At levels at or below 0.10 mg/m^3, nonspecific signs of neurotoxicity such as introversion, insomnia, and anxiety occur, along with increased urinary excretion of specific proteins and enzymes. At higher chronic levels of exposure, more pronounced neurologic effects such as short-term memory loss and changes in personality occur, and the neurotoxicity may progress to the point of overall body tremors indicative of C.N.S. damage. The nephrotoxic effects appear to be mediated by an autoimmune mechanism, and there are wide variations in individual susceptibility. No definitive thresholds have been established for the neurotoxicity associated with mercury vapor; however, such effects have not been reported at air concentrations of approximately 0.01 mg/m^3, and the early mild effects of mercury vapor exposure are reversible.[8]

Information on the effects of inhalation exposure to inorganic mercury compounds is very limited in humans. However, occupational exposure to mercuric oxide has been reported to cause peripheral nervous system damage and a syndrome similar to amyotrophic lateral sclerosis, which is reversible. Upon ingestion, mercurous (Hg^+) mercury is probably converted to the mercuric (Hg^{+2}) form, which is capable of penetrating the blood-brain and placental barriers (approximately 10 times less effectively than mercury vapor) and is accumulated in the kidney. It is known that chronically ingested mercuric compounds may produce signs typical of mercurialism in adults and acrodynia (a disease beginning with pains in the lower limbs and progressing to more general nervous and psychic disorders) in children.[8]

The primary effect of methyl mercury exposure is neurotoxicity, manifest mainly as sensory and coordination dysfunction, e.g., narrowing of visual fields, paresthesia, and ataxia, rather than the psychic changes associated with exposure to mercury vapor. Relative to other forms of mercury, methyl mercury is the most readily absorbed and penetrates the blood-brain and placental barriers most effectively.[8] It is noteworthy that the TLV for methyl mercury is based on Swedish industrial experience and on an occupational study in which consistent signs of poisoning were not seen at concentrations of between 0.01 and 0.1 mg/m^3.[12]

Irritation: Mercury vapor and other airborne mercury compounds are not generally regarded as irritants.

Basis for the AALG: Since atmospheric mercury is considered to account for almost all of the mercury vapor retained by humans each day and the bulk of methyl mercury and inorganic mercury exposure is thought to occur via the diet,[8] the relative source contribution for air will be 80% for mercury vapor and 20% for inorganic mercury and methyl (alkyl) mercury.

Three separate AALGs have been calculated for mercury: one for mercury vapor, one for inorganic mercury (Hg^+ and Hg^{+2} compounds), and another for alkyl (primarily methyl) mercury compounds. It was considered most reasonable to base all of these AALGs on the TLV/42 since the TLVs were set at levels that would be expected to be close to the NOAELs for neurotoxicity in humans, based on occupational studies available and on consideration of differences in absorption of the various forms of mercury. These AALGs should be considered provisional since they are based on occupational limits and more study is needed on the potential effects of airborne mercury on the developing fetus. It should be noted that the major source of methyl mercury, contaminated fish, is addressed via other regulatory measures.[8,12]

AALG: systemic toxicity (TLV-based)
- mercury vapor—0.95 μg (Hg)/m^3 8-hour TWA
- inorganic mercury—0.48 μg (Hg)/m^3 8-hour TWA
- alkyl mercury—0.05 μg (Hg)/m^3 8-hour TWA

Note: In 1973 the U.S. EPA promulgated a NESHAP (national emission standards for hazardous air pollutants) for mercury.[8] This NESHAP has been under review, and new requirements for monitoring, reporting, and testing were recently added.[13] ATSDR has recently published a draft "Toxicological Profile for Mercury."[14]

References

1. ACGIH. 1986. "Documentation of the Threshold Limit Values and Biological Exposure Indices." 5:358–60.
2. NIOSH. 1985. "NIOSH Recommendations for Occupational Safety and Health Standards." *Morb. Mort. Wkly. Rpt.* suppl. 34:5s-31s.
3. NIOSH. 1973. "Criteria for a Recommended Standard . . . Occupational Exposure to Inorganic Mercury." DHEW (NIOSH) 73–11024.
4. OSHA. 1989. "Air Contaminants; Final Rule." *Fed. Reg.* 54:2332–2959.
5. U.S. EPA. 1980. "Ambient Water Quality Criteria for Mercury." EPA-440/5–80–058.
6. U.S. EPA. 1987. "Health Advisory for Mercury." Office of Drinking Water (3/31).
7. U.S. EPA. 1988. "Drinking Water Criteria Document for Inorganic Mercury" (draft). U.S. EPA, Washington, DC.
8. U.S. EPA. 1984. "Mercury Health Effects Update." EPA-600/8–84–013F.
9. U.S. EPA. 1985. "Drinking Water Criteria Document for Mercury " (draft). U.S. EPA, Washington, DC.
10. Mitsumori, K., K. Maita, T. Saito, S. Tsuda, and Y. Shikasu. 1981. "Carcinogenicity of Methyl Mercury Chloride in ICR Mice: Preliminary Note on Renal Carcinogens." *Cancer Letters* 12:305–310.
11. NTP. 1988. "Management Status Report." National Toxicology Program, Research Triangle Park, NC (1/21/88).
12. U.S. EPA. 1984. "Health Effects Assessment for Mercury." EPA/540/1–86–042.
13. U.S. EPA. 1987. "National Emission Standards for Hazardous Air Pollutants; Review and Revision of the Standards for Mercury; Final Rule; Review." *Fed. Reg.* 52:8723–28.
14. ATSDR. 1989. "Toxicological Profile for Mercury" (draft). Agency for Toxic Substances and Disease Registry (4/14).

MESITYL OXIDE

Synonyms: 4-methyl-3-pentene-2-one, isobutenyl methyl ketone, isopropylidene acetone

CAS Registry Number: 141–79–7

Molecular Weight: 98.14

Molecular Formula: $(CH_3)_2C = CHOCH_3$

AALG: systemic toxicity—0.12 ppm (0.48 mg/m^3) 8-hour TWA

Occupational Limits:
- ACGIH TLV: 15 ppm (60 mg/m^3) TWA; 25 ppm (100 mg/m^3) STEL
- NIOSH REL: 10 ppm (45 mg/m^3) TWA
- OSHA PEL: 15 ppm (60 mg/m^3) TWA; 25 ppm (100 mg/m^3) STEL

Basis for Occupational Limits: The ACGIH limits are based on irritant effects and the greater systemic toxicity of mesityl oxide compared with other saturated ketones.[1] The NIOSH limit is also based on irritant effects, and it is noted in the criteria document that ". . . there is reason to believe that eye irritation by mesityl oxide is more serious than that produced by the lower ketones."[2] OSHA recently adopted the TLV-TWA and TLV-STEL as part of its final rule limits.[3]

Drinking Water Limits: No drinking water limits were found for this substance.

Toxicity Profile

Carcinogenicity: No data were found implicating mesityl oxide as a carcinogen.

Mutagenicity: No data were found implicating mesityl oxide as a mutagen.

Developmental Toxicity: No data were found implicating mesityl oxide as a developmental toxin.

Reproductive Toxicity: No data were found implicating mesityl oxide as a reproductive toxin.

Systemic Toxicity: Data on the inhalation toxicity of mesityl oxide in animal models are limited. Based on LC_{50} values reported in RTECS—9000 mg/m^3 (4-hour) in rats and 10,000 mg/m^3 (2-hour) in mice—the acute toxicity of mesityl oxide is relatively low.[4] However, it is stated in the TLV documentation[1] that "the acute toxicity of mesityl oxide is considerably greater than that of comparable saturated ketones."

Smyth et al.[5] exposed male Wistar rats and male and female guinea pigs (n = 10/species/group) to 0, 50, 100, 250, or 500 ppm mesityl oxide (commercial grade) via inhalation for 8 hours per day, 5 days a week for 6 weeks. In general, the responses of the rats and guinea pigs were similar, although the rats were slightly more sensitive. Exposure to mesityl oxide at the highest concentration (only 10 exposures conducted) was associated with high mortality and pathology consisting of congestion and cloudy swelling of the kidney, liver congestion somewhat less often, and "light" congestion of the lungs. The authors concluded that death due to repeated exposure to mesityl oxide occurs mainly as a consequence of its "anesthetic action upon circulation and respiration." Exposure to 250 ppm mesityl oxide resulted in nose and eye irritation, poor growth, and albumin in the urine, and exposure at 100 or 250 ppm also resulted in histological changes in the liver and kidneys as described above. No effects were detected at 50 ppm.

Irritation: Silverman et al.[6] tested the sensory response of humans (n = 12) to a variety of organic vapors during exposures of 15 minutes, using the subjects as their own controls. The concentrations were determined nominally. Eye irritation was reported at 25 ppm and nasal irritation at 50 ppm. Many subjects also reported an unpleasant taste lasting 3–6 hours.

Mesityl oxide is reported to have a peppermintlike smell[1] and an odor threshold of 0.45 ppm (geometric mean of reported literature values).[7]

Basis for the AALG: Mesityl oxide is considered slightly soluble in water and has a vapor pressure of 8 mm Hg at 20°C. It is used as a solvent for a number of substances (cellulose esters and ethers, gums, resins, lacquers, inks, stains), in ore flotation and paint removers, and as an insect repellant. Based on the lack of data on the environmental fate and distribution and the expected contribution of various media to human exposure, 50% of the contribution to total exposure will be allowed from air.

Mesityl oxide has not been adequately evaluated for potential carcinogenic, mutagenic, or developmental effects, and data on its systemic toxicity are also limited. Based on results from limited human and animal studies, it appears that the occupational limits fall in the range of a human NOAEL for the working population. Therefore, the NIOSH REL (since it is more conservative) was selected as the basis for the AALG, which should be considered provisional. The REL was divided by 42 (4.2 for continuous exposure adjustment × 10 for interindividual variation), and a 50% relative source contribution was applied.

AALG: • systemic toxicity—0.12 ppm (0.48 mg/m^3) 8-hour TWA

References

1. ACGIH. 1986. "Documentation of the Threshold Limit Values and Biological Exposure Indices." 5:361.
2. NIOSH. 1978. "Criteria for a Recommended Standard . . . Occupational Exposure to Ketones." DHEW (NIOSH) 78–173.
3. OSHA. 1989. "Air Contaminants; Final Rule." *Fed. Reg.* 54:2332–2959.
4. NIOSH. 1988. SB4200000. "3-Penten-2-one, 4-Methyl." RTECS, on line.
5. Smyth, H.F., J. Seaton, and L. Fischer. 1942. "Response of Guinea Pigs and Rats to Repeated Inhalation of Vapors of Mesityl Oxide and Isophorone." *J. Ind. Hyg. Tox.* 24:46–50.
6. Silverman, L., H.F. Schulte, and M.W. First. 1946. "Further Studies on Sensory Response to Certain Industrial Solvent Vapors." *J. Ind. Hyg. Tox.* 28:262–66.
7. Amoore, J.E., and E. Hautala. 1983. "Odor as an Aid to Chemical Safety: Odor Thresholds Compared with Threshold Limit Values and Volatilities for 214 Industrial Chemicals in Air and Water Dilution." *J. Appl. Tox.* 3:272–90.

METHYL ACRYLATE

Synonyms: methyl propenate

CAS Registry Number: 96–33–3

Molecular Weight: 86.10

Molecular Formula: CHCOOCH$_3$

AALG: irritation (TLV/10)—1 ppm (3.5 mg/m^3) 8-hour TWA

Occupational Limits:
- ACGIH TLV: 10 ppm (35 mg/m^3) TWA; skin notation
- NIOSH REL: none
- OSHA PEL: 10 ppm (35 mg/m^3) TWA; skin notation

Basis for Occupational Limits: The occupational limits are based on prevention of lacrimation and other primary irritant effects.[1]

Drinking Water Limits: No drinking water limits were found for this substance.

Toxicity Profile

Carcinogenicity: In a study available only as an abstract, Klimisch and Reininghaus[2] exposed male and female Sprague-Dawley rats (n = 86/sex/group) to 0, 15, 45, or 135 ppm methyl acrylate (purity unspecified) via inhalation for 6 hours per day, 5 days a week for 2 years. Other than a temporary retardation of body weight gain at the highest exposure level, no systemic toxicity or tumorigenic effects were reported. The irritant properties of methyl acrylate were manifest as a dose-dependent atrophy of the neurogenic portion of the olfactory nasal epithelium with proliferation of the reserve cells to a multilayered epithelium. These changes were primarily manifest at the transition of respiratory to olfactory epithelium at the anterior portion of the olfactory mucosa. In addition, ocular effects in the form of dose-related corneal opacity and vascularization of the eyes were reported.

Given the limited reporting of results for this study and a lack of other evidence, IARC concluded that there was inadequate evidence of carcinogenicity of methyl acrylate to experimental animals.[3] Methyl acrylate is most appropriately placed in group D ("unclassified substance") under the EPA weight-of-evidence classification.

415

Mutagenicity: Methyl acrylate has been reported to be nonmutagenic in several strains of *E. coli* and *S. typhimurium* under various conditions both with and without the addition of an exogenous metabolic activation system. However, methyl acrylate has been reported to induce chromosomal aberrations in cultured Chinese hamster lung cells with the addition of an exogenous metabolic activation system. An increase in the number of micronuclei in polychromic erythrocytes from bone marrow was also observed in mice given methyl acrylate by both gavage and intraperitoneal injection. IARC considered the degree of evidence for genetic activity from short-term tests to be limited.[3]

Developmental Toxicity: No data were found implicating methyl acrylate as a developmental toxin.

Reproductive Toxicity: No data were found implicating methyl acrylate as a reproductive toxin.

Systemic Toxicity: Oral LD_{50}s of 826, 227, and 200 mg/kg have been reported in mice, rats, and rabbits, respectively. LC_{50} values (time not specified) of 12,800 and 7300 mg/m^3 have been reported in mice and rats, respectively.[3]

With the exception of the Klimisch and Reininghaus[2] study reviewed in the carcinogenicity section, no chronic or subchronic studies of methyl acrylate toxicity in laboratory animals were found. Treon et al.[4] exposed rats, guinea pigs, and rabbits (n = 2 to 5 animals/group) to 31, 95, 237, or 578 ppm methyl acrylate for varying lengths of time over a period of 7 days. At the two highest exposure levels, all rabbits and guinea pigs died and the signs observed were consistent with severe ocular and respiratory irritation. Exposure to 95 ppm methyl acrylate produced signs of slight irritation in the rabbits and no effects in the other species, and 31 ppm methyl acrylate produced no apparent effects in any of the species studied under the conditions of this exposure.

In a study designed to evaluate structure-toxicity relationships among acrylate esters,[5] it was reported that a single dose of 86 or 172 mg/kg methyl acrylate in corn oil via gavage to F344 rats resulted in a dose-dependent gastric toxicity in the form of intra- and intercellular mucosal and submucosal edema and superficial mucosal necrosis.

Irritation: Methyl acrylate is reported to be highly irritating to the skin and mucous membranes in occupationally exposed individuals (no exposure levels given).[1,3] It has also been judged to produce moderate to severe eye and skin irritation in standard animal tests (RTECS).[6]

Methyl acrylate is reported to have a ''sweet, but sharp, fruity'' odor[1] and an odor threshold of 14 ppb.[3]

Basis for the AALG: Methyl acrylate is volatile (vapor pressure 68.2 mm Hg at 20°C) and slightly soluble in water. It is used in the preparation of polymers, amphoteric surfactants, and vitamin B_1.[1,3] Since the AALG for methyl acrylate is based on irritant effects, no relative source contribution is factored into the calculation. The reason is that allocation of a proportion of exposure to a given source is not relevant when the effects of exposure are not cumulative, as is the case with irritant effects.

Based on consideration of the entire database available for methyl acrylate, which unfortunately is quite limited, it appears that this compound exerts its major effects at the portal of entry, as would be expected of a potent primary irritant. Thus, based on the available data, the AALG for methyl acrylate should be based on irritation. The TLV/10 was considered the most suitable basis for derivation of the AALG, since it is based on industrial experience and is long-standing. The study of Klimisch and Reininghaus[2] was considered unsuitable as a basis for AALG derivation because the limited reporting of the study precluded identification of a LOAEL or NOAEL. This AALG should be considered provisional due to the limited data available for this compound and it is strongly recommended that periodic review of the literature be conducted to update this assessment.

AALG: • irritation (TLV/10)—1 ppm (3.5 mg/m^3) 8-hour TWA

References

1. ACGIH. 1986. "Documentation of the Threshold Limit Values and Biological Exposure Indices." 5:369.
2. Klimisch, H. J., and W. Reininghaus. 1984. "Carcinogenicity of Acrylates: Long-Term Inhalation Studies on Methyl Acrylate and *n*-Butyl Acrylate in Rats." *Toxicologist* 4:53, abstract no. 211.
3. IARC. 1986. "Methyl Acrylate." *IARC Monog.* 39:99–112.
4. Treon, J. F., H. Sigmon, H. Wright, and K. V. Kitzmiller. 1949. "The Toxicity of Methyl and Ethyl Acrylate." *J. Ind. Hyg. Tox.* 31:317–26.
5. Ghanayem, B. I., R. R. Maronpot, and H. B. Matthews. 1985. "Ethyl Acrylate–Induced Gastric Toxicity. II. Structure-Toxicity Relationships and Mechanism." *Tox. Appl. Pharm.* 80:336–44.
6. NIOSH 1988. AT2800000. "Acrylic Acid, Methyl Ester." RTECS, on line.

METHYL ALCOHOL

Synonyms: methanol, wood alcohol

CAS Registry Number: 67–56–1

Molecular Weight: 32.04

Molecular Formula: CH_3OH

AALG: systemic toxicity (TLV/42)—2.3 ppm (3.1 mg/m^3) 8-hour TWA

Occupational Limits:
- ACGIH TLV: 200 ppm (260 mg/m^3) TWA; 250 ppm (310 mg/m^3) STEL; skin notation
- NIOSH REL: 200 ppm (260 mg/m^3) TWA; 800 ppm (1048 mg/m^3) ceiling (15 minutes)
- OSHA PEL: 200 ppm (260 mg/m^3) TWA; 250 ppm (310 mg/m^3) STEL; skin notation

Basis for Occupational Limits: The occupational limits for methanol are based on prevention of recurring headaches in workers as well as consideration of narcotic and ocular effects and sensory irritation.[1,2] OSHA recently adopted the ACGIH-STEL and skin notation as part of its final rule limits.[3]

Drinking Water Limits: No drinking water limits were found for this substance.

Toxicity Profile

Carcinogenicity: No data were found implicating methanol as a carcinogen.

Mutagenicity: Methanol has been reported to interfere with DNA repair in *E. coli*, induce mutations in *S. cerevisiae* without the addition of S9 fraction, and cause chromosomal aberrations in bone marrow cells following ingestion and intraperitoneal injection in mice. It was positive in a mouse micronucleus test.[4] However, negative results are listed in the Gene-Tox database as summarized in RTECS for the Syrian hamster embryo cell clonal assay, oncogenic transformation in Syrian hamster embryo cells by adenovirus SA7, *N. crassa* aneuploidy assay, and *in vitro* induction of sister chromatid exchange (cell type unspecified).[4]

Developmental Toxicity: Nelson et al.[5] exposed Sprague-Dawley rats (n = 15/group) to 0, 5000, 10,000, or 20,000 ppm methanol for 7 hours per day on gestation

419

days 1 to 19 for all exposure groups except the highest, which was exposed on gestation days 7 through 15. Exposure to methanol did not produce alterations in feed intake, water consumption, or body weight gain in any of the exposed dams. However, a slightly unsteady gait was noted in rats in the highest exposure group. Fetotoxicity occurred in the form of a dose-dependent decrease in fetal body weight, which was statistically significant in the two highest exposure groups for both male and female rat fetuses. There was an increased incidence of gross, skeletal, and visceral malformations in both the 10,000- and 20,000-ppm exposure groups; this incidence was statistically significant only at the highest dose level. Malformations most commonly observed included exencephaly, encephalocele, extra or rudimentary cervical ribs, and urinary or cardiovascular defects. NOAELs of 5000 and 10,000 ppm are suggested by this study for fetotoxicity and teratogenic effects, respectively.

Infurna and Weiss[6] evaluated the neonatal behavioral toxicity of prenatal methanol exposure using the endpoints of suckling behavior in one-day-old rat pups and nest-seeking behavior in ten-day-old rat pups. In this study Long-Evans rats (n = 10/group) were unexposed or received drinking water containing 2.0% methanol on either gestation days 15 to 17 or 17 to 19; consumption figures indicated that the dams in both exposure groups received approximately 2500 mg/kg methanol over a three-day exposure period. Parameters measuring fetal growth and development indicated that these exposure levels produced no adverse effects, nor was there any evidence of maternal toxicity. However, one-day-old pups from methanol-exposed dams took significantly longer than controls to begin suckling, and ten-day-old pups required significantly longer than control pups to locate nesting material from their home cages.

Reproductive Toxicity: In a report cited in RTECS,[4] intraperitoneal injection of methanol in mice at a dose of 5000 mg/kg was reported to interfere with spermatogenesis.

Systemic Toxicity: There is a substantial database on the effects of methanol exposure in humans in the form of individual case reports and descriptions of clusters of cases from both the clinical and occupational health literature, and these are well reviewed in the NIOSH criteria document.[2] However, based on its review of the literature, NIOSH concluded that it was "difficult to recommend an environmental limit based upon unequivocal scientific data" due to the lack of rigorously conducted comprehensive epidemiologic and clinical studies.

In humans, similar symptoms have been reported in cases of exposure via ingestion, dermal absorption, and inhalation. These symptoms have included dizziness, nausea, and vomiting, various types of visual disturbances, vertigo, unsteady gait, multiple neuritis with paresthesia, pains in the extremities, metabolic acidosis, and headaches. However, the most characteristic manifestations of methanol poisoning are visual disturbances and metabolic acidosis, with a latency period of between 12 and 48 hours. With respect to specific correlations of inhalation exposure and effect, based on the occupational literature it appears that concentrations between 1200 and 8300 ppm are required to produce the characteristic visual disturbances such as blurred and double vision, constricted visual fields, changes in color perception, and blindness. In such cases clinical examination has revealed findings including sluggish pupils, pallid optic discs, retinal edema, papilledema, and hyperemia of the optic discs with blurred edges and dilated veins. In addition, exposures between 200 and 350 ppm have been found to produce severe and recurrent headaches in exposed workers. It should also be noted that methanol is readily absorbed through intact skin and sufficient exposure via this route may result in poisoning.[2,7]

With respect to studies in animal models, it has in general proven difficult to identify suitable species for mechanistic studies due to early problems in inducing symptoms of poisoning in traditional rodent models similar to those observed in humans. However, evidence accumulated in recent years has resulted in the consensus judgement that various types of primates are the most useful animal models for mechanistic studies based on the similarity of metabolism between primates and humans. It has been shown that although the products of intermediary methanol metabolism are the same in humans, primates, and rodent models, the enzymes responsible are different at the first step. Initial oxidation of methanol to formaldehyde is catalyzed by a catalase-peroxidase system in nonprimates and by alcohol dehydrogenase in primates and humans. Formaldehyde is then oxidized to formic acid by formaldehyde dehydrogenase, and formic acid may then be either eliminated in the urine or further metabolized by various enzyme systems.[7]

Although there is a reasonable degree of consensus on the differences in metabolism of methanol between primates and nonprimates, there is as yet no clear delineation of the exact mechanism of methanol toxicity in either primates or humans. However, it appears that the effects observed in both primates and nonprimates are a consequence of a metabolite(s) rather than methanol itself. In humans this is supported by the long latency period between exposure and the observance of symptoms and the fact that ethanol, if administered soon enough, will ameliorate the toxic effects of methanol.[7]

Irritation: Methanol is potentially irritating to the eyes and mucous membranes only at concentrations greater than those known to produce systemic effects. Estimates of the odor threshold for methanol are highly variable and cover a wide range of concentrations (3.3–5900 ppm). On the basis of animal tests, methanol has been characterized as being mildly to moderately irritating to the eyes and skin.[2]

Basis for the AALG: Based on its volatility (vapor pressure = 92 mm Hg at 20°C), widespread use as a solvent, and ease of absorption through the skin, there is a large potential for exposure via both the inhalation and dermal routes.[1,2] Therefore, 50% of the contribution to total exposure will be allowed from air.

Two AALGs were calculated for methanol: one based on developmental toxicity using data from the study of Nelson et al.[5] and another based on systemic toxicity using the TLV. The TLV/42 was considered an appropriate basis for AALG derivation due to the absence of long-term studies in appropriate animal models or humans and because the TLV is long-standing.

The rat NOAEL of 5000 ppm for fetotoxicity from the study of Nelson et al.[5] was used as the basis for the AALG. This NOAEL was converted to a human equivalent dose according to the procedures described in Appendix C of the methodology document and divided by a total uncertainty factor of 500 (10 for interindividual variation $\times$ 10 for interspecies variation $\times$ 5 for the developmental uncertainty factor). Note that the midpoint of the range of rat weights (0.188 kg) was used in the calculations and the developmental uncertainty factor was applied based on the occurrence of developmental toxicity at levels below those producing overt maternal toxicity.

The recommended AALG is based on the TLV/42 since it is more conservative. However, this AALG should be considered provisional, since it is based on an occupational limit.

AALG: • systemic toxicity (TLV/42)—2.3 ppm (3.1 mg/m^3) 8-hour TWA
• developmental toxicity—3.9 ppm (5.1 mg/m^3) 24-hour TWA

References

1. ACGIH. 1986. "Documentation of the Threshold Limit Values and Biological Exposure Indices." 5:372.
2. NIOSH. 1976. "Criteria for a Recommended Standard . . . Occupational Exposure to Methyl Alcohol." DHEW (NIOSH) 76–148.
3. OSHA. 1989. "Air Contaminants; Final Rule." *Fed. Reg.* 54:2322–959.
4. NIOSH. 1988. PC1400000. "Methanol." RTECS, on line.
5. Nelson, B. K., W. S. Brightwell, D. R. MacKenzie, A. Khan, J. R. Burg, W. W. Weigel, and P. T. Goad. 1985. "Teratological Assessment of Methanol and Ethanol at High Inhalation Levels in Rats." *Fund. Appl. Tox.* 5:727–36.
6. Infurna, R., and B. Weiss. 1986. "Neonatal Behavioral Toxicity in Rats Following Prenatal Exposure to Methanol." *Teratology* 33:259–65.
7. Wimer, W. W., J. A. Russell, and H. L. Kaplan, Eds. 1983. *Alcohols Toxicology* (Park Ridge, NJ: Noyes Data Corporation), pp. 8–26.

METHYL BROMIDE

Synonyms: monobromomethane, bromomethane

CAS Registry Number: 74–83–9

Molecular Weight: 94.95

Molecular Formula: CH_3Br

AALG: systemic toxicity—2.4 $\mu g/m^3$ (0.62 ppb) 24-hour TWA

Occupational Limits:
- ACGIH TLV: 5 ppm (20 mg/m^3) TWA; skin notation
- NIOSH REL: reduce to lowest feasible level
- OSHA PEL: 5 ppm (20 mg/m^3) TWA; skin notation

Basis for Occupational Limits: The TLV is based on the prevention of serious neurotoxic effects and pulmonary edema.[1] The NIOSH recommendation was based on the known neurotoxicity of methyl bromide, potential carcinogenic effects based on rat gavage studies, and the known carcinogenic effect of related monohalomethanes.[2] OSHA recently adopted the ACGIH TLV.[3]

Drinking Water Limits: No drinking water limits were found for this substance.

Toxicity Profile

Carcinogenicity: Danse et al.[4] exposed male and female Wistar rats (n = 10/sex/group) to 0, 0.4, 2, 10, or 50 mg/kg methyl bromide in arachis oil by gavage 5 days a week for 13 weeks. There was a decrease in food consumption and body weight gain in the highest dose groups, but no effect on survival. Squamous cell carcinomas of the forestomach were reported in 7 of 10 male and 6 of 10 female rats in the highest dose group; forestomach papillomas were reported in 2 of 10 high-dose males. There was also a dose-related increase in hyperplasia and hyperkeratosis of the forestomach with the following incidence:

- *Males*
  control, 0 of 10
  0.4 mg/kg, 0 of 10
  2 mg/kg, 2 of 10
  10 mg/kg, 6 of 10
  50 mg/kg, 10 of 10

- *Females*
  control, 1 of 10
  0.4 mg/kg, 1 of 10
  2 mg/kg, 1 of 10
  10 mg/kg, 9 of 10
  50 mg/kg, 10 of 10

The effects on food consumption and body weight gain, as well as non-neoplastic lesions such as granulocytosis, decreased hemosiderosis, and increased hematopoiesis in the high-dose male and female rats, were considered secondary to the effects of methyl bromide on the forestomach.

In a report available only as an abstract, Hubbs and Harrington[5] (cited in IARC[6]) reported that exposure of rats (age, strain, sex, and numbers unspecified) to 25 or 50 mg/kg methyl bromide in peanut oil by gavage for 30, 60, 90, or 120 days resulted in ulceration and epithelial hyperplasia of the forestomach without evidence of malignancy.

Based on reported difficulty in distinguishing marked hyperplasia versus neoplasia in reviewing forestomach slides from the Danse et al. study, Boorman et al.[7] conducted a study to evaluate the regression of lesions following cessation of methyl bromide exposure. Male Wistar rats were gavaged with 50 mg/kg methyl bromide in peanut oil for 13 or 25 weeks with interim sacrifices at 13 and 25 weeks in the vehicle control groups; at 17, 21, and 25 weeks in the 13-week exposure group; and at 13, 17, 21, and 25 weeks in the 25-week exposure group. No forestomach lesions were found in vehicle controls killed at 13 and 25 weeks. In animals treated for 13 weeks there was inflammation, acanthosis, fibrosis, and a high incidence of pseudoepitheliomatous hyperplasia of the forestomach, and in those treated for 25 weeks the hyperplastic lesions were more severe and there was evidence of neoplasia in a single rat. In the stop-treatment groups there was progressive regression of the lesions following cessation of methyl bromide exposure and at 12 weeks postexposure inflammation and hyperplasia were no longer apparent, but adhesions, fibrosis, and mild acanthosis were still evident.

Based on the available evidence, it seems apparent that the observed forestomach neoplastic and non-neoplastic lesions are a consequence of the irritant properties of methyl bromide. Although it has been speculated that agents that induce a high degree of irritation (hence cell proliferation) set up an environment conducive to the development of neoplasia, the implications of these findings for human inhalation exposure are uncertain. It should be noted that two-year inhalation studies in mice are being conducted under the NTP with the quality assessment in progress.[8] By our judgement, methyl bromide is most appropriately placed in group D ("unclassified substance") under the EPA weight-of-evidence classification.

Mutagenicity: The literature on genotoxic effects of methyl bromide has been reviewed by IARC[6,9] and is summarized as follows: "Methyl bromide is mutagenic to bacteria and plants. It induces sex-linked recessive lethal mutations in *Drosophila melanogaster* and mutations in cultured mammalian cells. It does not induce unscheduled DNA synthesis in cultured mammalian cells. Methyl bromide alkylates DNA in liver and spleen of mice treated by various routes. It induces micronuclei in bone-marrow and peripheral blood cells of rats and mice.'' IARC concluded that the evidence for genetic activity of methyl bromide in short-term tests was sufficient.

Developmental Toxicity: The potential developmental toxicity of methyl bromide has been evaluated in both rats and rabbits exposed via inhalation[10,11] (cited in IARC[6]). Wistar rats (n = 36/group) were exposed to 0, 20, or 70 ppm methyl bromide (purity, >99.5%) under the following exposure regimens: pregestational— 7 hours per day, 5 days a week for 3 weeks prior to breeding; gestational—7 hours per day on days 1 through 19 of gestation; or combined pregestational and gestational exposure. Although there was a decrease in maternal body weight gain in all groups exposed to 70 ppm methyl bromide, there were no significant effects on fetal growth or treatment-related increases in skeletal or visceral malformations. New Zealand white rabbits (n = 24/group) were also exposed to 0, 20, or 70 ppm methyl bromide on gestation days 1 through 15. In the original protocol it was planned to expose the animals up to day 24 of gestation; however, due to excess maternal mortality in the high-dose group, which precluded an evaluation of fetal effects, exposure was stopped after day 15. No maternal toxicity or fetotoxicity were noted in animals exposed to 20 ppm methyl bromide.

Reproductive Toxicity: No multigeneration studies evaluating the reproductive toxicity of methyl bromide were found. However, the findings of testicular atrophy in rats and mice exposed to 160 ppm methyl bromide in the Eustis et al. study[12] should be noted.

Systemic Toxicity: There are many case reports in man of occupational fatalities associated with the use of methyl bromide as a fire extinguisher (older literature) and fumigant. Exposure was generally to high levels, although some cases were associated with higher-than-usual exposure with a history of chronic low level exposure.[13] Signs of toxicity reported include drowsiness, dyspnea, lacrimation, irritation, lung congestion and pulmonary edema, pneumonia, and kidney damage. In addition, neurological signs including tremors, ataxia, CNS depression, and convulsions have been observed.[6]

The signs of neurotoxicity observed following methyl bromide exposure in man are dependent on the concentration and duration of exposure. The narcotic and convulsive actions are generally associated with high-level short-duration exposures and signs such as headache, dizziness, anorexia, speech impairment, visual problems, and general weakness are associated with chronic low-level exposure.[13] The exact concentrations producing given effects are uncertain in most cases. However, estimates for acute exposure effects are generally in the range of 300 to 500 ppm and higher; chronic exposure levels have been estimated in the range of 100 to 500 ppm and lower.[1] Watrous[14] (cited in ACGIH[1]) reported that mild systemic toxicity consisting of nausea, vomiting, headache, skin lesions, and other symptoms resulted from exposure concentrations of less than 35 ppm for two weeks.

Neurotoxicity has also been observed in laboratory animals following repeated inhalation exposure. Anger et al.[15] (cited in IARC[6] and Alexeeff and Kilgore[13]) reported that rabbits exposed to 65 ppm methyl bromide for 7.5 hours per day for one month exhibited weight loss, decreased eye blink response, and reduced nerve conduction velocities in the sciatic and ulnar nerves. However no such effects were noted in rats exposed to 65 ppm and 55 ppm for one month and one year, respectively.

Ikeda et al.[16] exposed male Wistar rats (n = 12/group) to 0, 200, or 300 ppm methyl bromide for 4 hours per day, 5 days a week for 3 weeks. Animals in both exposure groups showed a decrease in body weight gain, and while those exposed to 200 ppm were asymptomatic, animals exposed at 300 ppm exhibited a decrease in spontaneous activity, three died, and one developed convulsions. Performance on the rotarod test, a measure of equilibrium function, was significantly decreased in animals exposed to 300 ppm at 12 days after the final exposure and in animals exposed to both 200 ppm and 300 ppm at 28 days after the final exposure. Alterations in circadian rhythm were also reported in animals exposed to 300 ppm. Consistent with observations reported in previous studies, no gross or microscopic lesions were found in the peripheral or central nervous system.

Irish et al.[17] exposed various numbers of rats, rabbits, guinea pigs, and monkeys to 0, 17, 33, 66, 100, or 220 ppm methyl bromide (purity, >99%) for 7.5 to 8 hours per day, 5 days a week for up to 6 months or until the majority of animals died or showed severe signs of toxicity. At 220 ppm, most of the animals died after a few exposures (rats, rabbits, guinea pigs—no monkeys used). It was noted that most guinea pigs showed marked pulmonary changes including congestion, edema, leucocytic infiltration, and hemorrhage. At 100 ppm, obvious signs of toxicity and some deaths occurred in all groups of animals with the guinea pigs (n = 11) apparently being more resistant than the rats (n = 30). While some rats survived to the end of the exposure period, their growth was retarded and their general condition was poor; pathologic lesions were found in the lungs of some but not all rats. Rabbits exposed to 100 ppm methyl bromide rapidly developed paralysis and the exposed monkey developed convulsions. At 66 ppm, no signs of toxicity or gross or microscopic lesions were detected in rats (n = 10 males, 12 females) or guinea pigs (n = 14 males, 10 females). Most rabbits (n = 42) developed rapidly increasing paralysis and signs of pulmonary irritation after 14 to 46 exposures, and deaths were common. The paralysis of fore- or hindquarters or both was preceded by weakness, but no morphological changes were observed in the nervous system; the authors concluded that this indicated that the paralysis was due to a "functional" abnormality. It should be noted that if exposure was stopped at the first signs of paralysis, the rabbits made a rapid and apparently complete recovery. Of the monkeys exposed to 66 ppm, several developed paralysis, but recovered 5 to 8 weeks after exposure ceased. At 33 ppm, no exposure-related toxicity developed in rats, monkeys, or guinea pigs after 6 months of exposure. Rabbits (n = 58) exposed to 33 ppm methyl bromide tended to develop either pulmonary irritation followed by pneumonia or the characteristic weakness followed by paralysis. Extensive pathologic examination of the nervous system from affected rabbits failed to reveal evidence of abnormality. At 17 ppm, the lowest exposure level used, rabbits (n = 6) tolerated the exposure for 6 months without the development of paralysis or other signs of toxicity and their tissues were essentially normal upon microscopic examination. NOAELs of 66 ppm in rats and guinea pigs, 33 ppm in monkeys, and 17 ppm in rabbits were suggested by this study. It should be noted that the monkeys used in these studies were apparently wild-caught, and a number of them developed or died of tuberculosis during the course of the studies.

Inhalation experiments in male and female B6C3F$_1$ mice and F344 rats exposed to 160 ppm methyl bromide for 6 hours per day, 5 days a week for 3, 10, or 30 exposure periods have demonstrated species- and sex-related differences in target organ toxicity.[12] There was a high degree of mortality and only female rats survived the entire exposure period with less than 50% mortality. In rats, neuronal necrosis was observed in the cerebral cortex, hippocampus, and thalamus, while in mice neuronal necrosis was found mainly in the internal granular layer of the cerebellum. Nephrosis was observed in mice, but not in rats, and was the major cause of mortality in mice. Although necrosis of the olfactory epithelium and myocardial degeneration (male mice only) occurred in both mice and rats, the lesions were more severe and extensive in the rats. Lesions of the adrenal cortex were also found in both mice and rats. However, the lesions were inner zone atrophy in mice (females) and cytoplasmic vacuolation in rats. Testicular degeneration was observed in both rats and mice.

Irritation: Pulmonary effects and lacrimation due to the irritant properties of methyl bromide have been reported in man. However, at lower concentrations neurotoxic effects and other forms of systemic toxicity have been reported, and irritation is not the critical toxic effect.[6] Methyl bromide is reported to have a ''sweetish'' odor[1] and an odor threshold of 16.8 ppm.[6]

Basis for the AALG: Methyl bromide is slightly soluble in water (0.9 g/L) and is a gas at room temperature and pressure.[1,6] Methyl bromide is currently used as a fumigant in pest control and as a methylating agent in industry. Its uses as a refrigerant and fire extinguisher have been discontinued due to the high incidence of associated injuries and fatalities.[1] Methyl bromide occurs naturally in sea water (from marine organisms) and both natural and manmade sources contribute to ambient air concentrations. Monitoring studies in 10 U.S. cities revealed concentrations in the range of 0.041 to 0.259 ppb.[6] Based on these considerations 80% of the contribution to total exposure will be allowed from air.

Although it is the oldest study reviewed, Irish et al.[17] is the strongest study available on which to base the AALG. The reason is that this study is the one longest in duration that established a clear dose-response relationship (LOAELs and NOAELs) for multiple animal species using the appropriate route of exposure. Results of other studies essentially confirm that although other organs may be affected, the principal target organs for methyl bromide toxicity are the lung and nervous system. Consequently, the NOAEL of 17 ppm methyl bromide in rabbits (the most sensitive species) from the Irish et al. study[17] was selected as the starting point for AALG derivation. The NOAEL of 17 ppm was adjusted for continuous exposure (65 mg/m^3 × 7.5/24 × 5/7) and converted to a human equivalent NOAEL according to the procedures described in Appendix C. A total uncertainty factor of 1000 (10 for interspecies variation × 10 for interindividual variation × 10 for less-than-lifetime exposure) was applied. Note that in the absence of weight data, a 4.1-kg rabbit breathing 0.2448 m^3 air/day was assumed based on data from the Federation of American Societies for Experimental Biology data book;[18] as with the assumed weights and breathing rates for rats and mice used by the EPA CAG, these figures represent a conservative assumption, i.e., an older, heavier animal.

It should be noted that the TLV was not considered a suitable basis for the AALG, since the database correlating known exposure levels of methyl bromide and effects in humans is relatively weak, and it is therefore difficult to determine how closely the TLV corresponds to a human LOAEL or NOAEL. In addition, it was felt that a more conservative approach was warranted due to the potential carcinogenic effects of methyl bromide, which have not yet been adequately evaluated. This AALG should be considered provisional pending availability of results from the NTP inhalation bioassay in mice.

AALG: • systemic toxicity—2.4 μg/m^3 (0.62 ppb) 24-hour TWA

References

1. ACGIH. 1986. "Documentation of the Threshold Limit Values and Biological Exposure Indices." 5:376–77.
2. NIOSH. 1984. "Monohalomethanes: Methyl Chloride, Methyl Bromide and Methyl Iodide." *Current Intelligence Bulletin* No. 43. DHHS (NIOSH) 86–122.
3. OSHA. 1989. "Air Contaminants; Final Rule." *Fed. Reg.* 54(12):2332–2959.
4. Danse, L. H. J. C., F. L. van Velsen, and C. A. van der Heijden. 1984. "Methyl Bromide: Carcinogenic Effects in the Rat Forestomach." *Tox. Appl. Pharm.* 72:262–71.
5. Hubbs, A. F. and D. D. Harrington. 1986. "Further Evaluation of the Potential Gastric Carcinogenic Effects of Subchronic Methyl Bromide Administration." In: Proceedings of the 36th Annual Meeting of the American College of Veterinary Pathologists and the Annual Meeting of the American Society for Veterinary and Clinical Pathology (Denver, Colorado, December 1985), p. 92.
6. IARC. 1986. "Methyl Bromide." *IARC Monog.* 41:187–212.
7. Boorman, G. A. et al. 1986. "Regression of Methyl Bromide–Induced Forestomach Lesions in the Rat." *Tox. Appl. Pharm.* 86:131–39.
8. NTP. 1989. "Chemical Status Report" (Research Triangle Park, NC: National Toxicology Program).
9. IARC. 1986. "Methyl Bromide." *IARC Monog.* suppl. 6:386–88.
10. Hardin, B. D. et al. 1981. "Testing of Selected Workplace Chemicals for Teratogenic Potential." *Scand. J. Work Environ. Health* 7 (suppl. 4):66–75.
11. Sikov, M. R. et al. 1980. "Teratologic Assessment of Butylene Oxide, Styrene Oxide and Methyl Bromide." Cincinnati: U.S. DHEW (Contract No. 210–78–0025).

12. Eustis, S. L. et al. 1988. "Toxicology and Pathology of Methyl Bromide in F344 Rats and B6C3F$_1$ Mice Following Repeated Inhalation Exposure." *Fund. Appl. Tox.* 11: 594–610.

13. Alexeeff, G. V. and W. W. Kilgore. 1983. "Methyl Bromide." *Residue Rev.* 88:101–53.

14. Watrous, R. M. 1942. "Methyl Bromide: Local and Mild Systemic Toxic Effects." *Ind. Med.* 11(12):575.

15. Anger, W. K. et al. 1981. "Neurobehavioral Effects of Methyl Bromide Inhalation Exposures." *Scand. J. Work Environ. Health* 7(suppl. 4):40.

16. Ikeda, T. et al. 1980. "Behavioral Effects in Rats Following Repeated Exposure to Methyl Bromide." *Tox. Lett.* 6(4–5):293.

17. Irish, D. D. et al. 1940. "The Response Attending Exposure of Laboratory Animals to Vapors of Methyl Bromide." *J. Ind. Hyg. Tox.* 22:218–30.

18. Federation of American Societies for Experimental Biology (FASEB). 1974. "Respiratory Frequency, Tidal Volume and Minute Volume: Vertebrates." In: P. L. Altman and D. S. Dittmen, Eds. *Biological Data Books*, 2nd ed., Vol. III, No. 208 (Bethesda, MD: FASEB), p. 1581.

METHYL CHLORIDE

Synonyms: chloromethane, monochloromethane

CAS Registry Number: 74–87–3

Molecular Weight: 50.49

Molecular Formula: CH_3Cl

AALG: carcinogenicity—0.15 ppm (0.30 mg/m^3) annual TWA

Occupational Limits:
- ACGIH TLV: 50 ppm (105 mg/m^3) TWA; 100 ppm (205 mg/m^3) STEL; skin notation
- NIOSH REL: reduce to lowest feasible level
- OSHA PEL: 50 ppm (105 mg/m^3) TWA; 100 ppm (205 mg/m^3) STEL

Basis for Occupational Limits: The ACGIH limits are based on neurotoxicity, whereas the NIOSH recommendation takes into account potential carcinogenic and teratogenic effects.[1-3] OSHA recently adopted the ACGIH TLV-TWA and TLV-STEL as part of its final rule limits.[4]

Drinking Water Limits: The ambient water quality criterion for chloromethane is 3.8 mg/L assuming a 70-kg man drinking 2 L of water per day and excluding consumption of fish and shellfish since bioconcentration was not considered to occur in the case of methyl chloride.[5]

Toxicity Profile

Carcinogenicity: Pavkov et al.[6,7] (cited in NIOSH[3] and IARC[8]) conducted a lifetime inhalation bioassay in which male and female B6C3F$_1$ mice and F344 rats (numbers per group unspecified) were exposed to 0, 51, 224, or 997 ppm methyl chloride for 6 hours per day, 5 days a week for 24 months. At the highest exposure level, growth and survival were adversely affected in both sexes of mice and rats. There was a statistically significant increase in malignant and nonmalignant tumors of the kidney (cortical adenomas and adenocarcinomas, papillary cystadenomas, cystadenocarcinomas, and tubular cystadenomas) only in male mice and only at the highest exposure level; no treatment-related increase in neoplasms was reported in rats. Non-neoplastic lesions in the form of functional limb muscle impairment, brain lesions (degeneration and atrophy of the granular layer of the cerebellum), atrophy

431

of the spleen, and hepatocellular necrosis were observed in male and female mice. Testicular damage was reported in both rats (decreased relative and absolute testicular weight and testicular tubular degeneration and atrophy) and mice (seminiferous tubule degeneration) at the highest exposure level.

In their review, IARC concluded that the evidence of carcinogenicity was inadequate based on this study. However, this conclusion was largely based on the fact that only an abstract of this study was available to the IARC Working Group at the time of the review.[8] It is the judgment here that methyl chloride is most appropriately placed in group C, possible human carcinogen, under the EPA weight-of-evidence classification.

Mutagenicity: Methyl chloride was mutagenic to *S. typhimurium* strains TA100 and TA1535 in the presence and absence of a metabolic activation system (rat liver S9 fraction) and induced a dose-dependent increase in 8-azoguanine-resistant mutants of *S. typhimurium* strain TM677. It caused chromatid breaks in *Tradescantia paludosa* pollen grains and DNA damage in mammalian cells in vitro, but not in vivo. Methyl chloride also induced mutations and SCEs in cultured mammalian cells (TK6 human lymphoblasts). It was positive in a dominant lethal assay in rats and induced oncogenic transformation in Syrian hamster embryo cells by adenovirus SA7.[8]

The evidence for genotoxicity of methyl chloride was judged sufficient by IARC when evaluated as to breadth of endpoints and phylogenetic complexity.[8]

Developmental Toxicity: Wolkowski-Tyl et al.[9] exposed C57BL/6 mice (n = 33/group) to 0, 100, 500, or 1500 ppm methyl chloride via inhalation for 6 hours per day on gestation days 6 through 17. There was a low, but statistically significant, incidence of heart defects (reduction in size and absence of atrioventricular valves, chordae tendineae, and papillary muscles) in the 500-ppm group. Maternal effects in the form of overt neurotoxicity were observed after 6 to 9 days in the high dose group, and treatment was stopped.

In a follow-up to the previous study, Wolkowski-Tyl et al.[10] exposed C57BL/6 mice (n = 74–77/group) to 0, 250, 500, or 750 ppm methyl chloride via inhalation for 6 hours per day on gestation days 6 through 17. The incidence of cardiac defects was 0.7% in the controls, 1.3% at 250 ppm, 2.5% at 500 ppm, and 4.3% at 750 ppm, with the incidence in the two highest exposure groups statistically significant compared with the controls. Neurotoxicity in the form of ataxia, tremors, hypersensitivity to touch or sound, and convulsions were observed in the 750-ppm dams.

F344 rats (n = 25/group) were also studied under the same protocol as mice in the initial study[9] except that exposure was from days 7 through 19 of gestation. However, no exposure-related skeletal or visceral abnormalities were observed, and decreased fetal body weights and skeletal maturity occurred only in the high-dose group, probably secondary to decreased maternal food consumption and body weight gain.

Reproductive Toxicity: Hamm et al.[11] exposed male F344 rats to 0, 150, 475, or 1500 ppm methyl chloride for 6 hours per day, 5 days a week for 10 weeks and then 6 hours per day, 7 days a week for an additional 2 weeks. At the end of the exposure, 10 animals per group were euthanized. The only treatment-related lesions found were in the high-exposure group and included severe bilateral testicular degeneration (10 of 10) and granulomas in the epididymis (3 of 10). The remaining 30 males were each mated to 2 unexposed females, and the following findings were reported:

- 1500 ppm: No litters were born.
- 475 ppm: Fewer males were proven fertile than controls, but the difference was not statistically significant (30% vs 45%). The number of females with copulation plugs producing litters was lower (statistically significant) than controls (31% vs 59%).
- 150 ppm: No effects on fertility or lesions were observed.

Systemic Toxicity: Methyl chloride has potent narcotic and anesthetic activity. In the TLV documentation, ''moderate exposure'' to methyl chloride is stated to be characterized by ocular symptoms such as mistiness, diplopia, and difficulty in accommodation, which may persist for several weeks, whereas ''serious exposure'' is associated with severe central nervous system effects.[1] Repko et al. reported that workers chronically exposed to concentrations ranging from 7.4 to 70 ppm (average 33.6 ppm) methyl chloride had a ''significant performance decrement'' compared with control workers[12] (cited in ACGIH[1]). In animal models, neurotoxicity is the most common response to methyl chloride exposure. Note the references to these signs in the studies reviewed in the previous sections.

Irritation: Methyl chloride is reported to have a ''sweet'' or ''ethereal'' odor and an odor threshold of 10 ppm.[1,7]

Basis for the AALG: Methyl chloride is a gas at room temperature (vapor pressure = 3800 mm Hg at 22°C) and is slightly soluble in water. The major uses of methyl chloride at present are in the production of methyl silicone polymers and resins and the manufacture of antiknock compounds for gasoline. Although methyl chloride has been detected in both ground and surface waters, it is considered to be ''probably the most abundant halocarbon in the atmosphere,'' and the principal sources of atmospheric methyl chloride are production by seaweeds and marine microorganisms and combustion of organic matter.[8] Based on these considerations, 80% of the contribution to total exposure will be allowed from air.

Three AALGs have been calculated for methyl chloride. The first one is based on carcinogenicity and should be considered provisional since only a single study described in an abstract was available and a clear positive response in only one sex and species, i.e., kidney tumors in male mice, was observed. In spite of this, it was considered prudent to calculate an AALG based on carcinogenicity since the evidence of mutagenicity was considered sufficient by IARC. The NOAEL of 225 ppm for kidney tumors in male mice (Pavkov et al.[6]) was adjusted for continuous exposure (225 ppm × 6/24 × 5/7) and converted to a human equivalent exposure level according to the procedures described in Appendix C; a total uncertainty factor of 1000 was applied (10 for interindividual variation × 10 interspecies variation × 10 for the database factor). Application of the database factor was considered appropriate because of the severity of the effect and the limited available description of study.

The developmental AALG was based on the NOAEL (adjusted for continuous exposure and converted to a human equivalent exposure level according to the procedures described in Appendix C) of 250 ppm for cardiac malformations in mice from the study of Wolkowski-Tyl et al.[10] with a total uncertainty factor of 500 (10 for interindividual variation × 10 for interspecies variation × 5 for the developmental uncertainty factor). Application of the developmental uncertainty factor was based on the occurrence of treatment-related malformations at levels not producing overt maternal toxicity.

The reproductive AALG was based on the NOAEL (adjusted for continuous exposure and converted to a human equivalent exposure level according to the procedures described in Appendix C) of 150 ppm for fertility effects in male F344 rats from the study of Hamm et al.[11] with a total uncertainty factor of 100 (10 for interindividual variation × 10 for interspecies variation).

Since it is the most conservative, the recommended AALG is the one based on carcinogenicity.

AALG: • carcinogenicity—0.15 ppm (0.30 mg/m^3) annual TWA
   • developmental—0.41 ppm (0.86 mg/m^3) 24-hour TWA
   • reproductive (male)—0.6 ppm (1.2 mg/m^3) 24-hour TWA

References

1. ACGIH. 1986. "Documentation of the Threshold Limit Values and Biological Exposure Indices." 5:380–81.
2. NIOSH. 1985. "NIOSH Recommendations for Occupational Safety and Health Standards." *Morb. Mort. Wkly. Rpt.* 34:5s–30s.
3. NIOSH. 1984. "Monohalomethanes—Methyl Chloride, Methyl Bromide, and Methyl Iodide. NIOSH Current Intelligence Bulletin No. 43" (9/27).
4. OSHA. 1989. "Air Contaminants; Final Rule." *Fed. Reg.* 54:2332–2959.
5. U.S. EPA. 1980. "Ambient Water Quality Criteria for Halomethanes." EPA 440/5–80–051.
6. Pavkov, K.L., W.D. Kerns, C.E. Chrisp, D.C. Thake, R.L. Persing, and H.H. Harroff. 1982. "Major Findings in a Twenty-Four Month Inhalation Toxicity Study of Methyl Chloride in Mice and Rats (Abstract No. 566)." *Toxicologist* 2:161.
7. Pavkov, K.L. 1982. "Final Report on a Chronic Inhalation Toxicology Study in Rats and Mice Exposed to Methyl Chloride. Vols. I-IV." Chemical Industry Institute of Toxicology/Battelle Columbus Laboratories. CIIT Docket #12712.
8. IARC. 1986. "Methyl Chloride." *IARC Monog.* 41:161–86.

9. Wolkowski-Tyl, R., M. Phelps, and J.K. Davis. 1983. "Structural Teratogenicity Evaluation of Methyl Chloride in Rats and Mice after Inhalation Exposure." *Teratology* 27:181–95.

10. Wolkowski-Tyl, R., A.D. Lawton, M. Phelps, and T.E. Hamm. 1983. "Evaluation of Heart Malformations in B6C3F$_1$ Mice Fetuses Induced by In Utero Exposure to Methyl Chloride." *Teratology* 27:197–206.

11. Hamm, T.E., T.H. Raynor, M.C. Phelps, C.D. Auman, W.T. Adams, J.E. Proctor, and R. Wolkowski-Tyl. 1985. "Reproduction in Fischer-344 Rats Exposed to Methyl Chloride by Inhalation for Two Generations." *Fund. Appl. Tox.* 5:568–77.

12. Repko, J.D. et al. 1976. "Behavioral and Neurological Effects of Methyl Chloride." DHEW (NIOSH) 77–125.

METHYLENE CHLORIDE

Synonyms: dichloromethane

CAS Registry Number: 75–09–2

Molecular Weight: 84.97

Molecular Formula: CH_2Cl_2

AALG: 10^{-6} 95% UCL—0.12 $\mu g/m^3$ (0.036 ppb) annual TWA

Occupational Limits:
- ACGIH TLV: 50 ppm (175 mg/m^3) TWA; Appendix A2, suspect human carcinogen
- NIOSH REL: lowest feasible limit
- OSHA PEL: 500 ppm (1735 mg/m^3) TWA; 1000 ppm (3470 mg/m^3 ceiling; 2000 ppm (6940 mg/m^3) peak (5 minutes in 2 hours)

Basis for Occupational Limits: The TLV is based on prevention of carboxyhemoglobinemia in the absence of exposure to carbon monoxide and on industrial experience, with consideration of carcinogenic effects in the form of the recent addition of methylene chloride to Appendix A2.[1] Carboxyhemoglobinemia due to methylene chloride exposure is a result of the body's ability to metabolize methylene chloride to CO.[1,2] In 1986, NIOSH recommended the REL be reduced to the lowest feasible limit in consideration of carcinogenic effects.[3] This recommendation supercedes the original RELs, 75 ppm as a TWA and 500 ppm as a ceiling limit.[2] Methylene chloride is in the process of 6(b) rulemaking, and the OSHA limits given above are transitional.[4]

Drinking Water Limits: The ambient water quality criterion for methylene chloride is 6 $\mu g/L$ on the basis of "qualitative, but not quantitative," data on carcinogenicity.[5] The 1-day and 10-day HAs for a 10-kg child are 13.3 mg/L and 1.5 mg/L, respectively. No suitable data were available to derive longer-term HAs, and a lifetime HA was not derived due to the B2 designation of methylene chloride under the EPA weight-of-evidence classification.[6] Methylene chloride is scheduled for regulation by 1989 under the 1986 amendments to the Safe Drinking Water Act.[7]

Toxicity Profile

Carcinogenicity: No excess mortality due to cancer was reported in two cohort studies[8,9] (cited in IARC[10]) and one proportional mortality study[11] (cited in IARC[10]). However, it was noted by the IARC working group that these studies had limited power to detect excess risk on account of the limited numbers of individuals with long-term exposure and/or adequate follow-up time.[10]

Dichloromethane has been tested for carcinogenicity by oral administration in drinking water in rats and mice[12,13] (cited in IARC[10]), via inhalation exposure in rats, mice, and hamsters[14,15] (cited in IARC[10]), and by intraperitoneal administration in the mouse lung adenoma assay[16] (cited in IARC[10]). Negative results were reported for the oral administration studies in mice and male rats, as well as for the inhalation study in male hamsters and the lung adenoma bioassay in mice. Inconclusive results were reported for the oral administration study in female rats and the inhalation study in female hamsters. These studies are reviewed in detail by IARC[10] and in the Health Assessment Document.[17] Additional studies are also reviewed in the ATSDR toxicological profile.[18] The NTP inhalation bioassay is reviewed in detail here since it forms the basis for the quantitative risk assessment presented in the IRIS database.[19]

In the NTP bioassay,[14] male and female F344/N rats were exposed to 0, 1000, 2000, or 4000 ppm methylene chloride, and male and female B6C3F$_1$ mice to 0, 2000, or 4000 ppm methylene chloride, for 6 hours per day, 5 days a week for 102 weeks. There was *clear evidence* of carcinogenicity in female rats and *some evidence* of carcinogenicity in male rats based on an increased incidence of benign neoplasms of the mammary gland in both sexes. Under NTP criteria, *clear evidence* of carcinogenicity is "demonstrated by studies that are interpreted as showing a chemically related increased incidence of malignant neoplasms, studies that exhibit a substantially increased incidence of benign neoplasms, or studies that exhibit an increased incidence of a combination of malignant and benign neoplasms where each increases with dose." *Some evidence* of carcinogenicity is "demonstrated by studies that are interpreted as showing a chemically related increased incidence of benign neoplasms, studies that exhibit marginal increases in neoplasms of several organs/tissues, or studies that exhibit a slight increase in uncommon malignant or benign neoplasms." The incidences of benign neoplasms of the mammary gland in males were 0 of 50 at 0 ppm, 0 of 50 at 1000 ppm, 2 of 50 at 2000 ppm, and 5 of 50 at 4000 ppm. The incidences in females were 5 of 50 at 0 ppm, 11 of 50 at 1000 ppm, 13 of 50 at 2000 ppm, and 23 of 50 at 4000 ppm. Non-neoplastic lesions related to methylene chloride exposure included hepatic hemosiderosis, cytomegaly, cytoplasmic vacuolization, necrosis, granulomatous inflammation, and bile duct fibrosis.

There was *clear evidence* of carcinogenicity in both sexes of mice based on increased incidences of alveolar/bronchiolar neoplasms (incidences in males: 2 of 50 at 0 ppm, 10 of 50 at 2000 ppm, and 28 of 50 at 4000 ppm; incidences in females: 1 of 50 at 0 ppm, 13 of 48 at 2000 ppm, and 29 of 48 at 4000 ppm) and hepatocellular neoplasms (incidences in males: 22 of 50 at 0 ppm, 24 of 49 at 2000 ppm, and 33 of 49 at 4000 ppm; incidences in females: 3 of 50 at 0 ppm, 16 of 48 at 2000 ppm, and 40 of 48 at 4000 ppm). Effects considered secondary responses to neoplasia in mice included dose-related increases in testicular atrophy in males and uterine and ovarian atrophy in females.

Methylene chloride has been placed in group B2, probable human carcinogen, under the EPA weight-of-evidence classification.[19]

Mutagenicity: Positive responses for methylene chloride have been reported in a number of assay systems and endpoints including the following: mutagenicity in *S. typhimurium* with and without metabolic activation, induction of mitotic recombination in *S. cerevisiae*, sex-linked recessive lethal test in *Drosophila*, chromosomal aberrations in Chinese hamster ovary cells, and sister chromatid exchanges in V79 cells.[20] In addition, methylene chloride has been reported to induce oncogenic transformation in cultured mammalian cells, and IARC has rated the degree of evidence for genetic activity based on short-term test results to be sufficient.[10] IARC also recently updated its assessment of the genotoxicity of methylene chloride.[21]

Developmental Toxicity: A case-control study of women working in the Finnish pharmaceutical industry revealed an association between exposure to methylene chloride (and a number of other solvents) and the occurrence of spontaneous abortion[22] (cited in IARC[10]).

Exposure to high levels of methylene chloride (1225 ppm) via inhalation on days 6 through 15 of gestation in mice and rats caused an increased incidence of minor skeletal anomalies in the fetuses and increased absolute liver weights in the dams[23] (cited in U.S. EPA[6,20] and IARC[10]). In another study in which groups of rats were exposed to 0 or 4500 ppm methylene chloride before or during gestation, or both, there was no increase in skeletal, external, or visceral abnormalities, although fetal body weights were reduced and both absolute and relative liver weights were increased in the dams[24] (cited in U.S. EPA[6,20] and IARC[10]). Neurobehavioral evaluation of offspring from rats that were allowed to deliver revealed alterations in spontaneous locomotor activities[25] (cited in U.S. EPA[6,20] and IARC[10]).

Reproductive Toxicity: Refer to "Developmental Toxicity," above.

Systemic Toxicity: Methylene chloride is metabolized to carbon monoxide and produces elevated levels of COHb in both animals and humans. Controlled exposure experiments in humans have revealed that COHb levels are related to both duration of exposure and concentration of methylene chloride; in nonsmoking subjects exposed to methylene chloride for 7.5 hours on 5 consecutive days, peak COHb percentages occurred on day 5 and were as follows: 50 ppm—2.9%, 100 ppm—5.7%, and 250 ppm—9.6%[26] (cited in NIOSH[2]). Note that in nonsmoking individuals unexposed to methylene chloride, the average COHb level is approximately

0.7%. Decreased performance of psychomotor tasks and auditory vigilance has been reported in subjects exposed to methylene chloride in the range of 300 to 750 ppm for 3–5 hours[27,28] (cited in NIOSH[2]). Mild intoxication in humans also produces central nervous system signs such as somnolence, lassitude, anorexia, and light headedness; deaths due to methylene chloride exposure have been attributed to cardiac injury and heart failure. However, in nonfatal cases recovery is apparently complete.[20] In the epidemiologic studies cited in "Carcinogenicity," above, no excess deaths due to heart or other circulatory system disease were found.[20]

It is relevant to note that simultaneous exposure to methylene chloride and carbon monoxide is considered to be a potentially serious occupational hazard, especially to smokers, because both contribute to the formation of COHb. The interaction of these two chemicals is considered to be additive, and a specific formula is used to determine to what degree the occupational limits should be lowered under conditions of simultaneous exposure to methylene chloride and carbon monoxide.[2]

Subchronic and chronic inhalation exposure to methylene chloride in laboratory animals is primarily associated with hepatotoxic and some renal effects.[20] For further detail, refer to the description of non-neoplastic lesions in the NTP bioassay.[14]

Irritation: Liquid methylene chloride is irritating if splashed in the eyes and may cause burns if it remains in contact with the skin.[1] Methylene chloride has been reported to have a "penetrating etherlike" odor,[10] and although concentrations of less than 100 ppm are usually not detectable, concentrations above 300 ppm are detectable to most people.[1]

Basis for the AALG: Methylene chloride is a volatile liquid (vapor pressure = 440 mm Hg at 25°C) and is soluble in water to the extent of 1% by weight.[1] It is used in paint and varnish remover, in aerosol products as a cosolvent or vapor pressure depressant, and in solvent degreasing, food and plastics processing, foam blowing metal cleaning, and food extraction.[10] Because the widespread environmental releases of methylene chloride and its use in consumer products have resulted in ubiquitous low levels in both air and water,[10] 50% of the contribution to total exposure will be allowed from air.

The AALG for methylene chloride is based on carcinogenicity using the quantitative risk assessment performed on the NTP inhalation bioassay data. This risk assessment is described in detail in the Addendum to the Health Assessment Document[29] and in the IRIS database.[19] The linearized multistage model was applied to four data sets to derive animal q_1^* values that were then converted to human q_1^*s; the data sets used were male rat mammary or subcutaneous tumors, female rat mammary tumors, and male and female mouse lung or liver adenomas/carcinomas (combined organ sites). CAG considered the q_1^* for humans derived from the female mouse data, i.e., 4.1×10^{-6} $(\mu g/m^3)^{-1}$, or 1.4×10^{-2} $(ppm)^{-1}$, to be the best upper bound slope estimate since it was the most conservative. Using this value of q_1^* and applying the relationship $E = q_1^* \times d$, where E is the specified level of extra risk and d is the dose, gives the estimates shown below. It is stated in the IRIS assessment that the "unit risk should not be used if the air concentration exceeds 2000 $\mu g/m^3$, since above this concentration the slope factor may differ from that stated." This AALG should be considered provisional since a revision of this risk assessment, which will incorporate pharmacokinetic and metabolism data, is pending.

AALG:
- 10^{-6} 95% UCL—0.12 $\mu g/m^3$ (0.036 ppb) annual TWA
- 10^{-5} 95% UCL—1.22 $\mu g/m^3$ (0.357 ppb) annual TWA

References

1. ACGIH. 1986. "Documentation of the Threshold Limit Values and Biological Exposure Indices." 5:390–92.
2. NIOSH. 1976. "Criteria for a Recommended Standard . . . Occupational Exposure to Methylene Chloride." DHEW (NIOSH) 76–138.
3. NIOSH. 1986. "Current Intelligence Bulletin 46: Methylene Chloride." DHHS (NIOSH), Cincinnati, OH.
4. OSHA. 1989. "Air Contaminants; Final Rule." *Fed. Reg.* 54:2332–2959.
5. U.S. EPA. 1980. "Ambient Water Quality Criteria for Halomethanes." EPA 440/4–79–029.
6. U.S. EPA. 1987. "Health Advisory for Dichloromethane." Office of Drinking Water (3/31).
7. Ohanian, EV. 1989. "National Primary Drinking Water Regulations for Additional Contaminants to be Regulated by 1989." In *Safe Drinking Water Act: Amendments, Regulations, and Standards,* E.J. Calabrese, C.E. Gilbert, and H. Pastides (Lewis Publishers: Chelsea, MI), pp. 71–82.
8. Ott, M.G., L.K. Skory, B.B. Holder, J.M. Bronson, and P.R. Williams. 1983. "Health Evaluation of Employees Occupationally Exposed to Methylene Chloride: Mortality." *Scand. J. Work Environ. Health* 9(suppl. 1):8–16.
9. Hearne, F.T., and B.R. Friedlander. 1981. "Follow-up of Methylene Chloride Study." *J. Occ. Med.* 23:660.
10. IARC. 1986. "Dichloromethane." *IARC Monog.* 41:43–85.
11. Friedlander, B.R., F.T. Hearne, and S. Hall. 1978. "Epidemiologic Investigation of Employees Chronically Exposed to Methylene Chloride. Mortality Analysis." *J. Occup. Med.* 20:657–66.
12. Serota, D.G., A.K. Thakur, B.M. Ulland, J.C. Kirschman, N.M. Brown, R.G. Coots, and K. Morgareidge. 1986. "A Two-Year Drinking Water Study of Dichloromethane on Rodents: I. Rats." *Food Chem. Tox.* 24:951–58.
13. Serota, D.G., A.K. Thakur, B.M. Ulland, J.C. Kirschman, N.M. Brown, R.G. Coots, and K. Morgareidge. 1986. "A Two-Year Drinking Water Study of Dichloromethane on Rodents: II. Mice." *Food Chem. Tox.* 24:959–63.
14. NTP. 1986. "Toxicology and Carcinogenesis Studies of Dichloromethane (Methylene Chloride) in F344/N Rats and B6C3F$_1$ Mice (Inhalation Studies)." NTP TR-306, National Toxicology Program, Research Triangle Park, NC.
15. Burek, J.D., K.D. Nitschke, T.J. Bell, D.L. Wackerle, R.C. Childs, J.E. Beyer, D.A.

Dittenber, L.W. Rampy, and M.J. McKenna. 1984. "Methylene Chloride: A Two-Year Inhalation Toxicity and Oncogenicity Study in Rats and Hamsters." *Fund. Appl. Tox.* 4:30–47.

16. Theiss, J.C., G.D. Stoner, M.B. Shimkin, and E.K. Weisburger. 1977. "Test for Carcinogenicity of Organic Contaminants of United States Drinking Waters by Pulmonary Tumor Response in Strain A Mice." *Cancer Res.* 116:361–67.

17. U.S. EPA. 1985. "Health Assessment Document for Dichloromethane (Methylene Chloride)." EPA-600/8–82/004F.

18. ATSDR. 1988. "Toxicological Profile for Methylene Chloride." ATSDR/TP-88/18.

19. U.S. EPA. 1989. "Dichloromethane; CASRN 75–09–2." IRIS (1/1/89).

20. U.S. EPA. 1984. "Health Effects Assessment for Methylene Chloride." EPA-540/1–86/028.

21. IARC. 1986. "Dichloromethane." *IARC Monog.* suppl. 6:228–30.

22. Taskinen, H., M.L. Lindbohm, and K. Hemminki. 1986. "Spontaneous Abortions among Women Working in the Pharmaceutical Industry." *Br. J. Ind. Med.* 43:199–205.

23. Schwetz, B.A., B.J.K. Leong, and P.J. Gehring. 1975. "The Effect of Maternally Inhaled Trichloroethylene, Perchloroethylene, Methyl Chloroform, and Methylene Chloride on Embryonal and Fetal Development in Mice and Rats." *Tox. Appl. Pharm.* 32:84–96.

24. Hardin, B.D., and J.M. Manson. 1980. "Absence of Dichloromethane Teratogenicity with Inhalation Exposure in Rats." *Tox. Appl. Pharm.* 52:22–8.

25. Bornschein, R.L., L. Hastings, and J.M. Manson. 1980. "Behavioral Toxicity in the Offspring of Rats Following Maternal Exposure to Dichloromethane." *Tox. Appl. Pharm.* 52:29–37.

26. Stewart, R.D., H.V. Forster, C.L. Hake, A.J. Lebrun, and J.E. Peterson. 1973. "Human Responses to Controlled Exposures of Methylene Chloride Vapor." Report No. NIOSH-MCOW-ENVM-MC-73-7. Medical College of Wisconsin, Department of Environmental Medicine, Milwaukee, WI.

27. Fodor, G.G., and G. Winneke. 1971. "Nervous System Disturbances in Men and Animals Experimentally Exposed to Industrial Solvent Vapors." In *Proceedings of the 2nd International Clean Air Congress,* H.M. England, Ed. (New York, NY: Academic Press), pp. 130–44.

28. Winneke, G. 1974. "Behavioral Effects of Methylene Chloride and Carbon Monoxide as Assessed by Sensory and Psychomotor Performance." In *Behavioral Toxicology— Early Detection of Occupational Hazards,* C. Xintaras, B.L. Johnson, and I. deGroot, Eds., DHEW(NIOSH) 74–126.

29. U.S. EPA. 1985. "Addendum to Health Assessment Document for Dichloromethane (Methylene Chloride)." EPA-600/8–82–004FA.

METHYL ETHYL KETONE

Synonyms: 2-butanone, MEK

CAS Registry Number: 78–93–3

Molecular Weight: 72.12

Molecular Formula: $CH_3C(O)CH_2CH_3$

AALG: developmental toxicity—0.13 ppm (0.39 mg/m^3) 24-hour TWA

Occupational Limits:
- ACGIH TLV: 200 ppm (590 mg/m^3) TWA; 300 ppm (885 mg/m^3) STEL
- NIOSH REL: 200 ppm (590 mg/m^3) TWA
- OSHA PEL: 200 ppm (590 mg/m^3) TWA; 300 ppm (885 mg/m^3) STEL

Basis for Occupational Limits: Both the NIOSH and ACGIH limits are based on the minimization of eye and upper respiratory tract irritation.[1,2] OSHA recently adopted the TLV-STEL as part of its final rule limits.[3]

Drinking Water Limits: The following HAs have been derived for MEK: 1-day (child)—75 mg/L, 10-day (child)—7.5 mg/L, longer-term (child)—2.5 mg/L, longer-term (adult)—8.6 mg/L, and lifetime—0.17 mg/L from an RfD of 0.0247 mg/kg/day using 20% as the relative source contribution from drinking water.[4]

Toxicity Profile

Carcinogenicity: No data were found implicating MEK as a carcinogen.

Mutagenicity: MEK has been found to be nonmutagenic to several strains of *S. typhimurium* and *E. coli* strain WP2,[4] but it is reported to cause chromosomal nondisjunction in *S. cerevisiae*.[5]

Developmental Toxicity: Schwetz et al.[6] exposed Sprague-Dawley rats (n = approximately 20 litters/group) to 0, 1126, or 2618 ppm MEK vapor for 7 hours per day on gestation days 6 through 15. There was a statistically significant decrease in fetal body weight and crown-rump length at the low, but not at the high, exposure level, and a dose-related increase in skeletal anomalies was observed. There was also a statistically significant increase in visceral abnormalities (dilated ureters and subcutaneous edema) in the high-dose group relative to controls. No significant maternal toxicity was reported in either exposure group compared with controls. It is relevant to note that the finding of skeletal variants at an exposure level of approximately 1000 ppm has been reported in another study[7] (cited in U.S. EPA[4]).

Reproductive Toxicity: No data were found implicating MEK as a reproductive toxin.

Systemic Toxicity: Data on chronic inhalation and ingestion exposure to MEK are reported to be lacking in both animals and humans.[4]

Takeuchi et al.[8] (cited in U.S. EPA[9]) exposed male Wistar rats (n = 8/group) to 0 or 200 ppm MEK via inhalation for 12 hours per day, 7 days a week for 24 weeks. A significant increase in motor and mixed nerve conduction velocities and a decrease in distal motor latency were observed after 4, but not 24, weeks. Saida et al.[10] (cited in U.S. EPA[9]) observed no paralysis (the only endpoint measured) in rats exposed continuously for 5 months to 1125 ppm MEK. In connection with the neurotoxic potential of MEK, it is relevant to note that Takeuchi et al. also found that combined exposure to 200 ppm MEK and 100 ppm *n*-hexane resulted in significantly greater neurotoxicity after 24 weeks than exposure to either chemical alone.

Cavender et al.[11] (cited in U.S. EPA[9]) exposed male and female F344 rats (n = 15/sex/group) to 0, 1250, 2500, or 5000 ppm MEK via inhalation for 6 hours per day, 5 days per week for 90 days. No effects were seen at 1250 ppm, and although SGPT activity was elevated in females exposed at 2500 ppm, the increase was not statistically significant. At the highest exposure level, mean body weight was depressed, and there were significant increases in liver weight and liver-to-body and liver-to-brain weight ratios as well as decreased SGPT and increased alkaline phosphatase, potassium, and glucose values in females.

Irritation: Nelson et al.[12] identified a LOAEL of 200 ppm for eye and nose irritation in 3 to 5 minute exposures of human volunteers. MEK is reported to have an acetone-like odor and an odor threshold (100% response) of 10 ppm.[1]

Basis for the AALG: MEK is soluble in water (295 mg/L at 25°C) and volatile with a vapor pressure of 100 mm Hg at 25°C. It is used as a solvent for gums, resins, cellulose acetate, and cellulose nitrate; in the production of paraffin wax and high grade lubricating oil; and in consumer products such as lacquers, varnishes, paint remover, and glues. Although specific information on its environmental fate is lacking, it is thought to volatilize slowly from water and soil and to be mobile in soil.[4] Because of the lack of data on the expected contribution of various media to human exposure, 50% of the contribution to total exposure will be allowed from air.

On account of the paucity of human data and the broad spectrum of effects associated with MEK exposure in animal studies, AALGs were calculated for multiple endpoints. The AALG for developmental toxicity is based on the rat LOAEL of 1126 ppm from the study of Schwetz et al.[6], adjusted for continuous exposure (3320 mg/m^3 × 7/24), and converted to a human equivalent exposure level according to the procedures described in Appendix C with a total uncertainty factor of 2500 (10 for interindividual variation × 10 for interspecies variation × 5 for the developmental uncertainty factor). Application of the developmental uncertainty factor was warranted by the finding of significant fetotoxic and teratogenic effects at exposure levels where no maternal toxicity was apparent.

Two AALGs were calculated for systemic toxicity because of the difference in the nature of toxic effects reported, i.e., subtle changes in motor nerve function vs decreased body weight gain and altered liver function. The rat NOAEL of 200 ppm from the Takeuchi et al. study[8] was adjusted for continuous exposure (590 mg/m^3 × 12/24) and converted to a human equivalent inhalation exposure level according to the procedures described in Appendix C. A total uncertainty factor of 1000 (10 for interindividual variation × 10 for interspecies variation × 10 for less-than-chronic exposure) was applied. The other AALG was based on the rat NOAEL of 2500 ppm from the study of Cavender et al.,[11] adjusted for continuous exposure (7350 mg/m^3 × 6/24 × 5/7), and converted to a human equivalent exposure level according to the procedures described in Appendix C. A total uncertainty factor of 1000 (10 for interindividual variation × 10 for interspecies variation × 10 for less than chronic exposure) was applied.

The AALG for irritation is based on the human LOAEL of 200 ppm from the study of Nelson et al.[12] with a total uncertainty factor of 50 (10 for interindividual variation × 5 for a LOAEL).

Because the Schwetz et al. study[6] is stronger than the Takeuchi et al. study[8] and the derived criteria are similar in magnitude, the recommended AALG is the one based on developmental toxicity.

AALG:
- developmental toxicity—0.13 ppm (0.39 mg/m^3) 24-hour TWA
- systemic toxicity—0.11 ppm (0.33 mg/m^3) 24-hour TWA; 0.5 ppm (1.5 mg/m^3) 24-hour TWA
- irritation—4.0 ppm (12 mg/m^3) 24-hour TWA

References

1. ACGIH. 1986. "Documentation of the Threshold Limit Values and Biological Exposure Indices." 5:395.
2. NIOSH. 1978. "Criteria for a Recommended Standard . . . Occupational Exposure to Ketones." DHEW (NIOSH) 78–173.
3. OSHA. 1989. "Air Contaminants; Final Rule." *Fed. Reg.* 54:2332–2959.
4. U.S. EPA. 1987. "Health Advisory—Methyl Ethyl Ketone." Office of Drinking Water (3/31).
5. NIOSH. 1987. EL6475000. "2-Butanone." RTECS, on line.
6. Schwetz, B.A., B.J.K. Leong, and P.J. Gehring. 1974. "Embryo- and Fetotoxicity of Inhaled Carbon Tetrachloride, 1,1-Dichloroethane, and Methyl Ethyl Ketone in Rats." *Tox. Appl. Pharm.* 28:452–64.

7. Deacon, M.M., M.D. Pilny, J.A. John, B.A. Schwetz, F.J. Murray, H.O. Yakel, and R.A. Kuna. 1981. "Embryo- and Fetotoxicity of Inhaled Methyl Ethyl Ketone in Rats." *Tox. Appl. Pharm.* 59:617–19.

8. Takeuchi, Y., Y. Ono, N. Hisanga, M. Iwata, M. Aoyama, J. Kitoh, and Y. Sugiura. 1983. "An Experimental Study of the Combined Effects of *n*-Hexane and Methyl Ethyl Ketone." *Br. J. Ind. Med.* 40:199–203.

9. U.S. EPA. 1984. "Health Effects Assessment for Methyl Ethyl Ketone." PB86–134145, EPA-540/1–86/003 (Springfield, VA: NTIS).

10. Saida, K., J.R. Mendell, and H.S. Weiss. 1976. "Peripheral Nerve Changes Induced by Methyl *n*-Butyl Ketone and Potentiation by Methyl Ethyl Ketone." *J. Neuropath. Exp. Neurol.* 35:207–25.

11. Cavender, F.L., H.W. Casey, H. Salem, J.A. Swenberg, and E.J. Gralla. 1983. "A 90-Day Vapor Inhalation Toxicity Study of Methyl Ethyl Ketone." *Fund. Appl. Tox.* 3:264–70.

12. Nelson, K.W., J.F. Ege, M. Ross, L.E. Woodman, and L. Silverman. 1943. "Sensory Response to Certain Industrial Solvent Vapors." *J. Ind. Hyg. Tox.* 25:282–85.

METHYL ISOBUTYL KETONE

Synonyms: hexone, MIBK

CAS Registry Number: 108–10–1

Molecular Weight: 100.16

Molecular Formula: $CH_3C(O)CH_2CH(CH_3)_2$

AALG: systemic toxicity (TLV-based)—0.12 ppm (0.49 mg/m^3) 8-hour TWA

Occupational Limits:
- ACGIH TLV: 50 ppm (205 mg/m^3) TWA; 75 mg/m^3 (300 mg/m^3) STEL
- NIOSH REL: 50 ppm (205 mg/m^3) TWA
- OSHA PEL: 100 ppm (410 mg/m^3) TWA

Basis for Occupational Limits: The occupational limits for MIBK are based on prevention of irritation and possible nephrotoxicity.[1,2] OSHA recently adopted the ACGIH TLV-TWA and TLV-STEL limits as part of its final rule limits.[3]

Drinking Water Limits: No drinking water limits were found for this substance.

Toxicity Profile

Carcinogenicity: No data were found implicating MIBK as a carcinogen.

Mutagenicity: No data were found implicating MIBK as a mutagen.

Developmental Toxicity: No data were found implicating MIBK as a developmental toxin.

Reproductive Toxicity: No data were found implicating MIBK as a reproductive toxin.

Systemic Toxicity: The acute toxicity of MIBK is relatively low, based on both oral and inhalation data. Oral LD$_{50}$s of 2080, 2671, and 1600 mg/kg have been reported in rats, mice, and guinea pigs, respectively, and an LC$_{50}$ of 23,300 mg/m^3 (time not specified) has been reported in mice.[4]

MacEwen et al.[5] (cited in ACGIH[1] and NIOSH[2]) studied the effects of inhalation exposure to MIBK in rats, mice, dogs, and monkeys for periods of up to 90 days. In a preliminary two-week study, rats, mice, dogs, and monkeys (n = 50, 40, 8, and 4/group, sex unspecified) were exposed to 100 or 200 ppm MIBK continuously; comparable controls were exposed to room air in identical chambers. The only significant differences between controls and treated animals for any species were higher absolute and relative kidney weights in rats at both exposure levels and higher absolute and relative liver weights in the highest exposure group. Histologic examination of the kidneys revealed "toxic nephrosis" of the renal tubules at both exposure levels. All other hematologic, clinical chemistry, and histologic examinations revealed no differences between treated and control animals. In further experiments, dogs, rats, and monkeys were exposed to 100 ppm MIBK continuously at 5 psi (approximately one-third atmosphere) for 90 days; comparable numbers of animals served as unexposed controls. Results of histologic examination and hematologic and clinical chemistry determinations revealed no differences between treated and control animals with the exception of pathologic changes in the kidneys of rats. These changes consisted of hyaline droplet degeneration of the proximal tubules with occasional foci of tubular necrosis. Relative and absolute kidney weights were also significantly increased. Examination of animals killed at various time periods after exposure ceased revealed that these effects were slowly reversible.

The effects of MIBK exposure in humans have not been evaluated in epidemiologic studies. However, in a case report of workers exposed to 500 ppm MIBK for 20 to 30 minutes per day near a centrifuge (with MIBK concentrations of 80 ppm elsewhere in the area), there were complaints of weakness, anorexia, headache, burning of the eyes, stomach ache, nausea, vomiting, and sore throat. A few workers also complained of insomnia, somnolence, heartburn, and intestinal pain, and four workers were reported to have slightly enlarged livers. However, clinical chemistry determinations were within normal limits[6] (cited in ACGIH[1] and NIOSH[2]). Five years later, improved work practices had resulted in a reduction of exposure to 100–105 ppm for 15–20 minutes during centrifuging (with MIBK concentrations of 50 ppm elsewhere in the room) and workers were required to wear respiratory protection. Some workers still complained of GI and CNS effects, and slight liver enlargement had persisted in two workers. However, other symptoms were no longer reported[7] (cited in ACGIH[1] and NIOSH[2]).

In another case report, workers exposed to approximately 100 ppm MIBK during a boot waterproofing process reported experiencing headaches and nausea, but another group of similarly exposed workers reported only respiratory irritation[8] (cited in ACGIH[1] and NIOSH[2]).

Irritation: At high concentrations MIBK is irritating to the mucous membranes of the eyes, nose, and throat.[1,2] Silverman et al.[9] (cited in ACGIH[1] and NIOSH[2]) studied the sensory effects of exposure to MIBK in human subjects (n = 12 males and females) exposed for 15 minutes to concentrations of up to 200 ppm. At a concentration of 200 ppm MIBK the odor was judged to be objectionable and irritation of the eyes, nose, and throat occurred. The highest concentration judged to be satisfactory for an 8-hour work day was 100 ppm. It should be noted that these concentrations were nominally, not analytically, determined, and the actual concentrations may have been lower.

Basis for the AALG: MIBK is somewhat water-soluble (1.91 g/100 mL) with a vapor pressure of 7.5 mm Hg at 25°C. It is used as a solvent for nitrocellulose and cellulose ethers, etc.[1] Based on lack of data on its environmental fate, distribution, and occurrence, 50% of the contribution to total exposure will be allowed from air.

Consideration of all the available data resulted in the conclusion that the long-standing TLV was a more appropriate basis than the available animal studies, since the latter were subchronic in duration, only one exposure level was tested, the implications of exposure at less than atmospheric pressure for animal extrapolation are unclear, and no NOAEL was identified for renal effects in rats (the most sensitive species tested). The TLV was treated as a human LOAEL and a total uncertainty factor of 210 was applied (4.2 for continuous exposure adjustment × 10 for interindividual variation × 5 for a LOAEL). The decision to treat the TLV as a surrogate human LOAEL was based on a case report citing continued complaints of CNS and GI tract effects even after reduction of exposure levels to the range of 50 to 105 ppm[7] (cited in ACGIH[1] and NIOSH[2]). This AALG should be considered provisional due to limitations in the available database.

AALG: • systemic toxicity (TLV-based)—0.12 ppm (0.49 mg/m³) 8-hour TWA

References

1. ACGIH. 1986. "Documentation of the Threshold Limit Values and Biological Exposure Indices." 5:402.
2. NIOSH. 1978. "Criteria for a Recommended Standard . . . Occupational Exposure to Ketones." DHEW (NIOSH) 78–173.
3. OSHA. 1989. "Air Contaminants; Final Rule." *Fed. Reg.* 54:2332–959.
4. NIOSH. 1988. SA9275000. "2-Pentanone, 4-Methyl." RTECS, on line.
5. MacEwen, J. D., E. H. Vernot, and C. C. Haun. 1971. "Continuous Exposure to Methylisobutylketone on Dogs, Monkeys and Rats" (Springfield, VA: National Technical Information Service), NTIS AD 730 291.
6. Linari, F., G. Perrelli, and D. Varese. 1964. "Clinical Observations and Blood Chemistry Tests Among Workers Exposed to the Effect of a Complex Ketone—Methyl Isobutyl Ketone." *Arch. Sci. Med.* pp. 226–37.
7. Armeli, G., F. Linari, and G. Martorano. 1968. "Clinical and Hematochemical Examinations in Workers Exposed to the Action of a Ketone (MIBK) After Five Years." *Lav. Um.* 20:418–24.
8. Elkins, H. B. 1959. *The Chemistry of Industrial Toxicology,* 2nd ed. (New York: John Wiley and Sons), p. 121.
9. Silverman, L., H. F. Schulte, and M. W. First. 1946. "Further Studies on Sensory Response to Certain Industrial Solvent Vapors." *J. Ind. Hyg. Tox.* 28:262–66.

METHYL METHACRYLATE

Synonyms: methyl 2-methyl-2-propanoate, MMA

CAS Registry Number: 80–62–6

Molecular Weight: 100.13

Molecular Formula:

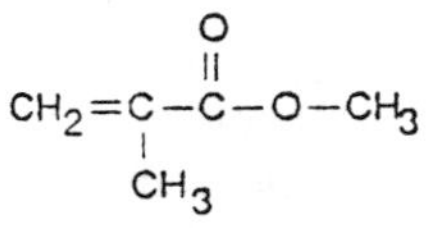

$$CH_2=C-\overset{\overset{\displaystyle O}{\|}}{C}-O-CH_3$$
$$|$$
$$CH_3$$

AALG: irritation (cumulative effects)—0.1 ppm (0.4 mg/m^3) 24-hour TWA

Occupational Limits:
- ACGIH TLV: 100 ppm (410 mg/m^3) TWA
- NIOSH REL: none
- OSHA PEL: 100 ppm (410 mg/m^3) TWA

Basis for Occupational Limits: The TLV was set based on protection against irritation and acute systemic effects. However, it is noted in the TLV documentation that "information is not available at this time to determine whether this level will be protective under conditions of chronic long-term exposures."[1]

Drinking Water Limits: In 1977, NAS calculated a SNARL in drinking water for methyl methacrylate (MMA) of 0.7 mg/L based on a two-year rat NOAEL of 2000 ppm (100 mg/kg/day) and an uncertainty factor of 1000.[2]

Toxicity Profile

Carcinogenicity: No treatment-related tumors were found in one lifetime study in which male and female Wistar rats were exposed to 0, 6, 60, or 2000 ppm MMA in drinking water[3] (cited in IARC[4]). However, IARC noted that there was insufficient reporting of survival data and histopathological examination. Negative results were also reported in a skin painting study using Wistar rats[5] (cited in IARC[4]). However, it was noted that insufficient sample sizes and short exposure duration rendered this study inadequate for carcinogenicity evaluation.

In an NTP bioassay,[6,7] male and female B6C3F$_1$ mice and male Fisher 344/N rats (n = 50/sex/group) were exposed to 0, 500, or 1000 ppm and female F344/N rats to 0, 250, or 500 ppm MMA via inhalation for 6 hours per day, 5 days per week for 102 weeks. No decrease in survival attributable to MMA exposure occurred in rats or mice; however, there was a decrease in body weight gain in treated rats after 80 weeks and treated mice after 20 weeks. No evidence of carcinogenicity was found for MMA in rats or mice of either sex; however, there was an increased incidence of serous and suppurative inflammation of the nasal cavity as well as degeneration of the olfactory sensory epithelium characterized by loss of neuroepithelial cells in exposed male and female rats. Inflammation of the nasal cavity was also observed in mice, in addition to hyperplasia of submucosal nasal epithelium and degeneration of the olfactory sensory epithelium; interstitial inflammation of the lung was noted in 1000 ppm male mice.

It is the judgment here that MMA is most appropriately placed in group D, unclassified substance, under the EPA weight-of-evidence classification.

Mutagenicity: In NTP studies, MMA was nonmutagenic to *S. typhimurium* in a variety of strains with and without metabolic activation but was mutagenic to cultured mammalian cells with and without metabolic activation. MMA was also found to induce SCEs and chromosomal aberrations in cultured CHO cells.[6]

Developmental Toxicity: Nicholas et al.[8] exposed Sprague-Dawley rats to 110 mg/L MMA for 17.2 (n = 22) or 54.2 (n = 27) minutes on gestation days 6 through 15; controls were either unexposed (n = 22) or exposed to air only (n = 26) for 54.2 minutes. There was a statistically significant decrease in fetal weight and fetal crown-rump length (dose-related) as well as delayed ossification in the exposed groups relative to both control groups. There was also a statistically significant increase in hematomas in the high-dose, but not the low-dose, fetuses. Maternal toxicity occurred in the form of decreased body weight gain, but for the low-dose group this was only statistically significant on day 15 (relative to untreated controls) and day 20 (relative to both controls).

Tansy and Kendall[9] (cited in ACGIH[1]) studied the effects of 60 hours of exposure to 116 or 400 ppm MMA via inhalation in pregnant mice and concluded that it did not produce teratogenic effects.

Reproductive Toxicity: No data were found implicating MMA as a reproductive toxin.

Systemic Toxicity: A study conducted by NIOSH[10] (cited in ACGIH[1] and IARC[4]) examined the health status of 91 exposed and 43 unexposed workers in five plants manufacturing polymethyl methacrylate; exposures were in the range of 16 to 200 mg/m³ (4 to 49 ppm) as an 8-hour TWA. Although no statistically significant effects were noted on symptomatology, blood pressure, respiratory function tests, blood counts, or blood and urine chemistries, it was stated in the TLV documentation that some of the findings were suggestive of alterations in skin and nervous system symptomatology and blood and urine chemistry values. It is relevant to note that animal studies indicate that the MMA monomer, not the polymer, is the toxicologically active entity.[6]

Nervous system effects, e.g., paresthesia of the fingers and siowed distal sensory conduction velocities from the digits, have been reported in individuals handling methacrylate putty.[6] MMA is also reported to be adipogenic in women and to cause disturbances in insulin, somatotropic hormone, and prolactin levels[11] (cited in NTP[6]).

In laboratory animals, the primary effects of MMA appear to be mainly manifest at the portal of entry, i.e., the nasal cavity as evidenced by lifetime inhalation studies in both rats and mice.[6,7] Tansy et al.[12] (cited in Chan et al.[7]) reported that exposure of adult male rats to 116 ppm MMA vapor for 7 hours per day, 5 days a week for 6 months resulted in loss of cilia in the trachea and decreased numbers of epithelial microvilli but produced no other exposure-related lesions.

Irritation: MMA is irritating to the skin, eyes, and mucous membranes in humans and has been reported to cause allergic dermatitis or stomatitis.[6] In a study cited in the TLV documentation, irritation was reported at concentrations of 170–250 ppm. MMA is reported to have a distinctive pungent smell with an odor threshold of less than 1 ppm.

Basis for the AALG: MMA is considered slightly soluble in water and has a vapor pressure of 30 mm Hg at 20°C. MMA polymerizes readily and is widely used in the manufacture of coatings, plastics, surgical implants, and in dentistry.[1] MMA has been detected in both ambient air and drinking water.[7] Because of a lack of information on its environmental fate and distribution and on the expected contribution of various media to human exposure, 50% of the contribution to total exposure will be allowed from air.

There are two possible bases for the AALG: the TLV and the mouse or rat LOAEL for effects on the olfactory mucosa. Neither is completely satisfactory. The TLV is not completely satisfactory because the exact rationale for selecting 100 ppm is not supported by adequate data correlating exposure and effect, and the animal LOAELs are not because the study was not designed, i.e., exposure doses were not selected, with the goal of establishing a LOAEL or NOAEL for noncarcinogenic effects.

For the TLV-based AALG, the TLV/42 is used since the TLV was set on the basis of acute systemic, as well as irritant, effects. For the LOAEL-based AALG, the rat LOAEL of 250 ppm with an adjustment for continuous exposure (250 ppm × 5/7 × 6/24) was used, rather than the mouse LOAEL of 250 ppm, because use of the former results in a more conservative criterion. The total uncertainty factor used was 500 (5 for a LOAEL × 10 for interindividual variation × 10 for interspecies variation). The AALG based on the rat LOAEL is recommended over the TLV-based AALG because it is more conservative.

AALG: systemic toxicity/cumulative irritant effects
- TLV-based—1.2 ppm (4.9 mg/m³) 8-hour TWA
- rat LOAEL—0.1 ppm (0.4 mg/m³) 24-hour TWA

References

1. ACGIH. 1986. "Documentation of the Threshold Limit Values and Biological Exposure Indices." 5:406–7.
2. Safe Drinking Water Committee. 1977. *Drinking Water and Health* (Washington, DC: National Academy of Science).
3. Borzelleca, J.F., P.S. Larson, G.R. Hennigar, E.G. Huf, E. Crawford, and R.S. Blackwell. 1964. "Studies on the Chronic Oral Toxicity of Monomeric Ethyl Acrylate and Methyl Methacrylate." *Tox. Appl. Pharm.* 6:29–36.
4. IARC. 1979. "Methyl Methacrylate and Polymethyl Methacrylate." *IARC Monog.* 19:187–211.
5. Oppenheimer, B.S., E.T. Oppenheimer, I. Danishefsky, A.P. Stout, and F.R. Eirich. 1955. "Further Studies of Polymers as Carcinogenic Agents in Animals." *Cancer Res.* 15:333–40.
6. NTP. 1986. "Toxicology and Carcinogenesis Studies of Methyl Methacrylate (CAS No. 80–62–6) in F344/N Rats and B6C3F$_1$ Mice." NTP TR-314, National Toxicology Program, Research Triangle Park, NC.
7. Chan, P.C., S.L. Eustis, J.E. Huff, J.K. Haseman, and H. Ragan. 1988. "Two-Year Inhalation Carcinogenesis Studies of Methyl Methacrylate in Rats and Mice: Inflammation and Degeneration of Nasal Epithelium." *Toxicology* 52:237–52.
8. Nicholas, C.A., W.H. Lawrence, and J. Autian. 1979. "Embryotoxicity and Fetotoxicity of Methyl Methacrylate Monomer in Rats." *Tox. Appl. Pharm.* 50:451–58.
9. Tansy, M.F., and F.M. Kendall. 1979. *Drug Chem. Tox.* 2:315–30.
10. Cromer, J., and K. Kronoveter. 1976. "A Study of Methyl Methacrylate Exposures and Employee Health." DHEW (NIOSH) 77–119.
11. Makarov, I., K. Makarenko, and N. Desjatnikova. 1981. "Adipogenic Effect of Some Industrial Poisons." *Gig. Trud. Prof. Zabol.* 12:29–31.
12. Tansy, M., F. Hohenleitner, D. White, R. Oberly, W. Landin, and F. Kendall. 1980. "Chronic Biological Effects of Methyl Methacrylate Vapor. III. Histopathology, Blood Chemistries, and Hepatic and Ciliary Function in the Rat." *Env. Res.* 21:117.

METHYL STYRENE

Synonyms: vinyl toluene, tolyethylene (includes meta- [CAS 100–80–1], ortho-[CAS 611–15–4], and para- [CAS 622–97–9] isomers)

CAS Registry Number: 25013–15–4

Molecular Weight: 118.19

Molecular Formula: $CH_3(C_6H_4)CH=CH_2$

AALG: systemic toxicity—0.16 ppm (0.76 mg/m^3) 24-hour TWA

Occupational Limits:
- ACGIH TLV: 50 ppm (240 mg/m^3) TWA; 100 ppm (480 mg/m^3) STEL
- NIOSH REL: none
- OSHA PEL: 100 ppm (480 mg/m^3) TWA

Basis for Occupational Limits: The occupational limits for methyl styrene are based on irritation and analogy with styrene for systemic toxicity.[1]

Drinking Water Limits: No drinking water limits were found for this substance.

Toxicity Profile

Carcinogenicity: Methyl styrene is currently under evaluation in the NTP bioassay program (inhalation studies, rats and mice); the technical report is scheduled for peer review.[2]

Mutagenicity: Positive results were reported in a mouse micronucleus test for the mixed isomers of methyl styrene. The meta-, ortho-, and para- isomers of methyl styrene have been reported to induce sister chromatid exchanges in cultured human lymphocytes; the para- isomer has also been reported to induce chromosomal aberrations in cultured human lymphocytes.[3]

Developmental Toxicity: No data were found implicating methyl styrene as a developmental toxin.

Reproductive Toxicity: No data were found implicating methyl styrene as a reproductive toxin.

Systemic Toxicity: An LC_{50} of 3020 mg/m^3 has been reported for methyl styrene in mice.[3] Data on the effects of subchronic and chronic inhalation exposure to methyl styrene are lacking in humans and limited in laboratory animals.

Wolf et al.[4] exposed Wistar-derived rats, guinea pigs, rabbits, and rhesus monkeys (numbers not specified; only female monkeys were used, but both sexes for all other species) to 0, 580, 1130, or 1350 ppm methyl styrene via inhalation for 7 hours per day, 5 days a week for approximately 20 weeks (92–100 exposures). No signs of toxicity or pathologic effects were noted in monkeys exposed at any level, and 580 ppm was the reported NOAEL for all other species. Growth depression was reported in both guinea pigs and rats, and decreased kidney weights in rabbits and guinea pigs at the middle and high exposure levels. The primary histopathologic changes observed were midzonal and centrilobular fatty degeneration of the liver, which were observed in both guinea pigs and rats exposed to 1130 and 1350 ppm methyl styrene.

Irritation: Methyl styrene is irritating to the eyes and mucous membranes. In a controlled human exposure study (time periods of exposure not given), subjects reported eye and nasal irritation at 400 ppm, strong objectionable odor at 300 ppm, and strong tolerable odor at 200 ppm. At 50 ppm, it was reported that the odor was detectable, but no irritation of mucous membranes occurred.[4]

Basis for the AALG: Methyl styrene is considered only very slightly soluble in water and has a vapor pressure of 1.1 mm Hg at 20°C. It is used primarily as a solvent and intermediate in chemical syntheses.[1] Because of the lack of data on environmental fate and distribution and the expected contribution of various media to human exposure, 50% of the contribution to total exposure will be allowed from air.

Two possible bases were considered in deriving the AALG: the TLV/42 and the rat NOAEL of 580 ppm from the Wolf et al. study.[4] In the case of the TLV, a divisor of 42, incorporating the 4.2 factor for continuous exposure adjustment, was considered appropriate since systemic toxicity has been demonstrated in animal models and there are positive data for chromosomal effects in human lymphocytes in vitro. The rat NOAEL of 580 ppm was adjusted for continuous exposure (580 ppm $\times$ 5/7 $\times$ 7/24) and converted to a human equivalent NOAEL (the midpoint of the range of rat weights from the original paper—212.5 g—was used in the calculations) according to the procedures described in Appendix C. A total uncertainty factor of 1000 was applied (10 for interspecies variation $\times$ 10 for interindividual variation $\times$ 10 for less-than-chronic exposure). Since the limit calculated on the basis of the Wolf et al. study,[4] 0.76 mg/m^3 (0.16 ppm), was more conservative than the one calculated on the basis of the TLV, 2.9 mg/m^3 (0.6 ppm), the former is recommended as the AALG. This AALG should be considered provisional pending availability of results from the NTP inhalation bioassay.

AALG: • systemic toxicity—0.16 ppm (0.76 mg/m^3) 24-hour TWA

References

1. ACGIH. 1986. ''Documentation of the Threshold Limit Values and Biological Exposure Indices.'' 5:630.
2. NTP. 1989. ''Management Status Report.'' National Toxicology Program, Research Triangle Park, NC (2/7).
3. NIOSH. 1988. WL5075000, WL5075800, WL5075900, WL5076000. ''Styrene, methyl; styrene, *m*-methyl; styrene, *o*-methyl; styrene, *p*-methyl.'' RTECS, on line.
4. Wolf, M.A., V.K. Rowe, D.D. McCollister, R.L. Hollingsworth, and F. Oyen. 1956. ''Toxicological Studies of Certain Alkylated Benzenes and Benzene.'' *AMA Arch. Ind. Heal.* 14:387–98.

NAPHTHALENE

Synonyms: napthalene, naphthene

CAS Registry Number: 91–20–3

Molecular Weight: 128.16

Molecular Formula:

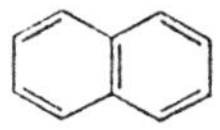

AALG: systemic toxicity—0.024 ppm (0.12 mg/m^3) 8-hour TWA (TLV-based)

Occupational Limits:
- ACGIH TLV: 10 ppm (50 mg/m^3) TWA; 15 ppm (75 mg/m^3) STEL
- NIOSH REL: none
- OSHA PEL: 10 ppm (50 mg/m^3) TWA; 15 ppm (75 mg/m^3) STEL

Basis for Occupational Limits: The occupational limits are based on the prevention of ocular effects, e.g., irritation, corneal injury, optical neuritis, and cataracts. It is acknowledged in the documentation that these limits may not be low enough to protect against hemolytic effects in hypersusceptible individuals.[1] OSHA recently adopted the ACGIH TLV-STEL of 15 ppm as part of its final rule limits.[2]

Drinking Water Limits: At the time the ambient water quality criteria were being established (1980), data were not sufficient to establish a health-based criterion for naphthalene.[3]

Toxicity Profile

Carcinogenicity: Data on naphthalene carcinogenicity are limited. Druckrey and Schmahl[4] (cited in U.S. EPA[5]) evaluated the carcinogenicity of naphthalene in the course of a study on anthracene carcinogenicity (naphthalene was used as the vehicle). BD I or BD III strain rats were exposed to either a total dose of 10 g naphthalene per rat orally (n = 28) or 0.82 g naphthalene per rat subcutaneously (time periods and dosing pattern not specified) and observed for over 1000 days; no tumors were found in either group. In another study[6] (cited in U.S. EPA[5]), naphthalene was not carcinogenic when applied to mouse skin. Naphthalene is currently on test in the NTP bioassay program (mice, inhalation) with pathology materials audit in progress.[7] Naphthalene has been placed in group D, unclassified substance, under the EPA weight-of-evidence classification.[5]

Mutagenicity: Negative results are listed in the Gene-Tox database, as summarized in RTECS, for naphthalene in two oncogenic transformation assays and in the Ames assay.[8]

Developmental Toxicity: Plasterer et al.[9] exposed CD-1 mice to 0 (n = 40) or 300 (n = 33) mg/kg/day naphthalene in corn oil via gavage on gestation days 8 through 14 in a study in which a number of other substances were also evaluated. At this dose level (which was selected to be at or just below the threshold of adult lethality), maternal mortality was significant (15%), and the average number of live young per litter was significantly reduced relative to controls; no teratogenic effects were observed.

Reproductive Toxicity: No data were found implicating naphthalene as a reproductive toxin.

Systemic Toxicity: Data on the inhalation toxicity of naphthalene in laboratory animals are lacking, and data on the ingestion toxicity of naphthalene are limited to subchronic studies. Shopp et al.[10] exposed random-bred male and female CD-1 mice (n = 40–112/sex/group) to corn oil alone or 5.3, 53, or 133 mg/kg naphthalene in corn oil (doses were 1/100, 1/10, and 1/4 the oral LD_{50} in male mice) by gavage for 90 consecutive days. With the exceptions of reduced spleen weights in female high-dose mice and a dose-related inhibition of aryl hydrocarbon hydroxylase in both males and females, the dose levels used did not result in any manifestations of toxicity as evaluated by serum enzyme and electrolyte levels, histopathology, and immunotoxicity tests.

The major toxic effects associated with naphthalene exposure in humans involve hemolytic anemia with associated jaundice and ocular effects. Susceptibility to naphthalene-induced hemolytic anemia is associated with glucose-6-phosphate dehydrogenase (G6PD) deficiency, a sex-linked genetic defect occurring most commonly among black males and males from certain peoples of Mediterranean descent. In addition, newborn infants also are sensitive to the hemolytic effects of naphthalene even in the absence of G6PD deficiency.[3]

The principal manifestations of naphthalene toxicity in occupationally exposed individuals are ocular effects, including optical neuritis and cataracts. Ghetti and Mariani[11] (cited in ACGIH[1] and U.S. EPA[3]) reported finding cataracts in 8 of 21 workers examined who were employed in a plant producing a dye intermediate from naphthalene (age distribution and incidences: 20–30 years, 2 of 4; 30–40 years, 3 of 5; 40–50 years, 2 of 8; and 50–60 years, 1 of 4). It is also noteworthy that cataracts and retinal damage have been produced experimentally in rabbits exposed to naphthalene (no inhalation data).[3]

Irritation: Naphthalene is reported to have a "strong coal tar" odor, and an odor threshold (geometric mean of reported literature values) of 0.084 ppm is listed by one source.[12] It has been reported that concentrations greater than 15 ppm naphthalene produce noticeable eye irritation.[1]

Basis for the AALG: Naphthalene is considered insoluble in water, and there is reportedly "appreciable volatilization" of the solid at room temperature (vapor pressure = 0.087 mm Hg at 25°C). It is used in the manufacture of a variety of materials, as a moth repellant, and in scintillation counters. Since data on the environmental fate and distribution of naphthalene and the expected contribution of various media to human exposure are lacking, 50% of the contribution to total exposure will be allowed from air.

Because nonoccupational inhalation data are lacking and the TLV is long-standing and based on an effect of major concern in the general population, the AALG is based on the TLV. However, since it has been acknowledged that this limit may not be protective of individuals with G6PD deficiency, a known high-risk group, it will be treated as a LOAEL, and the TLV will be divided by 210 (5 for a LOAEL × 10 for interindividual variation × 4.2 for continuous exposure adjustment). This limit should be considered provisional pending availability of results from the NTP inhalation bioassay.

AALG: • systemic toxicity—0.024 ppm (0.12 mg/m^3) 8-hour TWA (TLV-based)

References

1. ACGIH. 1986. "Documentation of the Threshold Limit Values and Biological Exposure Indices." 5:420.
2. OSHA. 1989. "Air Contaminants; Final Rule." *Fed. Reg.* 54:2332–2959.
3. U.S. EPA. 1980. "Ambient Water Quality Criteria for Naphthalene." EPA 440/5–80–059.
4. Druckrey, H., and D. Schmahl. 1955. "Cancerogene wirkung von anthracen" (German). *Die Naturwissenschaften.* 42:159.
5. U.S. EPA. 1984. "Health Effects Assessment for Naphthalene." EPA/540/1–86–014.
6. Schmeltz, I., et al. 1978. "Bioassays of Naphthalene and Alkyl Naphthalenes for Carcinogenic Activity. Relation to Tobacco Carcinogenesis." In *Carcinogenesis, Vol. 3: Polynuclear Aromatic Hydrocarbons,* P. Jones and K. Freudenthal, Eds. (New York, NY: Raven Press).
7. NTP. 1989. "Management Status Report." National Toxicology Program, Research Triangle Park, NC (2/7).
8. NIOSH. 1988. QJ0525000. "Naphthalene." RTECS, on line.
9. Plasterer, M.R., W.S. Bradshaw, G.M. Booth, M.W. Carter, R.L. Schuler, and B.D. Hardin. 1985. "Developmental Toxicity of Nine Selected Compounds Following Prenatal Exposure in the Mouse: Naphthalene, *p*-Nitrophenol, Sodium Selenite, Dimethyl

Phthalate, Ethylene Thiourea, and Four Glycol Ether Derivatives.'' *J. Tox. Environ. Health* 15:25–38.

10. Shopp, G.M., K.L. White, M.P. Holsapple, D.W. Barnes, S.S. Duke, A.C. Anderson, L.W. Condie, J.R. Hayes, and J.F. Borzelleca. 1984. ''Naphthalene Toxicity in CD-1 Mice: General Toxicology and Immunotoxicology.'' *Fund. Appl. Tox.* 4:406–19.

11. Ghetti, G., and L. Mariani. 1956. ''Eye Changes Due to Naphthalene.'' *Med. Lav.* 47:524.

12. Amoore, J.E., and E. Hautala. 1983. ''Odor as an Aid to Chemical Safety: Odor Thresholds Compared with Threshold Limit Values and Volatilities for 214 Industrial Chemicals.'' *J. Appl. Tox.* 3:272–90.

NICKEL AND COMPOUNDS

Synonyms: inorganic nickel salts, including nickel carbonyl (Ni(CO)$_4$, CAS 13463–39–3) and nickel subsulfide (Ni$_3$S$_2$, CAS 12035–72–2)

CAS Registry Number: 7440–02–0

Molecular Weight: 58.71

Molecular Formula: Ni

AALG:
- nickel subsulfide (10^{-6} 95% UCL)—0.4 ng (Ni)/m^3 annual TWA
- soluble nickel compounds (systemic toxicity)—0.36 ng (Ni)/m^3 24-hour TWA
- insoluble nickel compounds (systemic toxicity)—7.1 ng (Ni)/m^3 24-hour TWA

Occupational Limits:

- ACGIH TLV: 1 mg (Ni)/m^3 TWA metal and insoluble compounds; 0.1 mg (Ni)/m^3 TWA soluble inorganic compounds; 0.05 ppm (0.1 mg (Ni)/m^3) TWA nickel carbonyl
- NIOSH REL: 0.015 mg (Ni)/m^3 TWA inorganic nickel compounds; 1 ppb (7 µg (Ni)/m^3) TWA nickel carbonyl
- OSHA PEL: 1 mg (Ni)/m^3 TWA metal and insoluble compounds; 0.1 mg (Ni)/m^3 TWA soluble inorganic compounds; 1 ppb (7 µg (Ni)/m^3) TWA nickel carbonyl

Basis for Occupational Limits: In the TLV documentation, reference is made to epidemiologic and animal studies designed to assess the carcinogenicity of nickel, and these studies are briefly reviewed; however, it was concluded that "with the available knowledge, it is not felt that all forms of nickel are carcinogenic." It should be noted that, although nickel carbonyl is not designated as a carcinogen by ACGIH (i.e., listed in Appendix A1 or A2), the documentation states that the TLV is low enough to "minimize any potential carcinogenic effects." There is also a separate process TLV-TWA of 1 mg/m^3 with an A1a (recognized human carcinogen) designation for nickel sulfide roasting.[1]

Based on findings of hyperplasia and mild irritation of alveolar cells in rats exposed to 0.1 mg/m^3 nickel chloride 12 hours per day for 2 weeks and increased lung weights (considered a sign of mild irritation) in rats and guinea pigs exposed to 1.0 mg/m^3 nickel chloride for 6 months, the TLV for soluble nickel salts (e.g., chloride, sulfate, nitrate) was reduced to 0.1 mg/m^3. Note that the insoluble nickel salts include nickel oxides (NiO, Ni_2O_3), nickel sulfide, nickel carbonate, and trinickel disulfide.[1]

The health effects considered in establishing the NIOSH limit include dermal toxicity and lung and nasal cancer. Contrary to the position of the ACGIH, NIOSH concluded in its criteria document[2] that ''in the absence of evidence to the contrary, nickel metal and all inorganic nickel compounds, when airborne, should be considered carcinogens.'' It should be noted that the NIOSH and OSHA limit for nickel carbonyl of 1 ppb (7 $\mu g/m^3$) was set at the lowest detectable level.[3] OSHA recently adopted the ACGIH TLV for soluble nickel compounds as part of its final rule limits.[4]

Drinking Water Limits: The ambient water quality criterion for nickel is 0.632 mg/L for a 70-kg man consuming 2 L of water and 6.5 g of fish/shellfish per day.[5] The 1-day and 10-day HAs for a 10-kg child consuming 1 L of water per day are both 0.1 mg/L. The longer-term HA for a 70-kg adult is 0.35 mg/L, and the lifetime HA is 0.15 mg/L; both are derived from a reference dose calculated on the basis of a two-year rat feeding study.[6] Nickel is scheduled for regulation (MCL and MCLG) by 1989 under the 1986 amendments to the Safe Drinking Water Act.[7]

Toxicity Profile

Carcinogenicity: Human epidemiologic data indicate that at least some forms of nickel are carcinogenic to humans via inhalation, but the evidence for the ingestion route is inadequate. Nickel sulfide ore smelting and refining processes have been associated with an excess risk of lung, nasal, and laryngeal cancers in a number of studies, as well as buccal, pharyngeal, prostate, and kidney cancers in a few instances. The association is considered strong based on the consistency of findings in several countries, specificity of the tumor site (lung and nose), high relative risks, and a dose-response relationship based on length of exposure.[8] Both nickel refinery dust from pyrometallurgical nickel sulfide matte refineries and nickel subsulfide are classified as group A, human carcinogens, based on the EPA weight-of-evidence classification.[9,10] It should be noted that nickel refinery dust is a mixture of many nickel compounds, and it is uncertain which compounds in the dust are carcinogenic.[9] However, the major component of nickel refinery dust believed to be responsible for the observed carcinogenic effects is nickel subsulfide.

Nickel subsulfide has also been shown to be carcinogenic to F344 rats via inhalation, inducing primarily pulmonary adenomas and adenocarcinomas[11] (cited in U.S. EPA[8]). Other studies pertinent to the carcinogenicity of nickel refinery dust and nickel subsulfide in animal models are referenced and briefly reviewed in the IRIS summaries.[9,10]

Nickel carbonyl is classified as a group B2, probable human carcinogen, by the U.S. EPA on the basis of production of pulmonary adenomas in rats exposed via inhalation[12,13,14] (cited in U.S. EPA[8]). It has also been shown to induce malignant tumors in rats following intravenous injection, and biochemical studies indicate that it binds to DNA and inhibits RNA polymerase activities.[8]

Sunderman et al.[14] (cited in U.S. EPA[8]) evaluated the carcinogenicity of nickel carbonyl ($Ni(CO)_4$) in male Wistar rats using the following exposure groups:

1. solvent controls (50% ethanol/ether, n = 41)
2. 0.03 mg/L for 30 minutes, three times a week for one year (n = 64)
3. 0.06 mg/L for 30 minutes, three times a week for one year (n = 32)
4. a single 0.25 mg/L exposure (n = 80)

Of the nine animals surviving two years or more, four had lung tumors (one in group 2, one in group 3, and two in group 4), and no tumors were found in the controls. No pathological examination was carried out on animals that died during the interim period, and survival in all groups was poor. In a more elaborate follow-up to this study, Sunderman and Donnelly[12] (cited in U.S. EPA[8]) confirmed that malignant neoplasms of the lung occurred in all exposure groups, but not in the controls. Although the sample sizes were small and statistical analysis was not performed, the rarity of this tumor type in Wistar rats gives the finding significance. A LOAEL of 0.03 mg/L (30 mg/m^3) is suggested from the first study.

With respect to other nickel compounds, nickel oxide (NiO) and nickel metal have both been reported to induce injection site sarcomas, nickel (III) oxide (Ni_2O_3) has not been tested sufficiently to allow conclusions to be drawn as to its carcinogenicity, and nickel sulfate ($NiSO_4$) and nickel chloride ($NiCl_2$) have given negative responses in some injection studies. However, it should be noted that positive results have been reported for nickel sulfate and nickel chloride in some genotoxicity assays. Nickel acetate (a soluble nickel salt) has been reported to induce pulmonary tumors in strain A mice following intraperitoneal injection, as well as inducing oncogenic transformation in cultured mammalian cells and inhibiting DNA and RNA synthesis.[8]

At present only nickel carbonyl and nickel subsulfide mixtures can be considered carcinogenic based on the EPA weight-of-evidence classification. However, it is noted in the HAD[8] that ''the carcinogenic potential of other nickel compounds remains an important area for further investigation. Some biochemical and in vitro studies seem to indicate the nickel ion as a potential carcinogenic form of nickel and nickel compounds. If this is true, all nickel compounds might be potentially carcinogenic with potency differences related to their ability to enter and make the carcinogenic form of nickel available to a susceptible cell.'' At present, nickel (II) oxide, nickel sulfate hexahydrate, and nickel subsulfide are in the NTP bioassay program with prechronic studies completed and two-year studies in progress.[15]

Mutagenicity: The evidence for mutagenicity of nickel is summarized as follows in the HAD[8]:

It appears that nickel may induce gene mutations in bacteria and cultured mammalian cells; however, the evidence is fairly weak. In addition, nickel appears to induce chromosomal aberrations in cultured mammalian cells and sister chromatid exchanges in both cultured mammalian cells and human lymphocytes. However, the induction of chromosomal aberrations in vivo has not been observed. More definitive studies are needed to determine whether or not nickel is clastogenic. Nickel does appear to have the ability to induce morphological cell transformations in vitro and to interact with DNA resulting in cross-links and strand breaks. In aggregate, studies have demonstrated the ability of nickel compounds to induce genotoxic effects; however, the translation of these effects into actual mutations is still not clearly understood.

Developmental Toxicity: Data to assess the developmental toxicity of inhaled nickel compounds are very limited. Sunderman et al.[16] (cited in U.S. EPA[8]) exposed F344 rats (n = 10–13/group) to 0, 0.08, 0.16, or 0.30 mg/L nickel carbonyl via inhalation for 15 minutes on gestation day 7, 8, or 9. Significant maternal lethality occurred at the highest exposure level, and fetal viability and weight were decreased in the 0.08 mg/L day 8, 0.16 mg/L day 8, 0.16 mg/L day 7, 0.16 mg/L day 9, and all 0.30 mg/L exposure groups. Nickel carbonyl also induced highly specific ocular malformations—anophthalmia and microphthalmia—in the 0.16 mg/L day 7 and day 8 groups, but not the day 9 group. In another study using hamsters, Sunderman et al.[17] (cited in U.S. EPA[8]) reported that inhalation exposure to 0.16 mg/L nickel carbonyl via inhalation for 15 minutes on gestation day 4 or 5, but not on days 6, 7, or 8, resulted in a significant increase in fetotoxicity and malformations such as exencephaly and cystic lungs.

In other studies, certain nickel compounds—nickel chloride, nickel acetate, and nickel subsulfide—have been reported to be fetotoxic and/or teratogenic following ingestion or parenteral administration.[8]

Reproductive Toxicity: Inhalation studies assessing the effects of nickel on the reproductive processes of laboratory animals are lacking. Ingestion and parenteral administration studies in male rats indicate that certain nickel salts, e.g., nickel sulfate, may cause degenerative changes in the testes and epididymis as well as effects on spermatogenesis. Limited studies in female rats and hamsters suggest that nickel (as an implant) may decrease embryo viability and inhibit the implantation process.[8]

Essentiality: It is relevant to note here that nickel has been found to be an essential element in prokaryotic organisms and laboratory animals and may be essential for humans also. Nickel deficiency in rats and swine has been associated with increased mortality during the suckling period and decreased litter sizes.[8]

Systemic Toxicity: The most prominent effects associated with nickel are allergenicity and toxicity to the respiratory tract. It is well established that nickel may induce contact dermatitis in sensitized individuals in both occupational and nonoccupational settings. Clinical studies have shown that control of dietary nickel decreases the frequency and severity of the allergic response associated with this compound.[8]

In the HAD, two major categories of respiratory effects due to nickel are identified: (1) direct respiratory effects such as asthma due to primary irritation or an allergic response and (2) increased risk of chronic respiratory tract infections secondary to the impairment of alveolar macrophage function.

Nickel dust, nickel chloride, and nickel oxide have all been reported to induce pulmonary toxicity and/or alterations in alveolar macrophage function in animal models exposed subchronically via inhalation at levels at or near the occupational limits. Weischer et al.[18] (cited in U.S. EPA[19]) exposed Wistar rats (numbers not reported) to 0, 0.2, 0.4, or 0.8 mg/m^3 nickel oxide continuously for 120 days. They reported severe lung, liver, and kidney lesions and increased lung weights in animals at all three exposure levels. Bingham et al.[20] exposed male Wistar-derived rats (numbers not reported) to 0.120 mg/m^3 nickel oxide (NiO) or 0.109 mg/m^3 nickel chloride for 12 hours per day, 6 days a week for periods of up to several months (exact exposure period not given). In animals exposed to NiO for two weeks, there were significantly increased numbers of alveolar macrophages present in lung washes, increased focal infiltration by lymphocytes, and greater mucous production; exposure over longer periods of time resulted in thickening of respiratory bronchi and alveolar walls and subsidence of cellular infiltration. In animals exposed to nickel chloride for several months, the most pronounced effect was hyperplasia of the bronchial epithelium and marked mucous production.

In a study sponsored by NIOSH, Clary[21] (cited in NIOSH[2]) reported the effects of exposure to nickel chloride at 1 mg/m^3 via inhalation in male rats exposed 5 days per week for 3 or 6 months. Fibrosis of the alveolar ducts occurred in all animals exposed to nickel chloride, and signs of irritation in the form of proteinaceous material in alveolar spaces and proliferation of type II granular pneumocytes were observed. Johansson et al.[22] exposed male rabbits (n = 6/exposure group with 12 controls) to 1.0 mg/m^3 metallic nickel dust via inhalation for 6 hours per day, 5 days a week for 3 or 6 months. In animals exposed for 6 months, pronounced changes in the morphology of type II cells were noted (enlargement, presence of lamellar bodies), and the alveoli contained excessive numbers of foamy macrophages and granular phospholipid material. In addition, foci of pneumonia were present in all of the animals exposed for 6 months, compared with only one control, suggesting higher susceptibility to pulmonary infections.

Dunnick et al.[23] reported the results of subchronic inhalation studies in rats and mice designed to evaluate the comparative toxicity of a set of nickel compounds, with particular emphasis on the relationship of amount of nickel accumulated in the lungs and relative water solubility of the nickel compounds to their toxicity. Male and female B6C3F$_1$ mice and F344/N rats (n = 10/sex/group/species) were exposed to filtered air, nickel oxide (NiO), nickel subsulfide (Ni$_2$S$_3$), or nickel sulfate hexahydrate (NiSO$_4$ · 6H$_2$O) for 6 hours per day, 5 days a week for 13 weeks. The exposure levels used were (1) nickel sulfate—0.02, 0.05, 0.1, 0.2, and 0.4 mg (Ni)/m^3; (2) nickel subsulfide—0.11, 0.2, 0.4, 0.9, and 1.8 mg (Ni)/m^3; and (3) nickel oxide—0.4, 0.9, 2.0, 3.9, and 7.9 mg (Ni)/m^3. No treatment-related mortality occurred in rats or mice; however, there was a small, but statistically significant, decrease in mean body weight after 13 weeks in male rats exposed to 0.2, 0.4, and 1.8 mg (Ni)/m^3 nickel subsulfide. There was an exposure-related increase in lung weights in both rats and mice, which was statistically significant relative to controls at lower exposure levels in rats compared with mice for all three compounds. Time course studies indicated that equilibrium levels of nickel subsulfide and nickel sulfate hexahydrate were reached by 13 weeks, but nickel oxide lung levels increased throughout the study. Histopathologic lesions of the lung were associated with exposure to all three compounds and included alveolar macrophage hyperplasia and chronic active inflammation in both mice and rats and fibrosis in mice only. Alveolar macrophage hyperplasia occurred at lower exposure levels than chronic active inflammation for all three compounds; rats were more sensitive than mice, with significant (all or almost all animals affected) alveolar macrophage hyperplasia occurring at the lowest exposure level used for all three compounds in male and female rats. However, it should be noted that chronic active inflammation progressed to fibrosis of the lung only in mice at the two highest exposure levels of nickel subsulfide and nickel sulfate hexahydrate and not in nickel oxide–exposed mice. Atrophy of the olfactory nasal epithelium was also associated only with exposure to nickel sulfate hexahydrate and nickel subsulfide at higher exposure levels; lymphoid hyperplasia of the bronchial lymph nodes was observed following exposure to all three compounds. The order of toxicity (most to least) was associated with water solubility (most to least) and was as follows: nickel sulfate hexahydrate, nickel subsulfide, and nickel oxide.

Irritation: Although dusts and fumes of various nickel compounds may be irritating to the respiratory tract and have been reported to cause perforation of the nasal septum (also loss of sense of smell), other forms of toxicity generally occur at lower levels of exposure than those producing these effects.[2]

Basis of the AALG: Because nickel is ubiquitous in the environment and most data indicate that the bulk of nickel intake is via the diet, 20% of the contribution to total exposure will be allowed from air.[6] However, it should be noted that the relative source contribution from air may be higher for individuals living in the vicinity of nickel refineries. The exception to this is nickel carbonyl, for which 80% of the contribution to total exposure will be allowed from air since it is a gas.[1]

For purposes of this evaluation, those nickel compounds classified in groups A or B2 are treated as carcinogens, and their AALGs are based on this endpoint. This includes nickel subsulfide (based on human epidemiologic data) and nickel carbonyl (based on animal inhalation data).

It was concluded in the HAD[8] that four data sets from epidemiologic studies of nickel refinery workers were suitable for quantitative risk estimation and model testing. These studies are cited here and are those from which the actual data for quantitative risk assessment were principally taken. They include studies of workers at plants in Huntington, West Virginia[24] (cited in U.S. EPA[8]), Copper Cliff, Ontario[25] (cited in U.S. EPA[8]), Clydach, Wales[26] (cited in U.S. EPA[8]), and Kristiansand, Norway[27] (cited in U.S. EPA[8]). Workers at these plants have been under study for many years, and numerous additional reports pertaining to these sites are reviewed in the HAD. Since there is support for the use of both additive and multiplicative risk models, both types were used on all four data sets and the midpoint of the range, 2.4×10^{-4} $(\mu g/m^3)^{-1}$ (range: 1.1×10^{-5} to 4.6×10^{-4} $(\mu g/m^3)^{-1}$), was taken as the best estimate of q_1* for nickel refinery dust. The estimate of q_1* for nickel subsulfide was taken as twice that of nickel refinery dust, i.e., 4.8×10^{-4} $(\mu g/m^3)^{-1}$, based on the roughly 50% nickel subsulfide composition of refinery dust and demonstrations in animal studies that nickel subsulfide is the most carcinogenic nickel compound.[8] Applying the relation $E = q_1* \times d$, where E is the specified level of extra risk and d is the dose, gives the estimates shown below. It should be noted that although nickel oxide and nickel sulfate are important constituents of refinery dust, their carcinogenic potencies relative to nickel subsulfide have not been determined; thus, the above estimates do not apply to these two compounds.

Two provisional AALGs have been calculated for nickel carbonyl: one based on carcinogenicity and the other on developmental toxicity. These AALGs are designated as provisional because in our judgment the confidence in the database for this compound is low (the available studies suffer from design limitations). Nickel carbonyl is operationally treated as a class C carcinogen, i.e., the AALG is calculated on the basis of the LOAEL suggested from the studies of Sunderman and Donnelly[12–14] and cited in U.S. EPA.[8] The LOAEL of 0.03 mg/L (30 mg/m^3) in rats was adjusted for continuous exposure (30 mg/m^3 × 0.5/24 × 3/7 × 1/2), converted to a human dose as described in Appendix C, and divided by a total uncertainty factor of 10,000 (10 for a LOAEL × 10 for interindividual variation × 10 for interspecies variation × 10 for the database factor). A developmental AALG was calculated on the basis of the NOAEL for eye malformations in rats suggested from the study of Sunderman et al.[16] (cited in U.S. EPA[8]). Although this study suffers from design limitations, the highly specific nature of the malformation and its rarity justify calculation of a provisional AALG. The NOAEL of 0.08 mg/L (80 mg/m^3) in rats was converted to a human equivalent exposure level as described in Appendix C and divided by a total uncertainty factor of 500 (10 for interindividual variation × 10 for interspecies variation × 5 for the developmental uncertainty factor). These AALGs are not recommended for use since they are higher than those of other compounds that are likely to be less toxic.

The AALGs for soluble nickel compounds (nickel sulfate hexahydrate as the prototype) and insoluble nickel compounds (nickel oxide as the prototype) are based on systemic toxicity. These AALGs should be considered provisional on account of the short duration of the studies used as their basis (subchronic) and the possibility that nickel ion itself may be carcinogenic.

The AALG for soluble nickel compounds is based on the LOAEL of 0.02 mg (Ni)/m^3 as nickel sulfate hexahydrate in rats suggested by the study of Dunnick et al.[23] The LOAEL was adjusted for continuous exposure (0.02 mg/m^3 × 6/24 × 5/7), converted to a human equivalent exposure level as described in Appendix C, and divided by a total uncertainty factor of 5000 (10 for interindividual variation × 10 for interspecies variation × 10 for a less than chronic study × 5 for a LOAEL). Note that the assumed weight of the rat used in the calculations was 0.25 kg, based on data provided in the paper.

The AALG for insoluble nickel compounds is based on the LOAEL of 0.4 mg $(Ni)/m^3$ as nickel oxide in rats suggested by the study of Dunnick et al.[23] The LOAEL was adjusted for continuous exposure (0.4 mg/m^3 × 6/24 × 5/7), converted to a human equivalent exposure level as described in Appendix C, and divided by a total uncertainty factor of 5000 (10 for interindividual variation × 10 for interspecies variation × 10 for a less than lifetime study × 5 for a LOAEL). Note that the assumed weight of the rat used in the calculations was 0.25 kg, based on data provided in the paper.

AALG:
- nickel subsulfide
  10^{-6} 95% UCL—0.4 ng $(Ni)/m^3$ annual TWA
  10^{-5} 95% UCL—4.2 ng $(Ni)/m^3$ annual TWA
- nickel carbonyl
  carcinogenicity—0.024 μg/m^3 annual TWA
  developmental—210 μg/m^3 24 hour TWA
- soluble nickel compounds
  systemic toxicity—0.36 ng $(Ni)/m^3$ 24-hour TWA
- insoluble nickel compounds
  systemic toxicity—7.1 ng $(Ni)/m^3$ 24-hour TWA

References

1. ACGIH. 1986. "Documentation of the Threshold Limit Values and Biological Exposure Indices." 5:422–23.
2. NIOSH. 1977. "Criteria for a Recommended Standard . . . Occupational Exposure to Inorganic Nickel." DHEW (NIOSH) 77–164.
3. NIOSH. 1985. "NIOSH Recommendations for Occupational Safety and Health Standards." *Morb. Mort. Wkly. Rpt.* 34(suppl.):5s-31s.
4. OSHA. 1989. "Air Contaminants; Final Rule." *Fed. Reg.* 54:2332–2959.
5. U.S. EPA. 1980. "Ambient Water Quality Criteria for Nickel." EPA 440/5–80–060.
6. U.S. EPA. 1987. "Health Advisory for Nickel." Office of Drinking Water (3/31).
7. Ohanian, E. V. 1986. "National Primary Drinking Water Regulations for Additional Contaminants to be Regulated by 1989." Conference on the Safe Drinking Water Act: Amendments, Regulations and Standards. Sept. 23–24, 1986. Northeast Regional Environmental Public Health Center, University of Massachusetts, Amherst, MA.
8. U.S. EPA. 1986. "Health Assessment Document for Nickel and Nickel Compounds." EPA/600/8–83/012F.
9. U.S. EPA. 1988. "Nickel Refinery Dust; CASRN 00–02–0." IRIS (9/26/88).
10. U.S. EPA. 1988. "Nickel Subsulfide; CASRN 12035–72–2." IRIS (3/1/88).
11. Ottolenghi, A. D., J. K. Haseman, W. W. Payne, H. L. Falk, and H. N. MacFarland. 1974. "Inhalation Studies of Nickel Sulfide in Pulmonary Carcinogenesis of Rats." *J. Natl. Cancer Inst.* 54:1165–72.
12. Sunderman, F. W., and A. J. Donnelly. 1965. "Studies of Nickel Carcinogenesis: Metastasizing Pulmonary Tumors in Rats Induced by the Inhalation of Nickel Carbonyl." *Am. J. Pathol.* 46:1027–38.

13. Sunderman, F. W., A. J. Donnelly, B. West, and J. F. Kincaid. 1959. "Nickel Poisoning. IX. Carcinogenesis in Rats Exposed to Nickel Carbonyl." *Arch. Ind. Health* 20:36–41.

14. Sunderman, F. W., A. J. Donnelly, B. West, and J. F. Kincaid. 1957. "Nickel Poisoning. IV. Chronic Exposure of Rats to Nickel Carbonyl: A Report after One Year of Observation." *AMA Arch. Ind. Health* 16:480–85.

15. NTP. 1989. "Management Status Report." National Toxicology Program, Research Triangle Park, NC (2/7/89).

16. Sunderman, F. W., S. K. Shen, M. C. Reid, and P. R. Allpass. 1980. "Teratogenicity and Embryotoxicity of Nickel Carbonyl in Syrian Hamsters." *Teratogen. Carcinogen. Mutagen.* 1:223–33.

17. Sunderman, F. W., P. Allpass, and J. Mitchell. 1978. Ophthalmic Malformations in Rats Following Prenatal Exposure to Inhalation of Nickel Carbonyl." *Ann. Clin. Lab. Sci.* 8:499–500.

18. Weischer, C. H., H. Oldises, D. Hochrainer, and W. Koerdel. 1980. "Subchronic Effects Induced by Nickel-Monoxide Inhalation in Wistar Rats." *Dev. Toxicol. Environ. Sci.* 8:555–58.

19. U.S. EPA. 1984. "Health Effects Assessment for Nickel." PB86–134293, EPA/540/1–86/018.

20. Bingham, E., W. Barkley, M. Zerwas, K. Stemmer, and P. Taylor. 1972. "Responses of Alveolar Macrophages to Metals." *Arch. Environ. Health* 25:406–14.

21. Clary, J. J. 1977. "Report on Six Months Nickel Inhalation Study (1 mg/m^3) Using Rats and Guinea Pigs" (Cincinnati, OH: U.S. DHEW, PHS, CDC, NIOSH).

22. Johansson, A., P. E. R. Camner, and B. Robertson. 1981. "Effects of Long-Term Nickel Dust Exposure on Rabbit Alveolar Epithelium." *Environ. Res.* 25:391–403.

23. Dunnick, J. K., M. R. Elwell, J. M. Benson, C. H. Hobbs, F. F. Hahn, P. J. Haly, Y. S. Cheng, and A. F. Eidson. 1989. "Lung Toxicity after 13-Week Inhalation Exposure to Nickel Oxide, Nickel Subsulfide or Nickel Sulfate Hexahydrate in F344/N Rats and B6C3F$_1$ Mice." *Fund. Appl. Tox.* 12:584–94.

24. Enterline, P. E., and G. M. Marsh. 1982. "Mortality among Workers in a Nickel Refinery and Alloy Manufacturing Plant in West Virginia." *J. Natl. Cancer Inst.* 68:925–33.

25. Chovil, A., R. B. Sutherland, and M. Halliday. 1981. "Respiratory Cancer in a Cohort of Nickel Sinter Plant Workers." *Br. J. Ind. Med.* 38:327–33.

26. Peto, J., H. Cuckle, R. Doll, C. Herman, and L. G. Morgan. 1984. "Respiratory Cancer Mortality of Welsh Nickel Refinery Workers," in *Nickel in the Human Environment: Proceedings of a Joint Symposium*, March 1983, Lyon, France, IARC Scientific Publication No. 53.

27. Magnus, K., A. Anderson, and A. C. Hogetucit. 1982. "Cancer of Respiratory Organs among Workers of a Nickel Refinery in Norway." *Int. J. Cancer* 30:681–85.

NITRIC ACID

Synonyms: red fuming nitric acid, white fuming nitric acid

CAS Registry Number: 7697–37–2

Molecular Weight: 63.02

Molecular Formula: HNO_3

AALG: irritation (TLV-based)—8 ppb (21 $\mu g/m^3$) 8-hour TWA

Occupational Limits:
- ACGIH TLV: 2 ppm (5 mg/m^3) TWA; 4 ppm (10 mg/m^3) STEL
- NIOSH REL 2 ppm (5 mg/m^3) TWA
- OSHA PEL 2 ppm (5 mg/m^3) TWA

Basis for Occupational Limits: The occupational limits for nitric acid are based on prevention of irritation.[1,2] In the TLV documentation it is stated that the TLV is "sufficiently low to prevent irritation and corrosion effects on teeth, but possibly not to prevent potentiation of NO_2 effects."

Drinking Water Limits: No drinking water limits were found for this substance.

Toxicity Profile

Carcinogenicity: No data were found implicating nitric acid as a carcinogen.

Mutagenicity: Negative results are listed for nitric acid in the Gene-Tox database as summarized in RTECS for the oncogenic transformation of Syrian hamster embryo cells by adenovirus SA7.[3]

Developmental Toxicity: In a study cited in RTECS and unavailable in translation, oral exposure to 5275 mg/kg on days 1 through 21 of gestation in rats resulted in post-implantation mortality, fetotoxicity, and specific developmental abnormalities of the musculoskeletal system[4] (cited in NIOSH[3]).

Reproductive Toxicity: No data were found implicating nitric acid as a reproductive toxin.

Systemic Toxicity: Data on the effects of nitric acid exposure in experimental animals are for the most part limited to a few acute studies. Diggle and Gage[5] (cited in ACGIH[1] and NIOSH[2]) reported that a single exposure of rats to 25 ppm nitric acid (duration unspecified) resulted in no apparent adverse effects.

Gray et al.[6] (cited in NIOSH[2]) conducted comparative studies on the lethality of different forms of nitric acid as well as NO_2 using male rats. Median lethal concentrations (30 minutes) for red fuming nitric acid (RFNA), white fuming nitric acid (WFNA), and NO_2 were as follows: 310 ppm, 334 ppm, and 174 ppm; deaths were apparently due to pulmonary edema. Note that the RFNA and WFNA used contained 8% to 17% and 0.1% to 0.4% NO_2, respectively.

Both the authors and NIOSH concluded that these data suggest a possible synergistic effect between nitric acid vapor and NO_2, since RFNA has a higher NO_2 content by weight. It was also noted that although nitric acid appeared to be approximately half as toxic as NO_2 on an acute basis, the ratio may change at lower concentrations of nitric acid.

Gray et al.[7] (cited in NIOSH[2]) also conducted a chronic exposure study in which 90 rats, 30 mice, and 10 guinea pigs were exposed to 4 ppm RFNA for 4 hours per day, 5 days a week for 6 months. It was reported that there were no significant pathologic changes among treated animals compared to controls.

Irritation: The principal hazards associated with nitric acid exposure in humans are related to its irritant and caustic properties and include (1) chemical burns due to dermal contact, (2) corneal and other eye injuries due to "splash" accidents, (3) chemical burns around the mouth and mucous membranes following ingestion, and (4) pulmonary irritation following inhalation exposure.[1,2]

Chronic or continued exposure is reported to cause chronic bronchitis, which may progress to chemical pneumonitis[8] (cited in ACGIH[1]) and erosion of incisor and canine teeth may occur at levels (unspecified) not reported to cause other symptoms.[2]

Basis for the AALG: Nitric acid is miscible with water in all proportions and has a vapor pressure of 62 mm Hg at 25°C. Nitric acid is the commonest strong acid that is also a strong oxidizing agent, and it consequently has many applications, including etching and cleaning metals and synthesis of nitrates and nitro compounds.[1] Since the AALG for nitric acid is based on irritation, no relative source contribution is factored into the calculation. The reason is that allocation of a proportion of exposure to a given source is not relevant when the effects of exposure are not cumulative, as is the case with irritant effects.

Under practical occupational exposure conditions and in the limited animal studies available, it is difficult to separate the effects of exposure to NO_2 from those of nitric acid, since the two usually occur together. Given this difficulty and the fact that there are limited data indicating that there may be synergistic effects involving nitric acid and NO_2, it was considered prudent to apply a database factor of 5 and treat the TLV as a LOAEL, since it was the judgment of the TLV committee that the TLV might not be low enough to prevent potentiation of NO_2 effects. Thus a total uncertainty factor of 250 was applied (10 for interindividual variation × 5 for a LOAEL × 5 for the database factor). Given the limitations of the overall database for nitric acid, the AALG should be considered provisional.

AALG: • irritation (TLV-based)—8 ppb (21 μg/m³) 8-hour TWA

References

1. ACGIH. 1986. "Documentation of the Threshold Limit Values and Biological Exposure Indices." 5:428.
2. NIOSH. 1976. "Criteria for a Recommended Standard . . . Occupational Exposure to Nitric Acid." DHEW (NIOSH) 76–141.
3. NIOSH. 1988. QU5775000. "Nitric Acid." RTECS, on line.
4. *Z. Gesamte Hyg. Ihre Grenzegeb.* 29:667, 1983.
5. Diggle, W. M. and J. C. Gage. 1954. "The Toxicity of Nitrogen Pentoxide." *Br. J. Ind. Med.* 11:140–44.
6. Gray, E. L., F. M. Patton, S. B. Goldberg, and E. Kaplan. 1954. "Toxicity of the Oxides of Nitrogen—II. Acute Inhalation Toxicity of Nitrogen Dioxide, Red Fuming Nitric Acid and White Fuming Nitric Acid." *Arch. Ind. Hyg. Occup. Med.* 10:418–22.
7. Gray, E. L., S. B. Goldberg, and F. M. Patton. 1954. "Toxicity of the Oxides of Nitrogen—III. Effect of Chronic Exposure to Low Concentrations of Vapors from Red Fuming Nitric Acid." *Arch. Ind. Hyg. Occup. Med.* 10:423–25.
8. Fairhall, L. T. 1957. *Industrial Toxicology,* 2nd ed. (Baltimore: Williams and Wilkins), p. 83.

NITROFEN

Synonyms: 2,4-dichlorophenyl-4-nitrophenyl ether

CAS Registry Number: 1836–75–5

Molecular Weight: 284.10

Molecular Formula:

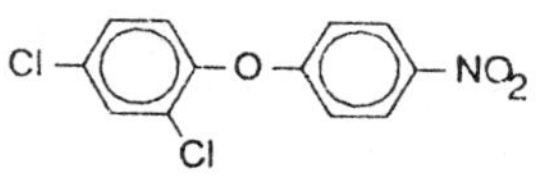

AALG: 10^{-6} 95% UCL—7.9 $\times$ 10^{-4} ppb (9.2 $\times$ 10^{-3} $\mu g/m^3$) annual TWA

Occupational Limits:
- ACGIH TLV: none
- NIOSH REL: none
- OSHA PEL: none

Drinking Water Limits: The Safe Drinking Water Committee of the NAS, using mouse tumor data from the 1979 NCI bioassay, estimated a lifetime risk of 4.4 $\times$ 10^{-5} with a 95% upper confidence limit of 5.6 $\times$ 10^{-5} for a 70-kg human consuming 1 L of water per day with a nitrofen concentration of 1 $\mu g/L$. The Safe Drinking Water Committee also derived ADIs based on developmental toxicity, using both a risk-based approach and an uncertainty factor approach assuming a 70-kg human consuming 2 L of water per day with a 20% relative source contribution from drinking water. The approaches yielded limits of 19 $\mu g/L$ and 1.1 $\mu g/L$, respectively.[1]

Toxicity Profile

Carcinogenicity: In a 1978 NCI bioassay,[2] groups of 50 male and female Osborne-Mendel rats received 2300 or 3656 ppm and 1300 or 2600 ppm technical-grade nitrofen in the diet, respectively, for 78 weeks and were observed for an additional 32 weeks (36 weeks in the case of the high-dose male rats). Groups of 50 male and female B6C3F$_1$ mice received 2348 or 4696 ppm nitrofen in the diet for 78 weeks and were observed for an additional 12 weeks. Controls (n = 20/sex/species) received the stock diet containing no nitrofen.

In the rats, there was a dose-related depression of body weight gain in both males and females. Insufficient numbers of male rats survived to allow evaluation of carcinogenicity; however, there was a statistically significant increased incidence of carcinoma of the pancreas (a rare tumor type) in female rats: controls, 0 of 20;

low-dose, 2 of 50; and high-dose, 7 of 50. In the mice, there was a dose-related depression of body weight gain in females and high-dose male mice relative to controls, but sufficient numbers of animals survived to allow evaluation of carcinogenicity. The incidence of hepatocellular carcinomas demonstrated a dose-related trend and was significantly elevated in both male and female mice. In addition, there was a significant positive dose-related trend in the incidence of hemangiosarcomas in both male and female mice compared to controls.

In another NCI bioassay,[3] male and female B6C3F$_1$ mice and F344 rats received 0, 3000, or 6000 ppm technical-grade nitrofen in the diet for 78 weeks with an additional 13-week observation period for the mice and 26-week observation period for the rats. Dose-related depression of body weight gain was observed in both sexes of rats and mice, but no treatment-related effects on survival were noted. No treatment-related increases in tumors occurred in the rats, but the incidences of hepatocellular adenomas and carcinomas were significantly increased in a dose-related trend in both male and female mice. The incidence of hepatocellular adenomas in mice was as follows: males—controls 1 of 20, low-dose 18 of 49, and high-dose 20 of 48; females—controls 0 of 18, low-dose 9 of 48, and high-dose 17 of 50. The incidence of hepatocellular carcinomas was as follows: males—controls 0 of 20, low-dose 13 of 49, and high-dose 20 of 48; females—controls 0 of 18, low-dose 5 of 48, and high-dose 13 of 50.

On the basis of these bioassay results, IARC[4] concluded that there was sufficient evidence that technical-grade nitrofen was a carcinogen in laboratory animals. Applying the criteria for the EPA weight-of-evidence classification for carcinogens, nitrofen is most appropriately placed in group B2, probable human carcinogen.

Mutagenicity: Nitrofen has been reported to induce mutations in *S. typhimurium* in both the presence and absence of a metabolic activation system in at least one study[4,5] and to induce oncogenic transformation in rat embryo cells.[5] However, it did not induce chromosomal aberrations in rat bone marrow cells or micronuclei in mouse bone marrow cells for in vitro assays.[4] The evidence for genetic activity of nitrofen based on short-term test results would be considered limited based on IARC criteria.

Developmental Toxicity: The developmental toxicity of nitrofen has been extensively evaluated in several rodent species (mouse, rat, hamster) using both the oral and dermal routes. Nitrofen is teratogenic in all three species by both routes, producing a variety of soft tissue anomalies including diaphragmatic hernias, heart defects, hydronephrosis, thyroid and gonad alterations, and effects on the development and secretion of the harderian glands at levels that do not cause maternal toxicity. Diaphragmatic hernias and heart defects often occur together to produce neonatal respiratory distress, cyanosis, and death. It is interesting to note that doses of up to 80 mg/kg/day on gestation days 6 to 18 produced no terata or fetal deaths in rabbits, although the number of live fetuses per litter was decreased. The developmental studies on nitrofen (up to 1983) have been well reviewed by Burke Hurt et al.,[6] and only those studies relevant to criteria derivation will be discussed here.

In one of a series of studies conducted by the U.S. EPA[7] (cited in Burke Hurt et al.[6]), groups of Sprague-Dawley rats were gavaged with nitrofen in corn oil on days 8 to 16 of gestation at doses between 0 and 25 mg/kg/day. Evidence of missing and decreased size of the harderian glands was found in pups from dams exposed to 12.5 mg/kg/day nitrofen and higher doses; a LOAEL of 1.39 mg/kg/day was identified from this study for the occurrence of diaphragmatic hernias in the offspring.

In another EPA study,[8] the postnatal development of the offspring of CD-1 mice gavaged with 0, 6.25, 12.5, 25, 50, 100, 150, or 200 mg/kg/day nitrofen (99.6% purity) on gestation days 7 to 17 was evaluated. No evidence of maternal toxicity was found at any of the levels tested. At the two highest exposure levels, all offspring were dead after 3 days, and 50% of the 100-mg/kg pups were also dead at that time; examination of dead and moribund pups indicated a significant incidence of cleft palate and diaphragmatic hernia. Growth rates of offspring were reduced in the 12.5-, 25-, 50-, and 100-mg/kg exposure groups, and various organ weights were reduced in all exposure groups, including the 6.25-mg/kg group (decreased lung and liver weights). In addition, there were dose-dependent alterations in the harderian glands at 25 mg/kg and higher doses, and puberty was delayed in surviving female offspring in the 50- and 100-mg/kg exposure groups. It is relevant to note that a cross-fostering experiment indicated that these effects were due to prenatal exposure to nitrofen.

Reproductive Toxicity: Results of several well-conducted reproductive toxicity studies[9–11] (cited in Safe Drinking Water Committee[1]) in rats have indicated that concentrations of nitrofen of 100 ppm or more in the diet for more than 12 weeks cause decreased neonatal survival in treated maternal, but not paternal, animals. In general, fertility and gestational indices were not adversely effected.

In a three-generation reproduction study, Ambrose et al.[9] fed male and female Wistar rats (n = 20/sex/group) diets containing 0, 10, 100, or 1000 ppm nitrofen and mated them after 11 weeks. No treatment-related effects on fertility or gestational indices or on parental or weanling weights were reported in any generation. However, in the highest exposure group, there was an increase in stillbirths and a decrease in pups surviving to 5 days, and no F_{1b} pups survived to weaning. In the 100-ppm group, there was an increased number of stillbirths in the offspring of the F_0 generation. No structural or histopathologic abnormalities were reported in the offspring, and a NOAEL of 10 ppm for reproductive effects was identified from this study. This dietary concentration was estimated to correspond to a dose of 1 mg/kg/day by the NAS.[1]

Systemic Toxicity: Ingestion data in animals indicate that the organ most sensitive to nitrofen-induced toxicity is the liver. Functional and histologic changes induced in the livers of mice, rats, and dogs include induction of mixed-function oxidase activity at lower doses and hypertrophy and hyperplasia accompanied by increased liver-to-body weight ratios at higher doses.[6] For example, in Beagle dogs maintained on a diet containing 2000 ppm nitrofen (approximately 15 mg/kg) for two years, increased liver-to-body weight ratios without any pathologic changes were reported.[9] In a subchronic feeding study in B6C3F$_1$ mice, hepatic mixed-function oxidase activity was induced at levels of 100 ppm (20 mg/kg/day) and higher, and increased liver-to-body weight ratios occurred at 1000 ppm (200 mg/kg/day). Although effects on the kidneys and reproductive system have also been reported in nitrofen-exposed animals, these alterations occur only at doses higher than those producing liver damage.[6]

Inhalation data on nitrofen are quite limited. It has been reported that a 1-hour exposure to 205,000 mg/m^3 nitrofen in rats results in respiratory distress and death due to hemorrhage and pulmonary edema. In another study, exposure of Wistar rats to 270 mg/m^3 nitrofen for 1 hour resulted in decreased body weight gain, but no pathologic changes or deaths were reported.[6]

Irritation: Nitrofen has been reported to be irritating to the eyes and skin of occupationally exposed individuals.[4]

Basis for the AALG: Nitrofen is used as a pre- and postemergence herbicide, has a low vapor pressure, is slightly soluble in water, and undergoes rapid photodegradation in solution.[4] Based on these considerations, 20% of the contribution to total exposure will be allowed from air.

AALGs were calculated for nitrofen based on both carcinogenicity, using data from the 1979 NCI bioassay,[3] and developmental toxicity, using data from the Gray et al. study.[7] Quantitative risk assessment was applied to the data sets for hepatocellular adenomas and carcinomas for both male and female mice (total of four data sets). Values of q_1^* obtained were all in the range of 2.0 to 3.85 $\times$ 10^{-3} (mg/kg/day)$^{-1}$, and the largest value (most conservative) of q_1^*, 3.85 $\times$ 10^{-3} (mg/kg/day)$^{-1}$, was taken as the best estimate. This animal q_1^* was adjusted for less-than-lifespan duration of the experiment as recommended by the U.S. EPA (i.e., multiplied by a factor: lifespan/length of experiment)^3, converted to a human q_1^*, and converted to a slope coefficient for the inhalation route. The basis for these conversions is described in Appendix C. It must be emphasized that, because of route-to-route differences, this derived estimate of q_1^*, 2.18 $\times$ 10^{-5} (μg/m^3)$^{-1}$, has a higher than usual degree of uncertainty associated with it. Using this q_1^* and applying the relation E = q_1^* $\times$ d, where E is the specified level of extra risk and d is the dose, gives the estimates shown below (when the 20% relative source contribution from air is applied). These AALGs should be considered provisional since they were derived from a non-inhalation study.

A LOAEL of 1.39 mg/kg/day, as reported in Burke Hurt et al.,[6] was identified from the study of Gray et al.[7] for diaphragmatic hernias in the offspring of Sprague-Dawley rats. This LOAEL was selected since it was the most conservative one available from an appropriately designed and conducted study. The LOAEL was converted to a human equivalent inhalation exposure level according to the procedures described in Appendix C, and a total uncertainty factor of 5000 was applied (10 for interindividual variation $\times$ 10 for interspecies variation $\times$ 5 for a LOAEL $\times$ 2 for the database factor $\times$ 5 for the developmental uncertainty factor). Application of the database factor was considered warranted based on the use of non-inhalation data. This AALG should be considered provisional since it is based on data from a non-inhalation study with limited reporting of results.

AALG: • 10^{-6} 95% UCL—7.9 × 10^{-4} ppb (9.2 × 10^{-3} µg/m³)
• 10^{-5} 95% UCL—7.9 × 10^{-3} ppb (9.2 × 10^{-2} µg/m³) annual TWA
• developmental toxicity—0.014 ppb (0.17 µg/m³) 24-hour TWA

References

1. Safe Drinking Water Committee. 1986. *Drinking Water and Health,* Vol. 6 (Washington, DC: National Academy Press), 368–82.
2. NCI. 1978. "Bioassay of Nitrofen for Possible Carcinogenicity." NCI-TR-24.
3. NCI. 1979. "Bioassay of Nitrofen for Possible Carcinogenicity." NCI-TR-184.
4. IARC. 1983. "Nitrofen." *IARC Monog.* 30:271–82.
5. NIOSH. 1988. KN8400000. "Ether, 2,4-Dichlorophenyl *p*-Nitrophenyl." RTECS, on line.
6. Burke Hurt, S. S., J. M. Smith, and A. W. Hayes. 1983. "Nitrofen: A Review and Perspective." *Toxicology* 29:1–37.
7. Gray, L. E., R. J. Kavlock, N. Chernoff, and J. Ferrell. 1982. "The Effects of Prenatal Exposure to the Herbicide Tok R on the Postnatal Development of the Harderian Gland of the Mouse, Rat and Hamster." *Toxicologist* 2:291.
8. Gray, L. E., R. J. Kavlock, N. Chernoff, J. Ostby, J. McLamb, and J. Ferrell. 1983. "Postnatal Developmental Alterations Following Prenatal Exposure to the Herbicide, 2,4-Dichloro-phenyl-*p*-nitrophenyl Ether. A Dose Response Evaluation in the Mouse." *Tox. Appl. Pharm.* 67:1.
9. Ambrose, A. M., P. S. Larson, J. F. Borzelleca, R. B. Smith, and G. R. Hennigar. 1971. "Toxicologic Studies on 2,4-Dichloro-*p*-phenyl Ether." *Tox. Appl. Pharm.* 19:263–75.
10. Kimbrough, R. D., T. B. Gains, and R. E. Linder. 1974. "2,4-Dichloro-phenyl-*p*-nitrophenyl Ether (TOK). Effects on Lung Maturation of Rat Fetus." *Arch. Env. Heal* 28:316–20.
11. O'Hara, G. P., P. K. Chan, J. C. Harris, S. S. Burke, J. M. Smith, and A. W. Hayes. 1983. "The Effect of Nitrofen [4-(2,4-Dichloro-phenoxy)nitrobenzene] on the Reproductive Performance of Male Rats." *Toxicology* 28:323–33.

PHENOL

Synonyms: carbolic acid, hydroxybenzene

CAS Registry Number: 108–95–2

Molecular Weight: 94.11

Molecular Formula: C_6H_5OH,

AALG: systemic toxicity (TLV-based)—0.024 ppm (0.09 mg/m^3) 8-hour TWA

Occupational Limits:
- ACGIH TLV: 5 ppm (19 mg/m^3) TWA; skin notation
- NIOSH REL: 5.2 ppm (20 mg/m^3) TWA; 15.6 ppm (60 mg/m^3) 15-minute ceiling
- OSHA PEL: 5 ppm (19 mg/m^3) TWA; skin notation

Basis for Occupational Limits: The TLV is based on the prevention of systemic toxicity and is stated to have a "sufficiently large margin of safety" for this purpose provided skin absorption is avoided.[1] The NIOSH and OSHA limits are also based on systemic toxicity; the health effects considered include skin and eye irritation and central nervous system, liver, and renal toxicity.[2,3]

Drinking Water Limits: Two ambient water quality criteria are given for phenol: 3.5 mg/L for health effects, assuming a 70-kg man consuming 2 L of water and 6.5 g fish/shellfish per day, and 0.3 mg/L based on organoleptic quality. The latter, more conservative limit, takes precedence.[4]

Toxicity Profile

Carcinogenicity: Skin painting studies reviewed in U.S. EPA[4,5] and Bruce et al.[6] suggest that phenol may be a promotor and/or a weak skin carcinogen in mice. However, design limitations in most of these studies do not allow evaluation of the effects of solvents or pretreatment with known carcinogens.

In an NCI bioassay[7] (cited in Bruce et al.[6]), male and female B6C3F$_1$ mice and F344 rats (n = 50/sex/group) received tap water or 2500 or 5000 ppm phenol in drinking water for 103 weeks. In low-dose, but not high-dose, male rats there was a statistically significant increase relative to controls in adrenal pheochromocytomas (an endocrine tumor producing large amounts of catecholamines) and leukemias or lymphomas. However, it was concluded that there was no evidence of carcinogenicity in mice or rats in this study on account of the lack of a dose-response relationship and the high spontaneous rate of tumors in the concurrent controls relative to historical controls. The only noncarcinogenic effect noted in this study was a dose-related decrease in body weight gain, thought to be due to decreased water consumption.

Phenol is most appropriately placed in group D, unclassified substance, based on the EPA weight-of-evidence classification, and it has been recommended that it be tested in an inhalation bioassay.[5]

Mutagenicity: Phenol was reported to induce frameshift mutations in *S. typhimurium* with metabolic activation in one study, but negative results were reported in another study using four different strains with and without metabolic activation. Negative results have also been reported in the *Drosophila* sex-linked recessive lethal test and the mouse bone marrow cell micronucleus test.

Phenol induces mutations in Chinese hamster V79 fibroblasts without metabolic activation, but the response is significantly enhanced with metabolic activation. Phenol also induces sister chromatid exchanges in cultured human lymphocytes, inhibits DNA synthesis in cultured human HeLa cells, and inhibits DNA replication synthesis and DNA repair synthesis in cultured human diploid fibroblasts.[6]

Developmental Toxicity: Data on the developmental toxicity of inhaled phenol are limited to a Russian study described in an abstract[8] (cited in Bruce et al.[6]). An increase in pre-implantation loss and early postnatal death was reported in mice exposed to 1.3 and 0.3 ppm phenol, but the exposure conditions were not described.

Gavage studies in CD rats, CD-1 mice, and Sprague-Dawley rats[9-11] (cited in Bruce et al.[6]) indicate that phenol causes dose-related fetotoxicity, primarily in the form of decreased fetal body weight per litter, but not teratogenic effects. In at least one species, CD rats, no evidence of maternal toxicity was found at fetotoxic doses.

Reproductive Toxicity: In early multigeneration studies, Heller and Pursell[12] (cited in U.S. EPA[5]) reported no effects in rats exposed to 100–1000 ppm phenol in drinking water for five generations, and no toxic effects in animals exposed to 3000–5000 ppm in drinking water for three generations. At greater than 8000 ppm, many offspring deaths were reported due to maternal behavioral abnormalities, and at 12,000 ppm, reproduction ceased.

Systemic Toxicity: Chronic inhalation studies in animal models and human epidemiologic data are lacking for phenol. However, a few subchronic inhalation studies have been conducted. Deichmann et al.[13] (cited in Bruce et al.[6]) studied the effects of exposure to 25 ppm phenol for 7 hours per day, 5 days a week in guinea pigs, rats, and rabbits. High mortality was reported in guinea pigs after 28 days. No overt signs of toxicity were seen in rabbits after 88 days, but pathological changes were reported in the lungs, liver, and kidneys. No overt signs or histological evidence of toxicity were found in rats after 74 days. Sandage[14] (cited in Bruce et al.[6]) reported no toxic effects in mice, rats, and monkeys exposed to 5 ppm phenol for 8 hours per day, 5 days a week for 90 days. It should be noted that these two subchronic studies and others reviewed in U.S. EPA[5] and Bruce et al.[6] are of limited value for criteria derivation because a dose-response relationship was not established.

The acute toxicity of phenol is similar in humans and animals regardless of the route of exposure, and phenol is rapidly absorbed via all major exposure routes. In acute poisoning in humans, the heart rate is initially increased and then slows and becomes irregular; blood pressure increases initially and rapidly decreases, and this is often followed by salivation and labored breathing. This progression of physiological events is due to the effect of phenol on the C.N.S., and death usually results from respiratory failure.[3]

In controlled exposure studies conducted in humans, Piotrowski[15] (cited in NIOSH[3]) reported no adverse effects in subjects exposed to 1.5–5.2 ppm phenol via skin absorption or inhalation for 8 hours with two 30-minute breaks (2.5 and 5.5 hours after the start of exposure). Urinary phenol concentrations returned to normal 16 hours after exposure ended. In another study, Mukhitov[16] (cited in NIOSH[3]) reported that six 5-minute inhalation exposures to 0.004 ppm phenol increased the sensitivity to light in each of three dark-adapted subjects and that exposure to 0.006 ppm phenol for 15 seconds elicited the formation of conditioned electrocortical reflexes.

Irritation: Phenol vapors and liquid phenol are highly irritating to the skin, eyes, and mucous membranes. Since phenol is readily absorbed via the skin, prevention of contact with liquid phenol is emphasized in the documentation for the occupational limits.[1-3]

Phenol is reported to have a ''characteristic sweet odor''[1] and an odor threshold (100% response) of 0.05 ppm.[6]

Basis for the AALG: The major use of phenol is in the manufacture of phenolic resins and other chemicals. It is also used as a disinfectant in germicidal paints and slimicides. Twenty percent of the contribution to total exposure is allowed from air based on the following considerations:

1. Phenol is highly water soluble and has the potential for wide distribution in water based upon its use patterns.[4,6]
2. Partially on account of its low volatility, phenol usually does not constitute a serious respiratory hazard in industry. Inhalation exposure is largely restricted to occupational settings and the general population in the immediate vicinity of a point source.[6]

Because of the absence of human epidemiologic data and chronic inhalation data in animal models as well as the limitations of subchronic inhalation studies, the best option for AALG derivation is the long-established TLV of 5 ppm, which should be divided by 42. Given the limitations of the available data referred to above, the AALG should be considered provisional.

In view of the abundance of positive mutagenicity data and the suggestive findings in the NCI bioassay, periodic review of the literature with respect to the chronic inhalation toxicity and carcinogenicity of phenol is strongly recommended.

AALG: • systemic toxicity (TLV-based)—0.024 ppm (0.09 mg/m^3) 8-hour TWA

References

1. ACGIH. 1986. "Documentation of the Threshold Limit Values and Biological Exposure Indices." 5:469–71.
2. NIOSH. 1985. "NIOSH Recommendations for Occupational Safety and Health Standards." *Morb. Mort. Wkly. Rpt.* 34(suppl):4s–31s.
3. NIOSH. 1976. "Criteria for a Recommended Standard . . . Occupational Exposure to Phenol." DHEW (NIOSH) 76–196.
4. U.S. EPA. 1980. "Ambient Water Quality Criteria for Phenol." EPA 440/5–80–066.
5. U.S. EPA. 1984. "Health Effects Assessment for Phenol." EPA 540/1–86/007, PB86–134186.
6. Bruce, R. M., J. Santodonato, and M. W. Neal. 1987. "Summary Review of the Health Effects Associated with Phenol." *Tox. Ind. Health* 3:535–68.
7. NCI. 1980. "Bioassay of Phenol for Possible Carcinogenicity." NCI TR-203.
8. Korshunov, S. F. 1974. "Early and Late Embryotoxic Effects of Phenol (Experimental Data)," in *Industrial Hygiene and Condition of Specific Functions of Working Women in the Petrochemical and Chemical Industry* (Russian), R. A. Malysheva, Ed. Wverdlovskii Nauchno-Issledovatel'skii Institut Okhrany Materinstva i Mladenchestva Mindrava RSFSR, Sverdlovsk, USSR, 149–53.
9. Jones-Price, C., T. A. Ledoux, J. R. Reel, L. Langhoff-Paschke, and M. C. Marr. 1983. "Teratologic Evaluation of Phenol (CAS No. 108–95–2) in CD-1 Mice. Laboratory Study: 9/18/80 to 1/12/81" (Springfield, VA: NTIS), PB85–104461/XAB.
10. Jones-Price, C., T. A. Ledoux, J. R. Reel, P. W. Fisher, L. Langhoff-Paschke, and M. C. Marr. 1983. "Teratologic Evaluation of Phenol (CAS No. 108–95–2) in CD Rats" (Springfield, VA: NTIS), PB83–247726.
11. Minor, J. L., and B. A. Becker. 1971. "A Comparison of the Teratogenic Properties of Sodium Salicylate, Sodium Benzoate and Phenol." *Tox. Appl. Pharm.* 19:373.
12. Heller, V. G., and L. Pursell. 1938. "Phenol-Contaminated Waters and Their Physiological Action." *J. Pharm. Exp. Ther.* 63:99–107.
13. Deichmann, W. B., K. V. Kitzmiller, and S. Witherup. 1944. "Phenol Studies, VII: Chronic Phenol Poisoning, with Special References to the Effects upon Experimental Animals of the Inhalation of Phenol Vapor." *Am. J. Clin. Pathol.* 14:273–77.
14. Sandage, C. 1961. "Tolerance Criteria for Continuous Inhalation Exposure to Toxic Material. I. Effects on Animals of 90-Day Exposure to Phenol, CCl$_4$ and a Mixture of Indole, Skatole, H$_2$S and Methyl Mercaptan" (Springfield, VA: NTIS), AD-268783.

15. Piotrowski, J. K. 1971. "Evaluation of Exposure to Phenol: Absorption of Phenol Vapor in the Lungs and through the Skin and Excretion of Phenol in the Urine." *Br. J. Ind. Med.* 28:172–78.

16. Mukhitov, B. 1964. "The Effects of Low Phenol Concentrations on the Organism of Man or Animals and Their Hygienic Evaluation," in *USSR Literature on Air Pollution and Related Occupational Diseases,* Vol. 9, B. S. Levine (trans.) (Springfield, VA: NTIS), TT-64–11574.

PHOSPHORIC ACID

Synonyms: orthophosphoric acid

CAS Registry Number: 7664–38–2

Molecular Weight: 98.00

Molecular Formula: H_3PO_4

AALG: irritation (TLV/10)—0.1 mg/m^3 (0.025 ppm) 8-hour TWA

Occupational Limits:
- ACGIH TLV: 1 mg/m^3 TWA; 3 mg/m^3 STEL
- NIOSH REL: none
- OSHA PEL: 1 mg/m^3 TWA

Basis for Occupational Limits: The occupational limits are based on irritation (industrial experience) and by analogy with comparable experience and data for sulfuric acid. It is stated in the TLV documentation that the TLV-TWA is ''. . . below the concentration that causes throat irritation among unacclimated workers and well below that which is well tolerated by acclimated workers.''[1]

Drinking Water Limits: No drinking water limits were found for this substance.

Toxicity Profile

Carcinogenicity: No data were found implicating phosphoric acid as a carcinogen.

Mutagenicity: Negative results are listed for phosphoric acid in the Gene-Tox database as summarized in RTECS for the oncogenic transformation of Syrian hamster embryo cells by adenovirus SA7.[2]

Developmental Toxicity: No data were found implicating phosphoric acid as a developmental toxin.

Reproductive Toxicity: No data were found implicating phosphoric acid as a reproductive toxin.

Systemic Toxicity: Data on the systemic effects of phosphoric acid are lacking in both humans and animal models. An oral LD$_{50}$ of 1530 mg/kg has been reported in rats.[2]

Irritation: Based on the severely limited available data, the principal toxic effects associated with phosphoric acid are based on its irritant effects. It was reported in a written communication to the TLV committee that at 0.8–5.4 mg/m^3 the fumes of phosphorus pentoxide were "noticeable, but not uncomfortable" and at 3.6–11.3 mg/m^3 the fumes caused coughing among unacclimated workers, but could be tolerated; however, at 100 mg/m^3 the fumes were "unendurable except to hardened workers."[1] The AIHA hygienic guide[3] (cited in ACGIH[1]) reports that phosphoric acid is "less hazardous" than nitric or sulfuric acids.

Basis for the AALG: Phosphoric acid is highly soluble in water and has a vapor pressure of 0.03 mm Hg at 20°C. It is widely used in a variety of applications, including fertilizer manufacture, in beverages, food, and detergents, in water treatment, and in pickling and rustproofing metals.[1] Since the AALG for phosphoric acid is based on irritation, no relative source contribution was factored into the calculation. The reason is that allocation of a proportion of exposure to a given source is not relevant when the effects of exposure are not cumulative, as is the case with irritant effects.

The AALG for phosphoric acid is based on irritation and derived from the TLV/10, since the limit is long-standing and protective against irritant effects based on industrial experience. This AALG should be considered provisional in consideration of the limited available data on phosphoric acid.

AALG: • irritation (TLV/10)—0.1 mg/m^3 (0.025 ppm) 8-hour TWA

References

1. ACGIH. 1986. "Documentation of the Threshold Limit Values and Biological Exposure Indices." 5:483.
2. NIOSH. 1988. TB6300000. "Phosphoric Acid." RTECS, on line.
3. American Industrial Hygiene Association. 1957. *Hygienic Guide Series—Phosphoric Acid* (Akron, OH: AIHA).

PHTHALIC ANHYDRIDE

Synonyms: 1,3-isobenzofurandione, PAN

CAS Registry Number: 85–44–9

Molecular Weight: 148.11

Molecular Formula:

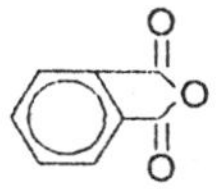

AALG: irritation (TLV/50)—0.05 ppm (0.3 mg/m^3) 8-hour TWA

Occupational Limits:
- ACGIH TLV: 1 ppm (6 mg/m^3) TWA
- OSH REL: none
- OSHA PEL: 1 ppm (6 mg/m^3) TWA

Basis for Occupational Limits: The ACGIH and OSHA limits are based on the irritant properties of phthalic anhydride relative to maleic anhydride.[1,2] OSHA recently adopted the TLV as part of its final rule limits.[2]

Drinking Water Limits: In 1977, the Safe Drinking Water Committee of the NAS concluded that insufficient data existed to recommend SNARLs for PAN.[3] No other drinking water limits were found for phthalic anhydride.

Toxicity Profile

Carcinogenicity: In an NCI cancer bioassay,[4] male and female F344 rats (n = 50/sex/group treated, n = 20/sex/group controls) were given 0, 7500, or 15,000 ppm phthalic anhydride in their diet for 105 weeks. Male and female B6C3F$_1$ mice (n = 50/sex/group treated, n = 20/sex/group controls) initially were given 0, 25,000, or 50,000 ppm phthalic anhydride in the diet for 32 weeks, but this was reduced to 12,500 ppm and 25,000 ppm for the male mice and 6250 ppm and 12,500 ppm for the female mice for the remaining 72 weeks of the study because of depressed body weight gain. Survival was not affected by phthalic anhydride administration for either the rats or mice. However, body weight gain was significantly lower relative to controls in the high-dose male rats and in both sexes of mice in both dose groups. With the exception of decreased body weight gain, no other abnormal signs or symptoms (or neoplastic or non-neoplastic lesions) attributable to phthalic anhydride administration were reported. It was concluded that under the conditions of this bioassay, phthalic anhydride was not carcinogenic to F344 rats or B6C3F$_1$ mice.

Mutagenicity: No data were found implicating phthalic anhydride as a mutagen.

Developmental Toxicity: No data were found implicating ingested or inhaled phthalic anhydride as a developmental toxin.

Reproductive Toxicity: No data were found implicating phthalic anhydride as a reproductive toxin.

Systemic Toxicity: Data on the systemic toxicity of phthalic anhydride are very limited. It has been reported to produce skin sensitization in guinea pigs. In a study in which rats were given oral doses of phthalic anhydride starting at 20 mg/kg/day and the dose doubled weekly, deaths due to severe nephrosis with the destruction of tubular epithelium were reported after 9 weeks; surviving animals exhibited gastric ulceration.[3] Oral LD$_{50}$s of 4020 and 2000 mg/kg have been reported in rats and mice, respectively.[5]

Irritation: Phthalic anhydride is reported to have a "mild" odor, but no specific odor threshold was found.[1]

Phthalic anhydride is a potent eye, skin, and mucous membrane irritant, although it is considered relatively less irritating than maleic anhydride and tetrachlorophthalic anhydride.[1]

In limited studies, animals were reported to exhibit signs of irritation when exposed to 30 mg/m^3. In workers exposed to phthalic anhydride, respiratory congestion and irritation have been reported as well as bronchitis, eye irritation, and bloody nasal discharge. These effects occurred at levels below 25 mg/m^3 (4.1 ppm), the limit of detection of the analytical method being used.[1-3]

Basis for the AALG: Phthalic anhydride is considered only slightly soluble in water and has a vapor pressure of less than 0.05 mm Hg at 20°C. According to the TLV documentation, PAN is used primarily in organic synthesis for the manufacture of a variety of resins, polyesters, dyes, and other substances.[1] PAN has been detected in finished drinking water.[3] Since the AALG is based on irritation, no relative source contribution from air was factored into the calculation of the AALG.

The AALG for phthalic anhydride is based on irritation using the TLV/50 (10 for interindividual variation × 5 for the database factor). Application of the database factor was considered warranted because the available inhalation data for PAN are limited, PAN may induce pulmonary sensitization, and the TLV is based on analogy. The TLV was chosen as the basis for AALG derivation, although it is based on analogy with other compounds, because (1) the TLV is long standing and (2) limited animal and human data indicate that the primary toxic effects associated with phthalic anhydride exposure are based on its irritant properties. The NCI bioassay results were not considered an appropriate basis for AALG derivation since the route of exposure was non-inhalation and the animals were only minimally affected by exposure to high levels of phthalic anhydride in the diet. Based on limitations of the data described above, this AALG should be considered provisional.

AALG: • irritation (TLV/50)—0.05 ppm (0.3 mg/m³) 8-hour TWA

References

1. ACGIH. 1986. "Documentation of the Threshold Limit Values and Biological Exposure Indices." 5:487.
2. OSHA. 1989. "Air Contaminants; Final Rule." *Fed. Reg.* 54:2332–2959.
3. Safe Drinking Water Committee. 1977. *Drinking Water and Health,* Vol. 1 (Washington, DC: National Academy Press), 755–56.
4. NCI. 1979. "Bioassay of Phthalic Anhydride for Possible Carcinogenicity." NCI Carcinogenesis Technical Report No. 159.
5. NIOSH. 1988. TI3150000. "Phthalic Anhydride." RTECS, on line.

PROPIONIC ACID

Synonyms: methylacetic acid, ethyl formic acid, carboxyethane

CAS Registry Number: 79–09–4

Molecular Weight: 74.09

Molecular Formula: CH_3CH_2COOH

AALG: irritation (TLV/10)—1.0 ppm (3.0 mg/m^3) 8-hour TWA

Occupational Limits:
- ACGIH TLV: 10 ppm (30 mg/m^3) TWA
- NIOSH REL: none
- OSHA PEL: 10 ppm (30 mg/m^3) TWA

Basis for Occupational Limits: The TLV for propionic acid is based mainly on analogy with acetic acid and is intended to prevent significant irritation of the eyes or respiratory passages.[1] OSHA recently adopted the ACGIH TLV as part of its final rule limits.[2]

Drinking Water Limits: No drinking water limits were found for this substance.

Toxicity Profile

Carcinogenicity: No data were found implicating propionic acid as a carcinogen.

Mutagenicity: No data were found implicating propionic acid as a mutagen.

Developmental Toxicity: No data were found implicating propionic acid as a developmental toxin.

Reproductive Toxicity: No data were found implicating propionic acid as a reproductive toxin.

Systemic Toxicity: Data on the systemic toxicity of propionic acid are limited to acute exposures. Propionic acid would appear to have a low order of toxicity based on the reported rat oral LD_{50} of 2600 mg/kg. Other reported LD_{50}s include:[3]

1. rat (intraperitoneal injection): 200 mg/kg
2. mouse (intravenous injection): 625 mg/kg
3. rabbit (dermal): 500 mg/kg

It has also been reported that propionic acid causes dyspnea, increased heart rate, partial paralysis, and frequent urination with recovery within 24 hours in rabbits given a single 2200 mg/kg intravenous injection, dogs given a single 550 mg/kg intravenous injection, and cats given a single 1000 mg/kg subcutaneous injection.[4] It was reported in the TLV documentation that an 8-hour exposure of rats to an ''atmosphere saturated with propionic acid'' resulted in no deaths.[1]

Irritation: Mild to moderate skin burns and mild eye redness as well as one case of an asthmatic-type response have been reported in workers occupationally exposed to propionic acid. In the results of an industrial hygiene survey communicated to the TLV committee, it was reported that no irritation was noted at TWA concentrations below 0.25 ppm (8-hour) with excursion concentrations of up to 2.1 ppm.[1] One source lists a nasal irritation threshold of 370 ppm for propionic acid.[5]

Propionic acid is reported to have a ''pungent'' odor[1] and odor thresholds of 0.16 ppm (calculated geometric mean of several literature citations) and 0.24 ppm (value given in a literature citation which also listed an irritation threshold) are listed by Amoore and Hautala.[5] These authors calculated an ''odor safety factor'' (TLV-TWA/odor threshold [geometric mean of all literature citations found, omitting extreme points and duplicate quotations]) for a number of compounds and developed an odor safety classification using letter designations A to E, in which class A compounds ''provide the strongest odorous warning of their presence at threshold limit value concentrations,'' whereas class E substances are ''practically odorless at the TLV concentration.'' Under this system, propionic acid was placed in class B, i.e., ''50–90% of distracted persons perceive warning of TLV.''

Basis for the AALG: Propionic acid is soluble in water and has a vapor pressure of approximately 4.1 mm Hg at 25°C.[5] Propionic acid and its salts are widely used as mold inhibitors, fungicides, herbicides, preservatives for grain and wood chips, and emulsifying agents; in flavors, perfumes, and drugs; in electroplating solutions; and in making cellulose propionate plastics. However, since the AALG for propionic acid is based on irritation, no relative source contribution from inhalation was factored into the calculation. The reason is that allocation of a proportion of exposure to a given source is not relevant when the effects of exposure are not cumulative, as is the case with irritant effects.

Since irritation is the primary effect associated with exposure to propionic acid in the absence of information on other toxicity endpoints, the AALG is based on the TLV/10. The TLV was chosen rather than the irritation threshold given in the literature because the basis for this latter figure was obscure and the TLV is backed by many years of industrial experience.

AALG: • irritation (TLV/10)—1.0 ppm (3.0 mg/m³) 8-hour TWA

References

1. ACGIH. 1986. "Documentation of the Threshold Limit Values and Biological Exposure Indices." 5:498.
2. OSHA. 1989. "Air Contaminants; Final Rule." *Fed. Reg.*: 54:2332–959.
3. NIOSH. 1988. UE5950000. "Propionic Acid." RTECS, on line.
4. Von Oettingen, W. F. 1960. "The Aliphatic Acids and Their Esters: Toxicity and Potential Dangers." *AMA Arch. Ind. Health* 21:100–13.
5. Amoore, J. E., and E. Hautala. 1983. "Odor as an Aid to Chemical Safety: Odor Thresholds Compared with Threshold Limit Values and Volatilities for 214 Industrial Chemicals in Air and Water Dilution." *J. Appl. Tox.* 3:272–90.

PROPYLENE

Synonyms: propene, methylethene, methylethylene

CAS Registry Number: 115–07–7

Molecular Weight: 42.09

Molecular Formula: $H_2C = CH\text{-}CH_3$

AALG: irritation/nasal epithelial changes—3.2 ppm (5.5 mg/m^3) 24-hour TWA

Occupational Limits:
- ACGIH TLV: no TLV recommended; listed in Appendix E of TLV booklet[1] as a "simple asphyxiant"
- NIOSH REL: none
- OSHA PEL: none

Basis for Occupational Limits: Simple asphyxiants are gases and vapors that when present at high concentrations in air, act to cause suffocation by limiting the available oxygen and are without other significant physiological effects.[1]

Drinking Water Limits: No drinking water limits were found for this substance.

Toxicity Profile

Carcinogenicity: In an NTP bioassay,[2] no evidence of carcinogenicity was found in male and female B6C3F$_1$ mice and F344/N rats exposed to air only or 5000 or 10,000 ppm propylene via inhalation 6 hours per day, 5 days a week for 103 weeks. It should be noted that the highest exposure level used, 10,000 ppm, was not the MTD, but was considered the highest level safe to work with on the basis of explosion hazard. The results are consistent with those reported in a study by Maltoni et al.[3] (cited in NTP[2]) in which Sprague-Dawley rats and Swiss mice were exposed to concentrations of up to 5000 ppm propylene via inhalation for 24 and 18 months, respectively.

Non-neoplastic changes occurred only in rats and consisted of squamous metaplasia of the respiratory epithelium, epithelial hyperplasia, and inflammatory changes. Squamous metaplasia occurred in both male and female rats; the greatest response was in the low exposure group, although a significant response was also observed in high-dose females. Epithelial hyperplasia occurred only in female rats; the maximal effect was at the highest concentration tested. However, inflammatory lesions in male rats were statistically significant only at the low dose (although of similar magnitude at the high dose).

It should be noted that the species difference reported in this study may be attributable to the stronger compensatory reflex apnea that occurs in B6C3F$_1$ mice compared with F344 rats in response to irritant gases.[4] This defense mechanism and species differences were studied by Chang et al.[5,6] (cited in Quest et al.[4]) using B6C3F$_1$ mice and F344 rats exposed to formaldehyde; the net effect of this reflex apnea is to decrease respiratory rates and minute volumes, resulting in decreased intake of formaldehyde in mice.

Mutagenicity: No data were found implicating propylene as a mutagen.

Developmental Toxicity: No data were found implicating propylene as a developmental toxin.

Reproductive Toxicity: No data were found implicating propylene as a reproductive toxin.

Systemic Toxicity: Although no systemic toxicity was observed in rats in the NTP bioassay, it was noted that focal inflammatory lesions occurred in the kidneys of mice exposed to propylene. However, the lesions were considered to be of minimal severity and their relationship to propylene exposure unclear.[4]

No adverse effects have been reported in humans exposed to concentrations of up to 4000 ppm (duration not reported), 20% of the lower flammability limit. Exposure to higher concentrations of propylene may cause incoordination, drowsiness, inability to concentrate, unconsciousness, and finally asphyxiation due to exclusion of oxygen[7] (cited in NTP[2]).

Irritation: Propylene is not generally recognized as an irritant; however, the response in rats in the NTP bioassay are suggestive of such an effect.

Basis for the AALG: There is evidence for widespread low levels of propylene in the atmosphere in both urban and rural areas; higher levels over metropolitan areas are attributable to automobile engine exhaust emissions and industrial activity.[2] In addition, since propylene is a gas at room temperature, 80% of the contribution to total exposure will be allowed from air.

The AALG is based on cumulative irritant effects as manifested by metaplastic and hyperplastic changes in the nasal epithelium of rats reported in the NTP lifetime inhalation bioassay.[2] The LOAEL of 5000 ppm was adjusted for continuous exposure (5000 $\times$ 5/7 $\times$ 6/24) and converted to a human equivalent exposure level for a 70-kg human breathing 20 m^3 of air per day (as described in Appendix C), and a total uncertainty factor of 500 was applied (10 for interindividual variation $\times$ 10 for interspecies variation $\times$ 5 for a LOAEL).

AALG: • irritation/nasal epithelial changes—3.2 ppm (5.5 mg/m^3) 24-hour TWA

References

1. ACGIH. 1987. ''TLVs—Threshold Limit Values for Chemical Substances and Physical Agents in the Work Environment with Intended Changes for 1987'' (Cincinnati, OH: ACGIH).
2. NTP. 1985. ''Toxicology and Carcinogenesis Studies of Propylene in F344/N Rats and B6C3F$_1$ Mice (Inhalation Studies)'' (Research Triangle Park, NC: National Toxicology Program), TR-272.
3. Maltoni, C., A. Ciliberti, and D. Carretti. 1982. ''Experimental Contributions in Identifying Brain Potential Carcinogens in the Petrochemical Industry.'' *Ann. NY Acad. Sci.* 381:216–49.
4. Quest, J. A., J. E. Tomaszewski, J. K. Haseman, G. A. Boorman, J. F. Douglas, and W. J. Clarke. 1984. ''Two-Year Inhalation Toxicity Study of Propylene in F344/N Rats and B6C3F$_1$ Mice.'' *Tox. Appl. Pharm.* 76:288–95.
5. Chang, J. C. F., E. A. Gross, J. A. Swenberg, and C. S. Barrow. 1983. ''Nasal Cavity Deposition, Histopathology, and Cell Proliferation after Single or Repeated Formaldehyde Exposures in B6C3F$_1$ Mice and F344 Rats.'' *Tox. Appl. Pharm.* 68:161–76.
6. Chang, J. C. F., W. H. Steinhagen, and C. S. Barrow. 1981. ''Effect of Single or Repeated Formaldehyde Exposure on Minute Volume of B6C3F$_1$ Mice and F344 Rats.'' *Tox. Appl. Pharm.* 61:451–59.
7. MCA. 1974. ''Chemical Safety Data Sheet SD-59: Propylene.'' (Washington, D.C.: Manufacturing Chemist's Association).

PROPYLENE OXIDE

Synonyms: methyloxidrane, propene oxide, 1,2-propylene oxide

CAS Registry Number: 75–56–9

Molecular Weight: 58.08

Molecular Formula: CH_3-CH-CH_2 O $\quad CH_3 - CH - CH_2$
$\quad\quad\quad\quad\quad\quad\quad\quad\quad\quad\quad\quad\quad\quad\quad\quad \diagdown \diagup$
$\quad\quad\quad\quad\quad\quad\quad\quad\quad\quad\quad\quad\quad\quad\quad\quad\quad O$

AALG: 10^{-6} 95% UCL—0.09 ppb (0.21 $\mu g/m^3$) annual TWA

Occupational Limits: • ACGIH TLV: 20 ppm (50 mg/m^3) TWA
  • NIOSH REL: none
  • OSHA PEL: 20 ppm (50 mg/m^3) TWA

Basis for Occupational Limits: The TLV for propylene oxide is based on its lesser toxicity than, and by analogy with, ethylene oxide. The TLV was reported to be currently under review in the 1986 TLV documentation but was not listed in the "notice of intended changes" for either 1987–88 or 1988–89.[1] OSHA has stated that the health hazards associated with this compound are "primary skin, eye, and respiratory irritation, as well as central nervous system depression." OSHA recently adopted the TLV-TWA of 20 ppm for propylene oxide as part of its final rule limits.[2]

Drinking Water Limits: No drinking water limits were found for this substance.

Toxicity Profile

Carcinogenicity: Propylene oxide has been tested for carcinogenicity using several different routes of exposure and strains of rats and mice. It has been shown to produce a statistically significant increase in injection site fibrosarcomas in female NMRI mice[3] (cited in IARC[4]) and a dose-dependent increased incidence of squamous cell carcinomas of the forestomach in female Sprague-Dawley rats gavaged with 0, 15, or 60 mg/kg propylene oxide in "salad" oil (composition not specified) twice a week for 109 weeks. In addition, hyperplasia, papillomas, or hyperkeratosis of the forestomach occurred in 7 of 50 low-dose rats and 17 of 50 high-dose rats, but not in controls[5] (cited in IARC[4]).

Several inhalation bioassays of propylene oxide have also been conducted. Lynch et al.[6] (cited in IARC[4]) reported a statistically significant increase in adrenal pheochromocytomas in male F344 rats exposed to 0, 100, or 300 ppm propylene oxide 7 hours per day, 5 days a week for 104 weeks (incidences: controls, 8 of 78; 100 ppm, 25 of 78; and 300 ppm, 22 of 80). There was also an increased incidence of peritoneal mesotheliomas as well as inflammatory lesions of the respiratory tract and dose-dependent "complex epithelial hyperplasia" of the nasal cavity. Reuzel and Kuper[7] (cited in U.S. EPA[8]) exposed male and female Cpb:WU Wistar rats to 0, 30, 100, or 300 ppm propylene oxide 6 hours per day, 5 days a week for 124 and 123 weeks, respectively. A dose-related increase in multiple fibroadenomas of the mammary gland was reported in female rats, and the incidence of total fibroadenomas of the mammary gland was statistically significant in the highest dose group compared with controls. In addition, there was a dose-related increase in focal hyperplasia in the nasal turbinates, but no tumors were found.

In an NTP bioassay,[9] male and female B6C3F$_1$ mice and F344/N rats were exposed to room air or 200 or 400 ppm propylene oxide via inhalation 6 hours per day, 5 days a week for 103 weeks. There was a statistically significant increase in the incidence of hemangiosarcomas and hemangiomas of the nasal cavity in both male and female mice at the highest exposure level relative to controls. In addition, there was an increased incidence and severity of inflammation of the respiratory epithelium. In rats, there was an increased incidence of epithelial hyperplasia, squamous metaplasia and suppurative inflammation of the respiratory epithelium of the nasal turbinates. In the high-dose group, papillary adenomas involving the respiratory epithelium and underlying submucosal glands of the nasal turbinates occurred in 2 of 50 male and 3 of 50 female rats. In addition, there was a statistically significant combined incidence of C-cell adenomas and carcinomas of the thyroid gland in female rats.

IARC has concluded that there is sufficient evidence for carcinogenicity of propylene oxide in laboratory animals and inadequate evidence in humans.[4] Propylene oxide has been tentatively placed in group B2, probable human carcinogen, under the EPA weight-of-evidence classification.[8]

Mutagenicity: Propylene oxide has been found to be a direct-acting mutagen in *S. typhimurium* and *E. coli*, to induce forward mutations in *Schizosaccharomyces pombe* and reverse mutations in *Klebsiella pneumoniae* and *Neurospora crassa*, and to induce chromosomal aberrations in cultured human lymphocytes. Positive results have also been reported in a *Drosophila* sex-linked recessive lethal assay and a mouse micronucleus test (intraperitoneal route), and negative to equivocal results in mouse and rat dominant lethal tests.[8] Propylene oxide has also been reported to induce oncogenic transformation in Syrian hamster embryo cells by adenovirus SA7 (Gene-Tox database cited in RTECS[10]). IARC has concluded that there is sufficient evidence for genetic activity of propylene oxide on the basis of short-term tests.[4]

Developmental Toxicity: Hardin et al.[11] exposed New Zealand rabbits and Sprague-Dawley rats to 0 or 500 ppm propylene oxide on gestation days 7 to 19 or 1 to 19 in the case of rabbits, and three weeks prior to gestation or gestation days 1 to 16 or 7 to 16 in the case of rats. Rabbits appeared to be relatively insensitive to the effects of propylene oxide. However, in the rats there was evidence of both maternal toxicity (decreased body weight gain and increased organ weights) and embryotoxicity (increased resorptions and decreased number of corpora lutea, implants, and live fetuses) as well as decreased fetal body weights and crown-rump length and delayed ossification. No teratogenic effects were observed, although a number of minor variations were seen, e.g., wavy ribs.

Reproductive Toxicity: Hayes et al.[12] reported that in a two-generation reproduction study using F344 rats exposed to up to 300 ppm propylene oxide via inhalation, no adverse effects on reproductive function occurred. However, two-year inhalation exposure of cynomolgus monkeys to 0, 100, or 300 ppm propylene oxide has been reported to result in significantly decreased sperm counts and sperm motility in both treatment groups[13] (cited in U.S. EPA[8]).

Systemic Toxicity: Lifetime inhalation studies in laboratory animals have indicated that propylene oxide may cause inflammatory lesions of the lungs, nasal cavity, and trachea; a NOAEL of 30 ppm for such effects in rats is suggested by the study of Reuzel and Kuper[7] (cited in U.S. EPA[8]).

Propylene oxide is reported to be a mild central nervous system depressant in humans that is capable of producing incoordination, ataxia, and general depression. Sprinz et al.[14] (cited in U.S. EPA[8]) reported that two years of exposure to propylene oxide resulted in axonal dystrophy in the medulla oblongata of monkeys exposed to 300 ppm, but not 100 ppm, for 6 hours per day, 5 days a week.

Irritation: Propylene oxide has been reported to have an "ethereal" odor and a median detectable odor concentration of approximately 200 ppm.[1]

"Excessive" exposure to propylene oxide vapor is reported to be irritating to the skin, eyes, and respiratory tract. Direct contact with diluted or undiluted liquid propylene oxide may cause burns and necrosis of the skin and corneal burns.[1]

Basis for the AALG: Propylene oxide is highly volatile (vapor pressure = 439 mm Hg at 20°C) and is soluble in water to the extent of 40.5 wt %.[4] Propylene oxide is reported to be highly reactive; it is used as a fumigant and as an intermediate in the manufacture of polyols for urethane foams, propylene glycol, surfactants, and other products.[1] It is estimated that most of the propylene oxide produced in the United States is released to the atmosphere.[8] Based on these considerations, 80% of the contribution to total exposure is allowed from air.

The AALG for propylene oxide is based on carcinogenicity, using the linearized multistage model applied to the data for nasal cavity hemangiomas and hemangiosarcomas (combined) in male $B6C3F_1$ mice from the NTP bioassay[9] since this was considered by CAG to result in the best estimates of cancer risk for inhalation exposure. The human q_1^* of 3.74×10^{-6} $(\mu g/m^3)^{-1}$ was calculated assuming 50% absorption. Using this slope coefficient and the relationship $E = q_1^* \times d$, where E is the specified level of extra risk and d is the dose, gives the estimates shown below.

AALG: • 10^{-6} 95% UCL—0.09 ppb (0.21 µg/m^3) annual TWA
           • 10^{-5} 95% UCL—0.9 ppb (2.1 µg/m^3) annual TWA

References

1. ACGIH. 1986. "Documentation of the Threshold Limit Values and Biological Exposure Indices." 5:504.
2. OSHA. 1989. "Air Contaminants; Final Rule." *Fed. Reg.* 54:2332–2959.
3. Dunkelburg, H. 1981. "Carcinogenic Activity of Ethylene Oxide and Its Reaction Products 2-Chloroethanol, 2-Bromoethanol, Ethylene Glycol and Diethylene Glycol. I. Carcinogenicity of Ethylene Glycol in Comparison with 1,2-Propylene Oxide after Subcutaneous Administration in Mice" (German). *Zbl. Bakt. Hyg. I. Abt. Orig. B.* 174: 383–404.
4. IARC. 1985. "Propylene Oxide." *IARC Monog.* 36:227–43.
5. Dunkelburg, H. 1982. "Carcinogenicity of Ethylene Oxide and 1,2-Propylene Oxide upon Intragastric Administration to Rats." *Br. J. Can.* 46:924–33.
6. Lynch, D. W., T. R. Lewis, W. J. Moorman, J. A. Burg, D. H. Groth, A. Khan, L. J. Ackerman, and B. Y. Cockrell. 1984. "Carcinogenic and Toxicologic Effects of Inhaled Ethylene Oxide and Propylene Oxide in F344 Rats." *Tox. Appl. Pharm.* 76:69–84.
7. Reuzel, P. G. J., and C. F. Kuper. 1983. "Chronic (28-Month) Inhalation Toxicity/ Carcinogenicity Study of 1,2-Propylene Oxide in Rats." Report No. V82.215/280853 (Zeist, The Netherlands: Civo Industries TNO).
8. U.S. EPA. 1987. "Summary Review of the Health Effects Associated with Propylene Oxide: Health Issue Assessment." EPA-600/8–86/007F.
9. NTP. 1985. "Toxicology and Carcinogenesis Studies of Propylene Oxide (CAS No. 75–56–9) in F344/N Rats and B6C3F$_1$ Mice (Inhalation Studies)" (Research Triangle Park, NC: National Toxicology Program), TR-267.
10. NIOSH. 1988. TZ2975000. "Propane, 1,2-Epoxy." RTECS, on line.
11. Hardin, B. D., R. W. Niemeier, M. R. Sikov, and P. L. Hackett. 1983. "Reproductive-Toxicologic Assessment of the Epoxides Ethylene Oxide, Propylene Oxide, Butylene Oxide and Styrene Oxide." *Scand. J. Work Environ. Health* 9:94–102.
12. Hayes, W. C., H. D. Kirk, T. S. Gushow, and J. T. Young. 1988. "Effect of Inhaled Propylene Oxide on Reproductive Parameters in Fischer 344 Rats." *Fund. Appl. Tox.* 10:82–88.
13. Lynch, D. W., T. R. Lewis, W. J. Mooreman, P. S. Sabharwal, and J. A. Burg. 1983. "Toxic and Mutagenic Effects of Ethylene Oxide and Propylene Oxide on Spermatogenic Functions in Cynomolgus Monkeys." *Toxicologist* 3:60.
14. Sprinz, H., H. Matzke, and J. Carter. 1982. "Neuropathological Evaluation of Monkeys Exposed to Ethylene and Propylene Oxide" (Springfield, VA: NTIS), PB83–134817.

RESORCINOL

Synonyms: *m*-dihydroxybenzene, 1,3-benzenediol

CAS Registry Number: 106–46–3

Molecular Weight: 110.11

Molecular Formula:

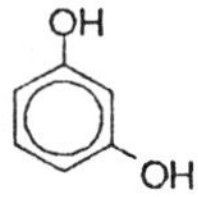

AALG: systemic toxicity (TLV-based)—0.12 ppm (0.54 mg/m^3) 8-hour TWA

Occupational Limits:
- ACGIH TLV: 10 ppm (45 mg/m^3) TWA; 20 ppm (90 mg/m^3) STEL
- NIOSH REL: none
- OSHA PEL: none

Basis for Occupational Limits: The TLV for resorcinol is based on analogy with phenol and catechol, together with a communication to the TLV committee stating that in a survey of workers occupationally exposed to resorcinol, no reports of irritation or discomfort occurred at exposure levels of approximately 10 ppm.[1] OSHA recently adopted the ACGIH limits for resorcinol as part of its final rule limits. Also, in its review, OSHA considered the ACGIH limits to be based on human NOAELs, probably because of the company survey results reported in a communication to the TLV committee.[2]

Drinking Water Limits: In 1980, the Safe Drinking Water Committee of the NAS recommended one-day and seven-day SNARLs of 11.7 and 0.5 mg/L, respectively, assuming a 70-kg man consuming 2 L of water per day with 100% contribution from drinking water.[3]

Toxicity Profile

Carcinogenicity: In a skin painting study in which the skin of groups of 50 female Swiss mice was painted twice weekly with 0.02 mL of 5, 25, or 50% resorcinol solution in acetone (control groups consisted of 150 untreated mice and 100 acetone-treated mice) for 100 weeks, there was no increase in tumors attributable to resorcinol in treated, compared with control, animals[4] (cited in IARC[5]). In another skin painting study in rabbits in which the dorsal skin was painted with 0.02 mL of a 5, 10, or 50% resorcinol solution, no evidence of toxicity or tumor production attributable to resorcinol was reported during lifetime observation[6] (cited in Safe Drinking Water Committee[3]). In addition, Van Duuren and Goldschmidt[7] (cited in Safe Drinking Water Committee[3]) have reported that resorcinol is not cocarcinogenic in mice in combination with benzo(a)pyrene.

Resorcinol is currently on test in the NTP bioassay program, with the quality assessment in progress for gavage studies in male and female rats and mice.[8]

Mutagenicity: In the Gene-Tox data base as summarized in RTECS, the following results are listed for resorcinol: negative in *Neurospora crassa* aneuploidy test, negative in an Ames assay, and inconclusive in a mammalian micronucleus test. In other citations in RTECS, resorcinol is reported to induce mutations in *S. typhimurium* with and without metabolic activation, mitotic recombination in *S. cerevisiae,* and chromosome aberrations in Chinese hamster ovary cells and human lymphocytes.[9]

Developmental Toxicity: No data were found implicating resorcinol as a developmental toxin.

Reproductive Toxicity: No data were found implicating resorcinol as a reproductive toxin.

Systemic Toxicity: Data on the inhalation toxicity of resorcinol in laboratory animals are limited to acute and subacute studies. No deaths, signs of toxicity, or histopathologic lesions were reported in female Harlan-Wistar albino rats (n = 6/group) given single exposures of up to 7800 mg/m^3 resorcinol for 1 hour or 2800 mg/m^3 for 8 hours and held for 14 days of observation; normal 14-day weight gains were reported for these animals. In other studies, no evidence of toxic effects were reported in rats, rabbits, or guinea pigs exposed to 34 mg/m^3 resorcinol 6 hours per day for 2 weeks with serial sacrifice over a period of several months. These latter studies were designed and conducted particularly to evaluate potential lung and tracheal damage.[10]

Although these studies appear to indicate a low order of inhalation toxicity for resorcinol at least in short-term exposures, their value for criteria derivation is limited because reporting on details of experimental protocols and results are lacking and a dose-response relationship was not established.

Data on the ingestion toxicity of resorcinol in laboratory animals are also limited. However, it is noteworthy that on the basis of acute toxicity tests, resorcinol is less toxic than either of the structurally related compounds, catechol or phenol.[1]

Irritation: Resorcinol may cause irritation, dermatitis, or corrosion if applied directly to the skin.[3] No complaints of irritation or discomfort were reported in a survey of 180 workers exposed to resorcinol at levels of approximately 10 ppm.[1]

Basis for the AALG: Resorcinol is considered soluble in water and has a vapor pressure of 1 mm Hg at 108.4°C. It is used in the manufacture of resins and other materials and as a rubber tackifier and cross-linking agent for neoprene.[1] Resorcinol was identified in treated effluents in one study.[3] Based on the lack of data on its environmental fate and distribution and the expected contribution of various media to human exposure, 50% of the contribution to total exposure will be allowed from air.

The AALG was derived from the TLV/42 because (1) this occupational limit is long-standing and (2) there is at least some evidence that this exposure level is without adverse effects in humans and that much higher levels (in short-term exposures) are without effect in laboratory animals. It was considered most appropriate to incorporate the factor of 4.2 to adjust for continuous exposure—for a total uncertainty factor of 42—since there are limited data indicating that resorcinol may be mutagenic and it has not been adequately tested for its ability to cause systemic toxicity. Because there is a lack of subchronic or chronic inhalation data for both animals and humans, this AALG should be considered provisional pending results of the NTP bioassay currently in progress.

AALG: • systemic toxicity (TLV/42)—0.12 ppm (0.54 mg/m^3) 8-hour TWA

References

1. ACGIH. 1986. "Documentation of the Threshold Limit Values and Biological Exposure Indices." 5:511.
2. OSHA. 1989. "Air Contaminants; Final Rule." *Fed. Reg.* 54:2332–2959.
3. Safe Drinking Water Committee. 1980. *Drinking Water and Health,* Vol. 3 (Washington, DC: National Academy Press), 227–31.
4. Stenback, F., and P. Shubik. 1974. "Lack of Toxicity and Carcinogenicity of Some Commonly Used Cutaneous Agents." *Tox. Appl. Pharm.* 30:7–13.
5. IARC. 1977. "Dihydroxybenzenes." *IARC Monog.* 15:155–75.
6. Stenback, F. 1977. "Local and Systemic Effects of Commonly Used Cutaneous Agents: Lifetime Studies of 16 Compounds in Mice and Rabbits." *Acta Pharm. Tox.* 41:417–31.
7. Van Duuren, B. L., and B. M. Goldschmidt. 1976. "Cocarcinogenic and Tumor-Promoting Agents in Tobacco Carcinogenesis." *J. Natl. Cancer Inst.* 51:703–5.
8. NTP. 1989. "Management Status Report." National Toxicology Program, Research Triangle Park, NC (2/7).
9. NIOSH. 1988. VG9625000. "Resorcinol." RTECS, on line.
10. Flickinger, C. W. 1976. "The Benzenediols: Catechol, Resorcinol and Hydroquinone—A Review of the Industrial Toxicology and Current Industrial Exposure Limits." *Am. Ind. Hyg. Assoc. J.* 37:596–606.

SODIUM HYDROXIDE

Synonyms: caustic soda, lye

CAS Registry Number: 1310–93–2

Molecular Weight: 40.01

Molecular Formula: NaOH

AALG: irritation (TLV-based)—0.04 mg/m^3 (0.024 ppm) ceiling

Occupational Limits:
- ACGIH TLV: 2 mg/m^3 ceiling
- NIOSH REL: 2 mg/m^3 ceiling
- OSHA PEL: 2 mg/m^3 TWA

Basis for Occupational Limits: The occupational limits for sodium hydroxide are based on the prevention of irritation to the upper respiratory tract.[1,2] Patty[3] has stated that " . . . from the irritant effects of caustic mists encountered in concentrations of 1 to 40 mg/m^3 of air, 2 mg sodium hydroxide per m^3 is believed to represent a concentration that is noticeably, but not excessively irritant."

Drinking Water Limits: No drinking water limits were found for this substance. However, it should be noted that NaOH is sometimes added during water treatment to adjust pH.

Toxicity Profile

Carcinogenicity: No data were found implicating NaOH as a carcinogen.

Mutagenicity: Sodium hydroxide is listed as negative for the oncogenic transformation of Syrian hamster embryo cells by adenovirus SA7 in the Gene-Tox database as summarized in RTECS.[4] It is stated in the "Assessment of Sodium Hydroxide as a Potentially Toxic Air Pollutant"[5] that "two bacterial assays have suggested sodium hydroxide is not mutagenic. However, an in vitro deoxyribonucleic acid (DNA) synthesis study and an in vivo study of insect spermatocytes suggested sodium hydroxide may be mutagenic."

Developmental Toxicity: No data were found implicating NaOH as a developmental toxin.

Reproductive Toxicity: No data were found implicating NaOH as a reproductive toxin.

Systemic Toxicity: The observed effects of NaOH in animals and humans are related to its caustic properties. An intraperitoneal LD_{50} of 40 mg/kg has been reported in mice.[4]

Ott et al.[6] reported the results of an epidemiologic study of workers chronically exposed to NaOH dust and found no relationship between exposure and the occurrence of cancer or excess mortality.

Irritation: Four basic hazards are usually identified based on the caustic properties of NaOH: (1) ocular injury due to "splash" accidents, e.g., corneal burns and opacity and possibly blindness, (2) alkali burns to the skin from dermal contact, (3) damage to the GI tract (e.g., esophageal stenosis and strictures) following ingestion, and (4) upper respiratory tract irritation and sometimes ulceration of the nasal passages following inhalation exposure to acid mists.[2]

In two case reports described in the NIOSH criteria document,[2] exposure to NaOH mists at concentrations in the range of 0.005 to 0.7 mg/m^3 resulted in "burning/redness" of the nose, throat, and eyes. However, other substances (e.g., Stoddard solvent) were present at levels as high as 780 mg/m^3. In another incident, workers briefly exposed to 0.24 to 1.86 mg/m^3 NaOH mist while cleaning ovens experienced irritation of the throat and lacrimation.[2]

Basis for the AALG: Sodium hydroxide is freely soluble in water and based on its low cost is widely used in industrial applications requiring a strong base.[1] Since the AALG for sodium hydroxide is based on irritant effects, no relative source contribution is factored into the calculation. The reason is that allocation of a proportion of exposure to a given source is not relevant when the effects of exposure are not cumulative, as is the case with irritant effects.

As is noted in the NIOSH criteria document, the effects of inhalation exposure to sodium hydroxide mist have not been systematically or reliably evaluated in either animal models or humans. Based on the weight of evidence of available data, it appears that the TLV might reasonably be expected to correspond to a human LOAEL for irritant effects. Therefore, the AALG was derived based on the TLV/50 (10 for interindividual variation × 5 for a LOAEL). This AALG should be considered provisional given the limitations of the available data.

AALG: • irritation (TLV-based)—0.04 mg/m^3 (0.024 ppm) ceiling

Note: The U.S. EPA recently completed an evaluation of NaOH with respect to the need to regulate NaOH emissions under the Clean Air Act.[5,7,8] This review was undertaken based on the production volume of NaOH and potential for adverse health effects. It was concluded that "given the paucity of data regarding systemic or acute health effects and the low potential for exposure to high concentrations of sodium hydroxide (due to its rapid atmospheric degradation), it is unlikely that routine emissions of sodium hydroxide pose a public health risk.''

References

1. ACGIH. 1986. ''Documentation of the Threshold Limit Values and Biological Exposure Indices.'' 5:535.
2. NIOSH. 1975. ''Criteria of a Recommended Standard . . . Occupational Exposure to Sodium Hydroxide.'' DHEW (NIOSH) 76–105.
3. Patty, F. A. 1949. *Industrial Hygiene and Toxicology* (New York: Wiley-Interscience), p. 561.
4. NIOSH. 1988. WB4900000. ''Sodium Hydroxide.'' RTECS, on line.
5. U.S. EPA. 1989. ''Assessment of Sodium Hydroxide as a Potentially Toxic Air Pollutant.'' *Fed. Reg.* 54:1140–43.
6. Ott, M. G., H. L. Gordon, and E. J. Schneider. 1977. ''Mortality Among Employees Chronically Exposed to Caustic Dust.'' *J. Occup. Med.* 19:813–16.
7. U.S. EPA. 1988. ''Sodium Hydroxide Preliminary Source Assessment.'' Office of Air Quality Planning and Standards, Research Triangle Park, NC.
8. U.S. EPA. 1988. ''Summary Review of Health Effects Associated with Sodium Hydroxide: Health Issue Assessment.'' Environmental Criteria and Assessment Office, Research Triangle Park, NC.

SODIUM SULFATE

Synonyms: disodium sulfate

CAS Registry Number: 7757–82–6

Molecular Weight: 142.04

Molecular Formula: Na_2SO_4

AALG:
- 150 $\mu g/m^3$ 24-hour TWA (as PM_{10})
- 50 $\mu g/m^3$ annual arithmetic TWA (as PM_{10}) (See "Basis for the AALG.")

Occupational Limits:
- ACGIH TLV: none
- NIOSH REL: none
- OSHA PEL: none

Drinking Water Limits: No drinking water limits were found for this substance. However, it should be noted that the National Secondary Drinking Water Standard for sulfate is 250 mg/L.[1]

Toxicity Profile

Carcinogenicity: No data were found implicating sodium sulfate as a carcinogen.

Mutagenicity: Negative results are listed for sodium sulfate in the Gene-Tox database as summarized in RTECS[2] for the oncogenic transformation of Syrian hamster embryo cells by adenovirus SA7.

Developmental Toxicity: Arcuri and Gautieri[3] reported that intraperitoneal injection of 60 mg/kg sodium sulfate in saline on gestation day 8 in CF-1 mice resulted in a statistically significant decrease in maternal body weight gain and an increase in pup skeletal abnormalities.

In a developmental toxicity screening test, gavage administration of 2800 mg/kg sodium sulfate to ICR/SIM mice on gestation days 8 through 12 did not adversely effect maternal health or neonatal survival and there was a significant increase in pup birth weight.[4] The significance of these findings for humans exposed via the inhalation route is uncertain.

Reproductive Toxicity: No data were found implicating sodium sulfate as a reproductive toxin.

Systemic Toxicity: Available data on sodium sulfate toxicity are limited. Sodium sulfate is of relatively low acute toxicity by the oral route with an LD_{50} of 5989 mg/kg reported in mice.[2]

Based on a request by an industrial entity to delist sodium sulfate (solution) from the list of toxic substances under Section 313 of SARA Title III, EPA conducted a review of the health and environmental effects of this compound. Based on limited available data on chronic toxicity (oral route, mice and chickens), it was concluded that Na_2SO_4 "did not pose a significant hazard of chronic toxicity except at high doses (i.e., greater than 10,000 ppm in chickens) where dehydration may occur due to the cathartic effect." It should be noted that sodium sulfate is used as a saline cathartic in humans with the usual therapeutic dose being 15 g, which corresponds to 214 mg/kg for a 70-kg person.[1]

Irritation: No data were found implicating sodium sulfate as an irritant.

Basis for the AALG: Sodium sulfate is soluble in "about 3.6 parts water" and occurs naturally as the minerals mirabilite and thenardite.[5] In its review of sodium sulfate, EPA concluded that this compound is released primarily to water and that land and air releases are very small relative to water releases.[1]

Based on its review of the health and environmental effects of Na_2SO_4, EPA[1] concluded that "Na_2SO_4 yielded no areas of concern" and that "there is no evidence which suggests that Na_2SO_4 is known or can reasonably be anticipated to cause health or environmental effects as described in section 313(d)(2)."

Given the very low order of toxicity and hazard associated with Na_2SO_4, and given the fact that since it is a crystalline inorganic solid, it would exist in air as a mist or dust, it is suggested that adherence to the current NAAQS for particulate matter would be protective of the health of the general public. The present primary (and secondary) standards for particulate matter are 150 $\mu g/m^3$ (no more than one exceedance expected per year) as a 24-hour TWA and 50 $\mu g/m^3$ annual arithmetic mean, both measured as PM_{10}. PM_{10} is an indicator that includes only those particles with an aerodynamic diameter of less than or equal to a nominal 10 μm. The basis for the particulate standard and the use of PM_{10} as an indicator are discussed in detail in EPA's "Ambient Air Quality Standards for Particulate Matter; Final Rules."[6]

AALG: • 150 $\mu g/m^3$ 24-hour TWA (as PM_{10})
  • 50 $\mu g/m^3$ annual arithmetic TWA (as PM_{10})

References

1. U.S. EPA. 1989. "Sodium Sulfate; Toxic Chemical Release Reporting; Community Right-to-Know." *Fed. Reg.* 54(32):7217–19.

2. NIOSH. 1988. WE1650000. "Sodium Sulfate (2:1)." RTECS, on line.

3. Arcuri, P. A. and R. F. Gautieri. 1973. "Morphine-Induced Fetal Malformations III: Possible Mechanisms of Action." *J. Pharm. Sci.* 62:1626–34.

4. Seidenberg, J. M., D. G. Anderson, and R. A. Becker. 1986. "Validation of an In Vivo Developmental Toxicity Screen." *Terat. Carcin. Mutagen.* 6:361–74.

5. Windholz, M., Ed. 1976. *The Merck Index,* 9th ed. (Rahway, NJ: Merck & Co., Inc.).

6. U.S. EPA. 1987. "Ambient Air Quality Standards for Particulate Matter; Final Rules." *Fed. Reg.* 52(125):24633–750.

STYRENE

Synonyms: phenyl ethylene, vinyl benzene, cinnamene

CAS Registry Number: 100–42–5

Molecular Weight: 104.14

Molecular Formula:

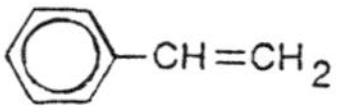

$$\langle\bigcirc\rangle\text{—CH}=\text{CH}_2$$

AALG: carcinogenicity—6.2 ppb (26.3 μg/m^3) annual TWA

Occupational Limits:
- ACGIH TLV: 50 ppm (215 mg/m^3) TWA; 100 ppm (425 mg/m^3) STEL; skin notation
- NIOSH REL: 50 ppm (215 mg/m^3) TWA; 100 ppm (425 mg/m^3) ceiling
- OSHA PEL: 50 ppm (215 mg/m^3) TWA; 100 ppm (425 mg/m^3) STEL; skin notation

Basis for Occupational Limits: The TLVs were set in consideration of central nervous system toxicity, irritation, and potential carcinogenic effects. The TLV-TWA was formerly 100 ppm, but was reduced to 50 ppm, "one-tenth the lowest concentration possibly causing lymphoid or hematopoietic tumors in female rats."[1] For the NIOSH limits, the TWA was based on nervous system effects and the ceiling limit on eye and respiratory tract irritation.[2] OSHA recently adopted the ACGIH limits as part of its final rule limits.[3]

Drinking Water Limits: NAS calculated a suggested no-adverse-effect level in drinking water of 0.9 mg/L based on a rat NOAEL from a 185-day study and using an uncertainty factor of 1000.[4] In the drinking water quality criteria document for styrene,[5] the following health advisories were derived from a rat NOAEL of 270 mg/kg using an uncertainty factor of 100: (1) 1-day HA, adults—94.5 mg/L, (2) 1-day HA, child—27 mg/L, (3) 10-day HA, adults—70 mg/L, (4) 10-day HA, child—20 mg/L, and (5) lifetime AADI—7 mg/L. Although there is limited data for the carcinogenicity of styrene, lung adenomas and carcinomas in female mice[6] (cited in U.S. EPA[5]) were used to derive a criterion of 0.014 μg/L at the 10^{-6} risk level. It is stated that this criterion is uncertain due to the short exposure duration and small number of animals per group used in that study.

Toxicity Profile

Carcinogenicity: There is some evidence for an association between styrene exposure in the styrene-butadiene rubber and styrene-polystyrene manufacturing industries and increased risk of lymphatic and hematopoietic tumors. However, the evidence is insufficient to designate styrene as a human carcinogen on account of limitations in the available studies such as small cohort sizes and concurrent exposure to multiple chemicals.

Jersey et al.[7] (cited in NIOSH[2]) exposed male and female Sprague-Dawley rats to 0, 600, or 1200 ppm styrene via inhalation for 6 hours per day, 5 days per week for 18.3 (males) or 20.7 (females) months with final sacrifice at 24 months; after 2 months the highest exposure level was lowered to 1000 ppm because of excessive mortality among the male rats. There was an increased combined incidence of leukemia and lymphosarcoma tumor types in female rats in both exposure groups, which was statistically significant when data from both exposure groups were combined and for each group compared individually to historical, but not concurrent, controls. In addition, there was a higher incidence of alveolar histiocytosis and increased liver weights in high-dose female rats. There was no increase in incidence of tumors attributable to treatment in the male rats; concurrent high mortality due to chronic murine pneumonia limits the conclusions that may be drawn from the male rat data.

In an NCI bioassay,[8] male and female B6C3F$_1$ mice and F344 rats received 150 or 300 mg/kg/day and 500, 1000, or 2000 mg/kg/day, respectively, by gavage in corn oil 5 days a week for 103 weeks (low-dose rats) or 78 weeks (all other groups). There was an increased incidence of lung adenomas in male mice compared with vehicle, but not historical, controls. NCI concluded that there was "suggestive" evidence of carcinogenicity in male B6C3F$_1$ mice, but the evidence was not "convincing" for either species.

The possible carcinogenicity of styrene has been investigated in several additional studies that are reviewed in the drinking water criteria document[5] and the NIOSH criteria document.[2] Regarding carcinogenicity, NIOSH states that "from the experimental animal investigations and from the epidemiological studies, there seems to be little basis to conclude that styrene is carcinogenic."[2] It seems most reasonable to tentatively place styrene in group C, possible human carcinogen, under the EPA weight-of-evidence classification.

Mutagenicity: Positive results are listed in the Gene-Tox database as summarized in RTECS for the following assays: (1) micronucleus test—in vitro human lymphocytes, (2) Ames assay, (3) *Drosophila* SLRL test, (4) *S. cerevisiae* gene conversion, and (5) in vivo cytogenetics—human lymphocytes. Negative results are listed for the following assays: (1) cell transformation—Syrian hamster embryo cells by adenovirus SA7, (2) gene mutation—V79 cell culture, and (3) in vitro unscheduled DNA synthesis-human fibroblasts.[9] The evidence for genetic activity of styrene in short-term tests was also recently reviewed by IARC.[10] Based on the

IARC classification scheme, it is our judgment that the evidence for genetic activity of styrene in short-term tests is sufficient.

Developmental Toxicity: Although there have been some reports of birth defects[11] (cited in NIOSH[2]) and increased incidence of spontaneous abortions[12] (cited in NIOSH[2]) in women occupationally exposed to styrene, evidence of developmental toxicity has not been found in several apparently well-conducted animal studies. These include a multigeneration study in which rats were exposed to styrene in drinking water[13] and a study in which rabbits were exposed to styrene via inhalation and rats were exposed via inhalation and gavage.[14]

In another study, however, Kankaanpaa et al.[15] exposed BMR/T6T6 mice to 0 or 250 ppm styrene for 6 hours per day on gestation days 6 through 16 of gestation and Chinese hamsters to 0, 300, 500, 750, or 1000 ppm styrene on days 6 through 18 of gestation. Embryotoxicity, in the form of significantly higher incidences of dead and resorbed fetuses, were observed in both species only in the highest exposure group relative to controls. Some degree of maternal toxicity was apparently observed, but no details were given and the authors did not attribute the embryotoxicity to maternal toxicity.

Reproductive Toxicity: No data were found implicating styrene as a reproductive toxin.

Systemic Toxicity: The best-documented form of systemic toxicity resulting from styrene exposure involves the central nervous system. Subjectively, this is manifest as complaints of headache, fatigue, dizziness, confusion, drowsiness, malaise, difficulty in concentrating, and a feeling of intoxication. Clinical signs include altered equilibrium, delayed reaction times, and abnormal EEGs. Alterations such as these have been reported in experimental studies at concentrations as low as 100 ppm[16,17] (cited in NIOSH[2]). Numerous studies, both clinical and experimental, documenting the C.N.S. toxicity of styrene are reviewed in the NIOSH criteria document.[2]

In addition to C.N.S. toxicity, there is also more limited evidence for other adverse effects attributable to styrene. These include peripheral neuropathy, abnormal pulmonary function, and alterations in liver function. The evidence for these effects is also reviewed in the criteria document.[2]

In a recent review of the toxicity of styrene, Bond et al.[16] concluded that many of the effects of styrene in humans and laboratory animals are similar. Further, it was noted that the major metabolic pathway in both humans and animal models involves oxidation of the vinyl group to styrene oxide with further metabolism to several products, at least two of which—mandelic and phenylglyoxylic acids—have been detected in both human and rodent urine subsequent to styrene exposure.

Irritation: The irritancy of styrene to the eyes, nose, skin, and respiratory tract is well documented in both clinical and experimental studies. A large proportion of subjects report upper respiratory tract irritation at exposures as low as 100 ppm, and some report eye irritation at concentrations of 20 or 50 ppm[16,17,19] (cited in NIOSH[2]). Styrene is reported to have an aromatic odor[1] and an odor threshold of 0.32 ppm (geometric mean of reported literature values).[20]

Basis for the AALG: Styrene is considered "very slightly" soluble in water and has a vapor pressure of 6.1 mm Hg at 25°C.[20] It is used in the production of polystyrene plastics, protective coatings, styrenated polyesters, copolymer resins, and as a chemical intermediate.[1] Styrene has been detected in both ambient air and finished drinking water in the United States.[21] Based on the lack of data on the environmental fate and distribution of styrene and the expected contribution of various media to human exposure, 50% of the contribution to total exposure is allowed from air.

Given the equivocal nature of the evidence for styrene carcinogenicity, two AALGs were calculated for styrene: one based on carcinogenicity and another based on systemic toxicity. Also, use of quantitative risk assessment was not deemed appropriate on account of the equivocal evidence of carcinogenicity. The AALG is instead based on the rat LOAEL of 600 ppm from the study of Jersey et al.[7] (cited in NIOSH[2]) using an uncertainty factor approach. The LOAEL was adjusted for continuous exposure (600 ppm × 5/7 × 6/24 × 20.7/24), and the total uncertainty factor used was 10,000 (10 for a LOAEL × 10 for interindividual variation × 10 for interspecies variation × 10 for the database factor). Use of the database factor was justified based on the severity of the effect and, in addition, data from human experimental studies[22,23] (cited in Amoore and Hautala[20]). Retention of 60% of the exposure dose was assumed. Given the weight-of-evidence ranking for styrene and the pending review of its carcinogenicity by EPA,[24] the AALG should be considered provisional.

An AALG was also calculated for systemic toxicity, specifically neurotoxicity, using the NIOSH REL-TWA as a basis. This limit was treated as a human LOAEL (REL/210; 10 for interindividual variation × 5 for a LOAEL × 4.2) based on the review in the criteria document,[2] in which it was stated that there were "effects such as slower reactions, subjective complaints related to CNS depression and abnormal EEGs at styrene concentrations around 50 ppm." This AALG should also be considered provisional.

AALG: • carcinogenicity—6.2 ppb (26.3 μg/m³) annual TWA
            • systemic toxicity—119 ppb (507 μg/m³) 8-hour TWA

References

1. ACGIH. 1986. "Documentation of the Threshold Limit Values and Biological Exposure Indices." 5:539.
2. NIOSH. 1984. "Criteria for a Recommended Standard . . . Occupational Exposure to Styrene." DHHS (NIOSH) 83–119.
3. OSHA. 1989. "Air Contaminants; Final Rule." *Fed. Reg.* 54:2332–2959.

4. Safe Drinking Water Committee. 1977. *Drinking Water and Health,* Vol. 1 (Washington, DC: National Academy Press).
5. U.S. EPA. 1985. "Drinking Water Criteria Document for Styrene" (draft) (Springfield, VA: NTIS), EPA-600/x-84–195–1, PB86–118056.
6. Ponomarkov, V. I., and L. Tomatis. 1978. "Effects of Long-Term Oral Administration of Styrene to Mice and Rats." *Scand. J. Work Environ. Health* 4(suppl. 2):127–35.
7. Jersey, G., M. Balmer, J. Quast, C. N. Park, D. J. Schuetz, J. E. Beyer, K. J. Olson, S. B. McCollister, and L. W. Rampy. 1978. "Two-Year Chronic Inhalation Toxicity and Carcinogenicity Study on Monomeric Styrene in Rats." Dow Chemical study for Manufacturing Chemists Association (12/6).
8. NCI. 1979. "Bioassay of Styrene for Possible Carcinogenicity." TR-185.
9. NIOSH. 1987. WL3675000. "Styrene." RTECS, on line.
10. IARC. 1986. "Styrene." *IARC Monog.* suppl. 6:498–501.
11. Hemminki, K., E. Franssila, and H. Vainio. 1980. "Spontaneous Abortion among Female Chemical Workers in Finland." *Int. Arch. Occup. Environ. Health* 45:123–26.
12. Holmberg, P. C. 1977. "Central Nervous Defects in Two Children of Mothers Exposed to Chemicals in the Reinforced Plastics Industry." *Scand. J. Work Environ. Health* 3:212–14.
13. Beliles, R. P., J. H. Butala, C. R. Stack, and S. Makris. 1985. "Chronic Toxicity and Three-Generation Reproduction Study of Styrene Monomer in the Drinking Water of Rats." *Fund. Appl. Tox.* 5:855–68.
14. Murray, F. J., J. A. John, M. F. Balmer, and B. A. Schwetz. 1978. "Teratologic Evaluation of Styrene Given to Rats and Rabbits by Inhalation or by Gavage." *Toxicology* 11:335–43.
15. Kankaanpaa, J. T. J., E. Elovaara, K. Hemminki, and H. Vainio. 1980. "The Effect of Maternally Inhaled Styrene on Embryonal and Foetal Development in Mice and Chinese Hamsters." *Acta Pharm. Tox.* 47:127–29.
16. Hake, C. L., R. D. Stewart, A. Wu, S. A. Graff, H. V. Forster, W. H. Keeler, A. J. Lebrun, P. E. Newton, and R. J. Soto. (undated). "Styrene—Development of a Biologic Standard for the Industrial Worker by Breath Analysis." NIOSH-MCOW-ENVM-STY-77–2. Milwaukee, Medical College of Wisconsin, NIOSH Contract No. HSM 99–72–84.
17. Oltramare, M., E. Desbaumes, C. Imhoff, and W. Michiels. 1974. "Toxicology of Monomeric Styrene—Experimental and Clinical Studies on Man" (French). Geneva, Editions Medicine et Hygiene.
18. Bond, J. A. 1989. "Review of the Toxicology of Styrene." *CRC Crit. Rev. Tox.* 19:227–49.
19. Stewart, R. D., H. C. Dodd, E. D. Baretta, and A. W. Schaffer. 1968. "Human Exposure to Styrene Vapor." *Arch. Environ. Health* 16:656–62.
20. Amoore, J. E., and E. Hautala. 1983. "Odor as an Aid to Chemical Safety: Odor Thresholds Compared with Threshold Limit Values and Volatilities for 214 Industrial Chemicals in Air and Water Dilution." *J. Appl. Tox.* 3:272–90.
21. IARC. 1979. "Styrene, Polystyrene and Styrene-Butadiene Copolymers." *IARC Monog.* 19:231–44.
22. Fiserova-Bergerova, V., and J. Teisinger. 1965. "Pulmonary Styrene Vapor Retention." *Ind. Med. Surg.* 34:620.
23. Bardodej, Z., and E. Bardodejova. 1970. "Biotransformation of Ethyl Benzene, Styrene and α-Methylstyrene in Man." *Am. Ind. Hyg. Assoc. J.* 31:206.
24. U.S. EPA. 1988. "Styrene; CASRN 100–42–5." IRIS (6/30/88).

SULFURIC ACID

Synonyms: hydrogen sulfate

CAS Registry Number: 7664–93–9

Molecular Weight: 98.08

Molecular Formula: H_2SO_4

AALG: irritation (TLV-based)—2 $\mu g/m^3$ 8-hour TWA

Occupational Limits:
- ACGIH TLV: 1 mg/m^3 TWA; 3 mg/m^3 STEL
- NIOSH REL: 1 mg/m^3 TWA
- OSHA PEL: 1 mg/m^3 TWA

Basis for Occupational Limits: The occupational limits for sulfuric acid are based on prevention of pulmonary irritation and damage to the teeth.[1,2] The ACGIH has recently added a TLV-STEL of 3 mg/m^3 for sulfuric acid.[3]

Drinking Water Limits: No drinking water limits were found for this substance.

Toxicity Profile

Carcinogenicity: No data were found implicating sulfuric acid as a carcinogen.

Mutagenicity: No data were found implicating sulfuric acid as a mutagen.

Developmental Toxicity: Murray et al.[4] exposed CF-1 mice and New Zealand white rabbits to filtered air (n = 40 mice and 20 rabbits) or 5 or 20 mg/m^3 sulfuric acid aerosol (n = 20 rabbits/group or 35 mice/group) for 7 hours per day on gestation days 6 through 15 and 6 through 18 in mice and rabbits, respectively. Slight maternal toxicity occurred only at the highest exposure level in both species and was evident as decreased maternal liver weights in mice and a trend toward a dose-related increase in subacute rhinitis and tracheitis in rabbits. There was no evidence for teratogenic effects or embryofetal toxicity in either species, but there was a significantly increased incidence of small non-ossified areas in the skull bones of rabbit fetuses in the highest exposure group. However, this was considered to be a minor variation in skeletal development by the authors.

Reproductive Toxicity: No data were found implicating sulfuric acid as a reproductive toxin.

Systemic Toxicity: In acute lethality studies in animals, it has been reported that 8-hour LC_{50}s range from 20 to 60 mg/m^3 sulfuric acid depending on the particle size and the age of the animals. Young animals are more susceptible, and at lethal concentrations 0.8-μm particles were less toxic than 2.6-μm particles.[1]

In the *Health Effects Assessment Document for Sulfuric Acid*,[5] it was stated that "since the major toxic effect of sulfuric acid is local irritation and since inhaled sulfuric acid is largely neutralized by NH_3 in the expired air, it is unlikely that significant systemic exposure to sulfuric acid occurs."

It should be noted that synergism has been demonstrated in laboratory animals exposed to sulfuric acid and many pollutants commonly found associated with automobile exhaust, e.g., SO_2, ozone, and metallic aerosols. In addition, when sulfuric acid is adsorbed onto airborne particulates, neutralization of sulfuric acid by expired NH_3 is reduced.[5]

Irritation: The predominant effects of both acute and chronic exposure to sulfuric acid in laboratory animals and humans are based on its irritant properties. An extensive literature exists on these effects and has been well reviewed by NIOSH[2,6] and Carson et al.[7]

Occupational exposure to sulfuric acid has been associated with dental erosion, eye irritation, and effects on the respiratory system. In a battery manufacturing plant where exposures had previously been estimated to be approximately 1.4 mg/m^3, Williams[8] (cited in U.S. EPA[5]) reported a slight increase in bronchitis among workers exposed to sulfuric acid compared to nonexposed workers from the same plant.

Dental erosion has been reported in workers exposed to sulfuric acid in battery manufacturing plants. Malcolm and Paul[9] (cited in U.S. EPA[5]) reported that the incidence and severity of dental erosion was greater in workers exposed to levels of between 3 and 16 mg/m^3 compared to workers exposed to levels between 0.8 and 2.5 mg/m^3. In more recent occupational studies, Jones and Gamble and Gamble et al.[10,11] (cited in U.S. EPA[5]) reported erosions of the teeth in workers exposed to an average level of 0.18 mg/m^3 (range 0 to 1.7 mg/m^3). It was further reported that exposure to 0.23 mg/m^3 for a period of 4 months was sufficient to initiate erosive changes in the teeth.

Amdur et al.[12] studied the responses of unacclimated human subjects to inhaled concentrations of sulfuric acid in the range of 0.35 to 5 mg/m^3 for periods of 5 to 15 minutes. It was reported that the lowest concentration detectable by odor, taste, or irritation was 1 mg/m^3. A concentration of 3 mg/m^3 was noticeable by all of the subjects and the sensation experienced was reported to be "like breathing dusty air." A concentration of 5 mg/m^3 was reported to be "very objectionable" to many of the subjects and a deep breath at this concentration usually induced coughing.

Based on their review of the literature, Carson et al.[7] estimated that "a lower exposure level that appears to be safe for man is in the range of 0.066 to 0.098

mg/m^3'' and also pointed out that this may be influenced by factors such as particle size, frequency and duration of exposure, and potential synergistic effects.

Based on its review of the literature published since the promulgation of the original sulfuric acid limit, NIOSH drew the following conclusions:

1. Data from human studies suggest that adverse pulmonary effects may occur at exposure levels equal to the current PEL ($1 \ mg/m^3$) and that the effects of exposure for extended periods are unknown.
2. It has been demonstrated that the concentration of ammonia in human airways is sufficient to neutralize low levels of sulfuric acid; however, it is not known whether such neutralization constitutes a significant protective mechanism.

Although the effects of exposure to sulfuric acid have been extensively studied in laboratory animals, extrapolation of these results to humans is complicated by the following considerations: (1) most of the species used have been nose-breathers rather than mouth-breathers, and (2) aerosol size, relative humidity, and other experimental factors which alter the effects of sulfuric acid also vary. Nonetheless certain general conclusions are relevant and may be drawn:

1. Results of studies in donkeys[13] (cited in NIOSH[6]) and guinea pigs[14] (cited in NIOSH[6]) suggest that exposure to $0.1 \ mg/m^3$ sulfuric acid may result in decreased pulmonary function and alterations in mucociliary clearance.
2. Long-term exposure of monkeys and guinea pigs indicated that the monkey is the more sensitive species, and slight but increasingly severe microscopic pulmonary lesions have been demonstrated in monkeys chronically exposed to sulfuric acid aerosols in the concentration range of 0.1 to $1.0 \ mg/m^3$ independent of particle size[15,16] (cited in NIOSH[6]).
3. Studies in animal models provide some evidence of additive and synergistic effects at low levels of sulfuric acid in combination with ozone, sulfur dioxide, and particulate carbon.

Alarie et al.[15] exposed the following groups of male and female Cynomolgus monkeys (n = 9/group) to sulfuric acid continuously for a period of 78 weeks: filtered air; $0.38 \ mg/m^3$, mass median diameter (MMD) 2.15 μm; $0.48 \ mg/m^3$, MMD 0.54 μm; $2.43 \ mg/m^3$, MMD 3.6 μm; and $4.79 \ mg/m^3$, MMD 0.73 μm. At the two highest concentrations, histological changes were observed in all exposed animals and included hyperplasia and hypertrophy of the bronchiolar epithelium and alveolar epithelium with focal thickening of the bronchiolar walls and alveolar septa as well as decreased functional capacity. In animals exposed to $0.48 \ mg/m^3$ there were no significant alterations in lung histopathology, but exposure to $0.38 \ mg/m^3$ sulfuric acid (higher MMD particles) resulted in slight bronchiolar epithelial hyperplasia in 5 of 9 exposed animals, slight thickening of the bronchiolar walls in 3 of 9 animals, and slight focal bronchiolar epithelial hyperplasia in 4 of 9 animals. Starting at 8 to 12 weeks after the beginning of exposure, a significant increase in respiratory rate was observed in animals exposed to $0.38 \ mg/m^3$ sulfuric acid. Respiratory rate remained elevated throughout the remainder of the study.

Basis for the AALG: Sulfuric acid is widely used as an industrial chemical in fertilizer manufacture, petroleum refining, electroplating, and acid cleaning; in storage batteries; and as a starting material for a variety of chemicals. Since it has a very low vapor pressure (less than 0.001 mm Hg at 20°C), it exists in air only as a mist or spray.[1] Since the AALG for sulfuric acid is based on irritation, no relative source contribution from inhalation was factored into the calculation. The reason is that allocation of a proportion of exposure to a given source is not relevant when the effects of exposure are not cumulative, as is the case with irritant effects.

It is clear from the available data that derivation of an AALG for sulfuric acid is highly complex due to (1) potential additive or synergistic effects with commonly occurring criteria pollutants such as ozone, particulate matter, and sulfur dioxide; (2) the variation in toxicity expected with variation in particle size; and (3) the effects of relative humidity on particle size. In connection with this latter issue it should be noted that owing to the hygroscopic nature of sulfuric acid, there is an increase in particle size, which is inversely related to relative humidity. NIOSH[6] has summarized the differential toxicity based on particle size as follows: ''For aerosols containing particle sizes that can penetrate to the lung, at least two mechanisms of action have been demonstrated. Aerosol particles that deposit in the upper lung appear to be more acutely harmful because reflexive bronchoconstriction occurs. Somewhat smaller aerosol particles, however, appear to cause greater alterations in pulmonary function and eventually in microscopic lesions. Exposure to very large particles would not lead to either of these effects. This all suggests that any future revision of the occupational exposure limit should consider aerosol particle size.''

Further complicating the derivation of an AALG for sulfuric acid is the fact that no single available animal or human study provides a suitable basis for criteria derivation, largely due to the considerations outlined above. It is for this reason that the occupational limit was selected as a starting point in AALG derivation. It is clear on the basis of animal studies and human occupational studies that the OEL of 1 mg/m^3 should be treated as a human LOAEL. In addition, a database factor of 10 is incorporated to account for the effects of synergism with other chemicals and the effects of particle size. Therefore, the AALG for sulfuric acid is based on the OEL of 1 mg/m^3 divided by a total uncertainty factor of 500 (10 for interindividual variation × 5 for a LOAEL × 10 for the database factor). It should be noted that the AALG of 2 μg/m^3 is below the level of 8–10 μg/m^3 at which Finklea et al.[17] (cited in NIOSH[6]) considered adverse effects in susceptible individuals likely and well below the lower level of safe exposure for man of 66–98 μg/m^3 estimated by Carson et al.[6]

This AALG should be considered provisional. It is strongly recommended that a concerted review of the available literature be undertaken for the purpose of developing a criterion for sulfuric acid that could be related to both particle size and relative humidity.

AALG: • irritation (TLV-based)—2 μg/m^3 8-hour TWA

References

1. ACGIH. 1986. "Documentation of the Threshold Limit Values and Biological Exposure Indices." 5:544 (1987).
2. NIOSH. 1974. "Criteria for a Recommended Standard . . . Occupational Exposure to Sulfuric Acid." DHEW (NIOSH) 74–128.
3. ACGIH. 1988. "Threshold Limit Values and Biological Exposure Indices for 1988–89" (Cincinnati, OH: ACGIH).
4. Murray, F. J., B. A. Schwetz, K. D. Nitschke, A. A. Crawford, J. F. Quast, and R. E. Staples. 1979. "Embryotoxicity of Inhaled Sulfuric Acid Aerosol in Mice and Rabbits." *J. Env. Sci. Health* C13(3):251–66.
5. U.S. EPA. 1984. "Health Effects Assessment Document for Sulfuric Acid." EPA/540/1-86/031.
6. NIOSH. 1981. "Review and Evaluation of Recent Literature—Occupational Exposure to Sulfuric Acid." DHHS (NIOSH) 82–104.
7. Carson, B. L., B. L. Herndon, H. V. Ellis, L. H. Baker, and E. Horn. 1981. "Sulfuric Acid Health Effects." Contract No. 68–03–2928 (Kansas City, MO: Midwest Research Institute), EPA/460/3-81-025, NTIS PB82–113135.
8. Williams, M. K. 1970. "Sickness Absence and Ventilatory Capacity of Workers Exposed to Sulfuric Acid Mist." *Br. J. Ind. Med.* 27:61–66.
9. Malcolm, D., and E. Paul. 1961. "Erosion of the Teeth Due to Sulfuric Acid in the Battery Industry." *Br. J. Ind. Med.* 18:63–69.
10. Jones, W., and J. Gamble. 1984. "Epidemiological-Environmental Study of Lead Acid Battery Workers. Environmental Study of Five Lead Acid Battery Plants." *Env. Res.* 35:1–10.
11. Gamble, J., W. Jones, J. Hancock, and R. Meckstroth. 1984. "Epidemiological-Environmental Study of Lead Acid Battery Workers. III. Chronic Effects of Sulfuric Acid on the Respiratory System and Teeth." *Env. Res.* 35:30–52.
12. Amdur, M. O., L. Silverman, and P. Drinker. 1952. "Inhalation of Sulfuric Acid Mist by Human Subjects." *AMA Arch. Ind. Hyg. Occ. Med.* 6:305–13.
13. Schlesinger, R. B., M. Halpern, R. E. Albert, and M. Lippmann. 1979. "Effect of Chronic Inhalation of Sulfuric Acid Mist upon Mucociliary Clearance from the Lungs of Donkeys." *J. Environ. Pathol. Toxicol.* 2:1351–67.
14. Amdur, M. O., M. Dubriel, and D. A. Creasia. 1978. "Respiratory Response of Guinea Pigs to Low Levels of Sulfuric Acid." *Environ. Res.* 15:418–23.
15. Alarie, Y., W. M. Busey, A. A. Krumm, and C. E. Ulrich. 1973. "Long-Term Continuous Exposure to Sulfuric Acid Mist in Cynomolgus Monkeys and Guinea Pigs." *Arch. Env. Health* 27:16–24.
16. Alarie, Y. C., A. A. Krumm, W. M. Busey, C. E. Ulrich, and R. J. Kantz. 1975. "Long-Term Exposure to Sulfur Dioxide, Sulfuric Acid Mist, Fly Ash and their Mixtures—Results of Studies in Monkeys and Guinea Pigs." *Arch. Environ. Health* 30:254–62.
17. Finklea, J. F., J. Moran, J. H. Knelson, D. B. Turner, and L. E. Niemeyer. 1975. "Estimated Changes in Human Exposure to Suspended Sulfate Attributable to Equipping Light-Duty Motor Vehicles with Oxidation Catalysts." *Environ. Health Perspect.* 5:1037–47.

TETRACHLOROETHYLENE

Synonyms: perchloroethylene, tetrachloroethene, PCE

CAS Registry Number: 127–18–4

Molecular Weight: 165.84

Molecular Formula: $Cl_2C=CCl_2$

AALG: 10^{-6} 95% UCL—0.12 ppb (0.86 $\mu g/m^3$) annual TWA

Occupational Limits:
- ACGIH TLV: 50 ppm (335 mg/m^3) TWA; 200 ppm (1340 mg/m^3) STEL
- NIOSH REL: minimize workplace exposure levels, limit numbers of workers exposed
- OSHA PEL: 25 ppm (170 mg/m^3) TWA

Basis for Occupational Limits: The TLV-TWA is set to prevent the discomfort and subjective complaints arising from prolonged exposure, e.g., dizziness and headaches. The TLV-STEL is set to prevent anesthetic effects.[1] The NIOSH recommendation is based on consideration of carcinogenic effects.[2] OSHA, as part of its final rulemaking effort for air contaminants, recently lowered the PEL from 100 ppm as a TWA to 25 ppm as a TWA in consideration of carcinogenic effects.[3]

Drinking Water Limits: The ambient water quality criterion for PCE is 0.80 $\mu g/L$ at the 10^{-6} risk level assuming consumption of 2 L of water and 6.5 g fish/shellfish per day.[4] Since data were insufficient to calculate a 1-day HA, it was recommended that the 10-day HA of 2.0 mg/L for a 10-kg child be used as a conservative estimate for the 1-day health advisory. The longer-term HAs for a 70-kg adult and a 10-kg child are 5.0 mg/L and 1.4 mg/L, respectively. Since PCE has been designated as a group B2 carcinogen by the U.S. EPA, a lifetime HA would not normally be recommended. However, because the EPA's Scientific Advisory Board, Halogenated Organics Subcommittee, considers PCE to be a group C carcinogen, a lifetime HA of 10 $\mu g/L$ was calculated.[5] An MCLG and MCL will be proposed pending evaluation of the NTP bioassay data.[6]

Toxicity Profile

Carcinogenicity: Evidence from epidemiologic studies is inadequate to evaluate the carcinogenicity of PCE because, although an association between cancer risk and

531

employment in the dry-cleaning industry has been reported, either the solvents used have not been reported or there was exposure to multiple solvents.[7]

PCE has been shown to induce hepatocellular carcinomas in both sexes of B6C3F$_1$ mice when administered by gavage[8] (cited in U.S. EPA[7]) or via inhalation[9] (cited in U.S. EPA[10]) and is classified as a group B2, probable human carcinogen, by the U.S. EPA.[10] The inhalation bioassay will be reviewed in detail here since this data was used for quantitative risk assessment.

In the NTP inhalation bioassay,[9] B6C3F$_1$ mice and F344 rats (n = 50/sex/ group/species) were exposed to 0, 100, or 200 ppm and 0, 200, or 400 ppm PCE (99.9% purity), respectively, 6 hours per day, 5 days a week for 2 years. PCE exposure did not result in consistent decreases in body weight gain in rats or mice; however, mortality was increased in female rats at the highest exposure level, male mice at both exposure levels, and female mice at the lowest exposure level. It was thought that early deaths may have been attributable to the development of neoplasms. There was a statistically significant, dose-related increase in the combined incidence of hepatocellular adenomas and carcinomas in both male and female mice as follows: controls—males 16 of 49, females 4 of 48; low dose—males 31 of 49, females 17 of 50; and high dose—males 40 of 50, females 38 of 50. Non-neoplastic lesions observed in both males and females included liver degeneration, hepatocellular necrosis, hepatocellular nuclear inclusion, and renal tubular cell karyomegaly. Findings in the rats included the following:

1. a marginally statistically significant increase in mononuclear cell leukemia in male rats at both exposure levels
2. a not statistically significant, but dose-related, increased incidence in renal tubular cell adenomas/carcinomas (a tumor type with a very low spontaneous incidence) in male rats
3. a statistically significant increased incidence of mononuclear cell leukemia in female rats

However, in a document presented to the EPA by the Halogenated Solvent Industry Alliance, the findings in male rats were questioned on the basis of the criteria used in staging the leukemias and "alleged deficiencies in animal handling and identification."[10]

It is relevant to note that negative results were reported in rats exposed to PCE by gavage in the NCI bioassay[8] (cited in U.S. EPA[7]) and in an earlier inhalation bioassay conducted by Dow Chemical Company[11] (cited in U.S. EPA[7]). However, these studies are insufficient to evaluate PCE carcinogenicity in rats due to excess mortality in the former study and the low exposure levels used in the latter study.[7]

Mutagenicity: PCE has generally produced negative or weak responses (which usually occurred at cytotoxic doses) in assays designed to evaluate gene mutations, chromosomal aberrations, and effects on DNA. However, the epoxide of PCE has been found to be mutagenic in bacterial assays and is generally recognized to be responsible (possibly along with other reactive metabolites) for the carcinogenic potential of PCE.[7] The genotoxic potential of PCE has also been recently reviewed by IARC.[12]

Developmental Toxicity: The inhalation teratogenicity of PCE has been evaluated in mice, rats, and rabbits[13-15] (cited in U.S. EPA[7]), and its postnatal effects on the offspring of rats exposed during gestation have also been investigated[16,17] (cited in U.S. EPA[7]). Based on these studies, it was concluded that the teratogenic potential of PCE is not significant, although effects related to delayed development were observed at levels producing maternal toxicity. It was concluded in the HAD that minor behavioral changes noted in the offspring of rats exposed to PCE were probably due to maternal nutritional deprivation.[7]

Reproductive Toxicity: Beliles et al.[13] (cited in U.S. EPA[7]) exposed mice and rats to 0, 100, or 500 ppm PCE via inhalation for 7 hours per day on 5 consecutive days and examined their sperm for abnormalities at the end of 1, 4, or 10 weeks. Negative results were reported for rats; however, mice exposed to 500 ppm PCE showed 19.7% abnormal sperm after 4 weeks relative to 6.0% for the control animals.

Systemic Toxicity: In humans and animal models, neurotoxicity appears to be the most sensitive endpoint, with renal and hepatic effects generally reported only after prolonged exposure periods (animals) or high exposure levels (humans). At levels at or near 100–200 ppm in humans, short-term exposures are reported to result in dizziness, confusion, nausea, and headaches. Longer-term exposures enhance these symptoms and may result in unconsciousness, and higher levels of exposure may cause disorientation, irritability, ataxia, and sleep disturbances.[7]

Irritation: In humans, short-term exposures in the range of 100 to 200 ppm have been reported to cause irritation of the eyes and mucous membranes.[7] PCE is reported to have an ''ethereal'' odor; odor thresholds have been reported to be as low as 5 ppm and as high as 50 ppm.[1]

Basis for the AALG: PCE is soluble in water to the extent of only 0.015 g/mL at 20°C and has a vapor pressure of 19 mm Hg at 25°C. Its principal uses are in the dry cleaning industry and in metal degreasing.[1] PCE has been identified in both ambient air and drinking and surface waters. Levels detected in both media are highly variable depending on proximity to a point source, season, time of day, etc. Based on these considerations, 50% of the contribution to total exposure will be allowed from air. Because PCE can be found in relatively high concentrations in ground water in contamination situations, the relative source contribution from air may be lower in specific exposure situations.[7]

The AALG is based on carcinogenicity. The quantitative risk assessment for PCE was performed on data from the NTP inhalation bioassay as described in the Addendum to the HAD.[10] Upper-bound slope estimates, q_1^*, were determined for 6 sets of dose-tumor incidence data (leukemia in male and female rats, liver carcinomas in male and female mice, and liver adenomas and carcinomas combined in male and female mice) using three different methods to determine metabolized dose. The first method used metabolized doses derived directly from a single exposure balance study, and the second and third methods used doses derived from physiologically based pharmacokinetic (PB-PK) models. The estimates presented here are based on the first method as recommended by CAG since "there are uncertainties with the parameters used in the PB-PK model that were estimated from the same empirical data set used in the direct estimation, and the model cannot be considered to give a better estimate." The range of upper-bound slope estimates using the six data sets and the direct method of calculating metabolized dose was 2.9×10^{-7} to 9.5×10^{-7} $(\mu g/m^3)^{-1}$ with a geometric mean of 5.78×10^{-7} $(\mu g/m^3)^{-1}$. Note that the use of the direct method of calculating metabolized dose assumes that the total amount metabolized over a 72-hour period from a single exposure is comparable to the steady-state metabolized dose under the NTP multiple exposure pattern. Using the geometric mean of 5.78×10^{-7} $(\mu g/m^3)^{-1}$ as the best estimate for the q_1* and applying the relation $E = q_1* \times d$, where E is the specified level of extra risk and d is the dose, gives the estimates shown below. It is relevant to note that the former best estimate of the upper-bound slope q_1*, based on the 1977 NCI gavage bioassay data for hepatocellular carcinomas in male and female B6C3F$_1$ mice, 4.8×10^{-7} $(\mu g/m^3)^{-1}$, falls within the range calculated using the NTP inhalation bioassay data. Since the basis for this quantitative risk assessment is currently under review,[18] the AALG should be considered provisional.

AALG: • 10^{-6} 95% UCL—0.12 ppb (0.86 $\mu g/m^3$) annual TWA
  • 10^{-5} 95% UCL—1.2 ppb (8.6 $\mu g/m^3$) annual TWA

References

1. ACGIH. 1986. "Documentation of the Threshold Limit Values and Biological Exposure Indices." 5:464–65.
2. NIOSH. 1985. "NIOSH Recommendations for Occupational Safety and Health Standards." *Morb. Mort. Wkly. Rpt.* 34(suppl.) 34:5s-31s.
3. OSHA. 1989. "Air Contaminants; Final Rule." *Fed. Reg.* 54:2332–2959.
4. U.S. EPA. 1980. "Ambient Water Quality Criteria for Tetrachloroethylene." EPA 440/5–80–073.
5. U.S. EPA. 1987. "Health Advisory for Tetrachloroethylene." Office of Drinking Water (3/31).
6. U.S. EPA. 1985. "National Primary Drinking Water Regulations; Volatile Synthetic Organic Chemicals; Final Rule and Proposed Rule." *Fed. Reg.* 50:46880–901.
7. U.S. EPA. 1985. "Health Assessment Document for Tetrachloroethylene (Perchloroethylene)." EPA/600/8–82/005F.
8. NCI. 1977. "Bioassay of Tetrachloroethylene for Possible Carcinogenicity." DHEW (NIOSH) 77–813.
9. NTP. 1985 (August). "NTP Technical Report on the Toxicology and Carcinogenesis Studies of Tetrachloroethylene in F344/N Rats and B6C3F$_1$ Mice (Inhalation Studies)." TR-311.

10. U.S. EPA. 1986. "Addendum to the Health Assessment Document for Tetrachloro-ethylene (Perchloroethylene)." EPA-600/8–82/005FA.

11. Rampy, L. W., J. F. Quast, M. F. Balmer, B. J. K. Leong, and P. J. Gehring. 1978. "Results of a Long-Term Inhalation Toxicology Study on Rats of a Perchloroethylene (Tetrachloroethylene) Formulation." Toxicology Research Laboratory, Dow Chemical Co., Midland, MI.

12. IARC. 1986. "Tetrachloroethylene." *IARC Monog.* suppl. 6:514–16.

13. Beliles, R. P., D. J. Brusick, and F. J. Mecler. 1980. "Teratogenic-Mutagenic Risk of Workplace Contaminants: Trichloroethylene, Perchloroethylene and Carbon Disulfide." U.S. DHEW, Contract No. 210–77–0047.

14. Schwetz, B. A., B. J. K. Leong, and P. J. Gehring. 1975. "The Effect of Maternally Inhaled Trichloroethylene, Perchloroethylene, Methyl Chloroform and Methylene Chloride on Embryonal and Fetal Development in Mice and Rats." *Tox. Appl. Pharm.* 32:84–96.

15. Tepe, S. J., M. K. Dorfmueller, R. G. York, and J. M. Manson. 1982. "Teratogenic Evaluation of Perchloroethylene in Rats." Unpublished.

16. Manson, J. M., S. J. Tepe, B. Lowrey, and L. Hastings. 1982. "Postnatal Evaluation of Offspring Exposed Prenatally to Perchloroethylene." Unpublished.

17. Nelson, B. K., B. J. Taylor, J. V. Setzer, and R. W. Hornung. 1980. "Behavioral Teratology of Perchloroethylene in Rats." *J. Environ. Pathol. Toxicol.* 3:233–50.

18. U.S. EPA. 1988. "Tetrachloroethylene; CASRN 127–18–4." IRIS (3/1/80).

TITANIUM DIOXIDE

Synonyms: titanium oxide

CAS Registry Number: 13463–67–7

Molecular Weight: 79.90

Molecular Formula: TiO_2

AALG: • 150 $\mu g/m^3$ 24-hour TWA (as PM_{10})
  • 50 $\mu g/m^3$ annual arithmetic TWA (as PM_{10})

Occupational Limits: • ACGIH TLV: 10 mg/m^3 TWA (total dust containing <1% quartz)
  • NIOSH REL: none
  • OSHA PEL: 10 mg/m^3 TWA (total dust containing <1% quartz)

Basis for Occupational Limits: The TLV documentation states that "there has been no evidence of danger to health from inhalation of TiO_2 in concentrations that do not exceed 10 mg/m^3."[1] Further, it is indicated that a number of investigators consider titanium dioxide to be in the category of a "nuisance" or inert dust, i.e., a dust that tends to remain in the lung tissue and although potentially able to produce adverse effects at higher concentrations, is not generally associated with induction of a proliferative response. OSHA recently proposed to adopt the TLV-TWA for total dust as part of its final rule limits.[2]

Drinking Water Limits: No drinking water limits were found for this substance.

Toxicity Profile

Carcinogenicity: The potential carcinogenicity of titanium dioxide has been investigated in two well-designed studies in which both the ingestion and inhalation route were used.[3,4] In an NCI bioassay, male and female B6C3F$_1$ mice and F344 rats (n = 50/sex/species/group) were exposed to 0, 2.5, or 5% TiO_2 in the diet for 103 weeks. No neoplastic lesions attributable to titanium dioxide exposure were reported, and it was concluded that TiO_2 was not carcinogenic to mice or rats under the conditions of this study.[3]

Lee et al.[4] exposed male and female CD rats (n = 100/sex/group) to 0, 10, 50, or 250 mg/m^3 (nominal concentrations, approximately 84% particles respirable

[<13 μm MMD]) titanium dioxide dust via inhalation for 6 hours per day, 5 days per week for 24 months. Interim kills and examination of 5, 5, and 10 rats from each treatment group were conducted at 3, 6, and 12 months, respectively. Titanium dioxide exposure did not adversely affect growth or survival or result in neoplastic or non-neoplastic lesions at sites other than the lung. At the lowest exposure level, the authors considered that the response observed in the rats met the biological criteria for a nuisance dust, which they list as (1) reversibility of lesions, (2) insignificant collagen fiber formation, and (3) preservation of alveolar structure. Lung lesions observed at the lowest exposure level included infiltration of "dust-laden" macrophages in the alveolar ducts and adjacent alveoli with Type II pneumocyte hyperplasia. At the 50- and 250-mg/m^3 exposure levels, there was a statistically significant dose-related increase in relative and absolute lung weights, and the histologic lesions observed included accumulation of dust in macrophages, foamy macrophage response, Type II pneumocyte hyperplasia, alveolar proteinosis, alveolar bronchiolarization, cholesterol granulomas, focal pleurisy, and dust deposition in the tracheobronchiolar lymph nodes. At the highest exposure level, there was also a statistically significant increase in bronchiole alveolar adenomas in both male and female rats (control: males, 2 of 79, and females, 0 of 79; 250 mg/m^3: males, 12 of 79, and females, 13 of 79) and squamous cell carcinomas in female rats (control: 0 of 79; 250 mg/m^3: 13 of 79). The authors concluded that the tumors occurring in rats at the highest exposure level were of limited biological relevance to humans based on the "excessive dust loading and overwhelmed clearance mechanisms" in the lungs and further noted that the tumors observed in rats differed with respect to type, location, and development compared with tumor types commonly observed in man.

Mutagenicity: Negative results were listed in the Gene-Tox database as summarized in RTECS for titanium dioxide in a cell transformation assay (Syrian hamster embryo cells by adenovirus SA7).[5]

Developmental Toxicity: No data were found implicating titanium dioxide as a developmental toxin.

Reproductive Toxicity: No data were found implicating titanium dioxide as a reproductive toxin.

Systemic Toxicity: In the clinical literature there have been some reports indicating that high levels of exposure to TiO_2 resulted in significant dust deposition in the lung and slight pulmonary fibrosis. However, a causative relationship between TiO_2 exposure and fibrosis could not be established on account of the presence of other particulates (TiO_2-coating materials) in the lung biopsy material.[4]

In the inhalation study described in "Carcinogenicity" above, the authors noted that the pulmonary response at 10 mg/m^3 satisfied the biological criteria for a nuisance dust. At 50 and 250 mg/m^3, a dose-dependent increase in dust cell accumulation, foamy macrophage response, type II pneumocyte hyperplasia, alveolar bronchiolarization, cholesterol granulomas, focal pleurisy, and dust deposition in the tracheobronchiolar lymph nodes were noted.[4]

Irritation: Titanium dioxide dust is capable of causing irritation of the eyes, skin, and respiratory tract at high exposure levels. However, the levels required to produce these effects are well in excess of those typically present in ambient air.[2,6]

Basis for the AALG: Titanium dioxide occurs as a crystalline solid; the major crystalline forms are anatase and rutile. Both forms are used as ceramic colorants and for welding-rod coatings; the anatase form is also used as a white pigment in enamels and exterior paints. Titanium dioxide is considered insoluble in water and, because of its physical properties, would exist as a dust in ambient air.[1] Based on the lack of information on its environmental fate and distribution and the expected contribution of various media to human exposure, 50% of the relative source contribution will be allowed from air.

In the derivation of an AALG for titanium dioxide, two options were considered: to use data from the lifetime rat inhalation study or to default to the particulate standard. The exposure level of 10 mg/m^3 from the Lee et al. study[4] was treated as a NOAEL based on biological significance, adjusted for continuous exposure (10 mg/m^3 $\times$ 5/7 $\times$ 6/24), and converted to a human equivalent exposure level according to the procedures described in Appendix C; a total uncertainty factor of 100 (10 for interindividual variation $\times$ 10 for interspecies variation) was applied. The resulting criterion was 20 μg/m^3 as a 24-hour TWA.

Since TiO$_2$ meets the biological criteria for an inert dust at exposure levels likely to be encountered by the general public and would exist as a dust or mist in air, it is suggested that adherence to the current NAAQS for particulate matter would be protective of the health of the general public. The present primary (and secondary) standards for particulate matter are 150 μg/m^3 (no more than one exceedance expected per year) as a 24-hour TWA and 50 μg/m^3 as an annual arithmetic mean, both measured as PM$_{10}$. PM$_{10}$ is an indicator that includes only those particles with an aerodynamic diameter of less than or equal to a nominal 10 micrometers. The basis for the particulate standard and the use of PM$_{10}$ as an indicator are discussed in detail in U.S. EPA.[7]

It is relevant to note that following a review of the data on TiO$_2$ toxicity, EPA granted a petition to delete TiO$_2$ from the list of toxic chemicals subject to release reporting under SARA title 313.[8]

AALG: • 150 μg/m^3 24-hour TWA (as PM$_{10}$)
• 50 μg/m^3 annual arithmetic TWA (as PM$_{10}$)

References

1. ACGIH. 1986. "Documentation of the Threshold Limit Values and Biological Exposure Indices." 5:576.

2. OSHA. 1989. "Air Contaminants; Final Rule." *Fed. Reg.* 54:2332–2959.

3. NCI. 1979. "Bioassay of Titanium Dioxide for Possible Carcinogenicity." NCI-TR-97.

4. Lee, K. P., H. J. Trochimowicz, and C. F. Reinhardt. 1985. "Pulmonary Response of Rats Exposed to Titanium Dioxide by Inhalation for Two Years." *Toxicol. Appl. Pharmacol.* 79:179–192.

5. NIOSH. 1988. XR2275000. "Titanium Oxide." RTECS, on line.

6. Carson, B. L., H. V. Ellis, and J. L. McCann. 1986. "Titanium," in *Toxicology and Biological Monitoring of Metals in Humans* (Chelsea, MI: Lewis Publishers), 264–7.

7. U.S. EPA. 1987. "Ambient Air Quality Standards for Particulate Matter; Final Rule." *Fed. Reg.* 52(125):24633–750.

8. U.S. EPA. 1988. "Toxic Chemical Release Reporting; Community Right-to-Know; Titanium Dioxide." *Fed. Reg.* 53:23108–12.

TOLUENE

Synonyms: toluol, methylbenzene, phenylmethane

CAS Registry Number: 108–88–3

Molecular Weight: 92.13

Molecular Formula:

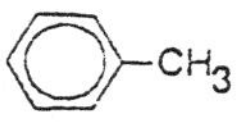

AALG: systemic toxicity (TLV-based)—0.38 ppm (1.4 mg/m^3) 8-hour TWA

Occupational Limits:
- ACGIH TLV: 100 ppm (375 mg/m^3) TWA; 150 ppm (560 mg/m^3) STEL
- NIOSH REL: 100 ppm (375 mg/m^3) TWA; 200 ppm (750 mg/m^3) ceiling (10 minutes)
- OSHA PEL: 100 ppm (375 mg/m^3) TWA; 150 ppm (560 mg/m^3) STEL

Basis for Occupational Limits: All of the occupational limits are set on the basis of prevention of the central nervous system depressant effects of toluene, e.g., narcosis, incoordination, and subjective symptoms such as fatigue and headache.[1–3] OSHA recently adopted the ACGIH TLVs as part of its final rule limits.[3]

Drinking Water Limits: The ambient water quality criterion for toluene is 14.3 mg/L based on the consumption of 2 L of water and 6.5 g of fish/shellfish per day.[4] The 1-day HA for a 10-kg child is 3.46 mg/L. Since suitable data were not available to calculate a 10-day or longer-term HA for toluene, it was recommended that the DWEL adjusted for a 10-kg child be used as a conservative estimate. The lifetime HA for toluene is 2.42 mg/L based on a NOAEL of 300 ppm from a lifetime inhalation bioassay (described in "Carcinogenicity," below).[5] The U.S. EPA has proposed an MCLG of 2.0 mg/L based on noncarcinogenic effects and assuming 20% contribution from drinking water.[6]

Toxicity Profile

Carcinogenicity: Negative results have been reported for toluene in two lifetime inhalation bioassays in rodents. In an inhalation bioassay[7] (cited in U.S. EPA[8]) conducted by the Chemical Industry Institute of Toxicology (CIIT), male and female F344 rats (n = 90/sex/group) were exposed to 0, 30, 100, or 300 ppm toluene

6 hours per day, 5 days a week for 24 months. There were no statistically significant differences in treated vs control animals with respect to neoplastic or proliferative lesions. However, it was pointed out in the HAD[8] that the highest dose used was less than the MTD. In this same study, in a battery of clinical tests that included hematology and urinalysis, there was a slight, but statistically significant, reduction in hematocrit and increase in mean corpuscular hemoglobin in female, but not male, rats in the 100- and 300-ppm exposure groups. It is relevant to note that in the derivation of the oral reference dose, the U.S. EPA considered 300 ppm to be a NOAEL in this study.[9] Negative results have also been reported in a recent NTP inhalation bioassay (galley copy of technical report in progress) in which both male and female B6C3F$_1$ mice and F344 rats were used.[10]

Additional studies have indicated that toluene did not produce tumors when applied to the skin of rodents and that it did not promote the development of skin tumors following initiation with 7,12-dimethylbenzanthracene in Swiss mice.[8] Toluene has been placed in group D, unclassified substance, under the EPA weight-of-evidence classification.[9,11]

Mutagenicity: Negative results have been reported for a variety of bacterial and mammalian in vitro mutagenicity assays with and without metabolic activation and for in vivo mutagenicity assays as well. Conflicting results have been reported in tests for the induction of chromosomal aberrations and sister chromatid exchanges in humans and animals.[8]

Developmental Toxicity: Results from several inhalation studies in mice[12] (cited in U.S. EPA[8]) and rats[13,14] (cited in U.S. EPA[8]) have indicated that toluene is not teratogenic in these species; however, fetotoxic effects, in the form of delayed skeletal development, were reported at levels where frank maternal toxicity was apparently not manifest.

Reproductive Toxicity: No data were found implicating toluene as a reproductive toxin.

Systemic Toxicity: The principal effect of toluene in animals and humans is central nervous system dysfunction. A consistent dose-response pattern for acute exposure to toluene in humans is evident from the aggregate results of a number of studies reviewed in the HAD:[8]

1. 10,000–30,000 ppm—onset of narcosis within a few minutes, longer exposures may be lethal
2. >4000 ppm—likely to cause rapid impairment of reaction time and coordination, narcosis and death possible with exposures exceeding one hour
3. 1500 ppm—probably not lethal for exposure periods of up to eight hours
4. 300–800 ppm—obvious incoordination expected during exposure periods of up to eight hours

5. 100–300 ppm—detectable signs of incoordination expected during exposure periods of up to eight hours
6. 50–100 ppm—subjective complaints, e.g., fatigue and headache, but probably no observable impairment of reaction time or coordination
7. >37 ppm—probably perceptible to most humans

In a study[15] (cited in U.S. EPA[8]) in which humans were exposed to toluene at various concentrations (0, 100, 300, 500, or 700 ppm) for periods of 20 minutes and their performance on tests of perceptual speed and reaction time measured, 100 ppm was established as a NOAEL.

Myelotoxic effects were attributed to toluene in a few early occupational studies, but recent investigations have not supported earlier findings and these are now attributed to concurrent exposure to benzene. Hepatotoxicity and renal toxicity following toluene exposure have been reported only at very high concentrations or in chronic abusers of toluene.[8]

The effects of toluene are similar in humans and laboratory animal models, and a number of relevant animal studies are reviewed in both the HAD[8] and Low et al.[16]

Irritation: Acute exposures to approximately 400 ppm toluene have been associated with lacrimation and irritation of the eyes and throat, and exposure to approximately 200 ppm has been reported to cause mild throat and eye irritation.[8] Toluene is reported to have a "typical hydrocarbon odor" and an odor threshold in air of 0.16–37 ppm.[1]

Basis for the AALG: Toluene is slightly soluble in water (515 mg/mL at 20°C) and has a vapor pressure of 28 mm Hg at 25°C.[1,16] Toluene is derived mainly from petroleum and is present in gasoline and various petroleum solvents. The principal uses of toluene are as a solvent and chemical intermediate.[1] Eighty percent of the contribution to total exposure will be allowed from air for toluene since it is considered to be the most prevalent hydrocarbon in the atmosphere and the atmosphere is the major reservoir for it.[8]

Although none of the studies cited in the HAD is sufficient by itself for criteria derivation, taken in aggregate they support a NOAEL for C.N.S. effects (e.g., muscular incoordination, slowed reaction times) of 100 ppm (the same as the TLV) in humans exposed for up to eight hours. Therefore, the AALG is based on the TLV; however, since subjective complaints have been reported at concentrations at and below 100 ppm, the TLV is treated as a human LOAEL and divided by a total uncertainty factor of 210 (4.2 × 5 for a LOAEL × 10 for interindividual variation). This AALG should be considered provisional since calculation of an inhalation reference dose by EPA is pending.[9]

AALG: • systemic toxicity (TLV-based)—0.38 ppm (1.4 mg/m^3) 8-hour TWA

References

1. ACGIH. 1986. "Documentation of the Threshold Limit Values and Biological Exposure Indices." 5:578–79.

2. NIOSH. 1973. "Criteria for a Recommended Standard . . . Occupational Exposure to Toluene" (Springfield, VA: NTIS), PB-222–219/8.
3. OSHA. 1989. "Air Contaminants; Final Rule." *Fed. Reg.* 54:2332–2959.
4. U.S. EPA. 1980. "Ambient Water Quality Criteria for Toluene." EPA 440/5–80–075.
5. U.S. EPA. 1987. "Health Advisory for Toluene." Office of Drinking Water (3/31).
6. U.S. EPA. 1985. "National Primary Drinking Water Regulations; Synthetic Organic Chemicals, Inorganic Chemicals and Microorganisms; Proposed Rule." *Fed. Reg.* 50: 46936–7025.
7. Chemical Industry Institute of Toxicology (CIIT). 1980. "A Twenty-Four Month Inhalation Toxicology Study in Fischer-344 Rats Exposed to Atmospheric Toluene. Executive Summary and Data Tables." Conducted by Industrial Bio-Test Laboratories, Inc., Decatur, IL, and Experimental Pathology Laboratories, Inc., Raleigh, NC, for CIIT, Research Triangle Park, NC (10/15/80).
8. U.S. EPA. 1983. "Health Assessment Document for Toluene." EPA-600/8–82–008F.
9. U.S. EPA. 1989. "Toluene; CASRN 108–88–3." IRIS (7/1/89).
10. NTP. 1989. "Chemical Status Report." National Toxicology Program, Research Triangle Park, NC (7/7/89).
11. U.S. EPA. 1984. "Health Effects Assessment for Toluene." EPA-540/1–86/033.
12. Hudak, A., and G. Ungvary. 1978. "Embryotoxic Effects of Benzene and Its Methyl Derivatives: Toluene and Xylene." *Toxicology* 11:55.
13. Litton Bionetics, Inc. 1978. "Teratology Study in Rats. Toluene. Final Report." Submitted to the American Petroleum Institute, Washington, DC in January 1978. LBI Project No. 20698–4 (Kensington, MD: Litton Bionetics, Inc.).
14. Tatrai, E., A. Hudak, and G. Ungvary. 1979. "Simultaneous Effect on the Rat Liver of Benzene, Toluene, Xylene and CCl$_4$." *Acta. Physiol. Acad. Sci. Hung.* 53:261.
15. Gamberale, F., and M. Hultengren. 1972. "Toluene Exposure. II. Psychophysiological Functions." *Work Environ. Health* 9:131–39.
16. Low, L. K., J. R. Meeks, and C. R. Mackerer. 1988. "Health Effects of the Alkylbenzenes. I. Toluene." *Tox. Ind. Health* 4:49–75.

1,1,1-TRICHLOROETHANE

Synonyms: methylchloroform, MC

CAS Registry Number: 71–55–6

Molecular Weight: 133.42

Molecular Formula: CH_3CCl_3

AALG: systemic toxicity (TLV-based)—6.67 ppm (36.4 mg/m^3) 8-hour TWA

Occupational Limits:
- ACGIH TLV: 350 ppm (1900 mg/m^3) TWA; 450 ppm (2450 mg/m^3) STEL
- NIOSH REL: 350 ppm (1900 mg/m^3) 15-minute ceiling
- OSHA PEL: 350 ppm (1900 mg/m^3) TWA; 450 ppm (2450 mg/m^3) STEL

Basis for Occupational Limits: The TLV-TWA is based on prevention of beginning anesthetic effects and objections to odor; the TLV-STEL is based on prevention of anesthesia.[1] The NIOSH recommendation includes the same considerations as the TLVs, but is also stated to be based on "cardiovascular and respiratory effects associated with chronic exposures in several animal species, and the absence of reported effects in man at concentrations below the proposed limit."[2] OSHA recently adopted the ACGIH TLV-STEL as part of its final rule limits.[3]

Drinking Water Limits: The ambient water quality criterion for MC is 18.7 mg/L assuming consumption of 2 L of water and 6.5 g fish/shellfish per day.[4] The following HAs have been recommended for MC: (1) 1-day child HA—140 mg/L, (2) 10-day child HA—insufficient data, recommend use of longer-term child HA, (3) longer-term child HA—35 mg/L, (4) longer-term adult HA—125 mg/L, and (5) lifetime HA—0.2 mg/L.[5] The final MCL for MC is 0.2 mg/L.[6]

Toxicity Profile

Carcinogenicity: Although negative results have been reported in several bioassays designed to evaluate the carcinogenicity of MC, problems with study design or conduct limit the conclusions that may be drawn from these results. For example, although no statistically significant increases in tumors occurred in treated versus control animals in a 1977 NCI gavage bioassay using male and female B6C3F$_1$ mice and Osborne-Mendel rats, high mortality in both species rendered this study unusable to draw any conclusions regarding MC carcinogenicity[7] (cited in U.S. EPA[8]). It should be noted that the technical report for an additional NTP gavage bioassay conducted in F344/N rats and B6C3F$_1$ mice is currently on hold pending evaluation of potentially serious data discrepancies.[8]

In an inhalation bioassay, Quast et al.[9] (cited in U.S. EPA[8]) exposed male and female Sprague-Dawley rats to 0, 875, or 1750 ppm MC for 6 hours per day, 5 days a week for 12 months and then observed them until 31 months. There was an increase in ovarian granulosa cell tumors in female rats significant only at the low dose, and it was thought that this finding might not be related to MC administration. The only non-neoplastic changes reported were an increased incidence of focal hepatocellular alterations in high-dose female rats. It should be noted that the highest dose was not at the MTD and that exposure was not for the entire life span of the animal.[8]

In a recently published inhalation bioassay, Quast et al.[10] exposed male and female B6C3F$_1$ mice and F344 rats (n = 80/sex/species/group) to 0, 150, 500, or 1500 ppm 1,1,1-trichloroethane formulation for 6 hours per day, 5 days a week for 2 years. Interim kills (n = 10 sex/species/group) were performed after 6, 12, and 18 months. It should be noted that the 1,1,1-trichloroethane formulation consisted of approximately 94% 1,1,1-trichloroethane, 5% stabilizers (butylene oxide, t-amyl alcohol, methyl butynol, nitroethane, and nitromethane), and less than 1% minor impurities. There was no exposure-related decrease in survival or in body weight gain in either sex of rats or mice with one exception: in female rats in the 1500-ppm group, there was a statistically significant decrease in body weight gain from months 11 to 24. No exposure-related increases in neoplastic or proliferative lesions were reported in either rats or mice, and the only non-neoplastic lesions observed were in the livers of male and female rats in the 1500-ppm group. These lesions were described as slight and were observed in animals at all the interim kills; however, the lesions were not discernable at the termination of the study on account of confounding changes related to normal aging processes. The lesions consisted of "accentuation of the normal hepatic lobular pattern" with altered hepatocyte cytoplasmic staining in the area surrounding the central vein and smaller cells in the portal region.

Under the EPA weight-of-evidence classification, MC has been placed in group D, unclassified substance.[11]

Mutagenicity: Methyl chloroform has been tested for its capability to induce mutations, chromosomal effects, DNA alterations, and oncogenic transformations using a number of test systems and a broad range of cell types and species. However, the U.S. EPA in its Health Assessment Document concluded that most of these tests were inadequate because they used standard protocols that did not take into account the high volatility and low water solubility of this compound. However, where adequate exposure has been assured, technical grade MC is weakly mutagenic to *S. typhimurium* and genotoxic to mouse hepatocytes. In addition, it has been reported to induce oncogenic transformation in F344 rat and Syrian hamster embryo cells. It should be noted that technical grade MC contains up to 3% 1,4-dioxane, a known animal carcinogen, as a stabilizing agent.[8]

Developmental Toxicity: Schwetz et al.[12] (cited in U.S. EPA[8]) exposed Sprague-Dawley rats and Swiss-Webster mice to 0 or 875 ppm MC via inhalation 7 hours per day on gestation days 6 through 15. No teratogenic effects attributable to MC exposure were observed in either species. Mean and relative absolute liver weights were elevated in mice, but only absolute liver weights were elevated in rats. Fetuses from exposed mice were smaller as determined by both body weight and crown-rump measurements, but the differences were not statistically significant.

York et al.[13] (cited in U.S. EPA[8]) reported no significant signs of maternal toxicity or behavioral effects in pups from Long-Evans rats exposed to 2100 ± 200 ppm MC via inhalation before, during, or both before and during gestation. Although no teratogenic effects were reported, delayed development (decreased fetal body weight, delayed ossification, and delayed kidney development) occurred in pups from dams exposed either during gestation or before and during gestation.

It is relevant to note that although no teratogenic effects have been reported in inhalation studies, a higher incidence of persistent ductus arteriosus has been reported in weanling Sprague-Dawley rats from dams exposed to 10 ppm MC in drinking water throughout pregnancy and lactation.[14]

Reproductive Toxicity: Lane et al.[15] (cited in U.S. EPA[8]) reported no dose-related effects on fertility, gestation, and viability of offspring in a multigeneration study using ICR Swiss mice in which the parental generation had been exposed to 0, 100, 300, or 1000 mg/kg/day (nominal concentrations) MC in drinking water prior to mating. It should be noted that EPA considered this study inadequate to evaluate reproductive and teratogenic effects because the doses used were not high enough to produce overt toxic effects.[8]

Systemic Toxicity: Chronic inhalation studies with MC have been conducted by Quast et al.[9,10] and were reviewed in "Carcinogenicity," above. Based on the most recent of these studies that used lifetime exposure rather than chronic exposure with lifetime observation, a NOAEL of 500 ppm is suggested in rats, the most sensitive species used in these studies. However, it should be noted that technical grade MC (94% purity) was used in this latter study.

Subchronic inhalation studies using a variety of laboratory animals have suggested that the guinea pig is the most sensitive species to the effects of MC; fatty degeneration of the liver was observed after 1–3 months exposure to 3000 ppm, and significantly reduced growth at 650 ppm[16] (cited in U.S. EPA[17]). Work by Torkelson et al.[18] (cited in U.S. EPA[17]) suggested a NOAEL of 500 ppm in rats, guinea pigs, and hamsters exposed to MC for 0.5 hours per day, 5 days a week for 6 months.

At fatal exposure levels in humans (14,000–15,000 ppm), autopsy findings have included pulmonary edema and congestion of the lungs and liver.[2] Short-term exposure in the range of 5000 ppm and higher is associated with the onset of narcosis, and exposure in the range of 1900 to 2600 ppm may cause light-headed-ness; at around 1000 ppm, disturbances in equilibrium may occur.[8] The studies discussed in the next two paragraphs are relevant to exposure-effect correlations for longer-term exposures.

Kramer et al.[19] compared the health of 151 matched pairs of employees from two textile plants: one where MC was not used as a general cleaning solvent and one where it was. Exposed employees had worked in the plant for 1 to 6 years; personal monitoring and air sampling data concentrations of MC revealed concentrations in the range of 2 to 217 ppm. Physical examination and clinical measures of cardiac, renal, and hepatic function revealed no significant differences in the exposed versus nonexposed workers.

Stewart et al.[20] (cited in U.S. EPA[17]) reported that human subjects exposed to 500 ppm MC under controlled conditions for 7.5 hours per day, 5 days a week for 3 weeks complained of fatigue, irritation, and headaches. However, no abnormal-ities were reported in hematologic parameters, urine clinical chemistries, or tests of pulmonary function.

The relative hazard and major effects of MC in humans have been summarized as follows in the HAD:[8]

> The likelihood of adverse health effects resulting from chronic exposure to the ambient air levels commonly encountered appears to be extremely low based on presently available data. It is estimated that the no-observed-effect-level (NOEL) for short-term exposure of humans to MC is in the range of approximately 350 to 500 ppm (1890 to 2700 mg/m³). This NOEL is many orders of magnitude higher than the highest levels of MC (20 ppb; 0.108 mg/m³) measured in the ambient air of urban areas . . . In the range of the NOEL no significant abnormal blood chemistry or organ function dec-rements have been noted. The main health effects are symptoms of neurological dysfunction observed at higher exposure levels. These symptoms were qualitatively diagnosed by subjects' impaired performance of clinical-level cognitive and manual tasks. More extensive human and laboratory animal data would be needed before firm conclusions about adverse health responses to low-level exposures to MC could be drawn.

Irritation: MC has been reported to have a characteristic odor and an odor threshold of 100 ppm. Irritation of the throat was reported at 500 ppm in a study by Stewart et al.[20] (cited in U.S. EPA[17]).

Basis for the AALG: MC is considered almost insoluble in water and is volatile with a vapor pressure of 100 mm Hg at 20°C. It is used industrially in cold cleaning and degreasing processes.[1] Since inhalation is considered to be the principal route of exposure,[8] 80% of the contribution to total exposure will be allowed from air.

The AALG for MC is based on the TLV/42 for the following reasons:

1. The aggregate database for short-term human exposure to MC indicates a NOAEL in the range of 350 to 500 ppm.
2. Epidemiologic studies of workers exposed below the level of the TLV for

extended periods revealed no adverse effects detectable upon extensive physical examination and clinical testing.

3. Well-conducted animal inhalation studies have not provided evidence that MC poses a carcinogenic risk.

AALG: • systemic toxicity (TLV-based)—6.67 ppm (36.4 mg/m^3) 8-hour TWA

References

1. ACGIH. 1986. "Documentation of the Threshold Limit Values and Biological Exposure Indices." 5:382–83.
2. NIOSH. 1976. "Criteria for a Recommended Standard . . . Occupational Exposure to 1,1,1-Trichloroethane (Methyl Chloroform)." DHEW (NIOSH) 76–184.
3. OSHA. 1989. "Air Contaminants; Final Rule." *Fed. Reg.* 54:2332–2959.
4. U.S. EPA. 1980. "Ambient Water Quality Criteria for Chlorinated Ethanes." EPA-440/5–80–029.
5. U.S. EPA. 1987. "Health Advisory for 1,1,1-Trichloroethane." Office of Drinking Water (3/31).
6. U.S. EPA. 1987. "Drinking Water Regulations; Public Notification; Final Rule." *Fed. Reg.* 52:41533–550.
7. NCI. 1977. "Bioassay of 1,1,1-Trichloroethane for Possible Carcinogenicity." NCI TR-3.
8. U.S. EPA. 1984. "Health Assessment Document for 1,1,1-trichloroethane (Methyl Chloroform)." EPA-600/8–82–003F.
9. Quast, J. F., L. W. Rampy, M. F. Balmer, B. J. K. Leong, and P. J. Gehring. 1978. "Toxicological and Carcinogenic Evaluation of 1,1,1-Trichloroethane Formulation by Chronic Inhalation in Rats" (Midland, MI: Dow Chemical Company).
10. Quast, J. F., L. L. Calhoun, and L. E. Frauson. 1988. "1,1,1-Trichloroethane Formulation: A Chronic Inhalation Toxicity Study in Fischer 344 Rats and B6C3F$_1$ Mice." *Fund. Appl. Tox.* 11:611–25.
11. U.S. EPA. 1989. "1,1,1-Trichloroethane; CASRN 71–55–6." IRIS (6/1/89).
12. Schwetz, B. A., B. J. K. Leong, and P. J. Gehring. 1975. "The Effect of Maternally Inhaled Trichloroethylene, Perchloroethylene, Methyl Chloroform and Methylene Chloride on Embryonal and Fetal Development in Mice and Rats." *Tox. Appl. Pharm.* 32:84–96.
13. York, R. G., B. Sowry, L. Hastings, and J. Manson. 1982. "Evaluation of Teratogenicity and Neurotoxicity with Maternal Inhalation Exposure to Methyl Chloroform." *J. Tox. Environ. Heal.* 9:251–66.
14. Dapson, S. C., D. E. Hutcheon, S. H. Gilani, and J. R. Caputi. 1987. "Persistent Ductus Arteriosus in Weanling Rats Maternally Exposed to Methyl Chloroform." Submitted to New England Journal of Medicine.
15. Lane, R. W., B. L. Riddle, and J. F. Borzelleca. 1982. "Effects of 1,2-Dichloroethane and 1,1,1-Trichloroethane in Drinking Water on Reproduction and Development in Mice." *Tox. Appl. Pharm.* 63:409–21.
16. Adams, E. M., H. C. Spencer, V. K. Rowe, and D. D. Irish. 1950. "Vapor Toxicity of 1,1,1-Trichloroethane (Methyl Chloroform) Determined by Experiments on Laboratory Animals." *Arch. Ind. Hyg. Occup. Med.* 1:225–36.

17. U.S. EPA. 1986. "Health Effects Assessment for 1,1,1-Trichloroethane." EPA-540/ 1–86/005.

18. Torkelson, T. R., F. Oyen, D. D. McCollister, and V. K. Rowe. 1958. "Toxicity of 1,1,1-Trichloroethane as Determined on Laboratory Animals and Human Subjects." *Am. Ind. Hyg. Assoc. J.* 19:353–62.

19. Kramer, C. G., M. G. Ott, J. E. Fulkerson, and N. Hicks. 1978. "Health of Workers Exposed to 1,1,1-Trichloroethane: A Matched-Pair Study." *Arch. Env. Heal.* 33:331–42.

20. Stewart, R. D., C. L. Hake, A. Wu, S. A. Graff, H. V. Forster, A. J. LeBrun, P. E. Newton, and R. J. Soto. 1975. "1,1,1-Trichloro-ethane: Development of a Biologic Standard for the Industrial Worker by Breath Analysis." NIOSH-MCOW-ENVM-1,1,1-T-75–4.

TRICHLOROETHYLENE

Synonyms: 1,1,2-trichloroethylene, trichloroethene, TCE

CAS Registry Number: 79–01–6

Molecular Weight: 131.40

Molecular Formula: $Cl_2C = CHCl$

AALG: 10^{-6} 95% UCL—0.07 ppb (0.39 $\mu g/m^3$) annual TWA

Occupational Limits:
- ACGIH TLV: 50 ppm (270 mg/m^3) TWA; 200 ppm (1080 mg/m^3) STEL
- NIOSH REL: 25 ppm (134 mg/m^3) TWA
- OSHA PEL: 50 ppm (270 mg/m^3) TWA; 200 ppm (1080 mg/m^3) STEL

Basis for Occupational Limits: The TLV-TWA is set to control subjective symptoms such as headaches, fatigue, and irritability, while the STEL is based on the prevention of incoordination and other early anesthetic effects; these limits also have a wide margin of safety for the prevention of hepatotoxicity.[1] The NIOSH REL is based on consideration of carcinogenic effects, and at the time the 25-ppm level was recommended, it was thought that it could be achieved using existing engineering control technology.[2] OSHA recently adopted the ACGIH TLVs as part of its final rule limits.[3]

Drinking Water Limits: The ambient water quality criterion for TCE is 2.7 $\mu g/L$ at the 10^{-6} risk level assuming consumption of 2 L of water and 6.5 g fish/shellfish per day.[4] One-day, ten-day, and longer-term HAs were not calculated for TCE due to lack of suitable data; however, a lifetime HA of 0.26 mg/L was calculated based on an RfD from a noncarcinogenic LOAEL, and this level of exposure for a lifetime is associated with a 1×10^{-4} risk of cancer.[5] The NAS Safe Drinking Water Committee calculated one-day and ten-day SNARLs of 105 mg/L and 15 mg/L, respectively.[6] The final MCL for TCE is 0.005 ppm.[7]

Toxicity Profile

Carcinogenicity: Several epidemiologic studies have failed to find evidence of an association between TCE and cancer; however, all of these studies suffer from limitations that render them inadequate for evaluation of the carcinogenicity of TCE. These studies are reviewed in the Health Assessment Document.[8]

Results of animal bioassays with TCE are somewhat conflicting. In an NCI bioassay[9] (cited in U.S. EPA[8]) in which male and female B6C3F$_1$ mice were gavaged with TCE in corn oil 5 days a week for 78 weeks and observed until 90 weeks, an increased incidence of hepatocellular carcinomas was found. The doses and incidences were as follows: males—controls, 1 of 20; 1169 mg/kg/day, 36 of 50; 2339 mg/kg/day, 31 of 48; females—controls, 0 of 20; 869 mg/kg/day, 4 of 50; 1739 mg/kg/day, 11 of 47. No carcinogenic response was observed in Osborne-Mendel rats in the same bioassay.

NTP[10] (cited in U.S. EPA[8]) later repeated the NCI bioassay using TCE not stabilized with epoxides, i.e., epichlorohydrin, and the results confirmed the earlier bioassay findings. In male and female B6C3F$_1$ mice gavaged 5 days per week for 103 weeks with corn oil alone or 1000 mg/kg/day, the incidences of hepatocellular carcinoma were 8 of 48 and 30 of 50 for males and 2 of 48 and 13 of 49 for females, respectively. In this same bioassay, male and female F344 rats were gavaged with 0 (both untreated and vehicle control groups), 500, or 1000 mg/kg/day TCE 5 days a week for 103 weeks. Although there was a statistically significant increase in kidney adenocarcinomas in high-dose male rats, it was concluded that the evidence was inadequate for evaluation of TCE carcinogenicity since the high dose exceeded the maximally tolerated dose. In another study, Maltoni[11] (cited in U.S. EPA[8]) reported no carcinogenic response in Sprague-Dawley rats gavaged with 0, 50, or 250 mg/kg/day TCE 4 to 5 days per week for 52 weeks and observed for up to 140 weeks.

Bell et al.[12] (cited in U.S. EPA[8]) exposed male and female B6C3F$_1$ mice (n = 94–100/group) and Charles River rats to 0, 100, 300, or 600 ppm TCE via inhalation for 6 hours per day, 5 days a week for 24 months. There was an increased incidence of hepatocellular carcinomas in treated mice and no carcinogenic response in rats. However, problems with the conduct of the study (e.g., improper matching of control and treated groups, wide variation in exposure levels, and irregularities in the conduct of histopathologic examinations) severely limit the conclusions that may be drawn from the data, although it does provide supportive evidence for the carcinogenicity of TCE in the liver of B6C3F$_1$ mice.

In another inhalation study, Fukuda et al.[13] (cited in U.S. EPA[14]) exposed female ICR mice and Sprague-Dawley rats to 0, 50, 150, or 450 ppm TCE for 7 hours per day, 5 days a week for 104 weeks. There was an increased incidence of lung adenocarcinoma in mice in the two highest exposure groups and no carcinogenic response in the rats. Henschler et al.[15] (cited in U.S. EPA[8]) exposed male and female Han:NMRI mice, Wistar rats, and Syrian hamsters (n = 30/sex/species/group) to 0, 100, or 500 ppm TCE via inhalation for 6 hours per day, 5 days a week for 18 months and observed them until 30 months (mice and hamsters) or 36 months (rats). No carcinogenic response was observed in rats, hamsters, or male mice. However, there was a statistically significant increase in lymphomas in the female mice (controls, 9 of 29; low dose, 17 of 30; and high dose, 18 of 28). The authors speculated on the occurrence of immunosuppression, which enhanced susceptibility to tumor induction by endogenous viruses. There is some evidence that TCE may cause immunosuppression in another strain of mice also.

TCE has been classified as a B2, probable human carcinogen, by the U.S. EPA.[14] It should be noted that under this classification an increased incidence of a tumor that occurs with a high spontaneous background, e.g., mouse liver tumors, generally constitutes sufficient evidence of carcinogenicity unless there is other mitigating evidence.[16]

Mutagenicity: Based on an evaluation by the U.S. EPA, it was concluded that there is "suggestive" evidence that commercial grade TCE (which is often contaminated with other chemicals that may be mutagenic) is a weakly active indirect mutagen.[8] Purified TCE is reported to be weakly mutagenic in *S. typhimurium* and *S. cerevisiae* following metabolic activation. Positive results were reported in the mouse spot test, but negative findings were reported in the *Drosophila* sex-linked recessive lethal test and a mouse dominant lethal assay.[14] It was concluded in the Health Assessment Document that there was insufficient data to draw firm conclusions pertaining to the mutagenicity of pure TCE. The genotoxic potential of TCE has also been recently reviewed by IARC.[17]

Developmental Toxicity: Inhalation studies in mice and rats indicate that TCE is fetotoxic, but not teratogenic, only at maternally toxic dose levels. However, it is relevant to note that in an inhalation study in rabbits, an increased incidence (not statistically significant) of a rare anomaly—external hydrocephalus—was reported at an exposure level (500 ppm) at which maternal toxicity was not evident. It was concluded in the Health Assessment Document that at "low ambient levels that do not cause maternal toxicity, TCE would not pose a significant hazard to the developing conceptus," but the need for further research was emphasized.[8]

Reproductive Toxicity: Belilies et al.[18] (cited in U.S. EPA[8]) exposed male CD-1 mice (n = 12/group) to 0, 100, or 500 ppm TCE via inhalation for 7 hours per day for 5 days. Animals were killed 1, 4, and 10 weeks after exposure, and their sperm analyzed for abnormalities. A significantly increased incidence of abnormal sperm was seen in the 100-ppm animals after 4 weeks and in the 500-ppm animals after 1 week and 4 weeks. An increased incidence of sperm abnormalities was also reported in another study[19] (cited in U.S. EPA[8]) in which mice were exposed to TCE via inhalation.

Systemic Toxicity: With respect to the systemic toxicity of TCE, the correlation of exposure and effect is summarized as follows in the HAD:[8]

> There is no reliable information concerning the toxicological effects in humans of chronic exposure to levels of TCE below the TLV (50 ppm). Based upon acute human overexposure information and limited animal testing, it is unlikely that chronic exposure to TCE at levels found or expected in ambient air would result in liver or kidney damage. Such damage has not been generally found even when exposure greatly exceeds the TLV.

The first sign likely to be observed upon excessive exposure to TCE is central nervous system dysfunction. Psychomotor function and subjective complaints which have been studied in short-term controlled human studies, as well as data from accidental exposure and epidemiologic studies, indicate that nervous system function probably is affected by TCE concentrations ranging from 200 to 500 ppm (1076 to 2690 mg/m^3). Dose-response relationships for adverse health effects associated with TCE have not been established in man or experimental animals.

Irritation: TCE is reported to have a ''sweetish odor resembling chloroform'' and an odor threshold of 0.5 mg/L in water and 2.5–900 mg/m^3 in air.[1,5]

Basis for the AALG: TCE is considered relatively insoluble in water and has a vapor pressure of 58 mm Hg at 20°C. It is used primarily as a solvent and in vapor degreasing.[1] The AALG for TCE is based on carcinogenicity, and 50% of the contribution to total exposure is allowed from air since it is ubiquitous in the environment on account of large and dispersed releases of unconsumed TCE.[8]

The quantitative risk assessment for TCE presented in the HAD is based on data for hepatocellular carcinomas in male and female mice in the NTP and NCI bioassays. Data from inhalation studies, although supportive of the carcinogenicity of TCE in mice, are unsuitable for quantitative risk assessment for reasons explained in the ''Toxicity Profile,'' above. The q_1* values expressed as human equivalent values (mg metabolized dose/kg/day)$^{-1}$ ranged from 6.9×10^{-3} to 2.2×10^{-2}, with a geometric mean of 1.3×10^{-2}. In order to convert this q_1* to an equivalent inhalation figure, it was necessary to estimate the amount of TCE metabolized when humans are exposed to 1 μg/m^3 TCE in air. Based on data from a study by Monster et al.[20] (cited in U.S. EPA[8]), under continuous exposure to 1 μg/m^3 of TCE in air the body metabolic load is expected to be 9.9×10^{-5} mg/kg/day, and the corresponding inhalation q_1* is then 1.3×10^{-6} (μg/m^3)$^{-1}$. Applying the relationship $E = q_1* \times d$, where E is the specified level of extra risk and d is the dose, gives the estimates shown below. Further details on the application of the linearized multistage model to the NTP and NCI bioassay data and the use of interspecies and route-to-route extrapolation are discussed in the HAD for TCE. The AALG should be considered provisional since the carcinogenicity assessment in IRIS was recently withdrawn as a result of further review, and a new summary is in preparation.[21]

AALG: • 10^{-6} 95% UCL—0.07 ppb (0.39 μg/m^3) annual TWA
        • 10^{-5} 95% UCL—0.7 ppb (3.9 μg/m^3) annual TWA

References

1. ACGIH. 1986. ''Documentation of the Threshold Limit Values and Biological Exposure Indices.'' 5:595–97.

2. NIOSH. 1985. "NIOSH Recommendations for Occupational Safety and Health Standards." *Morb. Mort. Wkly. Rpt.* 34(suppl.):1s–31s.
3. OSHA. 1989. "Air Contaminants; Final Rule." *Fed. Reg.* 54:2332–2959.
4. U.S. EPA. 1980. "Ambient Water Quality Criteria for Trichloroethylene." EPA 440/5–80–077.
5. U.S. EPA. 1987. "Health Advisory for Trichloroethylene." Office of Drinking Water (3/31).
6. Safe Drinking Water Committee. 1980. *Drinking Water and Health*, Vol. 3 (Washington, DC: National Academy Press).
7. U.S. EPA. 1987. "Drinking Water Regulations; Public Notification; Final Rule." *Fed. Reg.* 52:41547.
8. U.S. EPA. 1985. "Health Assessment Document for Trichloroethylene." EPA/600/8–82/006F.
9. NCI. 1976. "Carcinogenesis Bioassay of Trichloroethylene." NCI-CG-TR-2.
10. NTP. 1982. "Carcinogenesis Bioassay of Trichloroethylene. CAS No. 79–01–6." NTP 81–84. NIH Publication No. 82–1799.
11. Maltoni, C. 1979. "Results of Long-Term Carcinogenicity Bioassays of Trichloroethylene Experiments by Oral Administration on Sprague-Dawley Rats." In *Archives of Research on Industrial Carcinogenesis, Vol. V: Experimental Research in Trichloroethylene Carcinogenesis,* C. Maltoni, G. Lefemine, and G. Cotti, Eds. (Princeton, NJ: Princeton Scientific Publishing Co.).
12. Bell, Z. G., K. J. Olson, and T. J. Benya. 1978. "Final Report of Audit Findings of the Manufacturing Chemists Association (MCA): Administered Trichloroethylene (TCE) Chronic Inhalation Study at Industrial Bio-Test Laboratories, Inc., Decatur, Illinois." Unpublished.
13. Fukuda, K., K. Takemoto, and H. Tsuruta. 1983. "Inhalation Carcinogenicity of Trichloroethylene in Mice and Rats." *Ind. Health* 21:243–54.
14. U.S. EPA. 1984. "Health Effects Assessment for Trichloroethylene." EPA/540/1–86/046.
15. Henschler, D., W. Romen, H. M. Elasser, D. Reichert, E. Eder, and Z. Radwan. 1980. "Carcinogenicity Study of Trichloroethylene by Long-Term Inhalation in Three Animal Species." *Arch. Toxicol.* 43:237–48.
16. U.S. EPA. 1986. "Guidelines for Carcinogen Risk Assessment." *Fed. Reg.* 51:33992–4003.
17. IARC. 1986. "Trichloroethylene." *IARC Monog.* suppl. 6:530–32.
18. Belilies, R. P., D. J. Brusick, and F. J. Meclar. 1980. Teratogenic-Mutagenic Risk of Workplace Contaminants: Trichloroethylene, Perchloroethylene and Carbon Disulfide." U.S. DHEW, Contract No. 210–77–0047.
19. Land, P. E., E. L. Owen, and H. W. Linde. 1979. "Mouse Sperm Morphology Following Exposure to Anesthetics during Early Spermatogenesis." *Anesthesiology* 51: S259.
20. Monster, A. C., G. Boersman, and W. C. Duba. 1976. "Pharmacokinetics of Trichloroethylene in Volunteers: Influence of Workload and Exposure Concentration." *Int. Arch. Occup. Environ. Health* 38:87–102.
21. U.S. EPA. 1989. "Trichloroethylene; CASRN 79–01–6." IRIS (7/1/89).

VANADIUM PENTOXIDE

Synonyms: vanadium oxide

CAS Registry Number: 1314–62–1

Molecular Weight: 181.88

Molecular Formula: V_2O_5

AALG: irritation (TLV/50)—1 $\mu g/m^3$ 8-hour TWA

Occupational Limits:
- ACGIH TLV: 0.05 mg/m^3 TWA as V_2O_5 respirable dust and fume
- NIOSH REL: 0.05 mg (V)/m^3 ceiling (15 minutes)
- OSHA PEL: 0.05 mg/m^3 TWA as V_2O_5 respirable dust and fume

Basis for Occupational Limits: The occupational limits for vanadium oxide are based on prevention of respiratory irritation.[1,2] NIOSH[2] considered that a ceiling limit for vanadium compounds as fume and dust was more appropriate than a TWA, since the primary effect associated with vanadium exposure is irritation, and based on the "belief that irritant effects are usually proportional to exposure concentration, rather than to exposure dose (as approximated by a TWA concentration), or in other words, are concentration-limiting rather than dose-limiting." OSHA recently adopted (final rule guidelines) the ACGIH TWA.[3]

Drinking Water Limits: No drinking water limits were found for vanadium pentoxide. In a 1977 review of the effects of vanadium, the Safe Drinking Water Committee[4] concluded that "the lack of data on acute or chronic oral toxicity is not surprising because of the extremely low absorption of vanadium from the gastrointestinal tract. Inhaled vanadium can produce adverse health effects, but the available evidence does not indicate that vanadium in drinking water is a problem." In a further review in 1980,[5] it was concluded that "because the essentiality of vanadium has not been proven, no requirements have been established. However, it may have nutritional significance. . . . Therefore, any contribution drinking water may give to the daily intake of vanadium may be beneficial."

Toxicity Profile

Carcinogenicity: No data were found implicating vanadium pentoxide as a carcinogen.

Mutagenicity: Positive results have been reported for vanadium pentoxide in the *B. subtilis* Rec assay (DNA repair) in the Gene-Tox database as summarized in RTECS.[6]

Developmental Toxicity: No data were found implicating vanadium pentoxide as a developmental toxin.

Reproductive Toxicity: No data were found implicating vanadium pentoxide as a reproductive toxin.

Systemic Toxicity: There are some indications that vanadium may be an essential element, but this has not been established with certainty.[5]

Available data on the inhalation toxicity of vanadium pentoxide in laboratory animals have been reviewed in the NIOSH criteria document.[2] In the most comprehensive study available, Pazynich[7] (cited in NIOSH[2]) exposed male white rats (n = 11/group) continuously to 0, 0.002, or 0.027 mg/m^3 vanadium pentoxide for 70 days. No significant effects were reported in the lowest exposure group, but in the highest exposure group there were alterations in motor chronaxy, decreased oxyhemoglobin in the venous blood, and an increase in leukocytes with "altered nuclei"; pathologic examination of tissues revealed damage to the lungs (vascular congestion, focal hemorrhages, indications of bronchitis), liver (acute vascular congestion in the central veins, accumulation of leukocytes in the portal areas, granular dystrophy of the hepatocytes), kidneys (granular dystrophy and necrosis of the convoluted tubular epithelium), and heart (congested blood vessels surrounded by focal hemorrhages in the myocardium).

Irritation: Relative to available animal studies, an extensive database exists on the effects of exposure to vanadium as vanadium pentoxide and in other forms in humans based on occupational epidemiologic studies, case reports, and clinical studies. The major effects reported in humans are irritant in nature and include upper respiratory tract irritation with mucus discharge; lower respiratory tract irritation with bronchitis, bronchospasm, coughing, and chest pain; eye irritation in the form of conjunctivitis and burning sensation; and skin irritation in the form of a sensation of heat or itching sometimes accompanied by rashes or eczema. In some cases a greenish-black discoloration of the tongue that disappears within a few months of cessation of exposure has also been reported.[2,8] In general, signs of respiratory irritation occur at lower concentrations than do signs of skin and eye irritation. In two occupational epidemiologic studies in which exposure concentrations were reported, exposure to vanadium ores for less than three years at concentrations of 0.01–2.12 mg (V)/m^3 was associated with eye and respiratory irritation[9] (cited in NIOSH[2]), and exposure to vanadium pentoxide and vanadates for 2.5 years at concentrations between 0.01 and 0.52 mg (V)/m^3 was associated with respiratory irritation and discoloration of the tongue[10] (cited in NIOSH[2]). In their review of the toxicity of inhaled vanadium, Carson et al.[8] reported that skin irritation may be associated with exposure to concentrations as low as 0.03 mg (V)/m^3 and eye irritation with concentrations as low as 0.018 mg (V)/m^3.

Zenz and Berg[11] conducted a controlled human exposure study that was strongly influential in lowering the TLV-TWA from 0.5 mg/m^3 to 0.05 mg/m^3. In this study, five human volunteers exposed to an average concentration of 0.2 mg/m^3 vanadium pentoxide respirable dust for 8 hours developed a persistent (7–10 days) productive cough without other systemic effects. Results of lung function tests conducted before, immediately after, and two weeks after exposure did not differ, nor were there any alterations in blood counts. Two volunteers exposed to 0.1 mg/m^3 vanadium pentoxide dust for 8 hours also developed, within 24 hours of exposure, slight coughing and mucus formation that increased after 48 hours and was gone completely after 4 days. No evidence of systemic toxicity or alterations in lung function tests or blood counts were reported.

Basis for the AALG: The primary sources of vanadium in air are the combustion of coal, crude oils, and undesulfurized heavy fuel oils. During the "heating season," concentrations of vanadium tend to be twice as high. In monitoring studies, concentrations of up to 1.32 μg (V)/m^3 have been reported in cities in the northeast and concentrations up to 0.022 μg (V)/m^3 have been reported in the west and midwest.[8] Since the AALG for vanadium oxide is based on irritant effects, no relative source contribution is factored into the calculation. The reason is that irritant effects are generally regarded as concentration-limited rather than dose-limited.

The AALG for vanadium pentoxide is based on irritant effects, since data from available human epidemiologic and controlled exposure studies indicate that this is the predominant critical effect associated with vanadium pentoxide exposure. Although no one human study can be considered suitable for AALG derivation, there is sufficient aggregate evidence that the TLV-TWA is an appropriate surrogate human NOAEL or LOAEL. Based on reports indicating that some individuals have experienced irritant effects at exposure concentrations below the TLV and the fact that systemic toxicity has been reported in at least one animal study, it was considered prudent to treat the TLV as a surrogate human LOAEL. Therefore, the TLV-TWA was divided by a total uncertainty factor of 50 (10 for interindividual variation × 5 for a LOAEL).

AALG: • irritation (TLV/50)—1 μg/m^3 8-hour TWA

References

1. ACGIH. 1986. "Documentation of the Threshold Limit Values and Biological Exposure Indices." 5:620.

2. NIOSH. 1977. "Criteria for a Recommended Standard . . . Occupational Exposure to Vanadium." DHEW (NIOSH) 77–222.

3. OSHA. 1989. "Air Contaminants; Final Rule." *Fed. Reg.* 54(12):2332–2959.

4. Safe Drinking Water Committee. 1977. *Drinking Water and Health,* Vol. 1. (Washington, DC: NAS).

5. Safe Drinking Water Committee. 1980. *Drinking Water and Health,* Vol. 3. (Washington, DC: NAS).

6. NIOSH. 1988. YW2450000 and YW2460000. "Vanadium Pentoxide, Dust and Fume." RTECS, on line.

7. Pazynich, V. M. 1966. "Experimental Data for Hygienic Determination of the Maximum Permissible Concentration of Vanadium Pentoxide in the Air." *Gig. Sanit.* 31: 8–12.

8. Carson, B. L., H. V. Ellis, and J. L. McCann. 1986. *Toxicology and Biological Monitoring of Metals in Humans* (Chelsea, MI: Lewis Publishers, Inc.), p. 276–282.

9. Vintinner, F. J. et al. 1955. "Study of the Health of Workers Employed in Mining and Processing of Vanadium Ore." *AMA Arch. Ind. Health* 12:642–53.

10. Lewis, C. E. 1959. "The Biological Effects of Vanadium. II. The Signs and Symptoms of Occupational Vanadium Exposure." *AMA Arch. Ind. Health* 19:497–503.

11. Zenz, C. and A. Berg. 1967. "Human Responses to Controlled Vanadium Pentoxide Exposure." *Arch. Env. Health* 14:709–12.

VINYL CHLORIDE

Synonyms: chloroethylene, chloroethene

CAS Registry Number: 75–01–4

Molecular Weight: 62.50

Molecular Formula: $CH_2 = CHCl$

AALG: 10^{-6} UCL—0.005 ppb (0.013 $\mu g/m^3$) annual TWA

Occupational Limits:
- ACGIH TLV: 5 ppm (12.8 mg/m^3) TWA; Appendix A1 recognized human carcinogen
- NIOSH REL: lowest reliably detectable level
- OSHA PEL: 1 ppm (2.56 mg/m^3) TWA; 5 ppm (12.8 mg/m^3) ceiling (15 minutes)

Basis for Occupational Limits: The occupational limits are based on the prevention of cancer, particularly angiosarcoma of the liver.[1,2] Regulations concerning work practices and worker protection are given in 29 CFR 1910.1017.[3]

Drinking Water Limits: The ambient water quality criterion for vinyl chloride is 2.0 $\mu g/L$ at the 10^{-6} risk level assuming a 70-kg human consuming 2 L of water and 6.5 g of fish/shellfish per day.[4] The following health advisories have been recommended for vinyl chloride: (1) 1-day child HA—insufficient data, recommend use of 10-day HA as a conservative estimate, (2) 10-day child HA—2.6 mg/L, (3) longer-term child HA—13 $\mu g/L$, and (4) longer-term adult HA—46 $\mu g/L$; since vinyl chloride is a known human carcinogen, no lifetime HA was recommended.[5] The MCLG for vinyl chloride is 0, and the final MCL is 0.002 ppm.[6,7]

Toxicity Profile

Carcinogenicity: A strong association exists between occupational exposure to vinyl chloride and the occurrence of cancer; the principal target organs are reported to be the liver, brain, lung, and lympho-hematopoietic system. However, a general nonspecific carcinogenic effect has also been suggested.[8] The most highly specific tumor type associated with vinyl chloride exposure is angiosarcoma of the liver (a rare tumor type in humans). Both IARC and the U.S. EPA have concluded that there is sufficient evidence for vinyl chloride carcinogenicity in humans based on epidemiologic data, and these studies are well reviewed in the ambient water quality criteria document.[4]

There is also sufficient evidence for vinyl chloride carcinogenicity in laboratory animals based on positive responses in multiple animal species (mice, rats, and hamsters) by both the oral and inhalation routes of exposure. These studies are reviewed in U.S. EPA;[4,9] only those studies relevant to quantitative risk assessment and AALG derivation are reviewed here.

The most comprehensive and in-depth investigation of vinyl chloride carcinogenicity conducted by a single institution was the ''Bologna project'' by Maltoni et al.[10] Multiple species and strains of laboratory animals were administered vinyl chloride by different routes and schedules of treatment, and starting at different ages. The major conclusions of this project are as follows:

1. Vinyl chloride caused tumors in all the animal species tested, i.e., male and female Sprague-Dawley rats, Wistar rats, Swiss mice, and Syrian golden hamsters.
2. Vinyl chloride is a multipotential carcinogen, i.e., it causes tumors of different types at different sites.
3. Angiosarcoma of the liver occurred in all species of animals tested.
4. Vinyl chloride is carcinogenic by inhalation and ingestion, and possibly by injection.
5. Obvious dose-response relationships in tumor production were observed in both ingestion and inhalation experiments.
6. Treatment duration and schedule markedly affected the neoplastic response, as did the species, strain, and sex of animals studied.
7. Neonatal animals were highly responsive to the effects of vinyl chloride and readily developed hepatocarcinomas and angiosarcomas.
8. Vinyl chloride acts as a transplacental carcinogen.
9. Vinyl chloride is carcinogenic at very low exposure levels, i.e., at 50 ppm and lower.
10. In experiments with male and female Sprague-Dawley rats (ingestion and inhalation), zymbal gland carcinomas, liver angiosarcomas, nephroblastomas, neuroblastomas, mammary gland adenocarcinomas, and forestomach papillomas were all statistically significant for at least one exposure level.
11. The exposure levels at which no increase in vinyl chloride–related tumors occurred were 5 ppm and 1 ppm by inhalation exposure (treated 4 hours per day, 5 days a week for 52 weeks) and 0.03 mg/kg by ingestion exposure (treated by gavage in olive oil once daily, 4 or 5 days per week for 59 weeks).

The most appropriate data for quantitative risk assessment purposes from the ''Bologna project'' are the result of the study in which male and female Sprague-Dawley rats (n = 60/treatment) were exposed to 0, 50, 250, 500, 2500, 6000, or 10,000 ppm vinyl chloride via inhalation for 4 hours per day, 5 days a week for 52 weeks (the longest exposure period used in this project) with lifetime observation; this study was the most appropriate for quantitative risk assessment purposes because of the large number of animals and wide range of dose levels employed. Significantly increased incidences of angiosarcoma of the liver as well as nephroblastomas of the kidney and carcinoma of the zymbal gland were reported. The overall combined tumor incidences (total number of rats with one or more tumors out of total number of animals at risk) were as follows: controls, 6 of 58; 50 ppm, 10 of 59; 250 ppm, 16 of 59; 500 ppm, 22 of 59; 2500 ppm, 32 of 59; 6000 ppm, 31 of 60; and 10,000 ppm, 38 of 61. A clear dose-response trend occurred in the incidence of angiosarcomas of the liver: controls, 0 of 58; 50 ppm, 1 of 60; 250 ppm, 3 of 59; 500 ppm, 6 of 60; 2500 ppm, 13 of 60; 6000 ppm, 13 of 59; and 10,000 ppm, 7 of 60.[10,11]

Vinyl chloride has been classified in group A, known human carcinogen, under the EPA weight-of-evidence classification.[5]

Mutagenicity: Although vinyl chloride is mutagenic to *S. typhimurium* without metabolic activation, the mutagenic response is markedly enhanced in the presence of a metabolic activation system. In addition, vinyl chloride has been reported to induce reverse mutations in *E. coli* strain K12, forward mutations in *Schizosaccharomyces pombe*, mitotic gene conversions in *S. cerevisiae*, and forward mutations in Chinese hamster ovary cells. All of these responses occurred in the presence of a metabolic activation system, indicating that a metabolite of vinyl chloride is predominantly responsible for the observed mutagenicity of vinyl chloride. Vinyl chloride is also positive in the *Drosophila* sex-linked recessive lethal assay, but negative in other *Drosophila* assays, including those designed to detect translocations, sex chromosome loss, and dominant lethal effects. In addition, negative results have been reported in a mouse inhalation dominant lethal assay and the mouse spot test.[5,8] In a recent IARC review of the genotoxic effects of vinyl chloride, it was also noted that exposure of workers to levels of vinyl chloride from 5 to 500 ppm was associated with the development of chromosomal aberrations in peripheral blood lymphocytes.[12]

Developmental Toxicity: Available studies indicate that vinyl chloride is not teratogenic in mice, rats, or rabbits exposed via inhalation during the period of major organogenesis at levels of up to 500 ppm, 6000 ppm, and 2500 ppm, respectively[13,14] (cited in U.S. EPA[5]). In some cases, inconclusive findings of increased fetal deaths or minor skeletal variations were reported; however, these findings occurred only at levels producing maternal toxicity.[8]

Some reports have cited an association between human exposure to vinyl chloride and adverse pregnancy outcomes. However, further more rigorous studies have not borne out this association, and ''evaluation of the parents of affected children in community studies has not shown a relationship between parental occupation in vinyl chloride industries and birth defects in the offspring or between proximity of the home to a vinyl chloride factory and birth defects in the offspring.''[15]

Reproductive Toxicity: In a recent review of the teratogenic risks associated with human male exposure to vinyl chloride, Uzych[16] concluded that ''data pertaining to the possible biological effects associated with potential vinyl chloride exposure are relatively sparse, conflicting and inconclusive'' and pointed out the need for further research in this area.

Systemic Toxicity: In humans, short-term exposure to high levels of vinyl chloride in the range of 40 to 900 ppm has been reported to result in dizziness, headaches, euphoria, and narcosis.[5]

Chronic inhalation exposure in animals and humans is associated with similar effects on numerous organ systems. In humans occupationally exposed to vinyl chloride, these effects have included hepatotoxicity in the form of hepatomegaly, serum enzyme alterations and fibrosis, splenomegaly, thrombocytopenia, thyroid insufficiency, pulmonary changes, acro-osteolysis (osteolysis involving the distal phalanges of the fingers and toes) accompanied by microvascular abnormalities, portal hypertension (attributed to abnormalities of the portal vein radials or hepatic sinusoids), functional disturbances of the C.N.S. with adrenergic sensory polyneuritis, angioneurosis (neuropathy affecting primarily the blood vessels), and Raynaud's syndrome (paroxysmal bilateral cyanosis of the extremities).[17]

Irritation: Vinyl chloride is reported to have an ethereal odor[1] that is detectable by some individuals at concentrations as low as 300 ppm.[17] Vinyl chloride is reported to be slightly irritating to the eyes and respiratory tract.[17]

Basis for the AALG: Major uses of vinyl chloride are in the manufacture of polyvinyl chloride resins and in organic synthesis.[1] Although vinyl chloride is a gas under standard conditions, it is degraded within a few hours when released to the ambient air. The major source of exposure to vinyl chloride for the general public is thought to be contaminated water.[5] Based on these considerations, 20% of the contribution to total exposure will be allowed from air.

The AALG for vinyl chloride is based on carcinogenicity, using data from the studies of Maltoni et al.[10] and Maltoni and Lefemine.[11] Quantitative risk assessment using the linearized multistage model was applied to two data sets: total tumors (as was originally done by U.S. EPA[4]) and liver angiosarcomas in Sprague-Dawley rats with incidence data for males and females combined, since separate incidence data for both sexes was not reported in the available literature. The larger (more conservative) q_1^*, 6.8×10^{-3} $(ppm)^{-1}$, was used as the basis for the estimates derived below. This animal q_1^* was converted to a human q_1^*, and the relation $E = q_1^* \times d$, where E is the specified level of extra risk and d is the dose, was applied to derive the estimates shown below.

AALG: • 10^{-6} UCL—0.005 ppb (0.013 $\mu g/m^3$) annual TWA
  • 10^{-5} UCL—0.05 ppb (0.13 $\mu g/m^3$) annual TWA

References

1. ACGIH. 1986. ''Documentation of the Threshold Limit Values and Biological Exposure Indices.'' 5:623–26.

2. NIOSH. 1978. "Vinyl Halide Carcinogenicity—Vinyl Bromide, Vinyl Chloride, Vinylidene Chloride. Current Intelligence Bulletin No. 28." DHEW (NIOSH) 79–146.

3. OSHA. 1988. 29 Code of Federal Regulations, Part 1910.1017.

4. U.S. EPA. 1980. "Ambient Water Quality Criteria for Vinyl Chloride." EPA 440/5–80–078.

5. U.S. EPA. 1987. "Health Advisory for Vinyl Chloride." Office of Drinking Water (3/31).

6. U.S. EPA. 1985. "National Primary Drinking Water Regulations; Volatile Synthetic Organic Chemicals; Final Rule and Proposed Rule." *Fed. Reg.* 50:46879–901.

7. U.S. EPA. 1987. "Drinking Water Regulations; Public Notification; Final Rule." *Fed. Reg.* 52:41533–550.

8. U.S. EPA. 1984. "Health Effects Assessment for Vinyl Chloride." EPA/540/1–86/036.

9. U.S. EPA. 1983. "Reportable Quantity Document for Vinyl Chloride." Prepared by the Environmental Criteria and Assessment Office, Cincinnati, OH, Office of Health and Environmental Assessment for the Office of Solid Waste and Emergency Response, Washington, DC.

10. Maltoni, C., G. Lefemine, A. Ciliberti, G. Cotti, and D. Carretti. 1981. "Carcinogenicity Bioassays of Vinyl Chloride Monomer: A Model of Risk Assessment on an Experimental Basis." *Env. Heal. Perspect.* 41:3–29.

11. Maltoni, C., and G. Lefemine. 1975. "Carcinogenicity Bioassays of Vinyl Chloride: Current Results." *Ann. NY Acad. Sci.* 246:195–218.

12. IARC. 1986. "Vinyl Chloride." *IARC Monog.* suppl. 6:566–69.

13. John, J. A., F. A. Smith, B. J. K. Leong, and B. A. Schwetz. 1977. "The Effects of Maternally Inhaled Vinyl Chloride on Embryonal and Fetal Development in Mice, Rats and Rabbits." *Tox. Appl. Pharm.* 39:497–513.

14. Radike et al. 1977. "Transplacental Effects of Vinyl Chloride in Rats." Annual Report. USPHS-ES-00159. Center for Study of the Human Environment, Department of Environmental Health, University of Cincinnati Medical Center, 183–85.

15. Reproductive Toxicology Center. 1986. "Reproductive Toxicity Review of Vinyl Chloride." REPROTOX, on line (Washington, DC: Columbia Hospital for Women Medical Center).

16. Uzych, L. 1988. "Human Male Exposure to Vinyl Chloride and Possible Teratogenic and Mutagenic Risks: A Review." *Hum. Tox.* 7:517–27.

17. NCI. 1978. "Vinyl Chloride: An Information Resource." DHEW (NIOSH) 79–1599.

VINYLIDENE CHLORIDE

Synonyms: 1,1-dichloroethylene, VDC, 1,1-dichloroethene

CAS Registry Number: 75–35–4

Molecular Weight: 96.94

Molecular Formula: $H_2C = CCl_2$

AALG: 10^{-6} 95% UCL—0.0025 ppb (0.01 $\mu g/m^3$) annual TWA

Occupational Limits:
- ACGIH TLV: 5 ppm (20 mg/m^3) TWA; 20 ppm (80 mg/m^3) STEL
- NIOSH REL: 1.0 ppm (4 mg/m^3) TWA; 5 ppm (20 mg/m^3) ceiling (15 minutes)
- OSHA PEL: 1.0 ppm (4 mg/m^3) TWA

Basis for Occupational Limits: It is stated in the TLV documentation that the ACGIH limits are "low enough to prevent overt toxicity," e.g., liver and kidney injury in exposed workers.[1] The NIOSH limits are based on consideration of carcinogenic effects in animal models.[1,2] OSHA recently adopted the NIOSH REL-TWA for vinylidene chloride as part of its final rule limits.[3]

Drinking Water Limits: The ambient water quality criterion for vinylidene chloride is 0.033 $\mu g/L$ at the 10^{-6} risk level assuming a 70-kg human consuming 2 L of water and 6.5 g fish/shellfish per day.[4] The following health advisories have been recommended for vinylidene chloride: (1) 1-day child HA—2.0 mg/L, (2) 10-day child HA—sufficient data unavailable, (3) longer-term child HA—1 mg/L, (4) longer-term adult HA—3.5 mg/L, and (5) lifetime HA—7 $\mu g/L$.[5] The MCLG and the final MCL are both set at 7 ppm (7 $\mu g/L$).[6,7]

Toxicity Profile

Carcinogenicity: Although no excess risk of cancer attributable to vinylidene chloride exposure was reported in two cohort studies[8,9] (cited in IARC[10]), it should be noted that these studies were inadequate to evaluate carcinogenic risk because of limited cohort size, short observation periods, and small numbers of deaths due to specific causes.[10]

As of 1985, the carcinogenicity of vinylidene chloride had been evaluated in 18 chronic animal studies, most of which used the inhalation route. In 16 of these 18 studies (reviewed in both the HAD[11] and by IARC[10]), there was no evidence for a treatment-related increase in tumors. However, it was pointed out in the HAD[11] that "these negative findings may be partially explained by study characteristics such as less than lifetime dosing, below maximum tolerated dose levels, and single dose studies, which individually or in combination, reduce the sensitivity of detecting a carcinogenic response. While the number of studies are many, the inadequacy of test conditions demonstrates the need for additional testing to elucidate the potential for human carcinogenicity."

In the remaining two studies, results indicative of a carcinogenic response were obtained; in one study, vinylidene chloride was demonstrated to be a tumor initiator in mouse skin[12] (cited in IARC[10] and U.S. EPA[11]). In the other study, Maltoni et al.[13] (cited in IARC[10] and U.S. EPA[11]) exposed male and female Swiss mice to 0, 10, or 25 ppm vinylidene chloride via inhalation 4 hours per day, 4 or 5 days a week for 52 weeks and held them for lifetime observation (126 weeks). In order to increase the power of this study, the authors added extra control (n = 90/sex) and 25 ppm (n = 120/sex) exposure groups. There was a statistically significant increase in kidney adenocarcinomas in male mice exposed to 25 ppm vinylidene chloride compared with controls in both sets of animals (first group: controls, 0 of 99; 10 ppm, 0 of 30; 25 ppm, 3 of 30; second group: controls, 0 of 87; 25 ppm, 25 of 120). In addition, there were statistically significant increases in mammary tumors in female mice and pulmonary adenocarcinomas in both male and female mice. It should be noted that male mice of several strains appeared particularly sensitive to the effects of vinylidene chloride, and the authors suggested that the occurrence of kidney tumors in these mice may be a strain-specific phenomenon.

Vinylidene chloride has been placed in group C, possible human carcinogen, under the EPA weight-of-evidence classification.[11] The following are additional supportive evidence for this classification: (1) genotoxic activity is observed in short-term tests with VDC, (2) a metabolite of VDC is known to alkylate and bind covalently with DNA, and (3) VDC is structurally related to the known human carcinogen, vinyl chloride.[14]

Mutagenicity: VDC has been shown to induce both frameshift and base pair substitution mutations in S. typhimurium. It is also mutagenic to E. coli WP2 uvrA and E. coli K12 and induces mitotic gene conversions and point mutations in S. cerevisiae D7. In all cases, the addition of an exogenous metabolic activation system is necessary for these responses to be manifest, indicating that the mutagenic activity of VDC is attributable to a metabolite rather than the parent compound.[10] In addition, VDC induced unscheduled DNA synthesis in isolated rat hepatocytes and in the kidneys of mice exposed to 50 ppm VDC for 6 hours. Negative findings have been reported in dominant lethal assays (inhalation) in both rats and mice, and VDC did not induce 8-azaguanine- or ouabain-resistant mutants in cultured Chinese hamster V79 cells. IARC has classified the degree of evidence for genetic activity of VDC in short-term tests to be sufficient.[10]

Developmental Toxicity: Short et al.[15] (cited in U.S. EPA[11]) exposed CD-1 mice and CD rats (n = 20/group/species for treated, n = 60/species for controls) to 0,

15, 30, 57, 144, or 300 ppm and 0, 15, 57, 300, or 449 ppm VDC, respectively, for 23 hours per day on gestation days 6 to 16. Maternal mortality was nearly 100% in the 144- and 300-ppm groups in mice and the 300- and 449-ppm groups in rats; maternal toxicity in the form of reduced weight gain and food consumption occurred in all groups of mice and rats except the 15-ppm mice. Fetal resorptions were significant at all exposure levels for both species, with the exception of the 15-ppm mice, and fetotoxicity was observed as soft tissue anomalies (rats only) and delayed ossification in all exposure groups. Further studies were conducted using CD-1 mice exposed to various concentrations of VDC during different time periods of gestation to evaluate VDC for possible behavioral effects in the offspring. No major adverse effects were reported to result from in utero exposure to VDC using a variety of behavioral tests.

No teratogenic effects were reported in a study in which Sprague-Dawley rats and New Zealand white rabbits were exposed to 0, 20 (rats only), 80, or 160 ppm VDC via inhalation 7 hours per day on gestation days 6 to 16 and 6 to 18, respectively. Effects on embryonal and fetal development (e.g., delayed ossification, wavy ribs) occurred only at exposure levels associated with maternal toxicity (decreased food consumption and weight gain)—80 and 160 ppm in rats and 160 ppm in rabbits.[16]

Although Short et al.[15] (cited in U.S. EPA[11]) concluded that VDC might be a weak teratogen, interpretation of their results is confounded by the occurrence of maternal toxicity. Based on the results of Murray et al.,[16] it appears that VDC is fetotoxic only at maternally toxic exposure levels.

Reproductive Toxicity: Nitschke et al.[17] (cited in U.S. EPA[11]) reported that VDC did not adversely effect reproductive function or neonatal development in Sprague-Dawley rats exposed to 50, 100, or 200 ppm VDC in drinking water in a three-generation reproduction study. However, mild dose-related lesions (fatty degeneration) occurred in the livers of adult VDC-treated rats.

Systemic Toxicity: In humans, vinylidene chloride is a central nervous system depressant, and repeated low-level exposure may result in liver and kidney injury.[10] In a cohort study, Ott et al.[8] (cited in IARC[10]) reported no differences in spirometry, clinical chemistries (including indicators of hepatic and renal injury), hematological parameters, and blood pressure measurements between vinylidene chloride workers and controls. Based on previous monitoring data, VDC concentrations in the work area ranged from <5 to 70 ppm.

In experimental animals, the principal target organs for both subchronic and chronic exposures are the liver and kidneys, whether exposure is via ingestion or inhalation. Pathologic changes observed in long-term animal studies include hepatocellular fatty degeneration, hypertrophy, necrosis, and centrilobular atrophy, as well as renal tubular degeneration. Based on information in the HAD, a NOAEL has not been established for VDC exposure in experimental animals either by ingestion or inhalation, and the lowest reported LOAEL is 5 ppm (significantly decreased body weight gain) in rats exposed continuously to VDC via inhalation for 90 days.

Irritation: VDC has a "mild, sweet" odor similar to chloroform that is detectable to most people at 1000 ppm and some at 500 ppm.[18,19] Liquid VDC is irritating to the eyes, and signs of nasal irritation have been reported in rats exposed to 200 ppm VDC via inhalation.[1]

Basis for the AALG: Vinylidene chloride is volatile (vapor pressure = 500 mm Hg at 20°C) and is considered only "sparingly" soluble in water. It polymerizes readily and is used almost solely as a copolymer in the production of films and coatings.[1] Since it has been estimated that exposure by both inhalation of ambient air and consumption of drinking water have the potential to contribute significantly to total VDC exposure,[11] 50% of the contribution to total exposure will be allowed from air.

    Although VDC is a possible (group C) human carcinogen, there is sufficient evidence for genetic activity of this compound based on IARC criteria. Thus, the AALG will be based on carcinogenicity, using the quantitative risk assessment presented in the HAD. The linearized multistage model was applied to the data for kidney adenocarcinomas in male Swiss mice observed in the Maltoni et al. study[13] (cited in IARC[10] and U.S. EPA[11]). The human equivalent upper-bound slope coefficient, q_1^*, calculated from this data set is 2.0×10^{-4} (ppb)$^{-1}$, or 5.0×10^{-5} $(\mu g/m^3)^{-1}$. Applying the relationship $E = q_1^* \times d$, where E is the specified level of extra risk and d is the dose, gives the estimates shown below. It is stated in the IRIS database[14] that "the unit risk should not be used if the air concentration exceeds 200 $\mu g/m^3$, since above this concentration the slope factor may differ from that stated."

AALG: • 10^{-6} 95% UCL—0.0025 ppb (0.01 $\mu g/m^3$) annual TWA
           • 10^{-5} 95% UCL—0.025 ppb (0.1 $\mu g/m^3$) annual TWA

References

1. ACGIH. 1986. "Documentation of the Threshold Limit Values and Biological Exposure Indices." 5:628–29.
2. NIOSH. 1978. "Vinyl Halides Carcinogenicity. Current Intelligence Bulletin No. 28." DHEW (NIOSH) 79–102.
3. OSHA. 1989. "Air Contaminants; Final Rule." *Fed. Reg.* 54:2332–2959.
4. U.S. EPA. 1980. "Ambient Water Quality Criteria for Dichloroethylenes." EPA 440/5–80–041.
5. U.S. EPA. 1987. "Health Advisory for 1,1-Dichloroethylene." Office of Drinking Water (3/31).
6. U.S. EPA. 1985. "National Primary Drinking Water Regulations; Volatile Synthetic Organic Chemicals; Final Rule and Proposed Rule." *Fed. Reg.* 50:46880–901.
7. U.S. EPA. 1987. "Drinking Water Regulations; Public Notification; Final Rule." *Fed. Reg.* 52:41547.
8. Ott, M. G., W. A. Fishbeck, J. C. Townsend, and E. J. Schneider. 1976. "A Health Study of Employees Exposed to Vinylidene Chloride." *J. Occ. Med.* 18:735–38.

9. Thiess, A. M., R. Frentzel-Beyme, and E. Penning. 1979. "Mortality Study of Vinylidene Chloride Exposed Persons," in *Proceedings of the 5th Medichem Congress*, San Francisco, CA, September 1977, C. Hien and D. J. Kilian, Eds. (San Francisco, CA: University of California at San Francisco), 3545–50.

10. IARC. 1986. "Vinylidene Chloride." *IARC Monog.* 39:195–226.

11. U.S. EPA. 1985. "Health Assessment Document for Vinylidene Chloride." EPA 600/8–83–031F.

12. Van Duuren, B. L., B. M. Goldschmidt, G. Loewengart, A. C. Smith, S. Melchionne, I. Seldman, and D. Roth. 1979. "Carcinogenicity of Halogenated Olefinic and Aliphatic Hydrocarbons in Mice." *J. Natl. Cancer Inst.* 63:1433–39.

13. Maltoni, C., G. Lefemine, G. Cotti, P. Chieco, and V. Patella. 1985. *Archives of Research on Industrial Carcinogenesis, Vol. III: Experimental Research on Vinylidene Chloride Carcinogenesis* (Princeton, NJ: Princeton Scientific Publishers).

14. U.S. EPA. 1989. "1,1-Dichloroethylene; CASRN 75–35–4." IRIS (4/1/89).

15. Short, R. D., J. L. Minor, J. M. Winston, B. Ferguson, and T. Unger. 1977. "Toxicity Studies of Selected Chemicals. Task II. The Developmental Toxicity of Vinylidene Chloride Inhaled by Rats and Mice during Gestation." Publication No. EPA-560/6–77–022. Prepared by Midwest Research Institute, Kansas City, MO, under Contract No. 68–01–3242. U.S. EPA, Office of Toxic Substances, Washington, DC.

16. Murray, F. J., K. D. Nitschke, L. W. Rampy, and B. A. Schwetz. 1979. "Embryotoxicity and Fetotoxicity of Inhaled or Ingested Vinylidene Chloride in Rats and Rabbits." *Tox. Appl. Pharm.* 49:189–202.

17. Nitschke, K. D., F. A. Smith, J. F. Quast, J. M. Norris, and B. A. Schwetz. 1983. "A Three-Generation Rat Reproductive Toxicity Study of Vinylidene Chloride in the Drinking Water." *Fund. Appl. Tox.* 3:75–79.

18. Torkelson, T. R., and V. K. Rowe. 1981. "Halogenated Aliphatic Hydrocarbons," in *Patty's Industrial Hygiene and Toxicology*, Vol. 2B, 3rd rev. ed., G. D. Clayton and F. E. Clayton, Eds. (New York, NY: Wiley Interscience), 3545–50.

19. U.S. EPA. 1984. "Health Effects Assessment for 1,1-Dichloroethylene." EPA 540/1–86/051.

XYLENES

Synonyms: This assessment includes all the xylene isomers:
- *o*-xylene (1,2-dimethylbenzene), CAS 95–47–6
- *m*-xylene (1,3-dimethylbenzene), CAS 108–38–3
- *p*-xylene (1,4-dimethylbenzene), CAS 106–42–3

CAS Registry Number: 1330–20–7

Molecular Weight: 106.18

Molecular Formula: $C_6H_4(CH_3)_2$

AALG: Limits for individual isomers are based on developmental toxicity:
- *o*-xylene—0.066 ppm (0.29 mg/m^3) 24-hour TWA
- *m*-xylene—0.66 ppm (2.9 mg/m^3) 24-hour TWA
- *p*-xylene—0.013 ppm (0.057 mg/m^3) 24-hour TWA

The limit for all xylene isomers combined is based on systemic toxicity (TLV-based):
- xylenes—1.2 ppm (5.2 mg/m^3) 8-hour TWA

Occupational Limits:
- ACGIH TLV: 100 ppm (435 mg/m^3) TWA; 150 ppm (655 mg/m^3) STEL
- NIOSH REL: 100 ppm (435 mg/m^3) TWA; 200 ppm (868 mg/m^3) ceiling (10 minutes)
- OSHA PEL: 100 ppm (435 mg/m^3) TWA; 150 ppm (655 mg/m^3) STEL

Basis for Occupational Limits: It was the consensus judgment of the TLV committee that at the recommended limits "irritant effects will be minimal and that no significant degree of narcosis or chronic injuries will result from continued occupational exposure at that level."[1] The NIOSH limits are based on prevention of irritant and narcotic effects.[2] OSHA recently adopted the ACGIH TLV-STEL for xylenes as part of its final rule limits.[3]

Drinking Water Limits: The following HAs have been calculated for xylenes: (1) 1-day child HA—12 mg/L, (2) 10-day child HA—data insufficient, recommend use of longer-term child HA, (3) longer-term child HA—7.8 mg/L, (4) longer-term adult HA—27.3 mg/L, and (5) lifetime HA—0.4 mg/L.[4] An MCLG of 0.44 mg/L has been proposed for xylenes.[5]

Toxicity Profile

Carcinogenicity: In an NTP bioassay,[6] B6C3F$_1$ mice and F344/N rats of both sexes were exposed to 0, 500, or 1000 mg/kg and 0, 250, or 500 mg/kg, respectively, xylenes in corn oil by gavage 5 days per week for 103 weeks. Technical-grade xylene (composition: 9% *o*-xylene, 60% *m*-xylene, 14% *p*-xylene, and 17% ethyl benzene) was used. No neoplastic or non-neoplastic lesions related to xylene administration were found in either sex of rats or mice. However, it was reported that hyperactivity lasting 5 to 30 minutes occurred in high-dose male mice following dosing from week 4 until the end of the study.

Maltoni et al.[7] (cited in NTP[6]) also evaluated the carcinogenicity of xylenes in rats exposed by gavage; although an increase in "malignant lesions" was reported in treated animals, the combining of tumors from multiple sites and the limited reporting of results render evaluation of their significance difficult.

The xylenes have been placed in group D, unclassified substance, under the EPA weight-of-evidence classification.[8]

Mutagenicity: In general, xylenes and the individual xylene isomers are considered nongenotoxic based on negative responses in the Ames assay with and without metabolic activation, in assays designed to measure DNA damage and chromosomal effects, and in the *Drosophila* sex-linked recessive lethal test.[6]

Developmental Toxicity: The developmental toxicity of both mixed xylenes and the individual xylene isomers by various exposure routes has been well reviewed by Hood and Ottley.[9] In general, fetotoxic effects have included delayed development, decreased fetal body weights, and altered enzyme activities. Although malformations have been reported at maternally toxic dose levels, it was concluded that there was no clear evidence for teratogenic effects.

Hudak and Ungvary[10] (cited in NTP[6] and U.S. EPA[11]) exposed CF4 rats continuously to 0 or 1000 mg/m^3 xylenes (10% *o*-xylene, 50% *m*-xylene, 20% *p*-xylene, 20% ethyl benzene) on gestation days 9 through 14. The only xylene-related effects were an increased incidence of fused sternebrae and extra ribs in the offspring.

In further studies, Ungvary et al.[12] exposed CFY rats to *o*-, *m*-, or *p*-xylene continuously at concentrations of 0, 150, 1500, or 3000 mg/m^3 on gestation days 7 through 14 (n = 15–30/group). Maternal toxicity, in the form of decreased food consumption in all 3000-mg/m^3 exposure groups and in the 1500-mg/m^3 *o*-xylene exposed group, was reported. In addition, there was increased mortality and decreased body weight gain in the 3000-mg/m^3 *m*-xylene exposed group, increased maternal liver-to-body weight ratio in all *o*-xylene exposed groups, and significantly lower placental weights in all *p*-xylene exposed groups. Significant fetal loss occurred in the 3000-mg/m^3 *p*-xylene exposed group, and the rate of pregnancy was decreased in dams exposed to 3000-mg/m^3 *o*- or *m*-xylene. Fetal body weight was significantly reduced in all 3000-mg/m^3 xylene exposure groups and in the 1500-

mg/m^3 *o*-xylene group. Retardation of skeletal growth occurred in all *p*-xylene exposed groups and in the 3000-mg/m^3 *o*-xylene exposed group. Although an increase in extra ribs was seen in the 3000-mg/m^3 *m*- and *p*-xylene offspring, this is considered a skeletal anomaly, and there were no other indications of external, visceral, or skeletal malformations. In summary, the following LOAELs and NOAELs are suggested by this study: a LOAEL of 150 mg/m^3 for maternal toxicity and a NOAEL of 150 mg/m^3 for fetotoxicity for *o*-xylene, NOAELs of 1500 mg/m^3 for both maternal and fetotoxicity for *m*-xylene, and LOAELs of 150 mg/m^3 for both maternal and fetotoxicity for *p*-xylene.

Reproductive Toxicity: No data were found implicating the xylenes or their isomers as reproductive toxins.

Systemic Toxicity: Based on case reports of occupationally exposed individuals, short-term exposure to xylene vapor may cause C.N.S. depression and minor reversible effects on the liver and kidneys. At high concentrations, inhalation of xylene may result in dizziness, staggering, drowsiness, and unconsciousness, and at very high concentrations, pulmonary edema, anorexia, nausea, vomiting, and abdominal pain may occur. It should be noted that data correlating exposure levels with these effects are lacking.[3,6]

Based on controlled human exposure studies and some animal studies, there is evidence for behavioral neurotoxicity due to xylene exposure. Gamberale et al.[13] (cited in U.S. EPA[4]) reported evidence of decreased performance on tests measuring choice reaction time and short-term memory in subjects exposed to 1300 mg/m^3 xylenes with 30 minutes of exercise on a bicycle ergometer. Savolainen et al.[14] reported adverse effects on body balance and manual coordination performance in subjects exposed to 391 mg/m^3 (90 ppm) *m*-xylene for 4 hours per day on 5 consecutive days and 1 day after the weekend; however, the authors also noted that tolerance against the observed effects developed during one working week.

Lifetime animal inhalation studies of xylene are lacking. However, in a year-long study, Tatrai et al.[15] (cited in U.S. EPA[11]) exposed male CFY rats to 0 or 4750 mg/m^3 *o*-xylene for 8 hours per day, 7 days a week. Liver toxicity in the form of hepatomegaly, altered enzyme patterns (cytochrome P-450, NADPH-cytochrome *c* reductase, aniline hydroxylase, and aminopyrine-N-demethylase activity were all increased) and increased bromosulfophthalein retention time (indicative of decreased clearance) was reported in treated animals. Histological examination revealed moderate proliferation of the smooth endoplasmic reticulum with increased numbers of peroxisomes and autophagous bodies as well as glycogen depletion and occasionally damaged mitochondria.

Irritation: Xylenes have been described as having an ''aromatic hydrocarbon odor,'' and an odor threshold of 1 ppm has been reported.[1,15,16] In controlled human exposure studies, 200 ppm xylene has been found to be definitely irritating to the eyes, nose, and throat.[1]

Basis for the AALG: The xylenes are considered insoluble in water and have a vapor pressure of 7–9 mm Hg at 20°C. Xylene is found in gasoline and is present in many petroleum solvents; its major uses are as a solvent and in chemical synthesis.[1] The xylenes have been identified in both drinking water and ambient air.[17] Based on the lack of data on its environmental fate and distribution and the expected contribution of various media to human exposure, 50% of the contribution to total exposure will be allowed from air.

The AALGs for the individual xylene isomers are based on developmental toxicity using data from the study of Ungvary et al.[12] In the case of o-xylene, 150 mg/m^3 was identified as a NOAEL for fetotoxicity, but was a LOAEL for maternal toxicity based on the endpoints studied. This exposure level was converted to a human equivalent exposure level according to the procedures described in Appendix C; the final AALG was the same whether the NOAEL for fetotoxicity was used (10 for interindividual variation × 10 for interspecies variation × 5 for developmental uncertainty factor) or the LOAEL for maternal toxicity was used (10 for interindividual variation × 10 for interspecies variation × 5 for a LOAEL) since the same uncertainty factor of 500 was used in both cases. In the case of m-xylene, a NOAEL of 1500 mg/m^3 was identified for fetotoxicity; this NOAEL was converted to a human equivalent exposure level according to the procedures described in Appendix C, and a total uncertainty factor of 500 was applied (10 for interindividual variation × 10 for interspecies variation × 5 for the developmental uncertainty factor). In the case of p-xylene, a LOAEL of 150 mg/m^3 was identified for fetotoxicity; this LOAEL was converted to a human equivalent exposure level according to the procedures described in Appendix C, and a total uncertainty factor of 2500 was applied (10 for interindividual variation × 10 for interspecies variation × 5 for developmental uncertainty factor × 5 for a LOAEL). Application of the developmental uncertainty factor was considered appropriate since exposure generally is to mixed xylene isomers and fetotoxicity generally occurred at or below exposure levels producing maternal toxicity.

The AALG for all xylene isomers combined is based on the TLV-TWA (TLV/42) since this limit is considered low enough to prevent systemic effects, e.g., narcotic and C.N.S. depressant effects.[1,2] This was considered appropriate even though the individual isomers are known to produce fetotoxic effects in laboratory animals since these effects occur at exposure levels also producing maternal toxicity. The AALG for the combined xylene isomers should be considered provisional since it is based on the occupational limit and because there has been at least one report[13] indicating that m-xylene might have adverse effects on body balance and coordination at levels below the TLV. In addition, a provisional designation is warranted because calculation of an inhalation reference dose by the U.S. EPA is pending.[8]

AALG: Limits for individual isomers are based on developmental toxicity:
- o-xylene—0.066 ppm (0.29 mg/m^3) 24-hour TWA
- m-xylene—0.66 ppm (2.9 mg/m^3) 24-hour TWA
- p-xylene—0.013 ppm (0.057 mg/m^3) 24-hour TWA

The limit for all xylene isomers combined is based on systemic toxicity (TLV-based):

- xylenes—1.2 ppm (5.2 mg/m^3) 8-hour TWA

References

1. ACGIH. 1986. "Documentation of the Threshold Limit Values and Biological Exposure Indices." 5:637–38.
2. NIOSH. 1975. "Criteria for a Recommended Standard . . . Occupational Exposure to Xylene." DHEW (NIOSH) 75–168.
3. OSHA. 1989. "Air Contaminants; Final Rule." *Fed. Reg.* 54:2332–2959.
4. U.S. EPA. 1987. "Health Advisory for Xylenes." Office of Drinking Water (3/31).
5. U.S. EPA. 1985. "National Primary Drinking Water Regulations; Synthetic Organic Chemicals, Inorganic Chemicals and Microorganisms; Proposed Rule." *Fed. Reg.* 50: 46935–7025.
6. NTP. 1986. "Toxicology and Carcinogenesis Studies of Xylenes (Mixed) (CAS 1330–20–7) in F344/N Rats and B6C3F$_1$ Mice (Gavage Studies)." (Research Triangle Park, NC: National Toxicology Program), TR-327.
7. Maltoni, C., B. Conti, G. Cotti, and F. Belpoggi. 1985. "Experimental Studies on Benzene Carcinogenicity at the Bologna Institute of Oncology: Current Results and Ongoing Research." *Am. J. Ind. Med.* 7:415–46.
8. U.S. EPA. 1989. "Xylenes; CASRN 1330–20–7." IRIS (7/1/89).
9. Hood, R., and M. Ottley. 1985. "Developmental Effects Associated with Exposure to Xylene: A Review." *Drug Chem. Tox.* 8:281–97.
10. Hudak, A., and G. Ungvary. 1978. "Embryotoxic Effects of Benzene and Its Methyl Derivatives: Toluene, Xylene." *Toxicology* 11:55–63.
11. U.S. EPA. 1984. "Health Effects Assessment for Xylene." EPA/540/1–86/006.
12. Ungvary, G., E. Tatrai, A. Hudak, G. Barcza, and M. Lorincz. 1980. "Studies on the Embryotoxic Effects of *ortho-*, *meta-* and *para-*Xylene." *Toxicology* 18:61–74.
13. Gamberale, F., G. Annwall, and M. Hultengren. 1978. "Exposure to Xylene and Ethylbenzene. III. Effects on Central Nervous Functions." *Scand. J. Work. Environ. Health* 4:204–11.
14. Savolainen, K., J. Kekoni, V. Riihimaki, and A. Laine. 1984. "Immediate Effects of *m-*Xylene on the Human Central Nervous System." *Arch. Tox.* suppl. 7:412–17.
15. Tatrai, E., G. Ungvary, I. R. Cseh, et al. 1981. "The Effect of Long-Term Inhalation of 0-Xylene on the Liver." *Ind. Environ. Xenobiotics, Proc. Int. Conf.* 293–300.
16. Clayton, G. D., and F. E. Clayton. 1981. *Patty's Industrial Hygiene and Toxicology,* 3rd ed. (New York, NY: John Wiley and Sons), 3291–3300.
17. U.S. EPA. 1985. "Drinking Water Criteria Document for Xylenes" (final draft) (Springfield, VA: NTIS), PB86–117942.

ZINC ACETATE

Synonyms: zinc diacetate, dicarbomethoxyzinc

CAS Registry Number: 557–34–6

Molecular Weight: 183.47

Molecular Formula: $Zn(O_2C_2H_3)_2$

AALG: analogy-based—6 $\mu g/m^3$ (0.9 ppb) 24-hour TWA

Occupational Limits:
- ACGIH TLV: none
- NIOSH REL: none
- OSHA PEL: none

Drinking Water Limits: At the time the ambient water quality criteria were being established (1980), data were not sufficient to derive a health-based criterion for zinc; a criterion of 5 mg Zn/L was recommended based on organoleptic considerations (bitter taste).[1] In 1985, the U.S. EPA concluded that "potential adverse health effects will not arise from zinc in drinking water" and thus no MCLG was proposed for zinc.[2]

Toxicity Profile

Carcinogenicity: No data were found implicating zinc acetate as a carcinogen. Zinc and its compounds have been placed in group D ("unclassified substance") under the EPA weight-of-evidence classification.[3]

Mutagenicity: No data were found implicating zinc acetate as a mutagen.

Developmental Toxicity: No data were found implicating zinc acetate *per se* as a developmental toxin. However, zinc itself has been implicated in certain studies. (Refer to reference 3 and the assessment for zinc sulfate).

Reproductive Toxicity: It is of interest to note that in mechanistic studies, zinc as zinc acetate has been found to be capable of at least partially reversing the deleterious effects of cadmium on spermatogenesis, steroidogenesis, and fertility in the male rat if given before or soon after cadmium treatment.[4]

Systemic Toxicity: Zinc is an essential trace element in humans and laboratory animals; it is a component of many important enzymes and is necessary for DNA, RNA, and protein synthesis; carbohydrate metabolism; utilization of nitrogen and sulfur; and cell division and growth. In addition, it is essential for spermatogenesis, oogenesis, and fetal nutrition. Zinc deficiency in animals and humans is associated with anorexia, decreased sense of taste, decreased growth, impaired wound healing, and other skin changes. It should be noted that excessive zinc intake may interfere with iron and copper metabolism and may aggravate marginal copper deficiency.[5]

No data were found on the effects of inhalation exposure to zinc acetate in humans or animal models, nor were any data available on the effects of subchronic or chronic exposure to zinc acetate via other routes.

Based on limited available acute data, zinc acetate is of relatively low oral toxicity with a reported LD_{50} of 2510 mg/kg in rats.[6] Jones et al.[7] studied the effects of pretreatment with a variety of heavy metals, including zinc as zinc acetate, on the intraperitoneal LD_{50} for that same metal in ICR mice. Although pretreatment with some metals increased the intraperitoneal LD_{50}, indicating a protective effect, no such effect was observed with zinc as zinc acetate. The intraperitoneal LD_{50}s with and without pretreatment were 22.8 mg/kg and 20.2 mg/kg, respectively.

Irritation: No data were found implicating zinc acetate as an irritant.

Basis for the AALG: Although no data were available on the specific use, occurrence, or release of zinc acetate, the major source of zinc exposure in humans is thought to be the diet.[3] Thus 20% of the contribution to total exposure will be allowed from air.

Derivation of an AALG for zinc acetate is problematic due to the absence of occupational limits, animal and human inhalation data, and subchronic or chronic exposure data by any route. Given the extreme paucity of data, only two approaches are viable: (1) use of an LD_{50}-derived criterion or (2) use of a criterion based on analogy with another zinc compound.

Based on the statistical analysis of the relationship of oral LD_{50}s to their corresponding chronic NOAELs for a large number of compounds, Layton et al.[8] recommended that the LD_{50} be multiplied by a factor of between 1.0×10^{-5} and 5.0×10^{-6} (equivalent to dividing by a factor between 100,000 and 200,000) to convert an LD_{50} to a corresponding surrogate animal NOAEL. In the case of zinc acetate the available rat oral LD_{50} of 2510 mg/kg was divided by 100,000, converted to a human equivalent surrogate NOAEL for the inhalation route according to the procedures described in Appendix C, and divided by a factor of 100 (10 for interspecies variation × 10 for interindividual variation).

In considering simple zinc compounds, e.g., salts and oxides, it would appear that the toxic properties would most likely be a consequence of the zinc moiety. Since two other zinc compounds have been evaluated, i.e., zinc oxide and zinc sulfate, it was deemed relevant to compare their physical properties and acute lethality data in order to evaluate the possibility of basing the zinc acetate AALG on analogy. This comparison is laid out in the following table:

Property	Zinc Acetate	Zinc Oxide	Zinc Sulfate
Molecular weight	183.47	81.38	161.43
Density	1.74	5.67	1.97
Melting point	237°C	1975°C	100°C
Water solubility	1 g/2.3 mL	insoluble	1 g/0.6 mL
Rat oral LD$_{50}$	2510 mg/kg	a	2950 mg/kg

[a]Mouse LD$_{50}$: 7900 mg/kg.[9]

It is apparent from this table that the physical properties and acute toxicity of zinc acetate and zinc sulfate are quite similar. Therefore, it was considered reasonable to base the AALG for zinc acetate on the AALG for zinc sulfate (which was based on a NOAEL from a subchronic ingestion study) with the inclusion of an additional database factor of 10 (analogy uncertainty factor). Note that the magnitude of the analogy uncertainty factor is arbitrary, since a scientific foundation for the numerical basis of such an uncertainty factor has not been established. For the reasons outlined above, it was considered more toxicologically sound to recommend the AALG based on analogy over the one based on the LD$_{50}$.

AALG:
- LD$_{50}$-derived—0.18 μg/m^3 (0.023 ppb) 24-hour TWA
- analogy-based—6 μg/m^3 (0.9 ppb) 24-hour TWA

References

1. U.S. EPA. 1980. "Ambient Water Quality Criteria for Zinc." EPA 440/5–80–079.
2. U.S. EPA. 1985. "National Primary Drinking Water Regulations. Synthetic Organic Chemicals, Inorganic Chemicals and Microorganisms; Proposed Rule." *Fed. Reg.* 50: 46936–47022.
3. U.S. EPA. 1984. "Health Effects Assessment for Zinc (and Compounds)." EPA 540/1–86/048.
4. Saksena, S., M. J. White, J. Mertzlufft, and I. Lau. 1983. "Prevention of Cadmium-Induced Sterility by Zinc in the Male Rat." *Contraception* 27:521–30.
5. Carson, B. L., H. V. Ellis, and J. L. McCann. 1986. *Toxicology and Biological Monitoring of Metals in Humans* (Chelsea, MI: Lewis Publishers, Inc.), pp. 290–98.
6. NIOSH. 1988. "Acetic Acid, Zinc(II) Salt." RTECS, on line.
7. Jones, M. M., J. E. Schoenheit, and A. D. Weaver. 1979. "Pretreatment and Heavy Metal LD$_{50}$ Values." *Tox. Appl. Pharm.* 49:41–44.
8. Layton, D. W., B. J. Mallon, D. H. Rosenblatt, and M. J. Small. 1987. "Deriving Allowable Daily Intakes for Systemic Toxicants Lacking Chronic Toxicity Data." *Reg. Tox. Pharm.* 7:96–112.
9. Windholz, M., Ed. 1976. *The Merck Index.* (Rahway, NJ: Merck & Co., Inc.).

ZINC OXIDE

CAS Registry Number: 1314–13–2

Molecular Weight: 81.38

Molecular Formula: ZnO

AALG: systemic toxicity (TLV-based)—4.8 $\mu g/m^3$ 8-hour TWA

Occupational Limits:
- ACGIH TLV: 5 mg/m^3 TWA—fume; 10 mg/m^3 TWA—total dust; 10 mg/m^3 STEL—fume
- NIOSH REL: 5 mg/m^3 TWA; 15 mg/m^3 ceiling (15 minutes)
- OSHA PEL: 5 mg/m^3 TWA—fume; 10 mg/m^3 TWA—total dust

Basis for Occupational Limits: The TLVs for zinc oxide fume are based on the prevention of metal fume fever, and it was considered that zinc oxide dust should be regarded as a nuisance dust.[1] The NIOSH limits apply to both zinc oxide dust and fume and are also based on the prevention of metal fume fever and other possible chronic respiratory effects.[2] OSHA recently adopted the ACGIH TLV-TWA for zinc oxide total dust as part of its final rule limits.[3]

Drinking Water Limits: At the time ambient water quality criteria were established (1980), data were not sufficient to derive a health-based criterion for zinc; a criterion of 5 mg/L was recommended based on organoleptic considerations (bitter taste).[4] In 1985, the U.S. EPA concluded that "potential adverse health effects will not arise from zinc in drinking water," and, thus, no MCLG for zinc was proposed.[5]

Toxicity Profile

Carcinogenicity: No data were found implicating zinc oxide as a carcinogen.

Mutagenicity: Zinc oxide has been reported to produce DNA damage in *E. coli*.[6]

Developmental Toxicity: No data were found implicating zinc oxide as a developmental toxin.

Reproductive Toxicity: No data were found implicating zinc oxide as a reproductive toxin.

Systemic Toxicity: *Essentiality:* Zinc is an essential element in humans and laboratory animals; it is a component of many important enzymes and is necessary for DNA, RNA, and protein synthesis, carbohydrate metabolism, utilization of nitrogen and sulfur, and cell division and growth. In addition, it is essential for spermatogenesis, oogenesis, and fetal nutrition. Zinc deficiency in animals and humans is associated with anorexia, decreased sense of taste, decreased growth, impaired wound healing, and other skin changes. It should be noted that excessive zinc intake may interfere with iron and copper metabolism and may aggravate marginal copper deficiency.[7]

Zinc oxide, either as a fume or dust, is the form of zinc individuals are most commonly exposed to in an occupational setting. Inhalation of zinc oxide, particularly the freshly generated fume, may result in metal fume fever (also called zinc chills or brass founders' ague); this condition is transient and characterized by fever, chills, muscular pain, nausea, and vomiting, with recovery occurring in 24 to 48 hours.[1] Metal fume fever is generally considered not to result in permanent effects, and although there is some evidence for the development of tolerance, it is apparently quickly lost.[8] Although there have been reports of metal fume fever occurring below concentrations of 5 mg/m^3, NIOSH in its 1975 criteria document considered these reports to be unsupported by firm data. However, it was also stated that "although the effects of zinc fume fever are transitory, the possibility of chronic respiratory effects resulting from zinc oxide inhalation, whether as a fume or in larger particulate form, cannot be dismissed."[2] In this connection it is relevant to note that nose-only exposure of guinea pigs to 5 mg/m^3 freshly generated zinc oxide particles (projected area diameter = 0.05 μm) for 3 hours per day on 6 consecutive days resulted in decreased vital capacity, functional residual capacity, and alveolar volume, which were still present 72 hours after the last exposure. In addition, there was histological evidence of a diffuse interstitial inflammatory response involving the respiratory bronchioles and alveolar ducts, and fibroblasts were present in the interstitial infiltrates. The authors concluded that their results suggested that the current TLV of 5 mg/m^3 may not be adequate to protect exposed workers.[9]

Studies on the effects of inhalation exposure to zinc oxide in laboratory animals are limited. Pistorius[10] (cited in U.S. EPA[8]) exposed rats (numbers and strain not specified) to 15 mg/m^3 zinc oxide (particulate size < 1 μm) via inhalation for 1, 4, or 8 hours per day for 84 days in one experiment and for 4 hours per day, 5 days a week for 1, 14, 28, or 56 days in another experiment. The only effect on lung function noted in the first experiment was a significant decrease in specific conductance and difference volume after two weeks that disappeared following continued exposure. In the second experiment, pulmonary zinc levels were highest after 1 and 14 days of treatment; after 28 and 56 days of treatment, inflammatory changes in the lungs (including infiltration of leukocytes and macrophages) were observed.

Irritation: Zinc oxide is not generally recognized as an irritant.

Basis for the AALG: Zinc oxide occurs as a white or yellowish-white powder, and zinc will volatilize to produce zinc oxide fume when heated to temperatures near its boiling point (907°C); zinc oxide powder also sublimes under normal pressure. Zinc oxide is used extensively in pigments, rubber, cosmetics, ointments, and electronic devices. Based on its use patterns and the fact that the general public is unlikely to be exposed extensively to zinc oxide fume, 20% of the contribution to total exposure will be allowed from air.

The AALG for zinc oxide (both as the fume and dust) is based on consideration of both the long-standing TLV-TWA for zinc oxide fume of 5 mg/m^3 and the NIOSH REL-TWA of 5 mg/m^3 for both zinc oxide fume and dust since respiratory effects from zinc oxide exposure would be the effect of major concern in both occupationally exposed individuals and in the general population. It was considered prudent to treat this figure as a human LOAEL based on the statement in the TLV documentation that "it is believed that if concentrations are kept below this level, the incidence of metal fume fever will be low and any attacks which occur will be mild" and in consideration of the histological findings of Lam et al.[9] in the lungs of guinea pigs exposed to zinc oxide particles at 5 mg/m^3 for relatively short time periods. Therefore, the figure of 5 mg/m^3 was divided by a total uncertainty factor of 210 (4.2 for continuous exposure adjustment × 10 for interindividual variation × 5 for a LOAEL). This AALG should be considered provisional given the suggestive nature of the Lam et al.[9] study and the lack of chronic studies pertinent to the evaluation of zinc oxide inhalation toxicity in both humans and animal models.

AALG: • systemic toxicity (TLV-based)—4.8 μg/m^3 8-hour TWA

References

1. ACGIH. 1986. "Documentation of the Threshold Limit Values and Biological Exposure Indices." 5:645–46.
2. NIOSH. 1975. "Criteria for a Recommended Standard . . . Occupational Exposure to Zinc Oxide." DHEW (NIOSH) 76–104.
3. OSHA. 1989. "Air Contaminants; Final Rule." *Fed. Reg.* 54:2332–2959.
4. U.S. EPA. 1980. "Ambient Water Quality Criteria for Zinc." EPA 440/5–80–079.
5. U.S. EPA. 1985. "National Primary Drinking Water Regulations; Synthetic Organic Chemicals, Inorganic Chemicals and Microorganisms; Proposed Rule." *Fed. Reg.* 50: 46936–7022.
6. NIOSH. 1988. ZH4810000. "Zinc Oxide." RTECS, on line.
7. Carson, B. L., H. V. Ellis, and J. L. McCann. 1986. *Toxicology and Biological Monitoring of Metals in Humans* (Chelsea, MI: Lewis Publishers), 290–98.
8. U.S. EPA. 1984. "Health Effects Assessment for Zinc (and Compounds)." EPA 540/1–86–048.
9. Lam, H. F., M. W. Conner, A. E. Rogers, S. Fitzgerald, and M. O. Amdur. 1985. "Functional and Morphologic Changes in the Lungs of Guinea Pigs Exposed to Freshly Generated Ultrafine Zinc Oxide." *Tox. Appl. Pharm.* 78:29–38.
10. Pistorius, D. 1976. "Fruhe reaktionen der rattenlunge auf zinkoxidhaltige." *Atemluft. Beitr. Silikose-Forsch.* 28:70.

ZINC SULFATE

Synonyms: zinc vitriol, white vitriol

CAS Registry Number: 7733–02–0

Molecular Weight: 161.43

Molecular Formula: $ZnSO_4$

AALG: systemic toxicity—5 ppb (33 $\mu g/m^3$) 24-hour TWA

Occupational Limits:
- ACGIH TLV: none
- NIOSH REL: none
- OSHA PEL: none

Drinking Water Limits: At the time the ambient water quality criteria were being established (1980), data were not sufficient to derive a health-based criterion for zinc; a criterion of 5 mg (Zn)/L was recommended based on organoleptic considerations (bitter taste).[1] In 1985 the U.S. EPA concluded that "potential adverse health effects will not arise from zinc in drinking water," and, thus, no MCLG for zinc was proposed.[2]

Toxicity Profile

Carcinogenicity: No data were found implicating zinc sulfate as a carcinogen. Zinc and its compounds have been placed in group D, unclassified substance, under the EPA weight-of-evidence classification.[3]

Mutagenicity: Negative results have been reported in the Gene-Tox database as summarized in RTECS for zinc sulfate in an oncogenic transformation assay (Syrian hamster embryo cells with adenovirus SA7) and the *S. cerevisiae* gene conversion assay.[4]

Developmental Toxicity: Kumar[5] (cited in U.S. EPA[3]) reported three cases of premature delivery and one case where an infant was stillborn in mothers who had supplemented their diets with 100 mg zinc sulfate per day during the last trimester of pregnancy. It is stated in the Health Effects Assessment Document[3] that "zinc supplementation for pregnant women has been recommended, but because of the known interaction between zinc and copper, excessive zinc supplementation for prolonged times could have an adverse effect on the fetus." It should be noted that the interaction between zinc and copper in the diet is discussed extensively in the Ambient Water Quality Criteria Document for Zinc.[1]

Reproductive Toxicity: No data were found implicating zinc sulfate as a reproductive toxin.

Systemic Toxicity: *Essentiality:* Zinc is an essential trace element in humans and laboratory animals; it is a component of many important enzymes and is necessary for DNA, RNA, and protein synthesis, carbohydrate metabolism, utilization of nitrogen and sulfur, and cell division and growth. In addition, it is essential for spermatogenesis, oogenesis, and fetal nutrition. Zinc deficiency in animals and humans is associated with anorexia, decreased sense of taste, decreased growth, impaired wound healing, and other skin changes. It should be noted that excessive zinc intake may interfere with iron and copper metabolism and may aggravate marginal copper deficiency.[6]

Data on the effects of inhalation exposure to zinc sulfate in humans and laboratory animals are lacking. A number of reports have appeared in the clinical literature that suggest that zinc supplementation (administered orally as zinc sulfate) may accelerate the healing of leg ulcers and be helpful in the treatment of chronic refractory rheumatoid arthritis; several studies in which controls were used suggest a human NOAEL of 2.14 mg/kg/day (figures derived by U.S. EPA assuming a 70-kg human).[3] It should be noted that these studies are of severely limited value for risk assessment since the number of patients studied was small, all had concurrent disease conditions, treatment generally lasted only a few months, and multiple dose levels were not employed.

Several studies have been conducted in laboratory animals in which zinc sulfate was administered either in the diet or in drinking water. Heller and Burke[7] (cited in U.S. EPA[3]) studied the effects of 2500 ppm zinc in the diet of rats (control n = 3 males, 1 female; treated n = 2 males, 2 females) over 3 generations and reported no signs of toxicity. Walters and Roe[8] (cited in U.S. EPA[3]) exposed male and female Chester Beatty stock mice (numbers not specified) to 0, 1000, or 5000 ppm $ZnSO_4$ in drinking water for one year. Although intercurrent disease (ectromelia) caused some mortality and necessitated the replacement of some animals during the first eight weeks of the study, zinc was reported to have no effect on weight gain or mortality. Aughey et al.[9] (cited in U.S. EPA[3]) reported that C3H mice (sex and numbers not specified) exposed to $ZnSO_4$ in drinking water (500 mg Zn/L, estimated dose 85 mg/kg/day) for up to 14 months showed histological evidence of hypertrophy of the adrenal cortex and pancreatic islets and changes characteristic of pituitary hyperactivity. Maita et al.[10] (cited in U.S. EPA[3]) exposed rats and mice (strain, sex, and numbers not specified) to 3000 or 30,000 ppm $ZnSO_4$ in the diet for 13 weeks; decreased food intake, retarded growth, and "hematologic abnormalities" were reported in the high-dose mice and rats. The estimated NOAELs of 95 mg/kg/day and 188 mg/kg/day in rats and mice, respectively, in this study appear to constitute the maximum no-effect level reported in the literature.

Irritation: Zinc sulfate is not generally regarded as an irritant.

Basis for the AALG: Since the major source of zinc is thought to be the diet, 20% of the contribution to total exposure will be allowed from air.

Given the lack of inhalation data in animals and humans for zinc sulfate and the limitations of available animal and human ingestion studies, derivation of the AALG becomes problematic. The best available animal study is that of Maita et al. because it used multiple dose levels and incorporated both a dose level that produced no adverse effects and one at which overt toxicity occurred. The NOAEL estimated by the U.S. EPA[3] for rats of 95 mg/kg/day was used as the starting point for criteria derivation since it was the most conservative figure. The ingestion NOAEL was converted to an equivalent inhalation NOAEL according to the procedures described in Appendix C, and a total uncertainty factor of 2000 (10 for interspecies variation $\times$ 10 for interindividual variation $\times$ 10 for a less-than-lifetime study $\times$ 2 for the database factor) was applied. Use of the database factor was justified based on use of the non-inhalation route of exposure in the study that served as a basis for criteria derivation. This AALG should be considered provisional because of the use of ingestion data from a subchronic study with limited reporting of study design and results.

AALG: • systemic toxicity—5 ppb (33 $\mu g/m^3$) 24-hour TWA

References

1. U.S. EPA. 1980. "Ambient Water Quality Criteria for Zinc." EPA 440/5-80-079.
2. U.S. EPA. 1985. "National Primary Drinking Water Regulations; Synthetic Organic Chemicals, Inorganic Chemicals and Microorganisms; Proposed Rule." *Fed. Reg.* 50: 46936-7022.
3. U.S. EPA. 1984. "Health Effects Assessment for Zinc (and Compounds)." EPA 540/1-86/048.
4. NIOSH. 1988. ZH5260000. "Zinc Sulfate." RTECS, on line.
5. Kumar, S. 1976. "Effects of Zinc Supplementation on Rats during Pregnancy." *Nutr. Rep. Int.* 14:221.
6. Carson, B. L., H. V. Ellis, and J. L. McCann. 1986. *Toxicology and Biological Monitoring of Metals in Humans* (Chelsea, MI: Lewis Publishers), 290-98.
7. Heller, V. G., and A. D. Burke. 1927. "Toxicity of Zinc." *J. Biol. Chem.* 74:85-93.
8. Walters, M., and F. J. C. Roe. 1965. "A Study of the Effects of Zinc and Tin Administered Orally to Mice over a Prolonged Period." *Food Cosmet. Toxicol.* 3:271-76.
9. Aughey, E., L. Grant, B. L. Furman, and W. F. Dryden. 1977. "The Effects of Oral Zinc Supplementation in the Mouse." *J. Comp. Pathol.* 87:1-14.
10. Maita, H., M. Hirano, K. Mitsumori, K. Takahashi, and Y. Shirasu. 1981. "Subacute Toxicity Studies with Zinc Sulfate in Mice and Rats." *Nippon Noyaku Gakkaishi* 6: 327-36.

ZINEB

Synonyms: zinc ethylenebisdithiocarbamate

CAS Registry Number: 12122–67–7

Molecular Weight: 275.73

Molecular Formula:

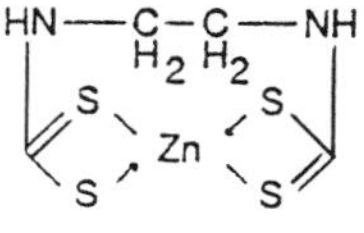

AALG: systemic toxicity—17.5 $\mu g/m^3$ 24-hour TWA

Occupational Limits:
- ACGIH TLV: none
- NIOSH REL: none
- OSHA PEL: none

Drinking Water Limits: In 1977 the Safe Drinking Water Committee of the NAS recommended a SNARL (no time period specified) of 35 $\mu g/L$ for zineb based on the standard 20% relative source contribution from drinking water and an alternate SNARL of 1.75 $\mu g/L$ based on a 1% contribution from drinking water.[1]

Toxicity Profile

Carcinogenicity: In 1976, IARC evaluated the evidence for zineb carcinogenicity.[2] It was reported that zineb caused an increased incidence of lung tumors in one of two strains of mice following oral administration[3] (cited in IARC[2]) as well as systemic reticulum cell sarcomas in mice given single subcutaneous injections of zineb and observed until 78 weeks of age[4] (cited in IARC[2]) and various sarcoma types in rats exposed to zineb in the form of a subcutaneous pellet implant[5] (cited in IARC[2]). However, no increased incidence of tumors was reported in two strains of mice exposed to zineb by gavage (464 mg/kg) and later in the diet (1298 mg/kg of diet) until 78 weeks of age[6] (cited in IARC[2]—note that this is the same study in which ethylene thiourea was found to produce liver tumors in mice) or in two more limited rat ingestion studies[5,7] (cited in IARC[2]).

IARC concluded that these studies were not sufficient to evaluate the carcinogenicity of zineb,[2] and no carcinogenicity studies were found for zineb that post-dated this IARC review. Based on this information, zineb is most appropriately placed in group D, unclassified substance, under the EPA weight-of-evidence classification.

It is significant to note that ethylene thiourea (ETU), which is classified as a B2, probable human carcinogen, occurs as a contaminant during zineb manufacture, a degradation product during storage and following application, and a metabolite following ingestion of zineb.[2] In addition, there are reports indicating that zineb on raw agricultural commodities would be expected to break down to ETU during cooking.[2] For further information on ETU carcinogenicity, refer to the chemical-specific evaluation of ETU.

Mutagenicity: The following short-term test results are reported for zineb in the Gene-Tox database as summarized in RTECS:[8] positive for *S. cerevisiae* homozygosis, negative for in vitro unscheduled DNA synthesis in human fibroblasts, and inconclusive in the *Drosophila* sex-linked recessive lethal assay. In addition, zineb is reported to cause chromosomal aberrations in mice given 10 mg/kg orally, gene conversion and mitotic recombination in *S. cerevisiae*, mutations in *B. subtilis* with metabolic activation, and sister chromatid exchanges in cultured human lymphocytes.[8] In one study, an increase in chromosomal aberrations in peripheral blood lymphocytes was reported in individuals occupationally exposed to zineb[9] (cited in IARC[2]).

Developmental Toxicity: Petrova-Vergieva and Ivanova-Tchemishanska[10] demonstrated a clear dose-response relationship for the occurrence of malformations (mainly of the brain, facial bones, limbs, and tail) in the offspring of albino rats (n = approximately 10/group) exposed to single oral doses of 0, 1, 2, 4, or 8 g/kg zineb on days 11 or 13 of gestation. A NOAEL of 1 g/kg zineb as a single oral dose was established in this study. In follow-up studies, groups of approximately 10 rats were given 0, 250, 500, or 1000 mg/kg zineb on gestation days 2 to 21 and groups of 4 rats were exposed to 100 mg/m^3 zineb powder via inhalation for 4 hours per day on gestation days 5 to 21, 6 to 21, or 7 to 21. No evidence of malformations or embryotoxicity was found in any of these follow-up studies. In addition, no effects on embryo viability, litter size, weight gain, or survival were found in rats exposed to 100 mg/m^3 zineb powder on days 4 to 21 of gestation and allowed to deliver their offspring, which were retained and studied for one month. The highest level of exposure used in the ingestion study was the NOAEL for the single-dose study, and the finding of no effects at this level was interpreted by the authors to indicate that the effects of zineb exposure in the offspring were not cumulative. It should be noted that maternal toxicity was either not evaluated or not reported in this study.

Reproductive Toxicity: In a report cited in RTECS, effects on the ovaries and fallopian tubes were noted in rats exposed to 8 g/kg zineb.[8] In another study, it was reported that first pregnancies were "retarded" and there was an increased incidence of sterility and fetal resorption in rats exposed to 100 mg/kg zineb for periods of up to 6 months[11] (cited in IARC[2]).

Systemic Toxicity: The acute toxicity of insoluble EBDC salts such as zineb is generally considered to be low; LD_{50}s in the range of 1 to 8 g/kg have been reported.[2] It has been suggested that the effects observed at high concentrations are at least partially attributable to the metal component.[1]

Smith et al.[7] exposed male and female albino rats (n = 10/ sex/group) to 0, 500, 1000, 2500, 5000, or 10,000 ppm zineb in the diet for two years. There was a clear dose-related trend in the occurrence of thyroid hyperplasia confirmed by histopathologic examination in all exposure groups. Renal pathology in the form of congestion, nephritis, and nephrosis was also seen at the highest exposure level. In addition, body weight gain was significantly decreased in female rats at levels of 5000 ppm and higher and in male rats at the highest exposure level. A LOAEL of 500 ppm for lifetime exposure of rats to zineb in the diet for thyroid hyperplasia is suggested by this study. It is of interest to note that levels of 10,000 ppm in the diet of dogs for one year were required to produce thyroid hyperplasia.

In the above study, the toxicity of zineb (zinc EBDC) was also compared to the sodium EBDC salt, and the zinc salt was reported to be much less potent in the induction of thyroid effects. Further studies revealed that this was most likely due to lower absorption (11–17%) of zineb from the gastrointestinal tract of rats.

In addition to its effects on the thyroid, it has also been suggested that zineb may exert effects on the immune system. It has been reported that treatment with 100 mg/kg zineb (length of time not specified) caused a reduction in antibody titers and phagocytic activity of leukocytes in both rats and rabbits, but no such effects were evident at 10 mg/kg/day[11] (cited in Safe Drinking Water Committee[1]).

Irritation: No data were found implicating zineb as an irritant.

Basis for the AALG: Based on its use as an agricultural fungicide, the major source of exposure to zineb for the general population is likely to be the diet.[12] Therefore, 20% of the contribution to total exposure will be allowed from air. It should be noted that a larger proportion of exposure might occur via inhalation in certain population subgroups such as pesticide applicators and mixer/loaders, and individuals living in the vicinity of fungicide application areas.

The AALG for zineb is based on thyroid hyperplasia in rats observed in the Smith et al.[7] study. The LOAEL of 500 ppm zineb in the diet was converted to a figure in mg/kg/day assuming a 350-g rat consuming 5% of its body weight per day as food as recommended by the U.S. EPA and described in Appendix C. This figure was converted to a human equivalent inhalation exposure level, and a total uncertainty factor of 1000 was applied (10 for interspecies variation × 10 for interindividual variation × 5 for a LOAEL × 2 database factor).

Since specific developmental abnormalities have been found in the offspring of rats exposed to single oral doses of zineb on day 11 or 13 of gestation, an AALG based on this endpoint was calculated for comparative purposes. The NOAEL for malformations in the offspring of rats of 1 g/kg was converted to a human equivalent exposure level according to the procedures described in Appendix C, and a total uncertainty factor of 1000 was applied (10 for interindividual variation × 10 for interspecies variation × 5 for the developmental uncertainty factor × 2 for the database factor). The NOAEL was not adjusted for continuous exposure since further studies indicated that the effect was not cumulative; application of the developmental uncertainty factor was justified by the fact that no evidence of maternal toxicity was reported in either of the studies at levels producing a frank increase in malformations.

AALG: • systemic toxicity—17.5 μg/m^3 24-hour TWA
• developmental toxicity—600 μg/m^3 24-hour TWA

Note: These AALGs should be considered provisional based on possible oncogenic hazards associated with zineb that have not yet been adequately defined and the use of a non-inhalation studies to derive these limits. The EBDC pesticides, including zineb, are currently the subject of an EPA Special Review with particular reference to potential oncogenic and teratogenic hazards.[12] It is strongly recommended that this compound be reevaluated when the results of this review become available.

References

1. Safe Drinking Water Committee. 1977. *Drinking Water and Health,* Vol. 1 (Washington, DC: National Academy Press), 650–660.
2. IARC. 1976. "Zineb." *IARC Monog.* 12:245–57.
3. Chernov, O. V., and I. I. Knitsenko. 1969. "Blastomogenic Properties of Some Derivatives of Dithiocarbamic Acid." *Vop. Onkol.* 15:71–74.
4. NTIS. 1968. *Evaluation of Carcinogenic, Teratogenic and Mutagenic Activities of Selected Pesticides and Industrial Chemicals, Vol. 1: Carcinogenic Study* (Washington, DC: U.S. Department of Commerce).
5. Andrianova, M. M., and I. V. Alekseev. 1970. "On the Carcinogenic Properties of the Pesticides Sevine, Maneb, Ciram and Cineb." *Vop. Pitan.* 29:71–74.
6. Innes, J. R. M., B. M. Ulland, M. G. Valerio, L. Petrucelli, L. Fishbein, E. R. Hart, A. J. Pallotta, R. R. Bates, H. L. Falk, J. J. Gart, M. Klein, I. Mitchell, and J. Peters. 1969. "Bioassay of Pesticides and Industrial Chemicals for Tumorigenicity in Mice: A Preliminary Note." *J. Natl. Cancer Inst.* 42:1101–14.
7. Smith, B. R., J. K. Finnegan, P. S. Larson, P. F. Sahyoun, M. L. Dreyfuss, and H. B. Haag. 1953. "Toxicologic Studies on Zinc and Disodium Ethylene Bisdithiocarbamates." *J. Pharm. Exp. Ther.* 109:159–66.
8. NIOSH. 1988. ZH3325000. "Zinc, (Ethylenebis(dithio-carbamate))." RTECS, on line.
9. Pilinskaya, M. A. 1974. "Results of Cytogenetic Examination of Persons Occupationally Contacting with the Fungicide Zineb." *Genetika* 10:140–46.
10. Petrova-Vergieva, T., and L. Ivanova-Tchemishanska. 1973. "Assessment of the Teratogenic Activity of Dithiocarbamate Fungicides." *Fd. Cosmet. Tox.* 11:239–44.
11. Ryazanova, R. A. 1967. "Effect of the Fungicides Ciram and Cineb on the Generative Function of Test Animals." *Gig. i Sanit.* 32:26–30.
12. U.S. EPA. 1987. "EBDC Pesticides; Initiation of Special Review." *Fed. Reg.* 27172–177.

APPENDICES

APPENDIX A: DRINKING WATER LIMITS

Introduction

This appendix supports the information on drinking water limits given in the chemical-specific assessments. The descriptive material provided here is not intended to be comprehensive in nature; in depth information can be found in the references cited at the end of the appendix. The drinking water limits described here include maximum contaminant level goals (MCLGs) and maximum contaminant levels (MCLs), health advisories (HAs), suggested no adverse response levels (SNARLs), and ambient water quality criteria (AWQC).

MCLGs and MCLs

Under the Safe Drinking Water Act (SDWA) of 1974, it is the responsibility of the U.S. Environmental Protection Agency (EPA) to promulgate national primary drinking water regulations (NPDWRs) for pollutants that may pose a threat to human health and are known, or have the potential, to occur in public water supplies. Previously, the sequence followed in NPDWR development (Figure A.1) was first to propose a recommended maximum contaminant level (RMCL, a non-enforceable health goal), followed by RMCL promulgation and proposal of the MCL (an enforceable standard), and finally MCL promulgation; a time period for public comment, meetings, and workshops was allowed between each step.[1] Recent amendments (1986) to the SDWA now require that EPA propose and promulgate MCLs and MCLGs (formerly called RMCLs) simultaneously.[2]

As indicated in Figure A.2, concurrent with the regulatory process the EPA Office of Drinking Water (ODW) prepares a chemical-specific criteria document (CD), which will serve as the technical support document for the health-based MCLG. In the CD, available data on the chemical of interest (including physical and chemical properties, toxicokinetics, human exposure, health effects in animals and humans, and mechanisms of toxicity) are reviewed and evaluated. In a separate section of the CD (''Quantification of Toxicological Effects''), the correlation

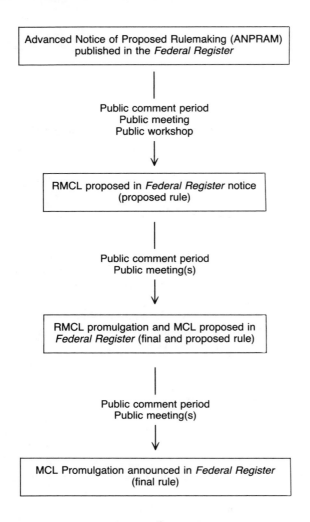

Figure A.1   Former NPDWR regulatory scheme. Adapted from EPA.[1]

between exposure and effect is discussed, and the quantitative basis for the MCLG is defined. The MCLG is based on health effects alone, with a sufficient margin of safety incorporated to protect susceptible individuals (members of high-risk groups).[2]

In addition to the CD, the ODW concurrently prepares other documents containing information on analytical methods, treatment technologies, human exposure potential, and cost-benefit analysis for the chemical under consideration. These documents are used to support the MCL, which is based on risk management considerations (i.e., technical and economic feasibility), as well as health effects, and is set as close to the MCLG as possible.[2]

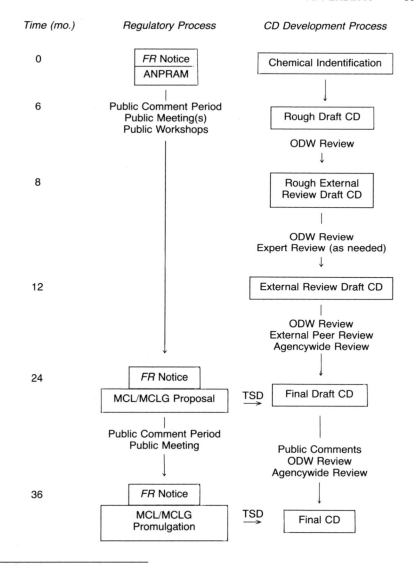

Time (mo.)	Regulatory Process	CD Development Process
0	FR Notice / ANPRAM	Chemical Indentification
6	Public Comment Period / Public Meeting(s) / Public Workshops	Rough Draft CD
		ODW Review
8		Rough External Review Draft CD
		ODW Review / Expert Review (as needed)
12		External Review Draft CD
		ODW Review / External Peer Review / Agencywide Review
24	FR Notice / MCL/MCLG Proposal	TSD → Final Draft CD
	Public Comment Period / Public Meeting	Public Comments / ODW Review / Agencywide Review
36	FR Notice / MCL/MCLG Promulgation	TSD → Final CD

FR—Federal Register
TSD—Technical Support Document
ANPRAM—Advanced Notice of Proposed Rulemaking

Figure A.2 Regulatory and criteria document development processes for MCL/MCLGs. *Source:* Ohanian.[2]

Table A.1 Three-Category Approach for Developing MCLGs

Evidence of Carcinogenicity	Classification	MCLG
Strong	EPA Group A or B IARC Group 1, 2A, or 2B	MCLG is set at zero
Equivocal	EPA Group C IARC Group 3	MCLG is (a) derived by DWEL approach with additional uncertainty factor or (b) set within 10^{-5} to 10^{-6} cancer risk range
Inadequate or Lacking	EPA Group D or E IARC Group 3	MCLG is derived using standard DWEL approach MCLG = (DWEL) × (%DWC)

Source: Ohanian.[2]
Note: DWEL—drinking water equivalent level
IARC—International Agency for Research on Cancer
DWC—drinking water contribution

A general overview of how MCLGs are set is given in Table A.1. If evidence that the chemical is carcinogenic is considered strong (EPA group A or B, or IARC groups 1, 2A, or 2B), the MCLG is set at zero, and in the CD a quantitative risk assessment is performed using mathematical models (e.g., linearized multistage model) and the appropriate epidemiologic or bioassay data. The upper 95% confidence limit of the slope, $q_1{}^*$, is usually used to derive low-dose estimates at the 10^{-4}, 10^{-5}, and 10^{-6} risk levels. The principles of quantitative risk assessment using animal or human data are discussed by Anderson and CAG[3] or in any of the health assessment documents issued by the EPA.

For substances with equivocal evidence of carcinogenicity (EPA group C or IARC group 3), the MCLG is derived in one of two ways: either on the basis of quantitative risk assessment using mathematical models with the MCLG set within the 10^{-5} to 10^{-6} cancer risk level range, or using a NOAEL (or LOAEL) with an uncertainty factor approach. This latter approach is described in Table A.2. If evidence for carcinogenicity is inadequate or lacking (EPA groups D or E, or IARC group 3), the MCLG is calculated using the NOAEL (or LOAEL) with an uncertainty factor approach.

Under the 1986 amendments to the SDWA, a number of chemicals have been scheduled for regulation (Table A.3). It should be noted that criteria documents and regulations are currently in effect for some of these substances. However, one of the provisions of the SDWA is that the NPDWRs are reviewed at least once every three years and revised regulations promulgated if necessary.[2]

Health Advisories

In addition to its function in developing MCLs and MCLGs under the SDWA, the U.S. EPA ODW also sponsors the Health Advisory Program. The purpose of health advisory documents and the Health Advisory Program is to provide "informal technical guidance to assist federal, state and local officials responsible for

Table A.2 Derivation of MCLGs Based on Noncarcinogenic Effects

Step 1	Calculation of reference dose (RfD)[a]
	definition: "The reference dose is an estimate (with an uncertainty spanning perhaps an order of magnitude) of the daily exposure to the human population (including sensitive subgroups) that is likely to be without an appreciable risk of deleterious health effects during a lifetime."[b]

$$RfD = \frac{NOAEL \ (or \ LOAEL)}{UF(s)} = mg/kg \ body \ weight/day$$

Step 1a	Selection of uncertainty factors (UFs)
	(a) 10 high-quality chronic or subchronic human NOAEL with good supporting toxicity data in other species
	(b) 100 high-quality chronic animal NOAEL (one or more species) or high-quality chronic or subchronic human LOAEL
	(c) 1000 high-quality chronic or subchronic animal LOAEL or when available data are incomplete or limited
	(d) An additional uncertainty factor from 1 to 10 may be included if a less-than-lifetime study was used, the adverse health effect is considered highly serious, or pharmacokinetic considerations altering dose exist, or to counterbalance beneficial effects (e.g., essential nutrients).
Step 2	Calculation of drinking water equivalent level (DWEL)
	definition: The DWEL is a medium-specific lifetime exposure level at which adverse, noncarcinogenic health effects are not expected to occur.

$$DWEL = \frac{(RfD) \times (body \ weight \ in \ kg)}{drinking \ water \ volume \ in \ L/day} = mg/L$$

A body weight of 70 kg and a drinking water volume of 2 liters/day are assumed for an adult.

Step 3	Calculation of the final MCLG
	MCLG = (DWEL) × (% DWC)

[a]The RfD was formerly termed the acceptable daily intake (ADI).
[b]Ohanian.[2]

protecting public health when emergency spills or contamination situations occur." Health advisories are health-based only guidelines for noncarcinogenic endpoints, not legally enforceable standards. The primary technical support documents for health advisories are the CDs prepared during the MCL/MCLG regulatory process.[4] The HA development process itself is shown in Figure A.3.

Health advisories may be calculated based on several different exposure periods—one-day, ten-day, longer-term (approximately 7 years or 10% of normal lifespan), or lifetime—for either a 70-kg man drinking 2 liters of water per day or a 10-kg child drinking 1 liter of water per day. A DWEL (drinking water equivalent level) approach similar to that described in Table A.2

$$HA = \frac{NOAEL \ (or \ LOAEL) \times body \ weight}{UF \times \_\ L/day} = \_ \ mg/L \ (or \ \mu g/L)$$

is used to derive HAs; it is considered most desirable to select an RfD (reference dose) study that is of approximately equivalent duration to the HA exposure period. Since HAs are based on noncarcinogenic effects, lifetime health advisories are not

Table A.3 Contaminants Scheduled for Regulation by 1989 Under the 1986 Amendments to the Safe Drinking Water Act

By 1987 (9 contaminants):[a]

Benzene	Carbon tetrachloride	p-Dichlorobenzene
1,2-Dichloroethane	1,1-Dichloroethylene	1,1,1-Trichloroethane
Trichloroethylene	Vinyl chloride	Fluoride

By 1988 (40 contaminants):

Chlorobenzene	0-Dichlorobenzene	cis-1,2-DCE[b]
trans-1,2-DCE[b]	Arsenic	Asbestos
Barium	Cadmium	Chromium
Copper	Lead	Mercury
Nitrate	Selenium	Acrylamide
Alachlor	Aldicarb	Carbofuran
2,4-D	Dibromochloropropane	1,2-Dichloropropane
Epichlorhydrin	Ethylbenzene	Ethylene dibromide
Heptachlor	Heptachlor epoxide	Lindane
Methoxychlor	PCBs	Pentachlorophenol
Styrene	Toluene	Toxaphene
2,4-TP	Xylene	Total coliforms
Giardia	Viruses	Turbidity

By 1989 (34 contaminants):

Methylene chloride	Trichlorobenzene	Antimony
Beryllium	Cyanide	Nickel
Sulfate	Thallium	Acrylonitrile
Adipates	Atrazine	Dalapon
Dinoseb	Diquat	Endothall
Endrin	Glyphosate	HCCP[c]
PAHs	Phthalates	Picloram
Simazine	1,2,2-Trichloroethane	2,3,7,8-TCDD (dioxin)
Vydate (Oxmyl)	Standard plate count	Legionella
Radium-226	Radium-228	Radon
Uranium	Photon radioactivity	Beta particle activity
Gross alpha particle activity		

Source: Ohanian.[2]

[a]Final MCLs were promulgated for these 9 chemicals in the *Federal Register* 52:41547-49, 10/28/87.
[b]DCE—dichloroethylene
[c]HCCP—hexachlorocyclopentadiene

recommended for known or probable human carcinogens (EPA group A or B), although HAs for shorter exposure periods are developed for them.[2] In addition, in calculating the lifetime health advisory, the issue of multimedia exposure is considered, and in the absence of specific data, 20% of the contribution to total exposure is assumed to come from drinking water for organic chemicals and 10% for inorganic chemicals. A list of the substances for which HAs have been developed is shown in Table A.4.

SNARLs

Suggested no adverse response levels (SNARLs) are health-based guideline levels for contaminants in drinking water established by the Safe Drinking Water Committee (SDWC) of the National Research Council (NRC), which is part of the

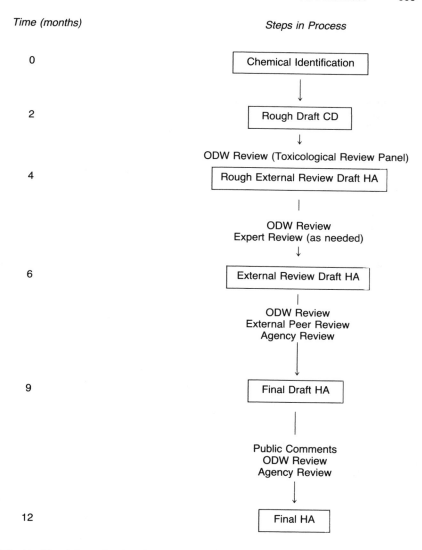

Figure A.3   Health advisory development process. *Source:* Ohanian.[4]

National Academy of Sciences (NAS). One provision of the SDWA of 1974 and its amendments in 1977 was that the NAS was mandated to conduct studies on the health effects of drinking water contaminants. Part of this responsibility included the evaluation of available data on drinking water contaminants for the purpose of assisting the EPA in the development of NPDWRs, identifying data gaps, and suggesting areas for future research.[5] SNARLs for specific chemicals evaluated by the SDWC are published in a series of volumes (eight as of 1988) entitled *Drinking Water and Health*.

Table A.4 Contaminants for Which Completed Health Advisories Have Been Issued as of 3/31/87

Acrylamide	Alachlor
Aldicarb	Arsenic
Barium	Benzene
Cadmium	Carbofuran
Carbon tetrachloride	Chlordane
Chlorobenzene	Cyanide
1,2-Dibromo-3-chloropropane	Chromium
o-,m-,p-Dichlorobenzenes	1,2-Dichloroethane
1,1-Dichloroethylene	cis-1,2-Dichloroethylene
trans-1,2-Dichloroethylene	1,2-Dichloropropane
2,4-Dichlorophenoxyacetic acid	p-Dioxane
2,3,7,8-TCDD (Dioxin)	Lead
Endrin	Epichlorhydrin
Ethylbenzene	Ethylene dibromide
Ethylene glycol	Heptachlor and heptachlor epoxide
Hexachlorobenzene	n-Hexane
Lindane	Mercury
Methoxychlor	Methyl ethyl ketone
Methylene chloride	Nickel
Nitrate/nitrite	Oxamyl
Pentachlorophenol	Perchloroethylene
Styrene	Toluene
Toxaphene	1,1,1-Trichloroethane
2,4,5-TP	Vinyl chloride
Xylenes	
Control of *Legionella* in plumbing systems	

Note: ODW is currently developing health advisories for several groups of substances including (1) pesticides specified by EPA's Office of Pesticide Programs as anticipated to be detected in water samples in the National Pesticide Survey, (2) certain unregulated volatile organic chemicals, and (3) munitions chemicals with the potential to contaminate drinking water during production, use, or disposal as specified by the Dept. of Army. Completion and availability of health advisories for the pesticides was announced in *Federal Register* 54:7598, (2/22/89).

In general, the basic approaches used by the SDWC in the development of SNARLs are the same as those used in the development of MCLGs. As with HAs, SNARLs may be calculated for both adults and children for one-day, ten-day, or chronic exposure periods. In the calculation of SNARLs, an arbitrary assumption of 20% contribution to total exposure from drinking water is made. Based on this it is pointed out that it is inappropriate to use SNARLs as "maximum contaminant intakes." In addition, SNARLs (like all other drinking water limits) have certain limitations, i.e., they do not provide a guarantee of absolute safety and do not generally take into account the potential for chemical interactions.[6]

Ambient Water Quality Criteria

The term *ambient water quality criteria*, as used in the chemical-specific assessments in this document, refers to the criteria for specific chemical pollutants or groups of pollutants published in a series of 65 documents (listed in Table A.5) by the U.S. EPA. Unlike other drinking water limits discussed here, the AWQCs

Table A.5 Chemicals and Classes of Chemicals for Which Ambient Water Quality Criteria Documents were Originally Published in 1980

Acenaphthene	Endosulfan
Acrolein	Endrin
Acrylonitrile	Ethylbenzene
Aldrin/dieldrin	Fluoranthene
Antimony	Haloethers
Arsenic	Halomethanes
Asbestos	Heptachlor
Benzene	Hexachlorbutadiene
Benzidine	Hexachlorocyclohexane
Beryllium	Hexachlorocyclopentadiene
Cadmium	Isophorone
Carbon tetrachloride	Lead
Chlordane	Mercury
Chlorinated benzenes	Napthalene
Chlorinated ethanes	Nickel
Chloroalkyl ethers	Nitrobenzene
Chlorinated naphthalene	Nitrophenols
Chlorinated phenols	Nitrosamines
Chloroform	Pentachlorophenol
2-Chlorophenol	Phenol
Chromium	Phthalate esters
Copper	Polychlorinated biphenyls
Cyanides	PAHs[a]
DDT	Selenium
Dichlorobenzenes	Silver
Dichlorobenzidine	Tetrachloroethylene
Dichloroethylenes	Thallium
2,4-Dichlorophenol	Toluene
Dichloropropanes/propenes	Toxaphene
2,4-Dimethylphenol	Trichloroethylene
Dinitrotoluene	Vinyl chloride
Diphenylhydrazine	Zinc

Note: These criteria documents were issued to satisfy the requirements of section 307(a)(1) of the Clean Water Act and as part of the conditions of the settlement agreement in the case of the National Resources Defence Council et al. vs. Train. Some of these documents have since been updated and reissued, e.g., arsenic, cadmium, chromium, copper, cyanide, lead, and mercury, and some new documents published, e.g., ammonia and chlorine.
[a]PAH—polynuclear aromatic hydrocarbons

were published pursuant to the Clean Water Act of 1977, section 304(a)(1), and apply to surface and ground waters, rather than finished drinking water from public water supplies, as is the case with the SDWA. In addition, criteria were also derived to protect aquatic life as well as human health.[7]

The AWQCs are guidelines with no regulatory impact which are intended to protect human health or aquatic life and do not take into account economic or technological feasibility. Criteria for the protection of aquatic life were derived by applying a set of guidelines to the database for each contaminant (minimum data requirements apply); the criteria specify both maximum and 24-hour average values. These criteria are not designed to protect all life stages of all species under all conditions; however, since some substances may be more toxic in marine waters than fresh waters or vise versa, provision was made to establish separate media-specific criteria if necessary. Details of the procedures and minimum data needed to derive aquatic life criteria are described in Appendix B of U.S. EPA.[7]

Criteria for the protection of human health are derived based on carcinogenic or toxic (noncarcinogenic) effects, or in a few cases based on organoleptic quality. The criteria are either based on the consumption of water and aquatic organisms (which are assumed to bioconcentrate the chemical from their aqueous environment) or solely from exposure to aquatic organisms. The consumption of 2 liters of water and 6.5 grams of fish/shellfish per day for a 70-kg man is assumed, and bioconcentration data for the specific chemical in question are used in the derivation of these criteria. As with other drinking water limits, the criterion is set at a specified level of risk based on mathematical models for carcinogens or using an NOAEL or LOAEL modified by appropriate uncertainty factors in the case of noncarcinogens.[7]

References

1. U.S. EPA. 1985. "National Primary Drinking Water Regulations; Synthetic Organic Chemicals, Inorganic Chemicals and Microorganisms; Proposed Rule." *Fed. Reg.* 50: 46936–7025.
2. Ohanian, E. V. 1989. "National Primary Drinking Water Regulations for Additional Contaminants to be Regulated by 1989." In E. J. Calabrese, C. E. Gilbert, and H. Pastides, *Safe Drinking Water Act—Amendments, Regulations, and Standards* (Chelsea, MI: Lewis Publishers), pp. 71–82.
3. Anderson, E. L., and Carcinogen Assessment Group, U.S. EPA. 1983. "Quantitative Approaches in Use to Assess Cancer Risk." *Risk Anal.* 3:277–95.
4. Ohanian, E. V. 1989. "Office of Drinking Water's Health Advisory Program." In E. J. Calabrese, C. E. Gilbert, and H. Pastides, *Safe Drinking Water Act—Amendments, Regulations, and Standards* (Chelsea, MI: Lewis Publishers), pp. 85–103.
5. Safe Drinking Water Committee, NAS. 1986. *Drinking Water and Health,* Vol. 6, preface (Washington, DC: National Academy Press).
6. Safe Drinking Water Committee, NAS. 1983. *Drinking Water and Health,* Vol. 5, introduction (Washington, DC: National Academy Press).
7. U.S. EPA. 1980. "Water Quality Criteria Documents; Availability." *Fed. Reg.* 45: 79318–79.

APPENDIX B: CHEMICAL-SPECIFIC WORKSHEET

Chemical _____   CAS No. _____

Synonyms _____

Molecular Weight _____

Molecular Formula:

Conversion (mg/m^3 to ppm):

Water Solubility (at _____ °C):

Vapor Pressure (at _____ °C):

Environmental Fate/Distribution/Occurrence Information:

Occupational Limits and Basis
TLV:

REL:

PEL:

Drinking Water Limits and Basis:

Availability of Toxicity Data by Endpoint (Source):

_____ Carcinogenicity   (notes on p. ____)
_____ Mutagenicity   (notes on p. ____)
_____ Developmental   (notes on p. ____)
_____ Reproductive   (notes on p. ____)
_____ Systemic   (notes on p. ____)
_____ Irritation   (notes on p. ____)

Principal & Supporting Studies by Endpoint (Source)

Basis for the AALG (calculations p.____):

Relative Source Contribution & Basis:

Data Sources
_____ TLV Documentation
_____ NIOSH Sources—CD, CIB, etc.
_____ Health Assessment Document/ Health Effects Assessment Document
_____ Ambient Water Quality Criteria Document
_____ Health Advisory
_____ IRIS
_____ Other Secondary Sources:
_____ Other Primary Sources:

APPENDIX C: DOSE CALCULATIONS, SCALING, AND PHYSIOLOGICAL PARAMETERS

I. Quantitative Risk Assessment

A. U.S. EPA quantitative risk assessments

In the calculation of AALGs for carcinogenic endpoints, whenever possible q_1^* values derived from quantitative risk assessments (QRAs) performed by the U.S. EPA Carcinogen Assessment Group (CAG) were utilized. These QRAs were generally available in either the health assessment document (HAD) for a particular compound or in the IRIS (Integrated Risk Information System) database. QRAs from IRIS were given precedence since they are more current. It should be noted that AALGs for carcinogens can be easily updated by checking the status of revisions to QRAs for a given compound in the IRIS database.

B. Quantitative risk assessment performed by the authors

In a few cases where sufficient animal bioassay data were available, but QRAs had not been performed by the U.S. EPA, the authors used EPA methodology as described by Anderson and CAG[1] and in the HADs to derive estimates for q_1^* and calculate AALGs as described in Chapter 3. In all cases, the linearized multistage model was applied to animal bioassay data using the program TOX_RISK.[2] Due to the more extensive review and evaluation, QRAs performed by the U.S. EPA should take precedence over QRAs performed by the authors should they become available, as for example in the IRIS database.

II. Calculation of AALGs for Noncarcinogenic Endpoints

A. Using an LOAEL/NOAEL from an animal inhalation study

1. Adjust the animal NOAEL/LOAEL (in units of mg/m^3) for continuous exposure as shown below. If the NOAEL/LOAEL is in units of ppm, first convert it to mg/m^3 as shown in Table C.1.

$$NOAEL_{adj} = NOAEL \times d/7 \times h/24 \times l_e/L_e,$$

Table C.1 Conversion of Exposure Levels in Units of ppm to Units of mg/m³

When the exposure level for a gas (NOAEL or LOAEL) used as a starting point in AALG derivation is expressed in units of ppm, this can be converted to mg/m³ on the basis of the Ideal Gas Law:

$$mg/m^3 = ppm \times \frac{g-mole}{22.4L} \times \frac{MW}{g-mole} \times \frac{273°}{T} \times \frac{P}{760\ mmHg} \times \frac{10^3\ L}{m^3} \times \frac{10^3\ mg}{g}$$

where

  ppm = concentration of the gas expressed on a volumetric basis L/10⁶ L
  MW = molecular weight of the gas in grams
  22.4 L = the volume occupied by 1 g-mole of any compound in the gaseous state at 0°C and 1 atm (760 mm Hg)
  T = actual temperature in degrees Kelvin
  P = actual pressure in mm Hg

Under standard conditions, i.e., at 25°C and 760 mm Hg, one g-mole of a perfect gas or vapor occupies 24.45 L, and this equation reduces to the following:

$$mg/m^3 = \frac{ppm \times MW}{24.45}$$

Source: Based on EPA.[3]

---

where d is the number of days per week exposed
h is the number of hours per day exposed
l_e is length of exposure
L_e is the length of the experiment
For example, if an NOAEL of 100 mg/m³ was identified from a study in which the animals were exposed for 6 hours per day, 5 days per week, for 13 weeks, and then observed for an additional 13 weeks, then the NOAEL$_{adj}$ would be calculated as NOAEL × 5/7 × 6/24× 13/26.

2. Another adjustment that can be made at this stage is to multiply the NOAEL$_{adj}$ by the absorption fraction (r) if such data are known to be available. For example, if it is known from other studies that the species in question absorbs 60% of the test agent, then the NOAEL$_{adj}$ would be multiplied by 0.6. In the absence of such data, r is assumed to be 1. It should be noted that this type of dosimetric adjustment is not appropriate if the critical effect (the one on which the AALG is based) is a portal-of-entry effect, which in this case would mean a pulmonary effect.

3. Convert the NOAEL$_{adj}$ (in mg/m³) to an animal NOAEL in units of mg/kg/day using the following formula:

$$NOAEL_{adj}\ (mg/kg/day) = \frac{NOAEL(mg/m^3) \times ABR\ (m^3/day)}{ABW\ (kg)}$$

where ABR is the animal breathing rate (in m³/day)
ABW is the animal body weight (in kg)
The ABR calculated based on the ABW is shown in Table C.2. Ideally, the ABR is calculated using weight data from the actual study, i.e., either the average weight or the midpoint of the range of weights is recommended here. In the absence of weight data from the actual study, default values may be used; those recommended are given in Table C.3.

Table C.2 Calculation of the Animal Breathing Rate (ABR) on the Basis of Animal Body Weight (ABW) Using a Body Weight–Surface Area Relationship

(1) The ABR is determined based on the body weight–surface area relationship described in Anderson and CAG[1] using either the average weight, or the midrange value of the weights for the control animals in the study from which the NOAEL was derived.

(2) In the absence of specific weight data, the default 30-gram mouse and 350-gram rat assumed by EPA[1] can be used. Other suggested default values may be found in Table C.3.

(3) The following basic data from the Federation of American Societies for Experimental Biology (FASEB) *Biological Data Book*[4] can be used in the species-specific equations shown under item (4) to determine the breathing rates for animals of various weights:
   —A 0.025-kg mouse breathes 35.5 liters per day.
   —A 0.113-kg rat breathes 105 liters per day.
   —A 0.092-kg hamster breathes 86.4 liters per day.
   —A 4.1-kg rabbit breathes 244.8 liters per day.
   —A 0.466-kg guinea pig breathes 230.4 liters per day.

(4) The equations shown below are derived from the above data under the assumption that body surface area is proportional to body weight to the 2/3 power, thus:

   For mice: $\text{ABR} = 0.0345 \times (\text{BW}/0.025)^{2/3}$
   For rats: $\text{ABR} = 0.105 \times (\text{BW}/0.113)^{2/3}$
   For hamsters: $\text{ABR} = 0.0864 \times (\text{BW}/0.092)^{2/3}$
   For rabbits: $\text{ABR} = 0.2448 \times (\text{BW}/4.1)^{2/3}$
   For guinea pigs: $\text{ABR} = 0.230 \times (\text{BW}/0.466)^{2/3}$

4. Convert the animal NOAEL_{adj} in mg/kg/day to a surrogate human NOAEL (NOAEL_{sh}) in units of mg/m^3:

$$\text{NOAEL}_{sh}\ (\text{mg/m}^3) = \frac{\text{NOAEL}_{adj}\ (\text{mg/kg/day}) \times \text{HBW (kg)}}{\text{HBR}\ (\text{m}^3/\text{day})}$$

where HBW is the human body weight (in kg)
HBR is the human breathing rate in m^3/day
In this work a 70 kg human breathing 20 m^3 of air per day was assumed, except in the case of developmental endpoints, where a 60 kg female breathing 20 m^3 of air per day was assumed.

5. Calculate the AALG by dividing the NOAEL_{sh} by appropriate multiplicative uncertainty factors (UFs) and adjust to account for the contribution of other sources of exposure, i.e., the relative source contribution (RSC), according to the guidelines in Chapter 3.

$$\text{AALG} = \frac{\text{NOAEL}_{sh}\ (\text{mg/m}^3) \times \text{RSC (\%)}}{\text{total UF}}$$

6. A number of assumptions are implicit in the calculations shown above. These assumptions and their limitations are outlined in Table C.4.

7. Updating the AALGs: It is strongly recommended that as inhalation reference concentrations (RfCs) become available in the IRIS database, that they be substituted for the AALGs calculated here with the RSC adjustment if appropriate. Note that as explained in Chapter 3, the RSC should not be applied if the critical effect is more concentration-dependent than

Table C.3 Default or Reference Values for Various Parameters in Humans and Laboratory Animals

Species	Lifespan (yrs)	Body Weight (kg)	Inhalation Rate[a] (m³/day)	Water Consumption (L/day)	Food Factor[b]
Mouse	2.0	0.03	0.039	0.0057	0.13
Rat	2.0	0.35	0.223	0.049	0.05
Hamster	2.4	0.14	0.13	0.027	0.083
Guinea pig	4.5	0.84	0.40	0.20	0.040
Rabbit	7.8	3.8	2.0	0.41	0.049
Dog	12.0	12.7	4.3	0.61	0.025
Rhesus monkey	18.0	8.0	5.4	0.53	0.040
Human	70.0	70.0	20	2	0.028

Note: The reference values given here are from EPA[5] with the exception of the inhalation rates for the 0.030-kg mouse and 0.35-kg rat, which were derived from the equations for animal breathing rate (ABR) in Table C.1, which came from Anderson and CAG[1]

[a]A correction factor for activity (2.11) is included in the inhalation rate given for humans, but not for the other species. The correction factor is based on the assumption of 20 L/min for 16 hours and 7.5 L/min for 8 hours per day.

[b]The food factor (f) is an empirically derived factor, $f = F/W$, which is the fraction of a species body weight consumed per day as food. An analogous factor, f_{water}, which is the fraction of species body weight consumed per day as water, also exists; the values given for this factor by EPA[6] are 0.078 and 0.17 for rats and mice, respectively.

time dependent, e.g., irritation. The reason for this recommendation is that the RfDs are backed by extensive peer review and represent an estimate based on more refined dosimetric adjustments.

B.   Using an NOAEL/LOAEL from a noninhalation animal study
1. If the NOAEL/LOAEL is not already in units of mg/kg/day, convert it to this basis, as described in Table C.5.
2. If the length of exposure (l_e) is less than the length of the experiment (L_e), convert the NOAEL to an $NOAEL_{adj}$ as follows:

$$NOAEL_{adj} = NOAEL \ (mg/kg/day) \times l_e/L_e$$

For example, if the NOAEL is 100 mg/kg/day and the animals received this dose daily for 12 months and were observed for an additional 6 months, then the $NOAEL_{adj}$ would be calculated as 100 mg/kg/day × 12/18 (see technical note under II.D).

3. Convert the animal $NOAEL_{adj}$ (in mg/kg/day) to a surrogate human NOAEL ($NOAEL_{sh}$) in units of mg/m³ using the following formula:

$$NOAEL_{sh} \ (mg/m^3) = \frac{NOAEL_{adj} \ (mg/kg/day) \times HBW \ (kg)}{HBR \ (m^3/day)}$$

where HBW is the human body weight (in kg)
HBR is the human breathing rate in m³/day
In this work, a 70 kg human breathing 20 m³ of air per day was assumed, except in the case of developmental endpoints, where a 60 = kg female breathing 20 m³ of air per day was assumed.

Table C.4 Assumptions Operative in Dosimetric Adjustments Described in This Appendix

1. Continuous exposure adjustment
   When an exposure level from a study in which a discontinuous exposure protocol was used is adjusted for continuous exposure, the operative assumption is that the derived exposure level approximates the original exposure level in terms of the resultant tissue concentrations. Since the toxicity of an exposure is a function of the concentration × time (C × T) curve, any factors which might affect this relationship will influence the validity of this assumption, and it has been emphasized that caution should be employed in the application of this type of adjustment.[7] In particular it should be noted that for effects which are more concentration-dependent as opposed to time-dependent, e.g., irritation, this type of dosimetric adjustment is not valid. Further cautions in the use of this type of adjustment with respect to portal-of-entry effects are discussed by EPA[7,a]

2. Absorption fraction
   If no adjustment is made for absorption (in the case of systemic or, with particular reference to the inhalation route, extrapulmonary effects), i.e., the absorption fraction is assumed to be 1, then it is implicitly assumed that absorption is complete, i.e., 100%. The following factors influence both the extent of absorption and time-to-steady-state blood levels, often differ between species, and should be considered in the application of dosimetric adjustments: (1) ventilation rates and airway mechanics, (2) blood flow rates (i.e., perfusion-limited), (3) metabolism, both pulmonary and extrapulmonary, (4) alveolar surface area, (5) thickness of the air:blood barrier, and (6) the blood:air and blood:tissue partition coefficients.[7]

3. Between species conversion of exposure levels
   When exposure levels are converted between species using adjustments for breathing rates and body weight, the underlying assumption is that mg/kg/day is an equivalent exposure between species after adjustment for breathing rate and body weight differences. This assumption is more defensible in the case of systemic effects, but is much more limited in the case of portal-of-entry (pulmonary) effects. The reasons for this are discussed extensively, and more appropriate dosimetric adjustments for portal-of-entry effects described, in a recent EPA document.[7,a] Note that when the ABR is calculated as shown in Table C.2 item (4), the ABW ratio is taken to the 2/3 power since many physiological parameters have been shown to be related to body surface area, which is proportional to the 2/3 power of body weight.[8]

[a]This document was published at the time this project was nearly completed.

4. Calculate the AALG by dividing the $NOAEL_{sh}$ by appropriate multiplicative uncertainty factors (UFs) and adjust to account for the contribution of other sources of exposure, i.e., the relative source contribution (RSC), according to the guidelines in Chapter 3.

$$AALG = \frac{NOAEL_{sh} \ (mg/m^3) \times RSC \ (\%)}{total \ UF}$$

5. A number of assumptions are implicit in the calculations shown above. These assumptions and their limitations are outlined in Table C.4. It is particularly significant to note the high degree of uncertainty associated with the type of route-to-route extrapolation described above. In cases where the AALG was derived from a noninhalation study, the AALG should be considered provisional and every effort should be made to update it as soon as animal or human inhalation data become available.

Table C.5 Conversion of Exposure Levels in ppm in the Diet or Water to mg of Test Article per kg of Body Weight per Day (mg/kg/day)

1. If the exposure levels are in units of ppm in drinking water, the following equation may be used:

$$mg/kg/day = \frac{ppm \ (mg/L) \times water \ intake \ (L/day)}{ABW \ (kg)}$$

where ppm is the exposure level in mg of the test agent per liter of water. The animal body weight (ABW) and water consumption data should ideally come from data reported in the study; alternatively default values, such as those given in Table C.3, may be used.

2. If the exposure levels are in units of ppm in the diet, the following equation may be used:

$$mg/kg/day = ppm \ (mg/kg) \times f$$

where ppm is the exposure level in the diet in terms of mg of test agent per kg of diet. The food factor (f) is an empirically derived factor, the fraction of a species body weight consumed per day as food (kg food consumed/kg of body weight/day). Values for f are species specific and are given in Table C.3.

3. In using the above equations certain potential sources of bias and error need to be kept in mind. These include, but may not be limited to, the following:
   a. If there is loss of the test article from the vehicle, e.g., volatilization of a compound from water, then the dose will be underestimated using these equations.
   b. If there is wastage of the food or water that is not accounted for and the dose in mg/kg/day is estimated using data from the original study, then the food or water consumption and hence the dose will be underestimated.
   c. Another limitation of using data from a study in which exposure is in ppm in the diet or drinking water is that bias may be introduced in estimation of the dose in mg/kg/day because there is a drop in food or water consumption and/or body weight gain that is not accurately monitored. In general, dose levels from gavage studies are more reliable than those from studies in which exposure is in ppm in the diet or water.

4. It has been pointed out that ppm in the diet or water is often assumed to be an equivalent exposure between species, and that this assumption is not justified due to the differences in calories/kg of diet, mainly related to moisture content differences, that also influence water consumption.[1,6] This is the basis for use of the empirically derived food factor described above.

Note: The basis for these equations and their use are described in Anderson and CAG[1] and EPA.[3]

C. An alternate approach for derivation of an AALG based on an LOAEL/NOAEL from a noninhalation study

   1. Starting with an LOAEL/NOAEL from a noninhalation study in units of mg/kg/day, the following equation may be used:

$$NOAEL_{alt} \ (mg/kg/day) = NOAEL \ (mg/kg/day) \times A_n/A_i$$

   where A_n is the efficiency of absorption via the noninhalation route (specific to the study used) and A_i is the efficiency of absorption via the inhalation route

   This approach was outlined in a draft EPA document.[3]

   2. The $NOAEL_{sh}$ is then calculated as described in B.3 above, i.e.,

$$NOAEL_{sh} \ (mg/m^3) = \frac{NOAEL_{alt} \ (mg/kg/day) \times HBW \ (kg)}{HBR \ (m^3/day)}$$

followed by calculation of the AALG as described in B.4, i.e.,

$$AALG = \frac{NOAEL_{sh}\ (mg/m^3) \times RSC\ (\%)}{total\ UF}$$

3. It should be noted that route-to-route extrapolation is best dealt with in the context of physiologically based pharmacokinetic modeling (PB-PK). However, in current practice the data and expertise required are often unavailable.

D. Technical note regarding adjustments for length of exposure (when shorter than the length of the experiment)
   1. Under some conditions, particularly in a study where the animals were exposed for only a portion of their lifespan but observed for a longer period, use of both the adjustment for less-than-lifetime exposure together with an uncertainty factor for less-than-lifetime exposure would be redundant and overly conservative.
   2. When l_e is less than L_e (length of exposure less than length of experiment), it is recommended that careful consideration be given to either not using or reducing the magnitude of the uncertainty factor for less-than-lifetime exposure.

References

1. Anderson, E. L., and the Carcinogen Assessment Group of the U.S. EPA. 1983. "Quantitative Approaches in Use to Assess Cancer Risk." *Risk. Anal.* 3:277–95.
2. Crump, K. S., R. B. Howe, and C. Van Landingham. 1987. *TOX RISK-Toxicology Risk Assessment Program* (Ruston, LA: Clement Associates, K. S. Crump Division).
3. U.S. EPA. 1987. "Interim Methods for the Development of Inhalation Reference Doses" (workshop draft). Environmental Criteria and Assessment Office (ECAO), U.S. EPA: Cincinnati, OH, ECAO-CIN-537. This document is superceded by a document of the same title with the following EPA no.: EPA 600/8–88/066F.
4. Federation of American Societies for Experimental Biology. 1974. *FASEB Biological Data Book,* 2nd ed, Vol. 3 (Bethesda, MD: FASEB).
5. U.S. EPA. "Reference Values for Risk Assessment." Prepared by the Office of Health and Environmental Effects, Environmental Criteria and Assessment Office, Cincinnati, OH, for the Office of Solid Waste, Washington, DC.
6. U.S. EPA. 1983. "Health Assessment Document for Acrylonitrile." EPA-600/8–82/007F.
7. U.S. EPA. 1989. "Interim Methods for Development of Inhalation Reference Doses." EPA 600/8–88/066F.
8. Calabrese, E. J. 1983. *Principles of Animal Extrapolation* (New York: John Wiley and Sons).

APPENDIX D.1:   CRITERIA FOR ASSESSING THE QUALITY OF INDIVIDUAL HUMAN EPIDEMIOLOGICAL STUDIES

A minimally acceptable study should meet the following criteria, which fundamentally represent good scientific practice. The study should have been reported or should be in press in the peer-reviewed literature.

1. The pertinent scientific background, such as reviews and supporting rationale upon which the study was based, should be given. Sponsorship and funding sources should be acknowledged.
2. The objectives of the study should be clearly stated and the study design described in relation to them. Underlying assumptions and limitations of the design also should be given.
3. The study population and comparison group description should include the specific population from which they were drawn and the method of selection. The rationale and criteria for inclusion/exclusion in the study should be given, particularly for exposure classifications. The appropriateness and limitations of the comparison group should be discussed. The extent to which the choice of subjects depended on existing or specially developed record systems, and implications of this upon the analysis, should be considered. The steps taken to ensure confidentiality of the subjects should be accounted for.
4. Methods of data collection should be described in detail, since these procedures will influence the derived interpretation and inferences. The validity (accuracy) and reliability (reproducibility) of the methods used to determine exposure should be stated. Response rates, including reasons for implications of differing rates, should be given. The direction and possible magnitude of any bias introduced into the study as a result of these rates should be described. The procedures used for following the study, methods to ensure completeness, and length

This material is reprinted directly from the document, *Interim Methods for the Development of Inhalation Reference Doses* (EPA/600/8–88/066F; April 1989). It was adapted from Interagency Regulatory Liaison Group,[1] Lebowitz,[2] and American Thoracic Society.[3]

of follow-up for each group or subgroup must be included. Other validity checks (e.g., avoiding bias by the independent ascertainment and classification of study variables, such as blind reading of histologic slides or clerical processing of data) also should be included.

5. Major demographic and anthropometric confounding factors should have been accounted for, such as age, sex, ethnic group, socioeconomic status, smoking status, and occupational exposure. Temperature, season, and day of the week are particularly important for acute studies of respiratory effects and also should be accounted for.

6. The procedures and statistical methods used to describe the data, estimate parameters, or test specific hypotheses should be presented. References and/or specific formulae also should be given for the statistical tests and for any programming procedures or packages that were applied.

The underlying assumptions and potential bias of the statistical methods should be stated. Explicit description of any method used to account for confounding factors (e.g., adjustment or matching) should be described explicitly. This includes methods to account for missing data, such as from nonresponse, attrition, or loss-to-follow-up. When reporting hypothesis tests, the measure of effect, statistical significance, power, and other criteria (e.g., one- vs. two-tailed test rationale) should be given. The point estimates and their standard errors and/or confidence intervals should be given when using estimation.

CRITERIA FOR CAUSAL SIGNIFICANCE

Statistical methods cannot establish proof of a causal relationship, but can define an association with a certain probability. The causal significance of an association is a matter of judgment that goes beyond any statement of statistical probability. To assess the causal significance or an air toxicant and a health effect, a number of criteria must be used, no one of which is pathognomonic by itself. These criteria include the following:

- Consistency (reproducibility) of the association. Causal inferences are strengthened when a variety of investigators have reproduced the findings under a variety of circumstances.
- Strength of the association. The larger the calculated relative risk, the greater the likelihood that the observed association is causal.
- Specificity of the association. Causality is more likely if a particular exposure is associated with only one illness and vice versa. This guideline rarely applies to air pollution research, in which all the diseases of major concern are multi-factorial.
- Temporal relationship of the association.
- Coherence of the association. An epidemiologic inference of causality is greatly strengthened when it conforms to knowledge concerning the biologic behavior of a toxin and its mechanism of action. This evidence may be obtained from clinical research or toxicologic studies.

References

1. Interagency Regulatory Liaison Group. 1981. "Guidelines for Documentation of Epidemiological Studies." *Am. J. Epidemiol.* 114:614–18.
2. Lebowitz, M. D. 1983. "Utilization of Data from Human Population Studies for Setting Air Quality Standards: Evaluation of Important Issues." *Environ. Health Perspect.* 52: 193–205.
3. American Thoracic Society. 1985. "Guidelines as to What Constitutes an Adverse Respiratory Health Effect, with Special Reference to Epidemiologic Studies of Air Pollution." *Am. Rev. Respir. Dis.* 131:666–68.

APPENDIX D.2: CRITERIA FOR ASSESSING THE QUALITY OF INDIVIDUAL ANIMAL TOXICITY STUDIES

A minimally acceptable study should meet the following criteria, which fundamentally represent good scientific practice.

1. All elements of exposure should be clearly defined.

   - The exposure amount, administration route, exposure schedule, and exposure duration must be described. Consideration should also be given to the concentration and time of exposure used vs. the expected level of human exposure.
   - If animal body weights, ages, or gender are not provided, consideration should be given to the uncertainty in appropriate default values.
   - Exposure information should include physicochemical characteristics of the substance used, such as purity, stability, pH, partition coefficient, particle size distribution, and vehicle. These properties can influence the local effects and the rate and extent of absorption, which can subsequently modify the toxic manifestations.
   - Exposure information should include description of generation and characterization technology used. The number of air changes, temperature, and relative humidity are exposure chamber characteristics which should be monitored. Cage (or other animal holder) rotation schedule should be described.
   - Animal care and holding procedures should be described.

2. Controls should be comparable with test animals in all respects except the treatment variable ("negative").

   - Concurrent controls must minimally include an "air-only" exposure group; if a vehicle is used, it is desirable that there be a "vehicle-only" group.

This material is reprinted directly from *Interim Methods for the Development of Inhalation Reference Doses* (EPA/600/8−88/066F; April 1989). This material was adapted from Society of Toxicology,[1] Muller et al.,[2] National Research Council,[3] James,[4] and Lu.[5]

- Historical control data can be useful in the evaluation of results, particularly where the differences between control and treated animals are small and are within anticipated incidences based on examination of historical control data.

3. Endpoints should answer the specific hypothesis addressed in the study, and the observed effects should be sufficient in number or degree (severity) to establish a dose-response relationship that can be used in estimating the hazard to the target species.

- The outcome of the reported experiment should be dependent on the test conditions and not influenced by competing toxicities.

4. The test performed must be valid and relevant to human extrapolation. The validity of using the test to mimic human responses must always be assessed. Issues to consider include the following:

- Does the test measure a toxicity directly or does it measure a response purported to indicate an eventual change (i.e., severity of the lesion)?
- Does the test indicate causality or merely suggest a chance correlation?
- Was an unproven or unvalidated procedure used?
- Is the test considered more or less reliable than other tests for that endpoint?
- Is the species a relevant or reliable human surrogate? If this test conflicts with data in other species, can a reason for the discrepancy be discerned?
- How reliable is high exposure (animal) data for extrapolation to low exposure (human scenario)?

5. Conclusions from the experiment should be justified by the data included in the report and consistent with the current scientific understanding of the test, the area of toxicology being tested, and the suspected mechanism of toxic action.
6. Due consideration in both the design and the interpretation of studies must be given for appropriate statistical analysis of the data.

- Statistical tests for significance can be performed only on those experimental units that have been randomized (some exceptions include weight-matching) among the dosed and concurrent control groups.
- Some frequent violations of statistical assumptions in toxicity testing include:
  —Lack of independence of observations.
  —Assuming a higher level of measurement than available (e.g., interval rather than ordinal).
  —Inappropriate type of distribution assumed.
  —Faulty specification of model (i.e., linear rather than nonlinear).
  —Heterogeneity of variance or covariance.
  —Large Type II error.

7. Subjective elements in scoring should be minimized. Quantitative grading of an effect should be used whenever possible.
8. Evidence of adherence to good laboratory practices is required unless excep-

tions have been negotiated (current testing) or considered (data obtained from studies carried out many years ago).

9. Peer review of scientific papers and of reports is extremely desirable and increases confidence in the adequacy of the work.
10. Reported results have increased credibility if they are reproduced by other researchers and supported by findings in other investigations.
11. Similarity of results to those of tests conducted on structurally related compounds should be considered.

References

1. Society of Toxicology, Task Force of Past Presidents. 1982. "Animal Data in Hazard Evaluation: Paths and Pitfalls." *Fund. Appl. Toxicol.* 2:101–107.
2. Muller, K. E., C. N. Barton, and V. A. Benignus. 1984. "Recommendations for Appropriate Statistical Practice in Toxicologic Experiments." *Neurotoxicology* 5:113–26.
3. National Research Council. 1984. *Toxicity Testing: Strategies to Determine Needs and Priorities* (Washington, DC: National Academy Press).
4. James, R. C. 1985. "Risk Assessment." In P. L. Williams, and J. L. Burson, Eds., *Industrial Toxicology: Safety and Health Applications in the Workplace* (New York: Van Nostrand Reinhold Company), pp. 369–98.
5. Lu, F. C. 1985. *Basic Toxicology: Fundamentals, Target Organs and Risk Assessment* (New York: Hemisphere Publishing).

APPENDIX E: BIBLIOGRAPHY OF NTP/NIEHS PAPERS RELEVANT TO EVALUATION OF TOXICOLOGY AND CARCINOGENESIS STUDIES

1. Anderson, M. W., D. G. Hoel, and N. L. Kaplan. 1980. "A General Scheme for the Incorporation of Pharmacokinetics in Low Dose Risk Estimation for Chemical Carcinogenesis. Example—Vinyl Chloride." *Toxicol. Appl. Pharmacol.* 55:154–61.
2. Ashby, J., and R. W. Tennant. 1988. "Chemical Structure, Salmonella Mutagenicity and Extent of Carcinogenicity as Indicators of Genotoxic Carcinogenesis Among 222 Chemicals Tested in Rodents by the U.S. NCI/NTP." *Mut. Res.* 204(1):17–115.
3. Bailer, A., and C. Portier. 1988. "An Illustration of Dangers of Ignoring Survival Differences in Carcinogenic Data." *J. Appl. Toxicol.* 8(3):185–89.
4. Bailer, A., and C. Portier. 1988. "Effects of Treatment Induced Mortality and Tumor Induced Mortality on Tests for Carcinogenicity in Small Samples." *Biometrics* 44(2): 417–32.
5. Boorman, G. A., C. A. Montgomery, Jr., S. L. Eustis, M. J. Wolfe, E. E. McConnell, and J. F. Hardisty. 1985. "Quality Assurance in Pathology for Rodent Carcinogenicity Studies." In H. Milman and E. Weisburger, Eds., *Handbook of Carcinogen Testing* (Park Ridge, NJ: Noyes Publications), pp. 345–57.
6. Boorman, G. A., D. A. Banas, S. L. Eustis, and J. K. Haseman. 1987. "Proliferative Exocrine Pancreatic Lesions in Rats. The Effect of Sample Size on the Incidence of Lesions." *Toxicol. Pathol.* 15:451–56.
7. Bristol, D. W. 1986. "National Toxicology Program Approach to Monitoring Study Performance." In *Managing Conduct and Data Quality of Toxicology Studies* (Princeton Scientific Publishing Co.), pp. 79–97.
8. Dinse, G. E., and J. K. Haseman. 1986. "Logistic Regression Analysis of Incidental Tumor Data from Animal Carcinogenicity Experiments." *Fundam. Appl. Toxicol.* 6: 44–52.
9. Haseman, J. K. 1983. "A Re-examination of False-Positive Rates for Carcinogenicity Studies." *Fundam. Appl. Toxicol.* 3:334–39.

This bibliography is part of the quarterly *Chemical Status Report* published by the NTP, Division of Research and Testing.

10. Haseman, J. K. 1983. "Patterns of Tumor Incidence in Two-Year Cancer Bioassay Feeding Studies in Fischer 344 Rats." *Fundam. Appl. Toxicol.* 3:1–9.
11. Haseman, J. K., J. E. Huff, and G. A. Boorman. 1984. "Use of Historical Control Data in Carcinogenicity in Rodents." *Toxicol. Pathol.* 12(2):126–35.
12. Haseman, J. K., D. D. Crawford, J. E. Huff, G. A. Boorman, and E. E. McConnell. 1984. "Results from 86 Two-Year Carcinogenicity Studies Conducted by the National Toxicology Program." *J. Toxicol. Environ. Health* 14:621–39.
13. Haseman, J. K. 1984. "Statistical Issues in the Design, Analysis and Interpretation of Animal Carcinogenicity Studies." *Environ. Health Perspect.* 58:385–92.
14. Haseman, J. K. 1985. "Issues in Carcinogenicity Testing: Dose Selection." *Fundam. Appl. Toxicol.* 5:66–78.
15. Haseman, J. K., J. E. Huff, G. N. Rao, J. E. Arnold, G. A. Boorman, and E. E. McConnell. 1985. "Neoplasms Observed in Untreated and Corn Oil Gavage Control Groups of F344/N Rats and (C57BL × C3H/HeN)F1 (B6C3F1) Mice." *J. Natl. Cancer Inst.* 75(5):975–84.
16. Haseman, J. K., J. S. Winbush, and M. W. O'Donnell. 1986. "Use of Dual Control Groups to Estimate False Positive Rates in Laboratory Animal Carcinogenicity Studies." *Fundam. Appl. Toxicol.* 7:573–84.
17. Haseman, J. K., E. C. Tharrington, J. E. Huff, and E. E. McConnell. 1986. "Comparison of Site-Specific and Overall Tumor Incidence Analyses for 81 Recent National Toxicology Program Carcinogenicity Studies." *Regul. Toxicol. Pharmacol.* 6:155–70.
18. Haseman, J. K., and J. E. Huff. 1987. "Species Correlation in Long-Term Carcinogenicity Studies." *Cancer Lett.* 37:125–32.
19. Haseman, J. K., J. E. Huff, E. Zeiger, and E. E. McConnell. 1987. "Comparative Results of 327 Chemical Carcinogenicity Studies." *Environ. Health Perspect.* 74:229–35.
20. Haseman, J. K. 1988. "Lack of Cage Effects on Liver Tumor Incidence in B6C3F1 Mice." *Fund. Appl. Toxicol.* 10:179–87.
21. Haseman, J. K. 1988. "Statistical Considerations in the Evaluation of Graded Responses in Carcinogenesis Studies." In H. Grice and J. Ciminera, Eds., *Carcinogenicity: The Design, Analysis, and Interpretation of Long-Term Animal Studies (ILSI Monograph)* (New York: Springer-Verlag), pp. 97–105.
22. Haseman, J. K. 1988. "Do Cage Effects Influence Tumor Incidence? An Examination of Laboratory Animal Carcinogenicity Studies Utilizing Fischer 344 Rats." *J. Appl. Toxicol.* 8(4):267–76.
23. Haseman, J. K., J. E. Huff, G. N. Rao, and S. L. Eustis. 1989. "Sources of Variability in Rodent Carcinogenicity Studies." *Fundam. Appl. Toxicol.* 12:793–804.
24. Hoel, D. G., N. L. Kaplan, and M. W. Anderson. 1983. "Implication of Nonlinear Kinetics on Risk Estimation in Carcinogenesis." *Science* 219:1032–37.
25. Hoel, D. G., J. K. Haseman, M. D. Hogan, J. E. Huff, and E. E. McConnell. 1988. "The Impact of Toxicity on Carcinogenicity Studies: Implications for Risk Assessment." *Carcinogenesis* 9(11):2045–52.
26. Huff, J. E. 1982. "Carcinogenesis Bioassay Results from the National Toxicology Program." *Environ. Health Perspect.* 45:185–98.
27. Huff, J. E., and J. A. Moore. 1984. "Carcinogenesis Studies Design and Experimental Data Interpretation/Evaluation at the National Toxicology Program." In J. Jarvisalo, P. Pfaffli, and H. Vainio, Eds., *Industrial Hazards of Plastics and Synthetic Elastomers*, pp. 43–64.
28. Huff, J. E., J. K. Haseman, E. E. McConnell, and J. A. Moore. 1986. "The National

Toxicology Program, Toxicology Data Evaluation Techniques, and Long-Term Carcinogenesis Studies.'' In E. E. Lloyd, Ed., *Safety Evaluation of Drugs and Chemicals* (Washington, DC: Hemisphere Publishing Company), pp. 411–46.

29. Huff, J. E., E. E. McConnell, J. K. Haseman, G. A. Boorman, S. L. Eustis, B. A. Schwetz, G. N. Rao, C. W. Jameson, L. G. Hart, and D. P. Rall. 1988. ''Carcinogenesis Studies: Results of 398 Experiments on 104 Chemicals from the U.S. National Toxicology Program.'' *Ann. NY Acad. Sci.* 534:1–30.

30. Huff, J. E., S. L. Eustis, and J. K. Haseman. 1989. ''Occurrence and Relevance of Chemically Induced Benign Neoplasms in Long-Term Carcinogenicity Studies.'' *Cancer Metastasis Review* 8:1–21.

31. Jameson, C. W. 1984. ''Analytical Chemistry Requirements for Toxicity Testing of Environmental Chemicals.'' In C. W. Jameson and D. B. Walters, Eds., *Chemistry for Toxicity Testing* (Boston, MA: Butterworth Publishers), pp. 3–14.

32. Kaplan, N., D. Hoel, C. Portier, and M. Hogan. 1988. ''An Evaluation of the Safety Factor Approach in Risk Assessment.'' In J. McLachlan, Ed., *Developmental Toxicology: Mechanisms and Risk,* Banbury Report 26, Cold Spring Harbor Laboratory, New York, pp. 335–46.

33. Luster, M. I., J. A. Blank, and J. H. Dean. 1987. ''Molecular and Cellular Basis of Chemically Induced Immunotoxicity.'' *Ann. Rev. Pharmacol. Toxicol.* 27:23–49.

34. Luster, M. I., A. E. Munson, P. T. Thomas, M. P. Holsapple, J. D. Fenters, K. L. White, Jr., L. D. Lauer, D. R. Germolec, G. L. Rosenthal, and J. H. Deans. 1988. ''Methods Evaluation. Development of a Testing Battery to Assess Chemical-Induced Immunotoxicity: National Toxicology Program's Guidelines for Immunotoxicity Evaluation in Mice.'' *Fundam. Appl. Toxicol.* 10:2–19.

35. Maronpot, R. R., J. K. Haseman, G. A. Boorman, S. L. Eustis, G. N. Rao, and J. E. Huff. 1987. ''Liver Lesions in B6C3F1 Mice: The National Toxicology Program, Experience and Position.'' *Arch. Toxicol. Suppl.* 10:10–26.

36. McConnell, E. E., H. A. Solleveld, J. A. Swenberg, and G. A. Boorman. 1986. ''Guidelines for Combining Neoplasms for Evaluation of Rodent Carcinogenesis Studies.'' *J. Natl. Cancer Inst.* 76(2):283–89.

37. Melnick, R. L., C. W. Jameson, T. J. Goehl, and G. O. Kuhn. 1987. ''Application of Microencapsulation for Toxicology Studies. I. Principles and Stabilization of Trichloroethylene in Gelatin-Sorbitol Microcapsules.'' *Fundam. Appl. Toxicol.* 8(4):425–31.

38. Melnick, R. L., C. W. Jameson, T. J. Goehl, R. R. Maronpot, B. J. Collins, A. Greenwell, F. W. Harrington, R. E. Wilson, K. E. Tomaszewski, and D. K. Agrawal. 1987. ''Application of Microencapsulation for Toxicology Studies. II. Toxicity of Microencapsulated Trichloroethylene in Fischer 344 Rats.'' *Fundam. Appl. Toxicol.* 8(4): 432–42.

39. Portier, C., and D. Hoel. 1983. ''Optimal Design of the Chronic Animal Bioassay.'' *J. Toxicol. Environ. Health* 12:1–19.

40. Portier, C., and D. Hoel. 1984. ''Design of Animal Carcinogenicity Studies for Goodness-of-Fit of Multistage Models.'' *Fundam. Appl. Toxicol.* 4:949–59.

41. Portier, C., and D. Hoel. 1987. ''Issues Concerning the Estimation of the TD50.'' *Risk Analysis* 7:437–47.

42. Portier, C. 1988. ''Life Table Analysis of Carcinogenicity Experiments.'' *J. Am. Coll. Toxicol.* 7(5):575–82.

43. Portier, C. 1988. ''Design of Long-Term Animal Carcinogenicity Experiments: Dose Allocation, Animal Allocation and Sacrifice Times.'' In D. Krewski and C. Franklin, Eds., *Statistical Methods in Toxicological Research* (Boca Raton, FL: CRC Press).

44. Portier, C. 1988. "Species Correlation of Chemical Carcinogens." *Risk Analysis* 8(4): 551–53.
45. Rall, D. P., M. D. Hogan, J. E. Huff, B. A. Schwetz, and R. W. Tennant. 1987. "Alternatives to Using Human Experience in Assessing Health." *Ann. Rev. Public. Health* 8:355–85.
46. Rao, G. N., W. W. Piegorsch, and J. K. Haseman. 1987. "Influence of Body Weight on the Incidence of Spontaneous Tumors in Rats and Mice of Long-Term Studies." *Am. J. Clin. Nutr.* 45:252–60.
47. Rao, G. N., R. L. Hickman, S. K. Sielkop, and G. A. Boorman. 1987. "Utero-Ovarian Infection in Aged B6C3F1 Mice." *Lab. Anim. Sci.* 37(2):153–58.
48. Rao, G. N., and J. J. Knapka. 1987. "Contaminant and Nutrient Concentrations of Natural Ingredient Rat and Mouse Diet Used in Chemical Toxicology Studies." *Fundam. Appl. Toxicol.* 9:329–38.
49. Rao, G. N. 1988. "Rodent Diets for Carcinogenesis Studies." *J. Nutr.* 118:929–31.
50. Rao, G. N., L. S. Birnbaum, J. J. Collins, R. W. Tennant, and L. C. Skow. 1988. "Mouse Strains for Chemical Carcinogenicity Studies: Overview of Workshop." *Fundam. Appl. Toxicol.* 10(3):385–94.
51. Reynolds, S. H., S. J. Stowers, R. M. Patterson, R. R. Maronpot, S. A. Aaronson, and M. W. Anderson. 1987. "Activated Oncogens in B6C3F1 Mouse Liver Tumors: Implications for Risk Assessment." *Science* 237:1309–16.
52. Shelby, M. D., and S. Stasiewicz. 1984. "Chemicals Showing No Evidence of Carcinogenicity in Long-Term, Two-Species Rodent Studies: The Need for Short-Term Test Data." *Environ. Mutagen.* 7:871–78.
53. Shelby, M. D. 1988. "The Genetic Toxicity of Human Carcinogens and Its Implications." *Mut. Res.* 204(1):3–15.
54. Solleveld, H. A., J. K. Haseman, and E. E. McConnell. 1984. "The Natural History of Body Weight Gain, Survival and Neoplasia in the Fischer 344 Rat." *J. Natl. Cancer Inst.* 72:929–40.
55. Stowers, S. J., R. R. Maronpot, S. H. Reynolds, and M. W. Anderson. 1987. "The Role of Oncogenes in Chemical Carcinogenesis." *Environ. Health Perspect.* 75:81–86.
56. Tennant, R. W., B. H. Margolin, M. D. Shelby, E. Zeiger, J. K. Haseman, J. Spalding, W. Caspary, M. Resnick, S. Stasiewicz, B. Anderson, and R. Minor. 1987. "Prediction of Chemical Carcinogenicity in Rodents from In Vitro Genetic Toxicity Assays." *Science* 236:933–41.

APPENDIX F: U.S. ENVIRONMENTAL PROTECTION AGENCY GENETIC TOXICOLOGY (GENE-TOX) PROGRAM PUBLICATION LIST

1. Introductory Papers

1. S. Greene and A. Auletta. 1980. "Editorial Introduction to the Reports of the Gene-Tox Program. An Evaluation of Bioassays in Genetic Toxicology." *Mut. Res.* 76:165–68.
2. M. D. Waters and A. Auletta. 1981. "The Gene-Tox Program: Genetic Activity Evaluation." *J. Chem. Inform. Comp. Sci. (ACS)* 21:35–38.

II. Phase I Panel Reports

1. David J. Bruisick, Vincent F. Simmon, Herbert S. Rosenkranz, Verne A. Ray, and Robert S. Stafford. 1980. "An Evaluation of the *Escherichia coli* WP2 *uvrA* Reverse Mutation Assay." *Mut. Res.* 76:169–90.
2. Walderico M. Generoso, Jack B. Bishop, David G. Gosslee, Gordon W. Newell, Ching-Ju Sheu, and Elizabeth Von Halle. 1980. "Heritable Translocation Test in Mice." *Mut. Res.* 76:191–215.
3. A. W. Hsie, D. A. Casciano, D. B. Couch, D. F. Krahn, J. P. O'Neill, and B. L. Whitfield. 1981. "The Use of Chinese Hamster Ovary Cells to Quantify Specific Locus Mutation and to Determine Mutagenicity of Chemicals. A Report of the Gene-Tox Program." *Mut. Res.* 86:329–54.
4. L. B. Russell, P. B. Selby, E. Von Halle, W. Sheridan, and L. Valcovic. 1981. "The Mouse-Specific Locus Test with Agents Other Than Radiations Interpretation of Data and Recommendations for Future Work." *Mut. Res.* 86:329–54.
5. L. B. Russell, P. B. Selby, E. Von Halle, W. Sheridan, and L. Valcovic. 1981. "Use of the Mouse Spot Test in Chemical Mutagenesis: Interpretation of Past Data and Recommendations for Future Work." *Mut. Res.* 86:355–79.

This publication list was current through 1988. Additional reports are published on an ongoing basis.

6. Samuel A. Latt, James Allen, Stephen E. Bloom, Anthony Carrano, Ernest Falke, David Kram, Edward Schneider, Rhona Schreck, Raymond Tice, Brad Whitfield, and Sheldon Wolff. 1981. "Sister-Chromatid Exchanges: A Report of the Gene-Tox Program." *Mut. Res.* 87:17–62.

7. Matthews O. Bradley, Bijoy Bhuyan, Mary C. Francis, Robert Langenbach, Andrew Peterson, and Eleizer Huberman. 1981. "Mutagenesis by Chemical Agents in V79 Chinese Hamster Cells: A Review and Analysis of the Literature. A Report of the Gene-Tox Program." *Mut. Res.* 87:81–142.

8. R. Julian Preston, William Au, Michael A. Bender, J. Grant Brewen, Anthony V. Carrano, John A. Heddle, Alfred F. McFee, Sheldon Wolff, and John S. Wassom. 1981. "Mammalian In Vivo and In Vitro Cytogenetic Assays: A Report of the U.S. EPA Gene-Tox Program." *Mut. Res.* 87:143–88.

9. Zev Leifer, Tsuneo Kada, Morton Mandel, Errol Zeiger, Robert Stafford, and Herbert S. Rosenkranz. 1981. "An Evaluation of Tests Using DNA Repair-Deficient Bacteria for Predicting Genotoxicity and Carcinogenicity. A Report of the U.S. EPA Gene-Tox Program." *Mut. Res.* 98:1–48.

10. Etta Kafer, Barry R. Scott, Gordon L. Dorn, and Robert Stafford. 1982. *"Aspergillus nidulans*: Systems and Results of Tests for Chemical Induction of Mitotic Segregation and Mutation. I. Diploid and Duplication Assay Systems. A Report of the U.S. EPA Gene-Tox Program." *Mut. Res.* 98:1–48.

11. Barry R. Scott, Gordon L. Dorn, Etta Kafer, and Robert Stafford. 1982. *"Aspergillus nidulans*: Systems and Results of Tests for Induction of Mitotic Segregation and Mutation. II. Haploid Assay Systems and Overall Response of All Systems. A Report of the U.S. EPA Gene-Tox Program." *Mut. Res.* 98:49–94.

12. Kathleen H. Larsen (Mavournin), Douglas Brash, James E. Cleaver, Ronald W. Hart, Veronica M. Maher, Robert B. Painter, and Gary A. Sega. 1982. "DNA Repair Assays As Tests for Environmental Mutagens. A Report of the U.S. EPA Gene-Tox Program." *Mut. Res.* 98:319–74.

13. Marvin S. Legator, Ernest Beuding, Robert Batzinger, Thomas H. Connor, Eric Eisenstadt, Michael G. Farrow, Gyula Ficsor, Abraham Hsie, John Seed, and Robert S. Stafford. 1982. "An Evaluation of the Host-Mediated Assay and Body Fluid Analysis. A Report of the U.S. Environmental Protection Agency Gene-Tox Program." *Mut. Res.* 98:319–74.

14. Milton J. Constantin and Elizabeth T. Owens. 1982. "Introduction and Perspectives of Plant Genetic and Cytogenetic Assays. A Report of the U.S. Environmental Protection Agency Gene-Tox Program." *Mut. Res.* 99:1–12.

15. Milton J. Constantin and Robert A. Nilan. 1982. "Chromosome Aberration Assays in Barley *(Hordeum vulgare)*. A Report of the U.S. Environmental Protection Agency Gene-Tox Program." *Mut. Res.* 99:13–36.

16. Milton J. Constantin and Robert Nilan. 1982. "The Chlorophyll-Deficient Mutant Assay in Barley *(Hordeum vulgare)*. A Report of the U.S. Environmental Protection Agency Gene-Tox Program." *Mut. Res.* 99:37–49.

17. G. P. Redei. 1982. "Mutagen Assay with Arabidopsis. A Report of the U.S. Environmental Agency Gene-Tox Program." *Mut. Res.* 99:234–55.

18. Te-Hsiu Ma. 1982. "Vicia Cytogenetic Tests for Environmental Mutagens. A Report of the U.S. Environmental Protection Agency Gene-Tox Program." *Mut. Res.* 99:257–71.

19. William F. Grant. 1982. "Chromosome Aberration Assays in Allium. A Re-

port of the U.S. Environmental Protection Agency Gene-Tox Program.'' *Mut. Res.* 99:273–91.

20. Te-Hsiu Ma. 1982. ''Tradescantia Cytogenetic Tests. (Root-Tip Mitosis, Pollen Mitosis, Pollen Mother-Cell Meiosis). A Report of the U.S. Environmental Protection Agency Gene-Tox Program.'' *Mut. Res.* 99:293–302.

21. J. Van't Hof and L. A. Schairer. 1982. ''Tradescantia Assay System for Gaseous Mutagens. A Report of the U.S. Environmental Protection Agency.'' *Mut. Res.* 99:303–15.

22. Michael J. Plewa. 1982. ''Specific-Locus Mutation Assays in *Zea mays*. A Report of the Environmental Protection Agency Gene-Tox Program.'' *Mut. Res.* 99:317–37.

23. Baldev K. Vig. 1982. ''Soybean (Glycine max (L.) merrill) As a Short-Term Assay for Study of Environmental Mutagens. A Report of the U.S. Environmental Protection Agency Gene-Tox Program.'' *Mut. Res.* 99:339–347.

24. Charles Heidelberger, Aaron E. Freeman, Roman J. Pienta, Andrew Sivak, John S. Bertram, Bruce C. Casto, Virginia C. Dunkel, Mary W. Francis, Takeo Kakunaga, John B. Little, and Leonard M. Schechtman. 1983. ''Cell Transformation by Chemical Agents: A Review and Analysis of the Literature. A Report of the U.S. Environmental Protection Agency Gene-Tox Program.'' *Mut. Res.* 114:283–385.

25. Andrew J. Wryobek, Laurie A. Gordon, James G. Burkhart, Mary W. Francis, Robert W. Kapp, Jr., Gideon Letz, Heinrich V. Malling, John C. Topham, and M. Donald Whorton. 1983. ''An Evaluation of the Mouse Sperm Morphology Test and Other Sperm Tests in Nonhuman Mammals. A Report of the U.S. Environmental Protection Agency Gene-Tox Program.'' *Mut. Res.* 115: 1–72.

26. Andrew J. Wyrobek, Laurie A. Gordon, James G. Burkhart, Mary W. Francis, Robert W. Kapp, Jr., Gideon Letz, Heinrich V. Malling, John C. Topham, and M. Donald Whorton. 1983. ''An Evaluation of Human Sperm As Indicators of Chemically Induced Alterations of Spermatogenic Function. A Report of the U.S. Environmental Protection Agency Gene-Tox Program.'' *Mut. Res.* 115:1–72.

27. N. Loprieno, R. Barale, E. S. Von Halle, and R. C. von Borstel. 1983. ''Testing of Chemicals for Mutagenic Activity with *Schizosaccharomyces pombe*. A Report of the U.S. Environmental Protection Agency Gene-Tox Program.'' *Mut. Res.* 115:215–23.

28. D. Clive, R. McCuen, J. F. S. Spector, C. Piper, and K. H. Mavournin. 1983. ''Specific Gene Mutations in L5178Y Cells in Culture. A Report of the U.S. Environmental Agency Gene-Tox Program.'' *Mut. Res.* 115:225–51.

29. John A. Heddle, Mark Hite, Barbara Kirkhart, Kathleen Mavournin, James T. MacGregor, Gordon W. Newell, and Michael F. Salamone. 1983. ''The Induction of Micronuclei As a Measure of Genotoxicity. A Report of the U.S. Environmental Protection Agency Gene-Tox Program.'' *Mut. Res.* 123:61–118.

30. W. R. Lee (Leader), S. Abrahamson, R. Valencia, E. S. Von Halle, F. E. Wurgler, and S. Zimmering. 1983. ''The Sex-Linked Recessive Lethal Test for Mutagenesis in *Drosophila melanogaster*. A Report of the U.S. Environmental Protection Agency Gene-Tox Program.'' *Mut. Res.* 123:183–279.

31. Ann D. Mitchell, Daniel A. Casciano, Martin L. Meltz, Douglas E. Robinson,

Richard H. C. San, Gary M. Williams, and Elizabeth S. Von Halle. 1983. "Unscheduled DNA Synthesis Tests. A Report of the U.S. Environmental Protection Agency Gene-Tox Program." *Mut. Res.* 123:363–410.

32. Herman E. Brockman, Frederick J. de Serres, Tong-Man Ong, David M. DeMarini, Alan J. Katz, Anthony J. F. Griffiths, and Robert S. Stafford. 1984. "Mutation Tests in *Neurospora crassa*. A Report of the U.S. Environmental Protection Agency Gene-Tox Program." *Mut. Res.* 133:87–134.

33. F. K. Zimmerman, R. C. von Borstel, E. S. Von Halle, J. M. Parry, D. Siebert, G. Zetterberg, R. Barale, and N. Loprieno. 1984. "Testing of Chemicals for Genetic Activity with *Saccharomyces cerevisiae*: A Report of the U.S. Environmental Protection Agency Gene-Tox Program." *Mut. Res.* 133:199–244.

34. R. Valencia, S. Abrahamson, W. R. Lee, E. S. Von Halle, R. C Woodruff, F. E. Wurgler, and S. Zimmering. 1984. "Chromosome Mutation Tests for Mutagenesis in *Drosophila melanogaster*. A Report of the U.S. Environmental Protection Agency Gene-Tox Program." *Mut. Res.* 133:61–88.

35. Sidney Green, Angela Auletta, Jill Fabricant, Robert Kapp, Madhu Manandhar, Ching-Ju Sheu, Janet Springer, and Brad Whitfield. 1985. "Current Status of Bioassays in Genetic Toxicology—the Dominant Lethal Assay. A Report of the U.S. Environmental Protection Gene-Tox Program." *Mut. Res.* 154:49–67.

36. Larry D. Kier, David J. Brusick, Angela E. Auletta, Elizabeth S. Von Halle, Vincent F. Simmon, Mary M. Brown, Virginia C. Dunkel, Joyce McCann, Kristien Mortelmans, Michael J. Prival, T. K. Rao, and Verne A. Ray. 1986. "The *Salmonella typhimurium*/Mammalian Microsome Mutagenicity Assay. A Report of the U.S. Environmental Protection Agency Gene-Tox Program." *Mut. Res.* 168:67–238.

37. S. Nesnow, M. Argus, H. Bergman, K. Chu, C. Frith, T. Helmes, R. McGaughy, V. Ray, T. J. Slaga, R. Tennant, and E. Weisburger. 1987. "Chemical Carcinogens: A Review and Analysis of the Literature of Selected Chemicals and the Establishment of the Gene-Tox Carcinogen Data Base. A Report of the U.S. Environmental Protection Agency." *Mut. Res.* 185:1–195.

III. Phase II Assessment Panel Reports

1. L. B. Russell, C. S. Aaron, F. de Serres, W. M. Generoso, K. L. Kannan, M. Shelby, J. Springer, and P. Voytek. 1984. "Evaluation of Mutagenicity Assays for Purposes of Genetic Risk Assessment. A Report of Phase II of the U.S. Environmental Protection Agency Gene-Tox Program." *Mut. Res.* 134:143–57.

2. David Brusick and Angela Auletta. 1985. "Developmental Status of Bioassays in Genetic Toxicology. A Report of Phase II of the U.S. Environmental Protection Agency Gene-Tox Program." *Mut. Res.* 153:1–10.

3. V. A. Ray, L. D. Kier, K. L. Kannan, R. T. Haas, A. E. Auletta, J. S. Wassom, S. Nesnow, and M. D. Waters. 1987. "An Approach to Identifying Specialized Batteries of Bioassays for Specific Classes of Chemicals: Class Analysis Using Mutagenicity and Carcinogenicity Relationships and Phylogenetic Concordant and Discordant Patterns. I. Composition and Analysis of the

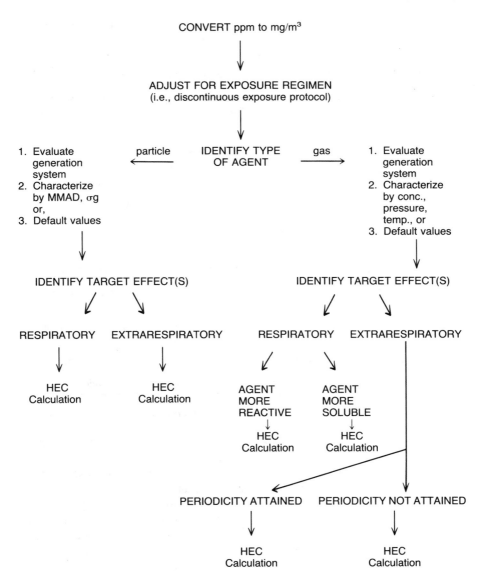

Figure G.1   Flowchart for HEC calculation. Adapted from Figure 4.1 (p. 4–11) of *Interim Methods for Development of Inhalation Reference Doses.*

on the lung (i.e., portal-of-entry or pulmonary effects), the equivalent dose across species is considered to be mass (mg) of toxic agent per unit (cm^2) surface area of the affected lung region (e.g., extrathoracic, tracheobronchial, pulmonary, thoracic, or total). However, in the case of particulates that exert extrapulmonary

Table G.1 Equations for Calculation of the HEC for Particulates That Exert Respiratory Effects

$$NOAEL_{HEC} \ (mg/m^3) = NOAEL_{ADJ} \ (mg/m^3) \times RDDR$$

where $NOAEL_{HEC}$ = the NOAEL human equivalent concentration
$NOAEL_{ADJ}$ = the NOAEL adjusted for exposure duration
$RDDR$ = $(RDD)_A/(RDD)_H$, the ratio of the regional deposited dose in animal species to that of humans for the region of interest for the toxic effect

$$RDD = \frac{10^{-6}YV_Tf}{S} \sum_{i=1}^{n} P_iE_i$$

where RDD = regional deposited dose (mg/cm^2 respiratory tract region/min)
n = number of size ranges
Y = exposure level (mg/m^3)
V_T = tidal volume (mL)
f = breathing frequency
S = regional surface area (cm^2) of toxic effect observed

For the ith size range of an exposure aerosol with a given particle diameter and σ_g,
P_i = the particulate mass fraction in that size range
E_i = the deposition efficiency for the species and respiratory tract region (i.e., extrathoracic, tracheobronchial and/or pulmonary, or total of interest)

(1)   The RDD is calculated and summed for regions of lung of interest, based on where the toxic effect is exerted. For example, for an effect confined to the nasal turbinate, only the RDD_{ET} (extrathoracic) would be calculated for both animals and humans.
(2)   Assumption: The equivalent dose across species is aerosol mass (mg) deposited per surface area (cm^2) of the relevant respiratory tract area.
(3)   Details concerning derivation of RDD values for humans and animals and the associated assumptions are described in the source document and its appendices.

Note: These are Equations 4.5 and 4.4, respectively, in the RfD$_i$ document.

effects, the equivalent dose across species is considered to be mass (mg) of particulate deposited per unit (kg) body weight. The equations used to calculate the HEC for particulates that exert their effects on the respiratory tract and for those that exert extrapulmonary effects are outlined in Tables G.1 and G.2, respectively.

In addition to pulmonary versus extrapulmonary effects, other important considerations for gases include the solubility of the gas in blood, reactivity with lung tissue, and periodicity. A gas is considered to display periodicity when alveolar blood concentrations are periodic with respect to time for most of exposure duration. In cases where the agent is a soluble gas with pulmonary effects or a gas that displays periodicity and exerts extrapulmonary effects, then blood-to-air partition coefficients become the most important factor to consider in dosimetric adjustments. For a gas that exerts extrapulmonary effects and does not display periodicity, the major factor used in dosimetric adjustments is the ratio of the alveolar ventilation rate divided by body weight of the animal species to the same parameters for humans. The equations used to calculate the HEC for gases that exert respiratory and extrarespiratory effects are outlined in Tables G.3 to G.5.

The major intent of this appendix was to provide an overview and introduction to the RfC. For details concerning the calculations and their experimental basis as

Table G.2 Equations for Calculation of the HEC for Particulates That Exert Extrarespiratory Effects

$$NOAEL_{HEC} \ (mg/m^3) = NOAEL_{ADJ} \ (mg/m^3) \times RDDR_{ER}$$

where $NOAEL_{HEC}$ = the NOAEL human equivalent concentration
$NOAEL_{ADJ}$ = the NOAEL adjusted for exposure duration
$RDDR_{ER}$ = $(RDD_{ER})_A/(RDD_{ER})_H$, the ratio of the dose available for uptake from the entire respiratory system of the animal model to that of humans

$$RDD_{ER} = \frac{10^{-6}YV_Tf}{S} \sum_{i=1}^{n} P_iE_i$$

where RDD_{ER} = extrarespiratory regional deposited dose (mg/kg of body weight per minute)
n = number of size ranges
Y = exposure level (mg/m^3)
V_T = tidal volume (mL)
f = breathing frequency
BW = body weight (kg)
P_i = the particulate mass fraction in the exposure size distribution (MMAD, σ_g)
E_i = the deposition efficiency of that size distribution (MMAD, σ_g) in the entire respiratory tract for the species of interest

(1) Assumption: For particulates that exert extrarespiratory (systemic) effects, the equivalent dose across species is mass of particulate (mg) deposited per unit body weight (kg).

(2) Assumption: 100% of the deposited dose to the entire respiratory tract is available for uptake to the systemic circulation. Distribution and clearance parameters are not yet incorporated into the equations. It is cautioned in the RfD_i document that "this assumption may result in slightly less conservative HEC estimates than using retained dose and accounting for differential uptake from various respiratory regions, but is more accurate than using exposure concentration."

(3) Technical details and exceptions and limitations of this approach are described in the RfD_i document.

Note: These are Equations 4.7 and 4.6, respectively, in the RfD_i document.

well as a complete discussion of the technical issues, the interested reader should consult the source document, *Interim Methods for Development of Inhalation Reference Doses*. It is emphasized in this document that the methodologies presented should be considered dynamic and extensive discussion is provided on both the limitations of the current approach and areas where additional research and clarification are needed.

Table G.3 Equations for Calculation of the HEC for Gases That Are More Reactive and Exert Respiratory Effects

$$\text{NOAEL}_{\text{HEC}} \; (\text{mg/m}^3) = \text{NOAEL}_{\text{ADJ}} \; (\text{mg/m}^3) \times \text{RGDR}$$

where $\text{NOAEL}_{\text{HEC}}$ = the NOAEL human equivalent concentration
$\text{NOAEL}_{\text{ADJ}}$ = the NOAEL adjusted for exposure duration
RGDR = $(\text{RGD})_\text{A}/(\text{RGD})_\text{H}$, the ratio of the regional gas dose in the animal species to that of humans for the region of interest for the toxic effect

$$\text{RDD} = \frac{10^{-6} Y V_\text{T} f}{S}$$

where RGD = regional gas dose
(mg/cm^2 respiratory tract region/min)
Y = exposure level (mg/m^3)
V_T = tidal volume (mL)
f = breathing frequency
S = regional surface area (cm^2) of toxic effect observed

(1)   The RGD is calculated and summed for regions of lung of interest, based on where the toxic effect is exerted. For example, for an effect confined to the conducting airways and alveoli (i.e., a "lung effect"), the RGD_{TH} (thoracic), which is the sum of the RGD_{TB} (tracheobronchial) and RGD_{PU} (pulmonary), would be calculated for both animals and humans.
(2)   This approach is recommended for gases and vapors that are very reactive and exert their toxic effect on the lung and is analogous to the approach for particulates that exert pulmonary effects.
(3)   Assumption: The equivalent dose across species is mass (mg) of toxic agent per surface area (cm^2) of the relevant respiratory tract area.
(4)   Assumption: The entire inspired concentration goes to the region of concern, whereas not all inspired gas is necessarily deposited.
(5)   Dosimetric issues with respect to gases as well as issues concerning reactivity and periodicity are discussed extensively in the RfD$_\text{i}$ document and its appendices.

Note: These are Equations 4.9 and 4.8, respectively, in the RfD$_\text{i}$ document.

Table G.4 Equation for Calculation of the HEC for Gases That Are More Soluble and Exert Respiratory Effects and for Gases That Exert Extrarespiratory Effects Where Periodicity Is Attained

$$\text{NOAEL}_{\text{HEC}} \ (\text{mg/m}^3) = \text{NOAEL}_{\text{ADJ}} \ (\text{mg/m}^3) \times \Gamma_A/\Gamma_H$$

where $\text{NOAEL}_{\text{HEC}}$ = the NOAEL human equivalent concentration
$\text{NOAEL}_{\text{ADJ}}$ = the NOAEL adjusted for exposure duration
Γ_A/Γ_H = the ratio of the blood to air partition coefficient of the chemical for the animal species to the human value, used only if $\Gamma_A \le \Gamma_H$

(1) For gases that are significantly more soluble in the blood relative to their reactivity with lung tissue, the above approach is recommended as a default "to account for uptake into the systemic circulation which may have decreased the amount of gas causing a direct effect in the lung and to account for the concentration available to the lung via blood circulation."

(2) The above equation was derived based on a PB-PK model derived and described in detail in an appendix to the RfD_i document.

(3) Assumption: "The toxic effects observed are related to the arterial blood concentration of the inhaled compound and that $\text{NOAEL}_{\text{HEC}}$s should be such that the human time-integrated arterial blood concentration is less than or equal to that of the exposed laboratory animal. This latter assumption is equivalent to assuming that time-average concentrations are equal to the equilibrium concentration adjusted for exposure duration."

(4) When $\Gamma_A > \Gamma_H$ or when Γ values are not known, a default value of $\Gamma_A/\Gamma_H = 1$ is recommended.

(5) Use of this equation is recommended if the conditions of periodicity are met during most of the exposure duration. It is stated that "if this condition is met for nine tenths of the time (e.g., periodic during the last 90 weeks of a 100-week experiment), then estimates of average concentrations will be in error by less than 10%."

(6) Issues relevant to the relationship of periodicity to partition coefficients and the influence of exposure duration and patterns are discussed in detail in the RfD_i document.

Note: This is Equation 4.10 in the RfD_i document.

Table G.5 Equation for Calculation of the HEC for Gases That Exert Extrarespiratory Effects Where Periodicity Is Not Attained

$$\text{NOAEL}_{HEC} \ (mg/m^3) = \text{NOAEL}_{ADJ} \ (mg/m^3) \times \frac{(V_A/BW)_A}{(V_A/BW)_H}$$

where NOAEL_{HEC} = the NOAEL human equivalent concentration
NOAEL_{ADJ} = the NOAEL adjusted for exposure duration

$\dfrac{(V_A/BW)_A}{(V_A/BW)_H}$ = the ratio of the alveolar ventilation rate (mL/min) divided by BW (kg) of the animal species to the same parameters for humans

(1)   This is a default approach intended for use when periodicity during 10% of exposure duration is suspected not to have been attained. It is associated with a higher degree of uncertainty relative to other approaches, and it is recommended that consideration be given to use of the modifying factor (MF).

(2)   The alveolar ventilation rate term is used to account for the volume of the respiratory tract where no gas exchange occurs, i.e., "physiologic dead space."

Note: This is Equation 4.11 in the RfD$_i$ document.

APPENDIX H: EXAMPLES OF LOAEL OR NOAEL SELECTION WHEN MULTIPLE RESPONSES ARE AVAILABLE

Effect Level	Species		
(mg/m^3)	Dog	Rat	Mouse
Example 1:			
LOAEL	100	120	
NOAEL	50	60	80
Comments:	Given the same critical effect, the proper choice is generally the highest dog NOAEL of 50 mg/m^3 since the potential experimental threshold in dogs (i.e., the potential LOAEL) may be below the highest NOAELs in both rats and mice.		
Example 2:			
LOAEL	120	100	90
NOAEL	90	75	
Comments:	Given the same critical effect, the proper choice is generally the mouse LOAEL of 90 mg/m^3, since the potential experimental threshold in mice may be less than the highest NOAELs for both dogs and rats. Judgment is needed in this example to insure that the adverse affects seen in all three species are truly minimal. For example, if any of the LOAELs in the species represented an increase in mortality, no firm basis for the development of a criterion exists. This is based on the general observation that mortality data are far removed quantitatively from chronic LOAELs and NOAELs, and thus, the data base has failed to establish the likely experimental threshold for the most sensitive endpoint		

Note: The responses, i.e., the numerical dose estimates, are adjusted to be dosimetrically equivalent between species prior to making comparisons such as those described in the above examples. These examples were taken directly from Appendix G of the document, *Interim Methods for the Development of Inhalation Reference Doses* (EPA/600/8-88/066F; April 1989).

Example 3:

LOAEL	75	80	90
NOAEL			

Comments: Given the same critical effect, the proper choice is generally the dog LOAEL of 75 mg/m^3, since by definition this represents the most sensitive species (see, however, the caution in example 2).

Example 4:

LOAEL	—	—	—
NOAEL	100	90	120

Comments: The proper choice is generally the highest rat NOAEL of 90 mg/m^3, since no assurance exists that the experimental threshold in rats is not below the highest NOAELs of both dogs and mice. This situation is unusual and should be judged carefully; since an LOAEL has not been determined, the criterion may be unduly conservative. Strict interpretation of this example might lead to strikingly lower criteria if other species are tested at much lower doses. Such criteria may not be appropriate.

CAS NUMBER INDEX

50–00–0	Formaldehyde
56–23–5	Carbon tetrachloride
64–19–17	Acetic acid
67–56–1	Methyl alcohol (methanol)
67–63–0	Isopropyl alcohol (isopropanol)
67–66–3	Chloroform
71–36–6	n-Butyl alcohol
71–43–2	Benzene
71–55–6	1,1,1-trichloroethane (methyl chloroform)
74–83–9	Methyl bromide
74–86–2	Acetylene
74–87–3	Methyl chloride
74–90–8	Hydrogen cyanide
75–01–4	Vinyl chloride
75–05–8	Acetonitrile
75–09–2	Methylene chloride (dichloromethane)
75–15–0	Carbon disulfide
75–21–8	Ethylene oxide
75–35–4	Vinylidene chloride (1,1-dichlorethylene)
75–56–9	Propylene oxide
75–65–0	t-Butyl alcohol
75–86–5	Acetone cyanohydrin
77–64–1	Acetone
78–59–1	Isophorone
78–84–2	Isobutyraldehyde
78–87–5	1,2-Dichloropropane
78–92–92	s-Butyl alcohol
78–93–3	Methyl ethyl ketone
79–01–6	Trichloroethylene (TCE)
79–06–1	Acrylamide
79–09–4	Propionic acid
79–10–7	Acrylic acid
80–62–6	Methyl methacrylate
84–74–2	Dibutyl phthalate
85–44–9	Phthalic anhydride
85–68–7	Butyl benzyl phthalate
91–20–3	Napthalene
94–36–0	Benzoyl peroxide
95–47–6	o-xylene (*See* Xylenes)
96–33–3	Methyl acrylate
96–45–7	Ethylene thiourea

98–82–8	Cumene
100–41–4	Ethyl benzene
100–42–5	Styrene
100–44–7	Benzyl chloride
100–51–6	Benzyl alcohol
101–68–8	Diisocyanate diphenyl methane
103–11–7	2-Ethylhexyl acrylate
105–60–2	Caprolactam
106–42–3	p-xylene (*See* Xylenes)
106–46–3	Resorcinol
106–89–8	Epichlorohydrin
106–99–0	1,3-Butadiene
107–02–8	Acrolein
107–05–1	Allyl chloride
107–06–2	Ethylene dichloride (1,2-dichloroethane)
107–13–1	Acrylonitrile
107–15–3	Ethylenediamine
107–21–1	Ethylene glycol
107–30–2	Chloromethyl methyl ether (CMME)
108–10–1	Methyl isobutyl ketone (MIBK)
108–31–6	Maleic anhydride
108–38–3	m-xylene (*See* Xylenes)
108–88–3	Toluene
108–90–7	Chlorobenzene
108–94–1	Cyclohexanone
108–95–2	Phenol
110–54–3	n-Hexane
111–42–2	Diethanolamine (DEA)
111–44–4	Bis-(2-chloroethyl)ether (BCEE)
115–07–7	Propylene
123–31–9	Hydroquinone
123–91–1	1,4-Dioxane
124–40–3	Dimethylamine
127–18–4	Tetrachloroethylene (perchloroethylene, PCE)
131–11–3	Dimethyl phthalate
140–88–5	Ethyl acrylate
141–32–2	Butyl acrylate
141–79–7	Mesityl oxide
302–01–2	Hydrazine
542–88–1	Bis-(chloromethyl)ether (BCME)
557–34–6	Zinc acetate
1306–19–0	Cadmium oxide (*See* Cadmium and compounds)
1309–64–4	Antimony trioxide (*See* Antimony and compounds)
1310–93–2	Sodium hydroxide
1313–13–9	Manganese dioxide (*See* Manganese and compounds)

1314–13–2	Zinc oxide
1314–60–9	Antimony pentoxide (*See* Antimony and compounds)
1314–62–1	Vanadium pentoxide
1315–04–4	Antimony pentasulfide (*See* Antimony and compounds)
1327–53–3	Arsenic trioxide (*See* Arsenic and compounds)
1330–20–7	Xylenes
1332–21–4	Asbestos
1345–04–6	Antimony trisulfide (*See* Antimony and compounds)
1836–75–5	Nitrofen
7439–92–1	Lead and compounds
7439–96–5	Manganese and compounds
7439–97–6	Mercury and compounds
7440–02–0	Nickel and compounds
7440–36–0	Antimony and compounds
7440–38–2	Arsenic and compounds
7440–43–9	Cadmium and compounds
7440–47–3	Chromium (IV) compounds (*See* Chromium and compounds)
7440–48–4	Cobalt and compounds
7440–50–8	Copper and compounds
7637–07–2	Boron trifluoride
7646–79–9	Cobaltous chloride (*See* Cobalt and compounds)
7647–01–0	Hydrochloric acid
7647–18–9	Antimony pentachloride (*See* Antimony and compounds)
7664–38–2	Phosphoric acid
7664–41–7	Ammonia
7664–93–9	Sulfuric acid
7697–37–2	Nitric acid
7726–95–6	Bromine
7733–02–0	Zinc sulfate
7757–82–6	Sodium sulfate
7782–50–5	Chlorine
7785–87–7	Manganese sulfate (*See* Manganese and compounds)
8018–01–7	Mancozeb
10025–91–9	Antimony trichloride (*See* Antimony and compounds)
10124–43–3	Cobaltous sulfate (*See* Cobalt and compounds)
10141–05–6	Cobaltous nitrate (*See* Cobalt and compounds)
10210–68–1	Cobalt carbonyl (*See* Cobalt and compounds)
12001–28–4	Crocidolite asbestos (*See* Asbestos)
12001–29–5	Chrysotile asbestos (*See* Asbestos)
12035–72–2	Nickel subsulfide
12122–67–7	Zineb
12172–73–5	Amosite asbestos (*See* Asbestos)
12427–38–2	Maneb
13463–39–3	Nickel carbonyl (*See* Nickel and compounds)
13463–67–7	Titanium dioxide

16065–83–3   Chromium (III) compounds (*See* Chromium and compounds)
16842–03–8   Cobalt hydrocarbonyl (*See* Cobalt and compounds)
25013–15–4   Methyl styrene (vinyl toluene)

Subject Index

AALs. *See* Ambient air levels
Acetic acid (ethanoic acid), 81–83
Acetone, 11, 28, 84–87
Acetone cyanohydrin
  (2-methyl-lactonitrile), 89–90
Acetonitrile (methyl cyanide), 91–93
Acetylene (ethine, ethyne, narcylen),
  95–96
ACGIH. *See* American Conference of
  Governmental Industrial Hygienists
Acroleic acid, 107–110
Acrolein (2-propenal), 97–100
Acrylamide monomer, 101–104
Acrylamide (propenamide, acrylamide
  monomer), 101–104
Acrylic acid (acroleic acid, ethylene
  carboxylic acid, 2-propenoic acid,
  vinyl formic acid), 107–110
Acrylonitrile (vinyl cyanide, propenitrile),
  10, 113–116
ADIs, 41
Agency for Toxic Substances and Disease
  Registry (ATSDR), 36, 46, 55, 61
  toxicological profiles, 26–28
Air toxics, 15–30
  chemical-specific documentation, 25–30
  documentation for occupational limits,
    23–24
  drinking water documents, 24–25
  Gene-Tox program, 20–21
  Integrated Risk Information System
    (IRIS), 18–20
  National Air Toxics Information
    Clearinghouse (NATICH), 18–19
  National Toxicology Program (NTP),
    21–22
  Registry of Toxic Effects of Chemical
    Substances (RTECS), 15–17
  regulation at the federal level, 4–5
  regulation at the state and local level,
    5–7
  REPROTOX database, 22
Allyl chloride (3-chloro-1-propene,
  3-chloropropene), 119–123

Alpha-chlorotoluene, 159–161
Ambient air level goals (AALGs). *See*
  Ambient air levels (AALs)
Ambient air levels (AALs), 6–12, 35–75
  AALG derivation for carcinogens,
    44–49
  AALG derivation for noncarcinogens,
    51–63
  adverse effects and thresholds of,
    39–40
  classification of toxicity, 36–37
  dose scaling and conversions, 40–42
  general issues, 37–44
  issues specific to certain noncarcinogenic
    endpoints, 63–71
  options for AALG derivation when only
    acute lethality data are available,
    71–75
  options for derivation when only acute
    lethality data are available, 71–75
  problems with approaches to AAL
    derivation, 8–12
  scope and limitations, 35–36
  secondary data sources, 15–30
  See also Air toxics
  use of genotoxicity data, 49–51
  use of OELs in derivation of, 7–8
Ambient water quality criteria documents
  (AWQCDs), 25
American Conference of Governmental
  Industrial Hygienists (ACGIH), 6–7,
  23–24
Ammonia (ammonia gas), 11, 125–126
Ammonia gas, 125–126
Antimony and compounds, 129–133
Arsenic and compounds, 10, 27–28,
  135–138
Asbestos, 28, 141–143
Association of Local Air Pollution Control
  Officials (ALAPCO), 18
ATSDR, 26–28; Agency for Toxic
  Substances and Disease Registry
Averaging times, guidelines for the
  selection of, 74–75

BACT. *See* Best available control technology
Barium, 28
BBP, 187–189
BCEE, 163–165
BCME, 167–169
Benzedine, 27
Benzene, 8–10, 27–28, 145–148
Benzene chloride, 229–232
1,4-Benzenediol, 361–364
Benzo(a)anthracene, 27
Benzo(a)fluoranthene, 27
Benzo(a)pyrene, 27–28
Benzoperoxide, 151–153
Benzoyl peroxide (benzoperoxide, benzoyl superoxide, dibenzoyl peroxide), 151–153
Benzoyl superoxide, 151–153
Benzyl alcohol (phenyl carbinol), 155–157
Benzyl chloride (alpha-chlorotoluene, chloromethylbenzene), 159–161
Beryllium, 27
Best available control technology (BACT), 6
Biethylene, 177–180
Bis(2-chloroethyl)ether (dichloroethyl ether, BCEE, 2,2-dichloroethyl ether), 27, 163–165
Bis(chloromethyl)ether (bisCME, BCME), 27, 167–169
BisCME, 167–169
Boron trifluoride (trifluoroborane), 171–172
Bromine, 173–174
Bromodichloromethane, 27
Bromomethane, 423–428
Butadiene, 177–180
1,3-Butadiene (butadiene, biethylene, erythene, divinyl), 177–180
n-Butanol, 191–193
1-Butanol, 191–193
2-Butanol, 195–196
t-Butanol, 199–201
2-Butanone, 443–445
Butyl 2-propenoate, 183–185
n-Butyl acrylate, 183–185
Butyl acrylate (n-butyl acrylate, butyl 2-propenoate), 183–185
sec-Butyl alcohol, 195–196

n-Butyl alcohol (n-butanol, normal butanol, 1-butanol), 191–193
s-Butyl alcohol (sec-butyl alcohol, secondary butanol, 2-butanol, 195–196
t-Butyl alcohol (t-butanol, 2-methyl-2-propanol), 199–201
Butyl benzyl phthalate (BBP), 187–189
n-Butyl phthalate, 267–271

Cadmium and compounds, 27–28, 203–206
CAG. *See* Carcinogen Assessment Group
Caloric demand, dose scaling and, 40–42
Caprolactam (2-oxohexamethylenimine), 209–212
Carbolic acid, 483–486
Carbon bisulfide, 213–216
Carbon disulfide (carbon bisulfide), 213–216
Carbon monoxide, 5
Carbon tetrachloride (tetrachloromethane, perchloromethane), 27–28, 219–222
Carboxyethane, 495–496
Carcinogen Assessment Group (CAG), 6, 41
Carcinogens, AALG derivation for, 44–49
CAS, 16
Caustic soda, 511–513
Chemical Abstracts Service (CAS), 16
Chemical Information System (CIS), 17, 21
Chemicals Manufacturers Association (CMA), 7
Chemical-specific documentation, 25–30
    ATSDR toxicological profiles, 26–28
    health assessment documents (HADs), 26
    IARC monographs, 25–26
Chlordane, 27–28
Chlorine (molecular chlorine), 223–227
3-Chloro-1-propene, 119–123
1-Chloro-2,3-epoxypropane, 297–300
Chlorobenzene (monochlorobenzene, MCB, phenylchloride, benzene chloride), 28, 229–232
Chloroethene, 27, 561–565
Chloroethylene, 561–565

Chloroform (trichloromethane), 27–28, 235–239
Chloromethane, 431–434
Chloromethylbenzene, 159–161
Chloromethyl methyl ether (dimethylchloroether, CMME), 241–243
3-Chloropropene, 119–123
Chromium and compounds, 27, 245–249
Chrysene, 27
Cinamene, 519–522
CIS, 17, 21
Clean Air Act of 1963, 4
CMME, 241–243
Cobalt and compounds, 251–255
Conversions, of AALGs, 40–42
Copper and compounds, 28, 257–259
Cresols, 28
Cumene (isopropyl benzene), 261–262
Current Intelligence Bulletins (CIBs), 24
Cyanide, 27–28
Cyclohexanone (pimelic ketone), 263–265

Database factor, guidelines for application of, 63
DBP, 267–271
DCP, 273–275
DEA, 277–279
Diamine, 349–352
1,2-Diaminoethane, 303–304
Dibenzoyl peroxide, 151–153
Dibutyl phthalate (n-butyl phthalate, DBP, di-n-butyl phthalate), 267–271
Dicarbomethoxyzinc, 579–581
Dichlormethane, 437–441
1,2-Dichloroethane, 27–28, 315–318
1,1-Dichloroethene, 27–28, 567–570
1,1-Dichloroethylene, 28, 567–570
Dichloroethyl ether, 163–165
2,2-Dichloroethyl ether, 163–165
2,4-Dichlorophenyl-4-nitrophenyl ether, 477–481
1,2-Dichloropropane (propylene dichloride, DCP), 27, 273–275
Diethanolamine (2, 2-iminodiethanol, DEA), 277–279
1,4-Diethylene dioxide, 293–296

Diisocyanate diphenyl methane (methylene bisphenyl isocyanate, MDI), 281–282
Dimethyl 1,2-benzenedicarboxylate, 289–291
Dimethylamine (N-methylmethanamine, DMA), 285–287
Dimethylchloroether, 241–243
Dimethyl phthalate (DMP, dimethyl 1,2-benzenedicarboxylate), 289–291
Di-n-butyl phthalate, 267–271
Dioxane, 293–296
p-Dioxane, 293–296
1,4-Dioxane (dioxane, p-dioxane, 1, 4-diethylene dioxide), 293–296
Disodium sulfate, 515–516
Dithane M-45, 389–391
Divinyl, 177–180
DMA, 285–287
DMP, 289–291
Documentation of the Threshold Limit Values (ACGIH), 23
Dose scaling, of AALGs, 40–42
Drinking Water and Health (NAS), 44
DWCDs, 46

Environmental Health Criteria (EHC), 28–28
    titles in series, 29–29
Environmental Protection Agency (EPA), 3–6, 8–12, 18, 20, 22, 25, 28, 36, 39–41, 44–46, 54–56, 60–64, 74–75
EPA. *See* Environmental Protection Agency
Epichlorohydrin (1-chloro-2,3-epoxypropane), 10, 297–300
Erythene, 177–180
1,2-Ethanediamine, 303–304
1,2-Ethanediol, 321–323
Ethanoic acid, 81–83
Ethine, 95–96
Ethoxycarbonylethylene, 305–309
Ethyl-2-propanoate, 305–309
Ethyl acrylate (ethoxycarbonylethylene, ethyl-2-propanoate), 305–309
Ethyl benzene (phenylethylene), 28, 311–314

Ethylene carboxylic acid, 107–110
Ethylenediamine (1,2-ethanediamine,
    1,2-diaminoethane), 303–304
Ethylene dibromide, 10
Ethylene dichloride (1,2-dichloroethane),
    315–318
Ethylene glycol (1,2-ethanediol), 11,
    321–323
Ethylene oxide (oxirane), 10, 325–328
Ethylene thiourea (2-imidazolidenethione),
    331–335
Ethyl formic acid, 495–496
2-Ethylhexyl 2 propenoate, 337–339
2-Ethylhexyl acrylate, 337–339
Ethyne, 95–96
ETU, 331–335, 396–397

Federal government, air toxics regulation
    and, 4–5
Federal Register (ATSDR), 27
Formaldehyde (HCHO), 10, 341–343
2,5-Furandione, 385–387

Gene-Tox program, 20–21, 36
Genotoxicity, 49–51
Glycol ethers, 28
Guidelines for Carcinogen Risk Assessment
    (EPA), 46
*Guidelines for Health Assessment of
    Suspect Developmental Toxicants*
    (EPA), 63
*Guidelines for Mutagenicity Risk
    Assessment* (EPA), 51

HADs. *See* Health assessment documents
HCHO, 341–343
HCN, 357–359
HEADs. *See* Health effects assessment
    documents
Health assessment documents (HADs),
    25–26, 36
Health effects assessment documents
    (HEADs), 26, 36
Hexachlorobenzene, 28
Hexachlorobutadiene, 28
Hexachlorocyclopentadiene, 28
n-Hexane, 11
Hexane, 345–347

Hexavalent chromium, 28
Hexone, 447–449
Hydrazine (diamine), 349–352
Hydrocarbons, 5
Hydrochloric acid (hydrogen chloride),
    353–355
Hydrocyanic acid, 357–359
Hydrogen chloride, 353–355
Hydrogen cyanide (hydrocyanic acid,
    HCN), 357–359
Hydrogen sulfate, 525–528
Hydroquinone, 361–364
Hydroxybenzene, 483–486

IARC. *See* International Agency for
    Research on Cancer
2-Imidazolidenethione, 331–335
2,2-Iminodiethanol, 277–279
Inhalation reference concentration (Rfc), 54
Inorganic nickel salts, 463–470
Integrated Risk Information System (IRIS),
    18–20, 36, 45
*Interim Methods for Development of
    Inhalation Reference Doses* (EPA), 44
International Agency for Research on
    Cancer (IARC), 17, 25–26, 36, 46,
    49–51
International Labor Organization, 29
International Program on Chemical Safety
    (IPCS), 28–29
IRIS. *See* Integrated Risk Information
    System
1,3-Isobenzofurandione, 491–493
Isobutenyl methyl ketone, 411–412
Isobutyl aldehyde, 365–367
Isobutyraldehyde (2-methyl propanol,
    isobutyl aldehyde), 365–367
Isophorone
    (3,5,5-trimethyl-2-cyclohexen-1-one),
    27, 369–372
Isopropanol, 375–377
Isopropyl alcohol (2-propanol,
    isopropanol), 375–377
Isopropylidene acetone, 411–412

Lead and compounds, 5, 27–28, 379–383
Lifespan, dose scaling and, 40–42
Lindane, 28

LOAEL. *See* Lowest observed adverse effect level
Local government, air toxics regulation in, 5–7
Lowest observed adverse effect level (LOAEL), 6, 54–63
Lye, 511–513

Maleic anhydride (2,5-furandione), 385–387
Mancozeb (manganese zinc complex ethylenebisdithiocarbamic acid, vondozeb, dithane M-45), 389–391
Maneb (manganese ethylene-1,2-bisdithiocarbamate), 393–398
Manganese and compounds, 28, 401–405
Manganese ethylene-1,2-bisdithiocarbamate, 393–398
Manganese zinc complex ethylenebisdithiocarbamic acid, 389–391
Maximum contaminant level goals (MCLGs), 24–25, 36
MC, 545–549
MCB (monochlorobenzene, MCB, phenylchloride, benzene chloride), 28, 229–232
MCLGs. *See* Maximum contaminant level goals
MDI, diisocyanate diphenyl methane (methylene bisphenyl isocyanate, MDI), 281–282
MEK, Methyl ethyl ketone (2-butanone, MEK), 28, 443–445
Mercury and compounds, 27–28, 407–410
Mesityl oxide (4-methyl-3-pentene-2-one, isobutenyl methyl ketone, isopropylidene acetone), 411–412
Methanol, 419–422
Methyl 2-methyl 2-propanoate, 451–453
2-Methyl-2-propanol, 199–201
4-Methyl-3-pentene-2-one, 411–412
Methylacetic acid, 495–496
Methyl acrylate (methyl propenate), 415–417

Methyl alcohol (methanol, wood alcohol), 419–422
Methylbenzene, 541–543
Methyl bromide (monobromomethane, bromomethane), 423–428
Methyl chloride (chloromethane, monochloromethane), 431–434
Methylchloroform, 545–549
Methyl cyanide, 91–93
Methylene bisphenyl isocyanate, 281–282
Methylene chloride (dichlormethane), 27–28, 437–441
Methylethene, 499–501
Methylethylene, 499–501
Methyl ethyl ketone (2-butanone, MEK), 28, 443–445
Methyl isobutyl ketone (hexone, MIBK), 447–449
2-Methyl-lactonitrile, 89–90
Methyl methacrylate (methyl 2-methyl 2-propanoate, MMA), 451–453
N-Methylmethanamine, 285–287
Methyloxidrane, 503–506
2-Methyl propanol, 365–367
Methyl propenate, 415–417
Methyl styrene (vinyl toluene, tolyethylene), 455–457
MIBK, 447–449
Mine Safety and Health Administration (MSHA), 24
MMA, 451–453
Molecular chlorine, 223–227
Monobromomethane, 423–428
Monochlorobenzene, 229–232
Monochloromethane, 431–434
Multimedia exposure, and relative source contribution, 73–74
Multispecies regression technique, 40

NAAQS. *See* National ambient air quality standards
Naphthalene (napthalene, naphthene), 11, 28, 459–462
Narcylen, 95–96
NAS. *See* National Academy of Sciences
NATICH. *See* National Air Toxics Information Clearinghouse
National Academy of Sciences (NAS), 35
Safe Drinking Water Committee, 44

National Air Toxics Information
 Clearinghouse (NATICH), 5, 18–19
National ambient air quality standards
 (NAAQS), 4–5
National Institute for Occupational Safety
 and Health (NIOSH), 15–17, 23–24,
 36
National Research Council (NRC), 12, 35
National Toxicology Program (NTP), 17,
 21–22
n-Hexane, 345–347
Nickel and compounds (inorganic nickel
 salts), 10, 27–28, 463–470
Nitric acid (red fuming nitric acid, white
 fuming nitric acid), 473–475
Nitrofen (2,4-dichlorophenyl-4-nitrophenyl
 ether), 477–481
Nitrogen dioxide, 5
Noncarcinogens,
 AALG derivation for, 51–63
 issues specific to endpoints, 63–71
No observed adverse effect level
 (NOAEL), 39, 54–63
Normal butanol, 191–193
NTP. *See* National Toxicology Program

Occupational exposure levels (OELs), 6–8
 use in derivation of ambient air levels,
 7–8
Occupational Safety and Health
 Administration (OSHA), 6
Octyl acrylate, 337–339
OELs. *See* Occupational exposure levels
 (OELs)
Office of Pesticides and Toxic Substances
 (OPTS), 20
Office of Testing and Evaluation (OTE),
 20
Orthophosphoric acid, 489–490
Oxidants, 5
Oxirane, 325–328
Ozone, 5

PAN, 491–493
Particulate matter, 5
PB-PK models, 44
PCBs, 10

PCE, 531–534
PELs, 6
Perchlorethylene, 531–534
Perchloromethane, 219–222
Permissible exposure limits (PELs), 6
Phenanthrene, 28
Phenol (carbolic acid, hydroxybenzene),
 11, 27–28, 483–486
Phenyl chloride, 229–232
Phenyl ethylene, 519–522
Phenylethylene (ethyl benzene), 311–314
Phenylmethane, 541–543
Phosphoric acid (orthophosphoric acid),
 489–490
Phthalic anhydride (1, 3-isobenzo-
 furandione, PAN), 491–493
Physiologically based pharmacokinetic
 models (PB-PK), 44
Polychlorinated aromatic hydrocarbons
 (PAHs), 28
Polycyclic biphenyls (PCBs), 28
Principal studies, in derivation of AALGs,
 42–44
2-Propanol, 375–377
Propenamide, 101–104
Propene, 499–501
Propene oxide, 503–506
Propenitrile, 113–116
2-Propenoic acid, 107–110
2-Propenol, 97–100
Propionic acid (methylacetic acid, ethyl
 formic acid, carboxyethane), 495–496
*Proposed Guidelines for Assessing Female
 Reproductive Risk* (EPA), 66–67
*Proposed Guidelines for Assessing Male
 Reproductive Risk* (EPA), 66–67
Propylene dichloride, 273–275
1,2-Propylene oxide, 503–506
Propylene oxide (methyloxidrane, propene
 oxide, 1,2-propylene oxide), 503–506
Propylene (propene, methylethene,
 methylethylene), 499–501

Reasonably available control technology
 (RACT), 6
Recommended exposure limits (RELs), 6
Red fuming nitric acid, 473–475
Registry of Toxic Effects of Chemical
 Substances (RTECS), 15–17, 36–38

Regulation. *See* Air toxics regulation
RELs, 6
Reportable Quantity Documents (RQDs),
    28
Reproductive toxicity, 66–68
Reproductive Toxicology Center (RTC), 22
REPROTOX database, 22
Resorcinol, 507–509
*Risk Assessment in the Federal
    Government, Managing the Process*
    (NRC), 12
RTECS. *See* Registry of Toxic Effects of
    Chemical Substances

SARA, 11–12; Superfund Amendments
    and Reauthorization Act
Scaling factor, 40
Scientifically defensible AALs,
    development of, 35–75
    See also Ambient air levels
Secondary butanol, 195–196
Selenium, 27
Sensory irritation, 68–71
Sodium cyanide, 28
Sodium hydroxide (caustic soda, lye),
    511–513
Sodium sulfate (disodium sulfate),
    515–516
STAPPA. *See* State and Territorial Air
    Pollution Program Administrators
State government, air toxics regulation
    and, 5–7
State and Territorial Air Pollution Program
    Administrators (STAPPA), 18
Styrene (phenyl ethylene, vinyl benzene,
    cinnamene), 519–522
Sulfur dioxide, 5
Sulfuric acid (hydrogen sulfate), 525–528
Superfund Amendments and
    Reauthorization Act (SARA), 11–12
Supporting studies, in derivation of
    AALGs, 42–44

TCE, 551–554
Tetrachloroethene, 531–524
Tetrachloroethylene (perchlorethylene,
    tetrachloroethene, PCE), 27, 531–534
Tetrachloromethane, 219–222

Threshold limit values (TLVs), 6–8, 23
*Threshold Limit Values and Biological
    Exposure Indices for 1989–90*
    (ACGIH), 23
Titanium dioxide (titanium oxide),
    531–534
Titanium oxide, 531–534
TLVs. *See* Threshold limit values
Toluene (toluol, methylbenzene,
    phenylmethane), 11, 27–28, 541–543
Toluol, 541–543
Tolyethylene, 455–457
Toxicity, AALG classification of, 36–37
1,1,1-Trichloroethane (methylchloroform,
    MC), 545–549
Trichloroethylene (1, 1, 2-trichloro-
    ethylene, trichloroethene, TCE), 27,
    551–554
1,1,2-Trichloroethylene, 551–554
Trifluoroborane, 171–172
3,5,5-Trimethyl-2-cyclohexen-1-one,
    369–372
Trivalent chromium, 28

Uncertainty factors, 6, 60
United Nations Environment Program, 29

Vanadium oxide, 557–559
Vanadium pentoxide (vanadium oxide),
    557–559
VDC, 567–570
Vinyl benzene, 519–522
Vinyl chloride (chloroethylene,
    chloroethene), 10, 27–28, 561–565
Vinyl cyanide, 113–116
Vinyl formic acid, 107–110
Vinylidene chloride (1,1-dichloroethylene,
    VDC, 1,1-dichloroethene), 567–570
Vinyl toluene, 455–457
Vondozeb, 389–391

White fuming nitric acid, 473–475
White vitriol, 587–589
Wood alcohol, 419–422
World Health Organization (WHO), 25, 29

Xylenes, 11, 28, 573–577

Zinc acetate (zinc diacetate,
    dicarbomethoxyzinc), 27–28, 579–581
Zinc diacetate, 27–28, 579–581
Zinc ethylenebisdithiocarbamate, 27–28,
    591–594
Zinc oxide, 27–28, 583–585

Zinc sulfate (zinc vitriol, white vitriol),
    27–28, 587–589
Zinc vitriol, 27–28, 587–589
Zineb (zinc ethylenebisdithiocarbamate),
    591–594

Name Index

Aaron, A.C., 266

ACGIH, 81–82, 87, 91, 93, 95, 97, 99, 101, 103, 107, 109–110, 113, 115, 119, 125–126, 129, 135, 137, 141–142, 145, 147–148, 151–153, 239, 243, 250, 255, 262, 265, 271, 275, 279, 282, 287, 291, 296, 300, 344, 347, 352, 355, 359, 364, 372, 377, 383, 387, 405, 410, 413, 417, 422, 428, 434, 441, 445, 449, 454, 457, 461, 470, 475, 486, 490, 493, 497, 501, 506, 509, 513, 522, 529, 534, 539, 543, 549, 554, 559, 564, 570, 577, 585

Adams, E.M., 119, 122, 549

ADIs, 41

Adrianova, M.M., 594

Agrawal, A.K., 102

Alarie, Y., 68, 529

Aldridge, W.N., 40

Alexeeff, G.V., 429

Ambrose, A.M., 481

Amdur, M., 83

Amdur, M.O., 529

American Industrial Hygiene Association, 490

Ames, S.R., 364

Amoore, J.E., 82, 93, 110, 276, 279, 301, 348, 356, 359, 373, 387, 413, 462, 497, 523

Anderson, B., 364

Anderson, E.L., 41–42, 48, 116, 133

Anger, W.K., 429

Arcuri, P.A., 517

Argus, M., 21

Armeli, G., 449

Ashby, J., 49

ATSDR, 146–148, 250, 359, 410, 442

Aughey, E., 589

Auletta, A., 20–21, 68, 71

Baldi, G., 82–83

Balin, P.N., 398

Bardodej, Z., 523

Barlow, S.M., 86

Barrett, J., 82

Barrow, C.S., 355

Beliles, R.P., 523, 535

Belilies, R.P., 555

Bell, Z.G., 555

Belyaeva, A.P., 131

Bernard, J.R., 78

Beyer, K. H., 279

Bierma, T., 9–11

Bigwood, E.J., 53

Bingham, E., 471

Bionetics Research Labs, 260

Bomski, H., 240

Bond, J.A., 523

Boorman, G.A., 428

Bornmann, G., 272

Bornschein, R.L., 442

Borxelleca, J.F., 454

Boxenbaum, H., 41

Boyland, E., 364

Brent, R.L., 40

Brieger, H., 131–132

Brown, C.C., 136, 138

Brown, H.S., 6–7, 57

Brown, N.A., 66

Bruce, R.M., 486

Bruckner, J.V., 86

Brusick, D., 49

Bryan, W.R., 59

Bucher, J.R., 372

Bull, R.J., 101

Bunker, J.W.M., 59

Bure, J.D., 441

Burke Hurt, S.S., 481

Bus, J.S., 348

Calabrese, E.J., 4, 6, 39–40, 42, 53, 73, 383

Carcinogen Assessment Group (CAG), 41

Carlson, A.J., 364

Carpenter, C.P., 72, 109

Carson, B.L., 131–132, 255, 529, 540, 560, 581

Cavender, F.L., 348, 446
Challen, P.J.R., 240
Chang, J.C.F., 501
Chan, P.C., 454
Chemical Industry Institute of Toxicology, 387, 544
Chernoff, N., 65, 399
Chernov, O.V., 594
Chovil, A., 471
Clary, J.J., 471
Claussen, U., 98
Clayton, G.D., 577
Connolly, W.M., 6–7, 57
Coon, R.A., 126, 287
Copper, E.R., 300
Cosmides, G.J., 30
Costlow, R.D., 64
Cote, I.L., 107
Cromer, J., 454

Danse, L.H.J.C., 428
Dapson, S.C., 549
Davies, T.A., 130
Deacon, M.M., 446
Dearfield, K.L., 102–103
Deichmann, W.B., 486
Deizell, E., 113
DePass, L.R., 107–109
Dickens, F., 387
Diggle, W.M., 475
Dourson, M.L., 49, 60–61
Druckrey, H., 461
Dunkelburg, H., 506
Dunnick, J.K., 471

Elder, R.L., 272, 292
Elfimova, E.W., 355
El Ghawabi, S.H., 359
Elkins, H.B., 356, 449
Enterline, P.E., 138, 471
Eustis, S.L., 429
Evans, R.D., 59

Fabre, R., 262
Fabricant, R., 68, 71
Fabro, S., 66
Fairhall, L.T., 475

Federation of American Societies for Experimental Biology, 429
Feron, V.J., 97–98, 100
Figueroa, W.G., 243
Finklea, J.F., 529
Fiserova-Bergerova, V., 523
Fitzhugh, O.G., 59
Flaga, C., 72
Flickinger, C.W., 509
Fodor, G.G., 442
Foster, P.M.D., 272
Freireich, E.J., 42
Friedlander, B.R., 441
Fukuda, K., 555

Gabel, B.E.G., 66
Gage, J.C., 109, 367
Gamberale, F., 544, 577
Gamble, J., 529
Gargus, J.L., 243
Gehan, E.A., 42
Gehring, P.J., 42
Gerarde, H.W., 262
Ghanayem, B.I., 417
Ghetti, G., 462
Giavini, E., 296
Gilman, J.P., 260
Gleason, R.P., 260
Goulding, E.H., 66
Gray, E.L., 475
Gray, L.E., 481
Gray, T.J.B., 272
Green, S., 20, 68, 71
Gross, W.G., 250
Groth, D.H., 130, 132–133

Haas, R.T., 21
Hake, C.L., 523
Hamilton, A., 359
Hamm, T.E., 435
Hardin, B.D., 121, 428, 442, 506
Harris, R.S., 59
Hartung, R., 279, 296
Hatch, T.F., 40
Haun, C.C., 352
Hausler, M., 121
Hayes, W.C., 506
Hayes, W.J., 59

Hazelton Laboratory, 92
Hearne, F.T., 435
Heindel, J.J., 22
Heller, V.G., 486, 589
Hemminki, K., 523
Henderson, Y., 355
Heppel, L.A., 276
Hermann, E.R., 40
Higgins, I., 138
Hill, T.A., 40–42
Holmberg,, P.C., 523
Hood, R., 577
Hubbs, A.F., 428
Hudak, A., 544, 577

IARC, 97–100, 102, 109, 114–115,
    120–121, 136, 146–147, 151–152,
    240, 243–244, 276, 296, 344, 352,
    364 377, 398, 417, 428, 435, 442,
    454, 481, 506, 509, 523, 535, 555,
    565, 571, 594
Ikeda, G.J., 102
Infurna, R., 422
Innes, J.R.M., 398, 594
International Program on Chemical Safety,
    355
Irish, D.D., 429

Jersey, G., 523
Johansson, A., 260, 471
John, J.A., 121, 301, 565
Johnson, E.M., 64, 66
Johnson, K.A., 101, 104
Johnstone, R.T., 296
Jones, M.M., 581
Jones-Price, C., 486
Jones, W., 529
Jorgenson, T.A., 240

Kane, L.E., 356
Kankaanpaa, J.T.J., 523
Kao, J., 66
Kapeghian, J.C., 125
Kapp, M., 68
Kavlock, R.J., 65
Kawabata, A., 300
Kawano, M., 272
Kerfoot, E.J., 255

Kerns, W.D., 344
Kier, L.D., 21
Kimbrough, R.D., 481
Klaasen, C.D., 61, 63
Klassen, C.D., 359
Klimisch, H.J., 417
Klimkina, N.V., 109
Klucik, I., 132
Kociba, R.J., 296
Konishi, T., 300
Konzen, R.B., 283
Korshunov, S.F., 486
Kramer, C.G., 550
Krasavage, W.J., 348
Krasovskij, G.N., 40, 59
Kucera, J., 86
Kuman, S., 589
Kuna, R.A., 146, 148

Lam, H.F., 585
Land, P.E., 555
Landua, E., 137
Lane, R.W., 549
Lange, A., 147
Larsson, K.S., 391, 399
Laskin, S., 244, 300
Layton, D.W., 72, 367, 581
Lee-Feldstein, A., 138
Lee, K.P., 540
Lehman, A.J., 59, 292
Lehmann, K.B., 240
Leong, B.K.J., 244
Leukroth, R.W., 40–42
Lewis, C.E., 560
Lijinsky, L., 108
Lijinskyy, W., 266
Linari, F., 449
Litton Bionetics, Inc., 544
Low, L.K., 544
Lu, M-H, 391, 399
Lundborg, M., 260
Lynch, D.W., 506

Maatsushita, T., 86
McCollister, D.D., 49, 59, 61
MacEwen, J.D., 352, 449
Machle, W., 355
McNamara, B.P., 59, 72

Magnus, K., 471
Maita, H., 589
Majka, J., 109
Makarov, I., 454
Malcolm, D., 529
Mallon, B.J., 72
Maltoni, C., 501, 555, 565, 571, 577
Mancuso, T.F., 250
Mannan, K.L., 21
Manson, J.M., 64, 535
Mantel, N., 41, 59
Marcon, L.V., 399
Marks, T.A., 300, 348
Massaro, E.J., 405
Matokhnyuk, L.A., 399
MCA, 501
Mikov, L.E., 292
Miller, R.R., 109–110
Minor, J.L., 486
Mitsumori,K., 410
Miyagaki, H., 348
Monster, A.C., 555
Motoc, F., 90
Mukhitov, B., 487
Murray, F.A., 240
Murray, F.J., 115, 146, 523, 529, 571

Nagymajtenyi, L., 137–138
NAS/NRC, 356
NAS Safe Drinking Water Committee, 91
NCI, 119, 240, 296, 481, 486, 493, 523, 534, 540, 549, 555, 565
Nelson, B.K., 422, 535
Nelson, K.W., 266, 377, 446
Nesnow, S., 21
Nettesheim, P., 250
Newman, L.M., 64
Nicholas, C.A., 454
Nikonorow, M., 272
NIOSH, 81–82, 86–87, 89–93, 95–96, 101–103, 109–110, 113–114, 119, 121–122, 125–126, 129–132, 135, 141–142, 145, 147, 151–153, 240, 250, 255, 262, 265–266, 272, 279, 282, 287, 292, 296, 300–301, 344, 348, 352, 355, 359, 364, 367, 372, 377, 383, 387, 391, 399, 405, 410, 413, 417, 422, 428, 431, 434, 445,
449, 457, 461, 470, 475, 481, 486, 490, 493, 497, 506, 509, 513, 517, 522- 523, 529, 534, 540, 544, 549, 555, 560, 565, 570, 577, 581, 585, 589, 594
Nishiyama, K., 272, 405
Nitschke, K.D., 571
Nogawa, K., 406
NTIS, 398, 594
NTP, 91, 276, 279, 291, 364, 372, 405, 410, 428, 441, 454, 457, 461, 471, 501, 506, 509, 534, 544, 555, 577

O'Berg, M., 113, 116
O'Brien, I.M., 283
Oglesby, F.L., 86–87, 364
Ohanian, E.V., 25, 73, 113, 435, 470
O'Hara, G.P., 481
Oishi, S., 272
Olefir, A.I., 399
Oltramare, M., 523
Onda, S., 272
Oppenheimer, B.S., 454
OSHA, 81–82, 91, 93, 97, 101–102, 107, 119, 125, 141, 145, 240, 243, 255, 260, 265, 275, 279, 282, 287, 296, 300, 344, 348, 352, 359, 372, 377, 383, 405, 410, 413, 422, 428, 431, 434, 445, 449, 461, 470, 493, 497, 506, 509, 522, 534, 540, 544, 555, 560, 565, 570, 577, 585
Othanian, E.V., 129
Ott, M.G., 148, 435, 513, 570
Ottolenghi, A.D., 470

Paladja, M., 21
Parkhurst, H.J., 250
Parnegguabu, L., 82
Patel, J.M., 99
Patty, F.A., 356, 513
Pavkov, K.L., 434
Pazynich, V.M., 560
Pedigo, N.G., 255
Pepelko, W.E., 43–44
Peto, J., 471
Petrova-Vergieva, T., 399, 594
Pilinskaya, M.A., 594
Pilny, M.K., 300

Pimental, J.C., 260
Piotrowski, J.K., 487
Pistorius, D., 585
Plasterer, M.R., 461
Ponomarkov, V.I., 523
Pozzani, U.C., 92
Princi, F., 250

Quast, J.F., 114–115, 549
Quest, J.A., 501

Racz, G., 364
Radike, 565
Raleigh, R.L., 86
Rall, D.P., 42, 59
Rampy, L.W., 535
Ramsey, J.C., 42
Ray, V.A., 21
Rengstorff, R.H., 266
Repko, J.D., 435
Reproductive Toxicology Center, 86, 102, 131, 137, 142, 565
Reuzel, P.G.J., 506
Rinsky, R.A., 148
Roe, F.J.C., 240, 364
Rogers, J.M., 65
Rosenblatt, D.H., 72
Rosenkranz, H.S., 21
Rowan, C.A., 6–7, 57
Rowe, V.K., 373

Safe Drinking Water Committee, 145, 240, 255, 355, 391, 398, 405, 454, 481, 493, 509, 523, 560, 594
Saida, K., 446
Saksena, S., 581
Salaman, M.H., 100
Sandage, C., 486
Saric, M., 406
Savolainen, K., 577
Schardein, J.L., 64, 66
Schaumburg, H., 348
Schlesinger, R.B., 529
Schmeltz, I., 461
Schmidt, L.H., 42
Schneiderman, M.A., 41
Schwetz, B.A., 240, 442, 445, 535, 549
Segal, A., 107

Seidenberg, J.M., 517
Serota, D.G., 441
Sheu, C., 68
Shiota, K., 272
Shopp, G.M., 462
Short, R.D., 387, 571
Shull, G., 66
Sikov, M.R., 428
Silverman, L., 373, 413, 449
Silverman, M., 119
Singh, A.R., 108, 272
Skalka, P., 364
Skipper, H.E., 42
Small, M.J., 72
Smith, B.R., 594
Smith, C.C., 272
Smolik, R., 147
Smyth, H.F., 109, 279, 373, 413
Sobotka, T.J., 399
Springer, J., 68
Sprinz, H., 506
Stara, J.F., 49, 60
Steinhagen, W.H., 287
Stenback, F., 509
Stewart, R.D., 442, 523, 549
Stockinger, H.E., 40
Sunderman, F.W., 470–471
Suzuki, Y., 406

Takeuchi, Y., 446
Tansy, M.F., 454
Taskinen, H., 442
Tatrai, E., 544, 577
Telford, I.R., 364
Tepe, S.J., 535
Theiss, A.M., 113
Theiss, J.C., 120, 442
Thiess, A.M., 571
Torkelson, T.R., 119, 121–122, 240, 296, 549, 571
Travis, C.C., 44
Treon, J.F., 266, 417

Ulrich, C.E., 405
Ulsamer, A.G., 344
Ungvary, G., 577
Union Carbide, 367
Upton, A.C., 40

U.S. EPA., 86, 92, 97, 98–99, 101–102, 104, 110, 113–116, 129, 131, 135–138, 141–143, 145–148, 240, 243, 250, 260, 266, 272, 276, 291, 296, 300–301, 344, 348, 352, 359, 372–373, 383, 391, 399, 405, 410, 431, 434, 442, 445–446, 461, 470–471, 486, 506, 513, 516–517, 523, 529, 534–535, 540, 544, 549, 555, 565, 570–571, 577, 581, 585, 589, 594
Uzych, L., 565

Van Duuren, B.L., 120, 122, 243, 300, 509, 571
Van Esch, G.J., 300
Venmen, B.C., 72
Vernot, E.H., 352
Vigliani, E.C., 82
Vintinner, F.J., 560
Von Oettingen, W.F., 497

Walden, R., 102
Walters, M., 589
Wand, R.C., 40–42
Ward, C.O., 147
Wassom, J.S., 21
Waters, M.D., 20–21
Watrous, R.M., 429
Watt, W.D., 130, 132

Webster, W.S., 405
Wehner, A.P., 255
Weil, C.S., 49, 59, 72, 377
Weill, C.S., 61
Weisburger, J.H., 48
Weischer, C.H., 471
Werner, J.B., 113
Wexler, P., 21, 30
Whitfield, B., 68
Wilhite, C.C., 92–93, 115
Williams, C.L., 359
Williams, G.M., 48
Williams, M.K., 529
Willis, J.H., 377
Wilson, J.G., 63
Wimer, W.W., 377, 422
Windholz, M., 517, 581
Windolz, M., 279
Winneke, G., 442
Withey, J.R., 43–44
Wolf, M.A., 262, 457
Wolfsie, J.H., 359
Wolkowski-Tyl, R., 435
Wong, O., 148
Woodside, M.D., 72
Woolrich, P.F., 283

York, R.G., 549

Zenick, H., 64, 103
Zenz, C., 560